Weighing the Odds

Weighing the Odds

A Course in Probability and Statistics

David Williams

CAMBRIDGE
UNIVERSITY PRESS

CAMBRIDGE UNIVERSITY PRESS
Cambridge, New York, Melbourne, Madrid, Cape Town, Singapore,
São Paulo, Delhi, Dubai, Tokyo

Cambridge University Press
The Edinburgh Building, Cambridge CB2 8RU, UK

Published in the United States of America by Cambridge University Press, New York

www.cambridge.org
Information on this title: www.cambridge.org/9780521006187

© Cambridge University Press 2001

First published 2001
Reprinted 2004

A catalogue record for this publication is available from the British Library

ISBN 978-0-521-80356-4 Hardback
ISBN 978-0-521-00618-7 Paperback

Transferred to digital printing 2010

For Sheila,
for Jan and Ben, for Mags and Jeff;
and, of course,
for Emma and Sam and awaited 'Bump'
from Gump, the Grumpy Grandpa

(I can *'put those silly sums away' now, Emma.)*

Contents

Preface

Probability and Statistics used to be married; then they separated; then they got divorced; now they hardly ever see each other. (That's a mischievous first sentence, but there is more than an element of truth in it, what with the subjects' having different journals and with Theoretical Probability's sadly being regarded by many – both mathematicians and statisticians – as merely a branch of Analysis.) In part, this book is a move towards much-needed reconciliation. It is written at a difficult time when Statistics is being profoundly changed by the computer revolution, Probability much less so.

This book has many unusual features, as you will see later and as you may suspect already! Above all, it is meant to be fun. I hope that in the Probability you will enjoy the challenge of some 'almost paradoxical' things and the important applications to Genetics, to Bayesian filtering, to Poisson processes, etc. The real-world importance of Statistics is self-evident, and I hope to convey to you that subject's very considerable intrinsic interest, something too often underrated by mathematicians. The challenges of Statistics are somewhat different in kind from those in Probability, but are just as substantial.

(a) *For whom is the book written?* There are different answers to this question. The book derives from courses given at Bath and at Cambridge, and an explanation of the situation at those universities identifies two of the possible (and quite different) types of reader.

At Bath, students first have a year-long gentle introduction to Probability and Statistics; and before they start on the type of Statistical theory presented here, they do a course in Applied Statistics. They are therefore familiar with the mechanics of elementary statistical methods and with some computer packages. Students with this type of background (for whom I provide reminders of some Linear Algebra, etc) will find in this book the unifying ideas which fit the separate Statistical methods into a coherent picture, and also ideas which allow those methods to be extended via modern computer techniques.

At Cambridge, students are introduced to Statistics only when they have a thorough grounding in Mathematics; and they (or at least, many of them) like to see how Linear Algebra, etc, may be applied to real-world problems. I believe it very important to 'sell' Probability and Statistics to students who see themselves as mathematicians; and even to try to 'convert' some of them.

As someone who sees himself as a mathematician (who has always worked in Probability theory and) who very much wishes he knew more Statistics and had greater wisdom in that subject, I hope that I am equipped to sell it to those who would find medians, modes, square-root formulae for standard deviation, and the like, a huge turn-off. (I know that meeting Statistics through these topics came very close to putting me off the subject for life.) You will see that throughout this book I am going to be honest. Mathematics students should know that Probability is just as axiomatic and rigorous as (say) Group Theory, to which it is connected in remarkable ways (see the very end of Chapter 9).

(b) Not only is it the case then that we look at a lot of important topics, but we also look at them seriously – which does not detract from our enjoyment.

So let me convince you just how serious this book is. (The previous sentence was written with a twinkle, of course. But seriously ... .)

I see great value in *both* Frequentist and Bayesian approaches to Statistics; and the last thing I want to do is to dwell to too great an extent on old controversies. However, it is undeniable that there is a profound difference between the two philosophies, even in cases where there is universal agreement about the 'answers'; and, for the sake of clarity, I usually present the two approaches separately. Where there seems to remain controversy (for example in connection with sharp hypotheses in Bayesian theory), I say very clearly what I think. Where I am uneasy about some aspect of the subject, and I *am* about several, I say that too.

In regard to Probability, I explain why the 'definition' of probability in terms of long-term relative frequency is fatally flawed from the point of view of logic (not just impracticality). I also explain very fully why Probability (the subject) only works if we do not attempt to define what probability means in the real world. This gives Mathematics a great advantage over the approach of Philosophy, a seemingly unreasonable advantage since the 'definition' approach seems at first more honest. It is worth quoting Wigner's famous (if rather convoluted) statement:

> The language of mathematics reveals itself [to be] unreasonably effective in the natural sciences ... a wonderful gift which we neither understand nor deserve. We should be grateful for it and hope that it will remain valid in future research and that it will extend, for better or for worse, to our pleasure even though perhaps to our bafflement, to wide branches of learning.

That for the quantum world we use a completely different Quantum Probability calculus will also be explained, the Mathematics of Analysis of Variance (ANOVA) having prepared the ground. It is important to realize that *the real world follows which Mathematics it chooses: we cannot insist that Nature follow rules which we think inevitable.*

(**c**) I said that, as far as Statistics is concerned, this book is being written at a difficult time. Statistics in this new century will be somewhat different in kind from the Statistics of most of last century in that a significant part of the *everyday practice* of Statistics (I am not talking about developments in research) will consist of applying Bayes' formula via MCMC (Monte-Carlo-Markov-Chain) packages, of which `WinBUGS` is a very impressive example. A package `MLwiN` is a more user-friendly package which does quite a lot of classical methods as well as some important MCMC work.

These packages allow us to study more realistic models; and they can deal more-or-less exactly with any sample size in many situations where classical methods can provide only approximate answers for large samples. (However, they can run into serious difficulty, sometimes on very simple problems.)

Many ideas from classical Frequentist Statistics – 'deviance', 'relative entropy', etc – continue to play a key rôle even in 'MCMC' Statistics. Moreover, classical results often serve to cross-check 'modern' ones, and those large-sample results do guarantee the desired consistency if sample sizes were increased. *A broad background in Statistics culture remains as essential as ever.* And the classical Principle of Parsimony ('Always use the **simplest** *acceptable* model') warns us not to be seduced by the availability of remarkable computer packages into using over-'sophisticated' (and, in consequence, possibly non-robust) models with many parameters. Of course, if Science says that our model *requires* many parameters, so be it.

In part, this book is laying foundations on which your Applied Statistics work can build. It does contain a significant amount of numerical work: to illustrate topics and to show how methods and packages work – or fail to work. It discusses some aspects of how to decide if one's model provides an acceptably good 'fit'; and indicates how to test one's model for robustness.

It takes enough of a look to get you started at the computer study of non-classical models. I hope that from such material you will learn useful methods and principles for real applications. The fact that I often 'play devil's advocate', and encourage you to think things out for yourself, should help. But the book is already much longer than originally intended; and because I think that each real example should be taken seriously (and not seen as a trite exercise in the arithmetic of the t-test or ANOVA, for example), I leave the study of real examples to parallel courses of study. *That parallel study of real-world examples should form the main set of 'Exercises' for the Statistics part of this book.* Mathematical exercises are less important; and this book contains very few exercises in which (for example) integration masquerades as Statistics. *In regard to Statistics, common sense and scientific insight remain, as they have always been, more important than Mathematics. But this must never be made an excuse for not using the right Mathematics when it is available.*

(d) The book covers a very limited area. **It is meant only to provide sufficient of a link between Probability and Statistics to enable you to read more advanced books on Statistics written by wiser people, and to show also that 'Probability in its own right'** (the large part of that subject which is 'separate' from Statistics) **is both fascinating and very important.** Sadly, to read more advanced Probability written by wiser people, you will need first to study a lot more Analysis. However, there is much other Probability at this level. See Appendix D.

This *textbook* is meant to *teach*. I therefore always try to provide the most intuitive explanations for theorems, etc, not necessarily the neatest or cleverest ones. The book is *not* a work of scholarship, so that it does not contain extensive references to the literature, though many of the references it mentions *do*, so you can follow up the literature with their help. *Appendix D, 'A small sample of the literature', is one of the key sections in this book. Keep referring to it.* The book is 'modern' in that it recognizes the uses and limitations of computers.

I have tried, as far as possible, to explain *everything* in the limited area covered by the book, including, for example, giving 'C' programs showing how random-number generators work, how statistical tables are calculated, how the MCMC 'Gibbs sampler' operates, etc. I like to know these things myself, so some of you readers might too. But that is not the main reason for giving 'C' details. However good a package is, there will be cases to which it does not apply; one sometimes *needs* to know how one can program things oneself.

In regard to 'C', I think that you will be able to follow the logic of the programs even if you are not familiar with the language. You can easily spot what things are peculiar to 'C' and what really matters. The same applies to WinBUGS . I have always believed that *the best way to learn how to program is to read programs*, not books on programming languages! I spare you both pointers and Object-Oriented Programs, so my 'C' in this book is rather basic. There is a brief note in Appendix A about *static* variables which can to some extent be used to obtain some advantages of OOP without the clumsiness.

Apology: I *do* omit explaining why the Gibbs sampler works. Since I have left the treatment of Markov chains to James Norris's fine recent book (which has a brief section on the Gibbs sampler), there is nothing else I can do.

(**e**) It is possible that in future *quantum computers* will achieve things far beyond the scope of computers of traditional design. At the time of writing, however, only very primitive quantum computers have actually been made. It seems that (in addition to doing other important things) quantum computers could speed up some of the complex simulations done in some areas of Statistics, perhaps most notably those of 'Ising-model' type done in image reconstruction. In Quantum Computing, the elegant Mathematics which underlies both ANOVA and Quantum Theory may – and probably will – find very spectacular application. See Chapter 10 for a brief introduction to Quantum Computing and other (more interesting) things in Quantum Probability.

(**f**) My account of Statistics is more-or-less totally free from statements of technical conditions under which theorems hold; this is because it would generally be extremely clumsy even to state those conditions. I take the usual view that the results will hold in most practical situations.

In Probability, however, we generally know both exactly what conditions are necessary for a result to hold and exactly what goes wrong when those conditions fail. In acknowledgement of this, I have stated results precisely. The very concrete Two-Envelopes Problem, discussed at several stages of the book, spectacularly illustrates the need for clear understanding, as does the matter of when one can apply the so-useful Stopping-Time Principle.

*On a first reading, you can (perhaps **should***) *play down 'conditions'.* And you should be told now that *every set and every function you will ever see outside of research in Mathematics will be what is called 'Borel'* (definition at 45L) – except of course for the non-Borel set at Appendix A6, p499! So, you can – on a first reading – always ignore 'Borel' if you wish; I usually say 'for a nice (Borel) function', which you can interpret as 'for *any* function'. (But the 'Borel' qualification is necessary for full rigour. See the Banach–Tarski Paradox at 43B for the most mind-blowing example of what can go wrong.)

(**g**) **Packages**. The only package used extensively in the book is WinBUGS. See also the remark on MLwiN above. For Frequentist work, a few uses of Minitab are made. Minitab is widely used for teaching purposes. The favourite Frequentist package amongst academic statisticians is S-PLUS. The wonderful Free Software Foundation has made available a package R with several of the features of S-PLUS at [190].

(**h**) **Note on the book's organization.** Things in this-size text are more important than things in small text. All subsections indicated by highlighting are to be read, and all exercises similarly indicated are to be done. You should, of course, do absolutely every exercise. A ▶ adds extra emphasis to something important, a ▶▶ to something very important, while a ▶▶▶ ... !

Do remember that 'the next section' refers to the next *section* not the next *sub*section. (A section is on average ten pages long.)

I wanted to avoid the 'decimal' numbering of theorems in the 'Theorem 10.2.14' style. So, in this book, 'equation 77(F1)' refers to equation F1 on page 77. If we were on page 77, that equation would be referred to merely as F1; but because of the way that LATEX deals with pages, an equation without a page number could be on any page within one page of the current one. It's easier for you to cope with that than for me to tinker further with the LATEX! Sometimes, the '□' symbol signifies a natural break-point other than the end of a proof. Forgive my writing 'etc' rather than 'etc.' throughout.

(i) Thanks. Much of this book was written at Bath, which is why that glorious city features in some exercises. My thanks to the Statistics (and Probability) group there for many helpful comments, and in particular to Chris Chatfield, Simon Harris, David Hobson, Chris Rogers and Andy Wood (now at Nottingham). Special thanks to Bill Browne (now at University College, London) and David Draper, both of whom, while MCMC enthusiasts, know a lot more than I do about Statistics generally. A large part of the book was written at Swansea, where the countryside around is even more glorious than that around Bath, and where Shaun Andrews, Roger Hindley and (especially) Aubrey Truman were of real help with the book.

Two initially-anonymous referees chosen by Cambridge University Press took their job very conscientiously indeed and submitted many very helpful comments which have improved the book. It is a pleasure to thank those I now know to be Nick Bingham and Michael Stein.

Anyone with any knowledge of Statistics will know the great debt I owe to Sir David Cox for his careful reading of the manuscript of the Statistics chapters and for his commenting on them with unsurpassed authority.

Richard Dawkins and Richard Tilney–Bassett prevented my misleading you on some topics in Genetics. Colin Evans, Chris Isham and Basil Hiley made helpful comments on Chapter 10.

In the light of all the expert advice, this really is an occasion where I really must stress that any remaining errors are, of course, mine. In particular, I must state that the book continued to evolve until after I felt that no further demands could be made on referees.

I typed the book in LaTeX: my thanks to Leslie Lamport and, especially, Donald Knuth. I am most grateful for willing help from Francis Burstall, Ryan Cheal and Andrew Swann (at Bath) and from Francis Clarke at Swansea in the cases where my LaTeX skill couldn't achieve the layout I wanted. Most of the diagrams I did in raw Adobe Postscript, some others with a 'C'-to-Postscript converter I wrote.

David Tranah of C.U.P. helped ensure that the book was actually written, suggested numerous improvements, and generally earned himself my strong recommendation to prospective authors. My thanks too to visionary artists, copy editors and other C.U.P. staff.

Malcolm and Pat, and Alun and Mair, helped me keep some semblance of sanity, and persuaded me that it was about time (for the sake of long-suffering Sheila – so worthy a chief dedicatee) that this book was finished. A big Thankyou to them.

For a fine rescue of my computer when all seemed to be lost, I thank Mike Beer, Robin O'Leary and, especially, Alun Evans.

For extremely skilled repair in 1992 on a machine in an even more desperate state, *me*, my most special thankyou of all to the team of miracle workers led by Dr Baglin at Addenbrooke's Hospital, Cambridge. It really is true that, but for them, this book would never have been written. Have a great new millennium, Addenbrooke's!

I am sad that the late great Henry Daniels will not see this attempt to make more widely known my lifelong interest in Statistics. We often discussed the subject when not engaged in the more important business of playing music by Beethoven and Brahms.

David Williams

Swansea, 2001

Please note that I use analysts', rather than algebraists', conventions, so

$$\mathbb{Z}^+ := \{0, 1, 2, \ldots\}, \qquad \mathbb{N} := \{1, 2, 3, \ldots\}.$$

($\mathbb{R}^+$ usually denotes $[0, \infty)$, after all, and $\mathbb{R}^{++}$ denotes $(0, \infty)$ in sensible notations.)

1

INTRODUCTION

Please, do read the Preface first!

Note. Capitalized 'Probability' and 'Statistics' (when they do not start a sentence) here refer to the *subjects*. Thus, Probability is the study of probabilities, and Statistics that of statistics. (Later, we shall also use 'Statistics' as opposed to 'statistics' in a different, technical, way.) 'Maths' is English for 'Math', 'Mathematics' made friendly.

1.1 Conditioning: an intuitive first look

One of the main aims of this book is to teach you to 'condition': to use conditional probabilities and conditional expectations effectively. Conditioning is an extremely powerful and versatile technique.

In this section and (more especially) the next, I want you to give your intuition free rein, using common sense and not worrying over much about rigour. All of the ideas in these sections will appear later in a more formal setting. Indeed, because fudging issues is *not* an aid to clear understanding, we shall study things with a level of precision very rare for books at this level; but that is for *later. For now, we use common sense,* and though we would later regard some of the things in these first two sections as 'too vague', 'non-rigorous', etc, I think it is right to begin the book as I do. **Intuition is much more important than rigour**, though we do need to know how to back up intuition with rigour.

In a subject in which it is easy to be misled by arguments which initially look plausible, our intuition needs constant honing. One is much more likely to make a mistake in elementary Probability than in (say) elementary Group Theory. We *all* make mistakes; and you should check out carefully everything *I* say in this book.

Always try to find counter-arguments to what I claim, shooting them down if I am indeed correct. Some notes on the need for care with intuition occupy the next subsection.

Since I shall later be developing the subject rather carefully, *it is important that you know from the beginning that it is one of the most enjoyable branches of mathematics.* I want you to get into the subject a little before we start building it up systematically. These first two sections contain some exercises for you – some of which (in the next section particularly) are a little challenging. In this connection, do remember that this is a 'first run through' this material: we shall return to it all later; and the later exercises when the book starts properly will give you more practice. If you peep ahead at the very extensive Section 4.1 for example, you will see one of the places where you get more practice with conditional probability.

Note. Several of our examples derive from games of chance. Probability did originate with the study of such games, and *their study is still of great value for developing one's intuition.* It also remains the case that *techniques developed for these 'frivolous' purposes continue to have great real-world benefit when applied to more important areas.* This is Wigner's 'unreasonable effectiveness of Mathematics' again.

A. Notes on the need for care with intuition. I mention some of the reasons why our intuition needs sharpening.

Aa. Misuse of the 'Law of Averages' in real-world discussions. All humans, even probabilists in moments of weakness, tend to misuse the 'Law of Averages' in everyday life. National Lotteries thrive on the fact that a person will think "I haven't won anything for a year, so I am more likely to win next week", which is, of course, nonsense. The (Un)Holy Lottery Machines behave *independently* on different weeks: they do not 'remember what they have done previously and try to balance out'.

A distinguished British newspaper even gave advice to people on how to choose a 'good' set of six numbers from the set $\{1, 2, \ldots, 49\}$ for the British National Lottery. (To win, you have to choose the same six numbers as the Lottery Machine.) It was said that the six chosen numbers should be 'randomly spread out' and should have an average close to 25. It is clear that the writer thought that Choice A of (say) the six numbers $8, 11, 19, 21, 37, 46$ is more likely to win than Choice B of the six numbers $1, 2, 3, 4, 5, 6$. Of course, on every week, these two choices have exactly the same chance of winning. Yet the average of all numbers chosen by the Lottery Machine over a year is very likely to be very close to 25; and this tends to 'throw' people.

One hears misuse of the 'Law of Averages' from sports commentators every week.

Some people tend to think that a new coin which has been tossed 6 times and landed 6 Heads, is more likely to land Tails on the next toss 'to balance things out'. They do not realize that what happens in future will swamp the past: if the coin is tossed a further

1000 times, what happened in the 6 tosses already done will essentially be irrelevant to the proportion of Heads in all 1006 tosses.

Now I know that *You* know all this. But (Exercise!) what do *you* say to the writer about the Lottery who says, "Since the average of all the numbers produced by the Lottery Machine over the year is very likely to be close to 25, then, surely, I would be better to stick with Choice A throughout a year than to stick with Choice B throughout the year." The British are of course celebrated for being 'bloody-minded', and the reality is that a surprisingly large number stick to Choice B!

Ab. Some common errors made in studying the subject. Paradoxically, *the most common mistake when it comes to actually doing calculations is to assume 'independence' when it is not present*, the 'opposite' of the common 'real-world' mistake mentioned earlier. We shall see a number of examples of this.

Another common error is to assume that things are equally likely when they are not. (The infamous 'Car and Goats' problem 15P even tripped up a very distinguished Math Faculty.) Indeed, I very deliberately avoid concentration on the 'equally likely' approach to Probability: it is an invitation to disaster.

Our intuition finds it very hard to cope with the sometimes perverse behaviour of ratios. The discussion at 179Gb will illustrate this spectacularly.

B. A first example on conditional probability. Suppose that 1 in 100 people has a certain disease. A test for the disease has 90% accuracy, which here means that 90% of those who do have the disease will test positively (suggesting that they *have* the disease) and 10% of those who do *not* have the disease will test positively.

One person is chosen at random from the population, tested for the disease, and the test gives a positive result. That person might be inclined to think: "I have been tested and found 'positive' by a test which is accurate 90% of the time, so there is a 90% chance that I have the disease." However, it is much more likely that the randomly chosen person does not have the disease and the test is in error than that he or she does have the disease and the test is correct. Indeed, we can reason as follows, using 'K' to signify 'thousand' (1000, not 1024) and 'M' for 'million'.

Let us suppose that there are 1M people in the population. Suppose that they are all tested. Then, amongst the 1M people,
about $1M \times 99\% = 990K$ would *not* have the disease, of whom
about $(1M \times 99\%) \times 10\% = 99K$ would test positively;
and
$1M \times 1\% = 10K$ *would* have the disease, of whom
about $(1M \times 1\%) \times 90\% = 9K$ would test positively.
So, a total of about $99K + 9K = 108K$ would test positively, of whom only 9K would actually have the disease. In other words, *only 1/12 of those who would test positively actually have the disease.*

Because of this, we say that the *conditional probability* that a randomly chosen person does have the disease *given* that that person is tested with a positive result, is 1/12.

[[*Note*. At 6G below, we shall modify the interpretation of conditional probability just given, in which we have imagined sampling without replacement of the entire population, to one valid in all situations where we are forced to imagine instead 'sampling with replacement' to ensure 'independence'. The numerical value of the conditional probability is here unaffected.]]

I do not want to get involved in 'philosophical' questions at this stage, but, provided you understand that this book contains such things only to the minimal extent consistent with clarity, I mention briefly now a point which will occur in other forms much later on.

C. Discussion: Continuation of Example 3B.

Now consider the situation where the experiment has actually been performed: a *real* person with an actual name – let's say it is Homer Simpson – has been chosen, and tested with a positive result. Can we tell Homer that the probability that he has the disease is 1/12? Do note that we are assuming that Homer is the person chosen at random; and that all we know about him is that his test proved positive. It is *not* the case (for example) that Homer is consulting his doctor because he fears he may have caught a sexually transmitted disease.

A Possible Frequency-School View. The problem is that there is no randomness in whether or not Homer has the disease: either he does have it, in which case the probability that he has it (conditional on any information) is 1; or he does not have it, in which case the probability that he has it (conditional on any information) is 0. All that we can say to Homer is that if every person were tested, then the fraction of those with positive results who would have the disease is 1/12; and in this sense he can be 11/12 'confident' that he does not have the disease. [The Note above on sampling with replacement applies here too, but does not really concern us now.] It is not very helpful to tell Homer only that the probability that he has the disease is either 0 or 1 but we don't know which.

The Bayesian-School View. If we take the contrasting view of the Bayesian School of Statistics, then we can interpret 'probability that a statement is true' as meaning 'degree of belief in that statement'; and then we *can* tell Homer that the probability (in this new sense) that he has the disease is 1/12.

Remarks. The extent to which the difference between the schools in this case is a matter of 'Little-endians versus Big-endians' is up to you to decide for yourself later. (The dispute in *Gulliver's Travels* was over which way up to put an egg in an eggcup. If your primary concern is with eating the egg, ...)

In this book, I am certainly happy to tell Homer that the conditional probability that he has the disease is 1/12. I shall sell him a copy of the book so that he can decide what that statement means to him.

The logic which we used in Example 3B is the uncontroversial and incontrovertible logic of Bayes' Theorem – a *theorem* – in Probability. We shall study the theorem as part of the full mathematical theory in Chapter 4. However, we shall develop many of its key ideas in this section.

D. Orientation: The Rules of Probability. Probability, the mathematical theory, is based on two simple rules: an Addition Rule and a Rule for Combining Conditional Probabilities, both of which we have in effect seen in our discussion of Example 3B. The Addition Rule is taken as *axiomatic* in the mathematical theory; the Rule for Combining Conditional Probabilities is really just a matter of *definition*. The subject is developed by application of logic to these rules. Especially, *no attempt is made to define 'probability' in the real world*. The *'long-term relative frequency' (LTRF)* idea is the key *motivation* for Probability, though, as we shall see, it is impossible to make it into a rigorous definition of probability.

The LTRF idea motivates the axioms; but once we have the axioms, we forget about the LTRF until its reappearance as a theorem, part of the Strong Law of Large Numbers.

▶ **E. LTRF motivation for Probability.** Let A be an event associated with some experiment $\mathcal{E}$, so that A might, or might not, occur when $\mathcal{E}$ is performed. Now consider a Super-experiment $\mathcal{E}^\infty$ which consists of an infinite number of independent performances of $\mathcal{E}$, 'independent' in that no performance is allowed to influence others. Write $N(A, n)$ for the number of occurrences of A in the first n performances of $\mathcal{E}$ within the Super-experiment. Then the LTRF idea is that

$$\frac{N(A, n)}{n} \text{ converges to } \mathbb{P}(A) \qquad (E1)$$

in some sense, where $\mathbb{P}(A)$ is the probability of A, that is, the probability that A occurs within experiment $\mathcal{E}$. If our individual experiment $\mathcal{E}$ consists of tossing a coin with probability p of Heads *once*, then the LTRF idea is that if the coin is thrown *repeatedly*, then

$$\frac{\text{Number of Heads}}{\text{Number of tosses}} \to p \qquad (E2)$$

in some sense. Here, A is the event that 'the coin falls Heads' in our individual experiment $\mathcal{E}$, and $p = \mathbb{P}(A)$. Since the coin has no memory, we believe (and postulate in mathematical modelling) that it behaves independently on different tosses.

The **certain event** Ω (Greek Omega), the event that 'something happens', occurs on every performance of experiment $\mathcal{E}$. The **impossible event** $\emptyset$ ('nothing happens') never occurs. The LTRF idea suggests that

$$\mathbb{P}(\Omega) = 1, \quad \mathbb{P}(\emptyset) = 0.$$

Note that in order to formulate and prove the Strong Law, we have to set up a model for the Super-experiment $\mathcal{E}^\infty$, and we have to be precise about 'in some sense'.

▶▶ **F. Addition Rule for Two Events.** If A and B are events associated with our experiment $\mathcal{E}$, and these events are *disjoint* (or *exclusive*) in that it is impossible for A and B to occur simultaneously on any performance of the experiment, and if

$A \cup B$ is the event that 'A happens or B happens',

then, of course,

$$N(A \cup B, n) = N(A, n) + N(B, n).$$

The appropriateness of the set-theoretic 'union' notation will become clear later. If we 'divide by n and let n tend to ∞' we obtain LTRF motivation – but not proof – of the Addition Rule for Two Events:

if A and B are disjoint, then $\mathbb{P}(A \cup B) = \mathbb{P}(A) + \mathbb{P}(B)$.   (F1)

Later, we take (F1) as an axiom.

  [[For example, if our individual experiment $\mathcal{E}$ is that of tossing a coin twice, then

$$\mathbb{P}(1 \text{ Head in all}) = \mathbb{P}(\text{HT}) + \mathbb{P}(\text{TH}),$$ (F2)

where, of course, HT signifies 'Heads on the 1st toss, Tails on the 2nd'.]]

  If A is any event, we write

A^c for the event 'A does not occur'.

Then A and A^c are disjoint, and $A \cup A^c = \Omega$: precisely one of A and A^c occurs within our experiment $\mathcal{E}$. Thus, $1 = \mathbb{P}(\Omega) = \mathbb{P}(A) + \mathbb{P}(A^c)$, so that

$$\mathbb{P}(A^c) = 1 - \mathbb{P}(A).$$

[[Hence, for our coin, $\mathbb{P}(\text{it falls Tails}) = q := 1 - p$.]]

▶ **G. The LTRF motivation for conditional probability.**   Let A and B be events associated with our experiment $\mathcal{E}$, with $\mathbb{P}(A) \neq 0$. The LTRF motivation is that we regard the *conditional probability* $\mathbb{P}(B \mid A)$ *that* B *occurs given that* A *occurs* as follows. Suppose again that our experiment is performed 'independently' infinitely often. Then (the LTRF idea is that) $\mathbb{P}(B \mid A)$ is the *long-term proportion of those experiments on which A occurs that B* (also) *occurs, in other words, that both A and B occur.*
In other words, if

$A \cap B$ is the event that 'A and B occur simultaneously', then we should have

$$\mathbb{P}(B \mid A) = \text{limit in some sense of } \frac{N(A \cap B, n)}{N(A, n)}$$

$$= \text{limit in some sense of } \frac{N(A \cap B, n)/n}{N(A, n)/n} = \frac{\mathbb{P}(A \cap B)}{\mathbb{P}(A)}.$$

in our experiment $\mathcal{E}^\infty$. In the mathematical theory, we *define*

$$\mathbb{P}(B \mid A) := \frac{\mathbb{P}(A \cap B)}{\mathbb{P}(A)}. \tag{G1}$$

Suppose that our experiment $\mathcal{E}$ has actually been performed in the real world, and that we are told only that event A has occurred. Bayesians would say that $\mathbb{P}(B \mid A)$ is then our 'probability as degree of belief that B (also) occurred'. Once the experiment has been performed, whether or not B has occurred involves no randomness from the Frequentist standpoint. A Frequentist would have to quote: 'the long-term proportion of those experiments on which A occurs that B (also) occurs is (whatever is the numerical value of) $\mathbb{P}(B \mid A)$'; and in this sense $\mathbb{P}(B \mid A)$ represents our 'confidence' that B occurred in an actual experiment on which we are told that A occurred.

With this Frequentist view of probability, we should explain to Homer that if the experiment 'Pick a person at random and test him or her for the disease' were performed *independently* a very large number of times, then on a proportion $11/12$ of those occasions on which a person tested positively, he or she would *not* have the disease. To guarantee the independence of the performances of the experiment, we would have to pick each person from the *entire* population, so that the same person might be chosen many times. It is of course assumed that if the person is chosen many times, no record of the results of any previous tests is kept. This is an example of sampling with replacement. We shall on a number of occasions compare and contrast sampling with, and sampling without, replacement when we begin on the book proper.

►► **H. General Multiplication Rule.** We have for any 2 events A and B,

$$\mathbb{P}(A \cap B) = \mathbb{P}(A)\mathbb{P}(B \mid A), \tag{H1}$$

this being merely a rearrangement of (G1).

For any 3 events A, B and C, we have, for the event $A \cap B \cap C$ that all of A, B and C occur simultaneously within our experiment $\mathcal{E}$,

$$\mathbb{P}(A \cap B \cap C) = \mathbb{P}\big((A \cap B) \cap C\big) = \mathbb{P}(A \cap B)\mathbb{P}(C \mid A \cap B),$$

whence

$$\mathbb{P}(A \cap B \cap C) = \mathbb{P}(A)\mathbb{P}(B \mid A)\mathbb{P}(C \mid A \cap B). \tag{H2}$$

The extension to 4 or more events is now obvious.

I. A decomposition result.   Let A and B be any two events. The events $G := A \cap B$ and $H := A^c \cap B$ are disjoint, and $G \cup H = B$. (*Clarification.* We are decomposing B according to whether or not A occurs. If B occurs, then either 'A occurs and B occurs' or 'A does not occur and B occurs'.) We have

$$\mathbb{P}(B) \;=\; \mathbb{P}(G) + \mathbb{P}(H) \;=\; \mathbb{P}(A)\mathbb{P}(B \mid A) + \mathbb{P}(A^c)\mathbb{P}(B \mid A^c). \qquad (\text{I1})$$

This, the simplest decomposition, is extremely useful.

Ia. Example 3B revisited.   In that example, let
    B be 'chosen person has the disease'
    A be 'chosen person tests positively'.
We want to find $\mathbb{P}(B \mid A)$. We are given that

$$\mathbb{P}(B) = 1\%, \quad \mathbb{P}(B^c) = 99\%, \quad \mathbb{P}(A \mid B) = 90\%, \quad \mathbb{P}(A \mid B^c) = 10\%.$$

We have, keeping the calculation in the same order as before,

$$\begin{aligned}
\mathbb{P}(A) &= \mathbb{P}(B^c \cap A) + \mathbb{P}(B \cap A) = \mathbb{P}(B^c)\mathbb{P}(A \mid B^c) + \mathbb{P}(B)\mathbb{P}(A \mid B) \\
&= (0.99 \times 0.10) + (0.01 \times 0.90) = 0.108 \quad (= 108K/1M).
\end{aligned}$$

We now know $\mathbb{P}(A \cap B)$ and $\mathbb{P}(A)$, so we can find $\mathbb{P}(B \mid A)$.

▶▶ **J. 'Independence means Multiply'.**   If A and B are two events, then we say that A and B are *independent* if

$$\mathbb{P}(A \cap B) \;=\; \mathbb{P}(A)\mathbb{P}(B), \qquad (\text{J1})$$

one of several assertions which we shall meet that 'Independence means Multiply'. If $\mathbb{P}(A) = 0$, no comment is necessary. If $\mathbb{P}(A) \neq 0$, then we may rearrange (J1) as

$$\mathbb{P}(B \mid A) \;=\; \frac{\mathbb{P}(A \cap B)}{\mathbb{P}(A)} \;=\; \mathbb{P}(B),$$

which says that the information that A occurs on some performance of $\mathcal{E}$ does not affect 'our degree of belief that B occurs' on that same performance.

If we consider the experiment 'Toss a coin (with probability p of Heads) twice', then, we believe that, since the coin has no memory, the results of the two tosses will be independent. (The laws of physics would be very different if they are not!) Hence we have

$$\mathbb{P}(\text{HT}) \;=\; \mathbb{P}(\text{Heads on first toss}) \times \mathbb{P}(\text{Tails on second}) \;=\; pq,$$

where q, the probability of Tails, is $1 - p$; and, using the Addition Rule as at 6(F2), we get the familiar answer that the probability of 'exactly one Head in all' is $pq + qp = 2pq$.

The Multiplication Rules for n independent events follow from the General Multiplication Rules at 7H similarly. If we toss a coin 3 times, the chance of getting HTT is, of course, pqq.

K. Counting.   I now begin a discussion (continued in the next section) of various 'counting' and 'conditioning' aspects of the famous binomial-distribution result for coin tossing. The 'counting' approach may well be familiar to you; but, in the main, I want you to condition rather than to count.

▶ **Ka. Lemma.** *For non-negative integers r and n with $0 \leq r \leq n$, the number $\binom{n}{r}$, also denoted by ^nC_r, of subsets of $\{1, 2, \ldots, n\}$ of size r is*

$$\binom{n}{r} = {}^nC_r = \frac{n!}{r!(n-r)!},$$

where, as usual,

$$n! := n(n-1)(n-2)\ldots 3.2.1, \qquad 0! := 1.$$

If $r < 0$ or $r > n$, we define $^nC_r := \binom{n}{r} := 0$.

Remarks. Here and everywhere, we adopt the standard convention in Maths that 'set' means 'unordered set': $\{1, 2, 3\} = \{3, 1, 2\}$. There are indeed $^4C_2 = 6$ subsets of size two of $\{1, 2, 3, 4\}$, namely, $\{1, 2\}$, $\{1, 3\}$, $\{1, 4\}$, $\{2, 3\}$, $\{2, 4\}$, $\{3, 4\}$. The empty set is the only subset of $\{1, 2, \ldots, n\}$ of size 0, even if $n = 0$.

The official rigorous proof would take one through the steps of the following lemma. Part (a) is not actually relevant for this purpose, but is crucial for other results.

▶ **Kb. Lemma.**   *(a) The number of ordered r-tuples $(i_1, i_2, \ldots, i_r)$ where each i_k is chosen from $\{1, 2, \ldots, n\}$ is n^r. (You will probably know from Set Theory that the set of all such r-tuples is the* Cartesian product $\{1, 2, \ldots, n\}^r$.)
(b) For $0 \leq r \leq n$, the number ^nP_r of ordered r-tuples $(i_1, i_2, \ldots, i_r)$ where each i_k is chosen from $\{1, 2, \ldots, n\}$ and $i_1, i_2, \ldots, i_r$ are distinct *is given by*

$$^nP_r = n(n-1)(n-2)\ldots(n-r+1) = \frac{n!}{(n-r)!}.$$

(c) The number of permutations of $\{1, 2, \ldots, n\}$ is $n!$.
(d) Lemma Ka is true.

Proof. In Part (a), there are n ways of choosing i_1, and, for each of these choices, n ways of choosing i_2, making $n \times n = n^2$ ways of choosing the ordered pair (i_1, i_2). For each of these n^2 choices of the ordered pair (i_1, i_2), there are n ways of choosing i_3; and so on.

In Part (b), there are n ways of choosing i_1, and, for each of these choices, $n-1$ ways of choosing i_2 (because we are now not allowed to choose i_1 again), making $n \times (n-1)$

ways of choosing the ordered pair (i_1, i_2). For each of these $n(n-1)$ choices of the ordered pair (i_1, i_2), there are $n-2$ ways of choosing i_3; and so on.

Part (c) is just the special case of Part (b) when $r = n$.

Now for Part (d). By Part (c), each subset of size r of $\{1, 2, \ldots, n\}$ gives rise to $r!$ ordered r-tuples $(i_1, i_2, \ldots, i_r)$ where $i_1, i_2, \ldots, i_r$ are the distinct elements of the set in some order. So it must be the case that $r! \times {}^nC_r = {}^nP_r$; and this leads to our previous formula for ${}^nC_r$.   □

L. 'National Lottery' Proof of Lemma 9Ka.   One can however obtain clearer intuitive understanding of Lemma 9Ka by using conditioning rather than counting as follows. Yes, a certain amount of intuition goes into the argument too.

A British gambler (who clearly knows no Probability) pays 1 pound to choose a subset of size r of the set $\{1, 2, \ldots, n\}$. (In Britain, $n = 49$, and $r = 6$.) The Lottery Machine later chooses 'at random' a subset of size r of the set $\{1, 2, \ldots, n\}$. If the machine chooses exactly the same subset as our gambler, then our gambler wins the 'jackpot'. It is clear that our gambler wins the 'jackpot' with probability $1/\binom{n}{r}$, and we can find $\binom{n}{r}$ from this probability.

Now, the probability that the first number chosen by the machine is one of the numbers in our gambler's set is clearly r/n. The *conditional probability* that the second number chosen by the machine is in our gambler's set *given* that the first is, is clearly $(r-1)/(n-1)$, because, given this information about the first, at the time the machine chooses its second number, there are $n-1$ 'remaining' numbers, $r-1$ of which are in our gambler's set. By Multiplication Rule 7(H1), the probability that the first *two* numbers chosen by the machine are in our gambler's set is

$$\frac{r}{n} \times \frac{r-1}{n-1}.$$

By extending the idea, we see that the probability that the machine chooses exactly the same set as our gambler is

$$\frac{r}{n} \times \frac{r-1}{n-1} \times \frac{r-2}{n-2} \times \cdots \times \frac{1}{n-r+1} = \frac{r!}{{}^nP_r} = \frac{r!(n-r)!}{n!}.$$

(In Britain, then, the probability of winning the jackpot is very roughly 1 in 14 million.) We have proved Lemma 9Ka.

For more on our gambler, see Exercise 17Rb below.

M. Binomial(n, p) distribution.   Now, we can combine the ideas of 8J with Lemma 9Ka.

▶ **Ma. Lemma.**   *If a coin with probability p of Heads is tossed n times, and we write Y for the total number of Heads obtained, then Y has the probability mass function of the binomial(n, p) distribution:*

$$\mathbb{P}(Y = r) = b(n, p; r) := \binom{n}{r} p^r (1-p)^{n-r}.$$

Proof. Because of the independence of the coin's behaviour on different tosses, we have for any outcome such as HTHHHTTHH ... with exactly r Heads and $n - r$ Tails,

$$\mathbb{P}(\text{HTHHHTTHH}\ldots) = pqppppqqpp\ldots = p^r q^{n-r},$$

where $q = 1 - p$. Now the typical result with r Heads in all is a sequence such as HTHHHTTHH ... in which the set of positions where we have H is a subset of $\{1, 2, \ldots, n\}$ of size r. Every one of these $\binom{n}{r}$ subsets contributes $p^r q^{n-r}$ to $\mathbb{P}(Y = r)$, whence the result follows. □

▶▶ **N. Stirling's Formula.** This remarkable formula, proved in 1730, states that

$$n! \sim \left(\frac{n}{e}\right)^n \sqrt{2\pi n} \quad \text{as } n \to \infty, \tag{N1}$$

with the precise meaning that, as $n \to \infty$, the ratio of the two sides of (N1) converges to 1. For $n = 10$, LHS/RHS $= 1.0084$. Much more accurate approximations may be found in Subsection 148D.

Na. Exercise. Use (N1) to show that, the probability $b(2n, \frac{1}{2}; n)$ that we would get n Heads and n Tails in $2n$ tosses of a fair coin satisfies

$$b(2n, \tfrac{1}{2}; n) \sim \frac{1}{\sqrt{\pi n}}. \tag{N2}$$

Check that when $n = 10$, the left-hand side is 0.1762 (to 4 places) and the right-hand side is 0.1784 (to 4 places).

Historically, approximation (N2) was a vital step on the route which eventually led to the general Central Limit Theorem of Chapter 5.

Nb. Heuristic explanation for Stirling's formula. (I promised that, as far as possible, I would not ask you to take things entirely on trust.) The Probability corresponding to this heuristic explanation can be found in Subsection 162F.

We have, with $x = n + y\sqrt{n}$,

$$n! = \int_0^\infty x^n e^{-x}\, dx = \int_{-\sqrt{n}}^\infty (n + y\sqrt{n})^n e^{-(n+y\sqrt{n})} \sqrt{n}\, dy$$

$$= \left(\frac{n}{e}\right)^n \sqrt{n} \int_{-\sqrt{n}}^\infty \left(1 + \frac{y}{\sqrt{n}}\right)^n e^{-y\sqrt{n}}\, dy.$$

Now (see Appendix A1, p495, and equation 496(ApA 2.4)), for any fixed y, we have, as $n \to \infty$,

$$\ln\left\{\left(1 + \frac{y}{\sqrt{n}}\right)^n e^{-y\sqrt{n}}\right\} = n \ln\left(1 + \frac{y}{\sqrt{n}}\right) - y\sqrt{n}$$

$$= n\left\{\frac{y}{\sqrt{n}} - \frac{1}{2}\frac{y^2}{n} + O\left(\frac{y^3}{n^{3/2}}\right)\right\} - y\sqrt{n} \to -\tfrac{1}{2}y^2,$$

(see Appendix A1, p495 for $O(\cdot)$ notation and also 496(ApA 2.4)). Hence for every $y \in \mathbb{R}$,

$$\left(1 + \frac{y}{\sqrt{n}}\right)^n e^{-y\sqrt{n}} \rightarrow e^{-\frac{1}{2}y^2},$$

and this suggests (but does not prove) that

$$A_n := \frac{n!}{(n/e)^n \sqrt{n}} \rightarrow \int_{-\infty}^{\infty} e^{-\frac{1}{2}y^2} \, dy = \sqrt{2\pi}, \tag{N3}$$

the famous last integral being evaluated at 146B.

It is significantly more difficult to make the above clear explanation into a rigorous proof than to provide the following rigorous proof (optional!) which is not an explanation.

Nc. Proof of Stirling's formula (optional!). This proof begins with an idea of H. Robbins, quoted in the books by Feller and by Norris. We have for $-1 < t < 1$,

$$\ln(1 + t) = \int_0^t \frac{1}{1 + s} \, ds = \int_0^t \left\{ 1 - s + s^2 - s^3 + \cdots \right\} ds$$

whence, for $0 < t < 1$,

$$\ln(1 + t) = +t - \tfrac{1}{2}t^2 + \tfrac{1}{3}t^3 - \cdots,$$
$$\ln(1 - t) = -t - \tfrac{1}{2}t^2 - \tfrac{1}{3}t^3 - \cdots,$$

Hence, for $0 < t < 1$,

$$t^{-1}\frac{1}{2}\ln\left(\frac{1+t}{1-t}\right) - 1 = \frac{1}{3}t^2 + \frac{1}{5}t^4 + \cdots \quad \text{(so is non-negative)}$$

$$\leq \frac{1}{3}t^2 + \frac{1}{3}t^4 + \cdots = \frac{t^2}{3(1 - t^2)}.$$

Hence (with $y = 1/t$), we have for $y > 1$,

$$0 \leq c(y) := y\frac{1}{2}\ln\left(\frac{y+1}{y-1}\right) - 1 \leq \frac{1}{3(y^2 - 1)}.$$

But if $a_n := \ln A_n$, where A_n is as at (N3), then (check!)

$$0 \leq a_n - a_{n+1} = c(2n + 1) \leq \frac{1}{12n} - \frac{1}{12(n+1)},$$

whence (a_n) is a decreasing sequence and $\left(a_n - (12n)^{-1}\right)$ is an increasing sequence. Hence, for some a,

$$a_n \rightarrow a, \quad A_n \rightarrow A := e^a,$$

and we have Stirling's formula in the form

$$n! \sim A\sqrt{n}\left(\frac{n}{e}\right)^n. \tag{N4}$$

It only remains to prove that $A = \sqrt{2\pi}$.

For an integer $k \geq 1$, integration by parts with $u = \cos^k \theta$, $v = \sin \theta$, shows that

$$I_{k+1} := \int_0^{\frac{1}{2}\pi} \cos^{k+1} \theta \, d\theta = \int u \, dv = [uv] - \int v \, du$$

$$= \int_0^{\frac{1}{2}\pi} k \cos^{k-1} \theta \sin^2 \theta \, d\theta = kI_{k-1} - kI_{k+1}$$

(using $\sin^2 = 1 - \cos^2$), whence, for $k \geq 1$,

$$I_{k+1} = \frac{k}{k+1} I_{k-1}.$$

It is immediately checked that $I_0 = \frac{1}{2}\pi$ and $I_1 = 1$. Hence, using also 12(N4),

$$I_{2n+1} = \frac{2n}{2n+1} I_{2n-1} = \frac{2n}{2n+1} \frac{2n-2}{2n-1} \cdots \frac{2}{3} I_1$$

$$= \frac{(n!2^n)^2}{(2n+1)!} \sim \frac{Ae}{2\sqrt{(2n+1)} \left(1 + \frac{1}{2n}\right)^{2n+1}} \sim \frac{A}{2\sqrt{2n}},$$

since $(1 + 1/m)^m \to e$. Similarly,

$$I_{2n} = \cdots = \frac{(2n)!}{(n!)^2 2^{2n}} \frac{1}{2} \pi \sim \frac{\pi}{A\sqrt{2n}}.$$

But since $0 \leq \cos \theta \leq 1$ for $\theta \in [0, \frac{1}{2}\pi]$, the definition of I_k implies that

$$I_{2n-1} \geq I_{2n} \geq I_{2n+1}.$$

Hence,

$$1 \geq \frac{I_{2n+1}}{I_{2n}} \sim \frac{A^2}{2\pi}, \quad 1 \geq \frac{I_{2n}}{I_{2n-1}} \sim \frac{2\pi}{A^2}.$$

The only possible conclusion is that $A = \sqrt{2\pi}$, as required. □

▶ **O. The Birthdays Problem.** Part (a) of the following problem is well known.

Oa. Exercise. Suppose that we have r people labelled $1, 2, \ldots, r$ in a room. *Assume* that their birthdays, correspondingly labelled, are equally likely to form any ordered r-tuple with numbers chosen from $\{1, 2, 3, \ldots, n\}$, where $n = 365$. We ignore leap years. In regard to the assumption, see Appendix B.
(a) Explain by 'counting' why the probability that no two have the same birthday is

$$\frac{{}^nP_r}{n^r} = \frac{n-1}{n} \times \frac{n-2}{n} \times \cdots \times \frac{n-r+1}{n}.$$

Explain also via the use of conditional probabilities. Check that the probability that no two have the same birthday is less than $\frac{1}{2}$ if $r = 23$. Two methods of doing this last step quickly with just a pocket calculator are given later in this subsection.

(b) Suppose that $r = 3$. There are three possibilities:

(i) no two have the same birthday,
(ii) some two have the same birthday and the third has a different birthday,
(iii) all three have the same birthday.

Calculate directly the probabilities of these possibilities, and check that they sum to 1.

(c) For general r, calculate the probability of the event that precisely one day in the year is the birthday of two of the people and no day is the birthday of three or more. Your answer should check with that to (b)(ii) when $r = 3$.

Ob. A very useful inequality.

$$1 - x \ \leq \ e^{-x} \quad \text{for } x \geq 0. \tag{O1}$$

To prove this, integrate $1 \geq e^{-y}$ $(y \geq 0)$ from 0 to x, getting $x \geq 1 - e^{-x}$ $(x \geq 0)$. Note that therefore, the probability that no two of our 23 people have the same birthday is at most

$$\exp\left(-\frac{1 + 2 + \cdots + 22}{365}\right) \ = \ \exp\left(-\frac{253}{365}\right) \ = \ 0.499998.$$

Oc. Beware of the 'independence' trap. Continue to ignore leap years, assuming that *every year* has 365 days. One might be tempted to argue that the probability that two people have different birthdays is $364/365$, that there are $\binom{23}{2} = 253$ pairs of people amongst 23 people, whence the chance that all 23 people have different birthdays is $(364/365)^{253} = 0.4995$. You can see that this argument is not correct by supposing that instead of 23 people, one had (say) 400. *You should always check out extreme cases to see if your argument might be flawed.*

The argument is essentially assuming that if, for a pair P of people, E_P denotes the event that the two people in the pair P have different birthdays, then the events E_P, where P ranges over the 253 pairs, are independent. But of course, if persons 1 and 2 have the same birthday and persons 2 and 3 have the same birthday, then persons 1 and 3 must have the same birthday. *You have been told already that the most common mistake made in the subject is to assume independence when it is not valid.*

The 'independence' argument does produce a good approximation for 23 people because (by the Binomial Theorem)

$$\left(\frac{364}{365}\right)^k \ = \ \left(1 - \frac{1}{365}\right)^k \ \approx \ 1 - \frac{k}{365} \quad \text{for } k = 1, 2, \ldots, 22,$$

whence (with the correct probability that no two have the same birthday on the left-hand side)

$$\frac{364}{365}\frac{363}{365}\cdots\frac{343}{365} \ \approx \ \left(\frac{364}{365}\right)^{1+2+\cdots+22} \ = \ \left(\frac{364}{365}\right)^{253}.$$

Indeed, mathematical induction shows that

$$1 - \frac{k}{365} \leq \left(\frac{364}{365}\right)^k,$$

whence the probability that no two have the same birthday is at most the 'independence' answer 0.4995.

P. The 'Car and Goats' Problem.
This has become notorious, and it is instructive. There are many other versions of essentially the same problem, which was much discussed long before its 'Monty Hall game show' incarnation around 1990.

Pa. The Problem. At the end of that American game show, a contestant is shown three closed doors. Behind one of the doors is a car; behind each of the other two is a goat. The contestant chooses one of the three doors. The show's host, who knows which door conceals the car, opens one of the remaining two doors which he knows will definitely reveal a goat. He then asks the contestant whether or not she wishes to switch her choice to the remaining closed door. *Should she switch or stick to her original choice?*

Wrong solution. There remain two closed doors, so the contestant has probability 1/2 of winning whether she switches or not.

Correct solution. A contestant who decides before the show definitely to stick with her original choice, wins the car if and only if her original choice was correct, that is, with probability 1/3. A contestant who decides before the show definitely to switch wins the car if and only if her original choice was wrong, that is, with probability 2/3. The fact that switching here doubles the chance of winning can be decided irrespective of what happens during the show.

Discussion. What is wrong with the Wrong solution? Well, just because there are only two possibilities, there is no reason at all to assign probability 1/2 to each of them. In this example, *the host's knowledge and strategy tilt the odds from the* 50 : 50 *situation.* Remember that the host's strategy precludes the possibility that the contestant first chooses a door with a goat and he then reveals a car; this causes the imbalance by shifting the weight of this possibility to the situation where the contestant chooses a door with a goat and the host reveals a goat. The following exercises clarify this.

Pb. Exercise: Having more goats sorts out our intuition. Suppose that there had been 1000 doors, 1 car and 999 goats. After the contestant's choice the host opens 998 doors each of which he knows will reveal a goat. By how much does switching now increase the contestant's chance of winning? Note that the '50 : 50 answer' now seems ridiculous to our intuition.

Pc. Exercise. Consider the following modification of the original problem. At the end of an American game show, a contestant is shown three closed doors. Behind one of the doors is a car; behind each of the other two is a goat. The contestant chooses one of the three doors. The show's host, who does *not* know which door conceals the car, opens one of the remaining two doors and *happens* to reveal a goat. He then asks the contestant whether or not she wishes to switch her choice to the remaining closed door. *Should she* *switch or stick to her original choice?* Show that for this problem, the probability of winning is $1/2$ irrespective of whether she switches or not. Generalize to 1000 doors.

Note. Mathematical models for the cases at Pa and Pc will be described at 73Ab.

▶▶ **Q. The 'Two-Envelopes Paradox'.** This is a *much* more important (apparent) paradox than the 'Car and Goats' one. Do not dismiss it as trivial. We shall discuss it seriously much later (399M), when it has important lessons for us.

Here is a very preliminary version of the problem, one which is not precisely posed. In this version, we shall assume that money is (non-negative and) real-valued rather than integer-valued in some units. In our later well-posed version with the same 'paradoxical' features, money will be integer-valued.

Someone shows you two envelopes and says that one contains twice as much money as the other. You choose one of the envelopes, and are then asked if you wish to swap it for the other.

(a) *'Conditioning' argument.* You might think as follows. "Suppose that I were to open the envelope I now have to reveal an amount x. Then the other envelope must contain either $\frac{1}{2}x$ or $2x$, so that on average (over many plays at this game), I would have $1\frac{1}{4}x$ after a swap. So it is best for me to swap." (Would you then swap a second time if you had the chance?!)

(b) *Symmetry argument.* Or you might think in this way. "Suppose that someone chose a number Z at random, and put Z in one envelope, $2Z$ in the other. If my first envelope were to contain Z, then, by swapping, I would gain Z; if my first envelope were to contain $2Z$, then I would lose Z by swapping. So, on average, it does not matter whether I swap."

That gives you the rough idea of what a properly-stated problem of 'two-envelopes' type will entail. At Subsection 399M, I give a precise statement of a problem with the 'paradoxical' features of this one, but one for which the issues are much clearer. I then give a rather extensive discussion of that problem because **I cannot let you imagine that Probability has real paradoxes: in spite of what may be claimed on the Internet, it doesn't.** In that subsection, I present the discussion in such a way that you will almost certainly be able to get the flavour of the correct explanation if you read that subsection now. That discussion will explain *any* type of two-envelope 'paradox'.

R. Miscellaneous Exercises

▶ **Ra. Exercise: Pooling blood samples.** This exercise, which is due to Dorfman and which has been used to great effect in practice, is to be found in Feller's masterpiece.

A certain disease, affecting only a small fraction p of the population, can be detected by an error-free blood test. To discover which individuals in a large group (of unrelated people from different regions) have the disease, the following strategy is adopted. The group is divided into blocks of x people. Samples from all people in a block are pooled (mixed together). If the pooled sample from a block is clear, then the x people in that block are clear; otherwise, every person in the block is tested separately.

Show that the expected total number of tests required per person is

$$T = \frac{1}{x} + 1 - (1 - p)^x.$$

(*Hint.* You need only consider a group consisting of x people, since each block situation is repeated. You certainly do one test on a block, and x more if and only if it is not the case that all people in the block are OK).

The Binomial Theorem allows us to approximate T by

$$T \approx \frac{1}{x} + 1 - (1 - xp).$$

Deduce that the best choice of x is about $1/\sqrt{p}$, yielding a minimum number of tests per person of about $2\sqrt{p}$.

If $p = 1/100$, use a pocket calculator to find the exact best value of x, and show that the strategy then leads to just over an 80% saving in number of tests compared with that of testing every individual separately from the beginning.

Rb. Exercise: Hypergeometric distributions (and the Lottery). Suppose that John chooses a subset S of size s ($0 \leq s \leq n$) at random from the set $\{1, 2, \ldots, n\}$, each of the $\binom{n}{s}$ subsets of size s having probability $1/\binom{n}{s}$ of being chosen. Suppose that, independently of John, Jane chooses a subset T of size t ($0 \leq t \leq n$) at random from the set $\{1, 2, \ldots, n\}$, each of the $\binom{n}{t}$ subsets of size t having probability $1/\binom{n}{t}$ of being chosen. Strictly speaking, the independence means that, for every subset S_0 of size s and every subset T_0 of size t,

$$\mathbb{P}(\text{John chooses subset } S_0 \text{ and Jane chooses subset } T_0)$$
$$= \mathbb{P}(\text{John chooses subset } S_0) \times \mathbb{P}(\text{Jane chooses subset } T_0)$$
$$= \frac{1}{\binom{n}{s}\binom{n}{t}}.$$

Let $|S \cap T|$ denote the number of elements in the set $S \cap T$. Prove (using the very substantial Hints only as a last resort) that

$$H(n, s, t; k) := \mathbb{P}(|S \cap T| = k)$$

may be expressed in the three equivalent ways:

(i) $\dfrac{\binom{s}{k}\binom{n-s}{t-k}}{\binom{n}{t}}$ (ii) $\dfrac{\binom{t}{k}\binom{n-t}{s-k}}{\binom{n}{s}}$ (iii) $\dfrac{\binom{n}{k}\binom{n-k}{s-k}\binom{n-s}{t-k}}{\binom{n}{s}\binom{n}{t}}$

We say that $|S \cap T|$ has a *hypergeometric* distribution. [*Hints.* For (i), we decide that we can pretend that John has chosen the subset $\{1, 2, \ldots, s\}$; then 'we want' Jane's subset to consist of a subset of size k $\{1, 2, \ldots, s\}$ and a subset of size $t - k$ in $\{s+1, s+2, \ldots, n\}$. In (iii), we imagine 'first choosing $K := S \cap T$, then $S \setminus K$, then $T \setminus S$'.]

Deduce that the probability that the (un)Holy Lottery Machine of Discussion 10L picks a total of k of our gambler's numbers is

$$\frac{\binom{r}{k}\binom{n-r}{r-k}}{\binom{n}{r}}.$$

It is always a good idea to think up several different methods in these combinatorial questions as a cross-check. The next exercise gives yet another method.

Rc. Exercise: A conditioning approach to the hypergeometric distribution. For the experiment of tossing a coin (with probability q of Tails) n times, calculate

$$\mathbb{P}(k \text{ Tails in first } s \text{ tosses} \mid t \text{ Tails in all});$$

and explain why this leads to the hypergeometric distribution. (*Note.* The fact that the answer does not involve q relates to the fact that t is a 'sufficient statistic' for q, as will be explained much later on.)

Rd. Exercise. If our answer (i) in Exercise 17Rb is correct, then

$$\sum_{k=0}^{s \wedge t} \binom{s}{k}\binom{n-s}{t-k} = \binom{n}{t} \qquad (\text{R1})$$

where we have used the standard notation $s \wedge t := \min(s, t)$, the minimum of s and t. The equation corresponds to $\sum_k H(n, s, t; k) = 1$. Prove equation (R1) by considering the coefficient of x^t in

$$(1 + x)^s (1 + x)^{n-s} = (1 + x)^n.$$

There is not really any need to specify the upper and lower limits for the sum in equation (R1) because the binomial coefficients are defined to be 0 for 'impossible' values. Even so, please answer the question: for what values of k is $H(n, s, t; k)$ strictly positive, and why?

1.2   Sharpening our intuition

Shortly, I shall look at some 'conditioning' approaches to Lemma 10Ma, the binomial-distribution result. First, I mention three puzzles, the first requiring only common sense, the next two some use of conditional probability. Stars indicate somewhat tricky things. All three puzzles are solved later *in the main text*.

A. Puzzle. Suppose that I keep throwing a fair die (perfectly cubical die) with scores $1, 2, \ldots, 6$ on its faces. Give a very accurate estimate that my accumulated score will at one stage be exactly 10000. *Hint.* What is the average gap between successive accumulated totals?

▶▶ **B. Puzzle: The 'Waiting for HH' problem.** Suppose that I keep tossing a fair coin.

(a) How many tosses are there on average before I get the first Head?

[*Solution.* You can do this by 'reversing' the way in which (I hope) you did Exercise A. But here's the way I now want you to do it. Let the answer be x. Then we have the equation

$$x = 1 + \left(\tfrac{1}{2} \times 0 \right) + \left(\tfrac{1}{2} \times x \right),$$

explained as follows:

the 1 is the first toss, which we have to make;

with probability $\tfrac{1}{2}$, the first toss produces Heads and we have 0 further tosses to make;

with probability $\tfrac{1}{2}$, the first toss produces Tails, in which case the whole process starts afresh and so, on average, I have to wait a further x tosses.

Here we used intuitively the decomposition result for average values (or 'expectations' or 'means') corresponding to 8(I1).]

(b) How many tosses on average do I make (from the beginning) until the first appearance of the HT pattern (Head immediately followed by a Tail)? (*Hint.* Using the result of part (a), you can do this immediately, without any further calculation.)

(c)* How many tosses on average do I make (from the beginning) until the first appearance of the HH pattern?

C. Puzzle: Eddington's 'Four Liars' Problem*. The great English astronomer Sir Arthur Eddington posed this problem. He got it wrong. See if you can do better by getting the correct answer $13/41$.

"Whenever A, B, C, or D makes a statement, he lies with probability 2/3, tells the truth with probability 1/3. D made a statement, C commented on D's statement, B on C's, and A on B's. What is the conditional probability that D spoke the truth given that

A affirms that B denies that C affirms that D lied?"

Hint. Why not start with the 'Two Liars' version, that of finding

$$\mathbb{P}(\text{D spoke the truth} \mid \text{C affirms that D lied}),$$

the answer to which is $\tfrac{1}{2}$?

		0	1	2	3	4	5	6
	0	1						
	1	1	1					
	2	1	2	1				
n	3	1	3	3	1			
	4	1	4	6	4	1		
	5	1	5	10	10	5	1	
	6	1	6	15	20	15	6	1

Table D(i): Pascal's triangle for $\binom{n}{r}$

D. Thinking about the binomial-distribution result 10Ma via conditioning.

I want to explain *two* useful methods for doing this. You will recall Pascal's triangle for calculating binomial coefficients.

The 'triangle' illustrates the recurrence relation

$$\binom{n}{r} = \binom{n-1}{r-1} + \binom{n-1}{r}. \tag{D1}$$

Suppose that a coin with probability p of Heads (and $q = 1 - p$ of Tails) is tossed n times. Let

A be the event 'Heads on first toss',

B be the event 'r Heads in all'

and let

$$b_1(n, p; r) := \mathbb{P}(B).$$

We want to prove that $b_1(n, r; p)$ equals the $b(n, r; p)$ of Lemma 10Ma.

Method 1. We have, 'obviously',

$$\mathbb{P}(A) = p; \quad \mathbb{P}(B \mid A) = b_1(n-1, p; r-1);$$
$$\mathbb{P}(A^c) = q; \quad \mathbb{P}(B \mid A^c) = b_1(n-1, p; r).$$

Using the decomposition result 8(I1), we therefore find that

$$b_1(n, p; r) = p\, b_1(n-1, p; r-1) + q\, b_1(n-1, p; r). \tag{D2}$$

This tallies exactly with recurrence relation (D1); and we could now deduce the required result by induction: 'Let S_m be the statement that $b_1(m, r; p) = b(m, r; p)$ for $0 \leq r \leq m$. Then S_{n-1} implies S_n, as we see by comparing (D2) with (D1). But S_0 is true ....'   □

We shall see later a more illuminating connection between Lemma 10Ma, Pascal's Triangle and the Binomial Theorem.

Method 2. We have

$$\mathbb{P}(A \mid B) \ = \ \frac{r}{n}$$

because 'it is the chance that the first toss is one of the r which produce Heads out of the n tosses'. Since

$$\mathbb{P}(A \cap B) \ = \ \mathbb{P}(A)\mathbb{P}(B \mid A) \ = \ \mathbb{P}(B)\mathbb{P}(A \mid B),$$

we have

$$b_1(n, p; r) \ = \ p \frac{n}{r} b_1(n-1, p; r-1) \ = \ p^2 \frac{n(n-1)}{r(r-1)} b_1(n-2, p; r-2)$$

$$= \ p^r \frac{n(n-1)\cdots(n-r+1)}{r(r-1)\cdots 1} b_1(n-r, p; 0).$$

But $b_1(n-r, p; 0)$, the probability of all Tails in $n-r$ tosses, is q^{n-r}; and we have the desired formula $b_1(n, p; r) = b(n, p; r)$. $\qquad\qquad\square$

Da. Exercise. For the 'hypergeometric' situation at Exercise 17Rb, let A and B be the events:

$$A : \text{'}1 \in S \cap T\text{'}, \quad B : \text{'}|S \cap T| = k\text{'}.$$

By considering $P(A)$, $\mathbb{P}(B \mid A)$, $\mathbb{P}(B)$, $\mathbb{P}(A \mid B)$, prove that

$$\frac{s}{n} \times \frac{t}{n} \times H(n-1, s-1, t-1; k-1) \ = \ H(n, s, t; k) \times \frac{k}{n}.$$

E. Solution to Problem 19B.
We have already seen that the answer to Part (a) is $x = 2$. The answer to Part (b) is therefore as easy as $2 + 2 = 4$. The reason is that if I wait for the first H (on average, 2 tosses) and then wait for the next T to occur (on average, a further 2 tosses), then I will have for the first time the pattern HT (after on average a total of 4 tosses).

Note that the time of the next H after the first H is not necessarily the time of the first HH. This explains the difference between the 'HT' and 'HH' cases.

Let $x = 2$ be the average wait for the first Head. Let y be the average total wait for HH. Then

$$y \ = \ x + 1 + (\tfrac{1}{2} \times 0) + (\tfrac{1}{2} \times y), \quad \text{whence } y = 6.$$

This is because I have to wait on average x tosses for the first Head. I certainly have to toss the coin 1 more time. With probability $\frac{1}{2}$, it lands Heads, in which case I have the HH pattern and 0 further tosses are needed; with probability $\frac{1}{2}$ it lands Tails and, conditionally on this information, I have to wait on average just as long (y) as I did from the beginning.

For a simulation study of this problem, see the last section of this chapter.

F. Important Note.   The correct intuition on which we relied in the above solution implicitly used ideas about the process 'starting afresh' after certain random times such as the time when the first Head appears. This is a trivial case of something called the '*Strong Markov Property*'. We shall continue in blissful ignorance of the fact that we are using this intuitively obvious result until we prove it in Subsection 129C.

G. Miscellaneous Exercises

▶ **Ga. Exercise.**   Let y_n be the average number of tosses before the first time I get n Heads in succession if I repeatedly toss a coin with probability p of Heads. Prove that

$$y_n = y_{n-1} + 1 + (p \times 0) + (q \times y_n)$$

whence

$$p\,y_n = y_{n-1} + 1, \quad y_0 = 0.$$

Deduce, by induction or otherwise, that

$$y_n = \frac{1}{q}\left(\frac{1}{p^n} - 1\right).$$

▶ **Gb. Exercise: Unexpected application to gaps in Poisson processes.**   Shopkeeper Mr Arkwright has observed that people arrive completely randomly at his shop, with, on average, $\lambda = 1$ person(s) arriving per minute. He will close the shop on the first occasion that $c = 6$ minutes have elapsed since the last customer arrived. Show that, on average, he keeps the shop open for approximately 6 hours and 42 minutes. Indeed, prove that the exact answer is $(e^{\lambda c} - 1)/\lambda$ minutes.

Hint. Use the previous exercise. God takes a very large integer N, and divides time into little intervals of length $\delta = 1/N$ minutes. At times $\delta, 2\delta, 3\delta, \ldots$, God tosses a coin with probability p of Heads; and if the coin falls *Tails*, causes a customer to arrive at the shop at that instant. What is the value of $q = 1 - p$ (with negligible error compared to $1/N$?) For some value of n which you must find, Mr Arkwright will close the shop the first time that God gets n Heads in a row. Apply Exercise Ga, and then let $N \to \infty$ to get the exact answer, using the well-known fact that

$$\left(1 + \frac{y}{k}\right)^k \to e^y \quad \text{as } k \to \infty.$$

▶ **Gc. Exercise: Which pattern first?: easy case.**   If I keep tossing a fair coin, what is the chance that I get
    (a) Pattern HH before Pattern HT,
    (b) Pattern TH before Pattern HH?

▶ **Gd. Exercise*: Which pattern first?: weird case.**   Consider the 8 possible patterns of length 3:

HHH, HHT, HTH, HTT, THH, THT, TTH, TTT.

Show that if I name *any* of these patterns, then you can name another one of them such that the chance that the pattern you name appears before the one I name in a sequence of tosses of a fair coin is strictly greater than $\frac{1}{2}$. Most people find this very counter-intuitive.

This is an excellent example to illustrate that sometimes you have to live with a problem before you can solve it.

1.3   Probability as Pure Maths

The way it is played these days, Probability is a branch of Pure Maths, derived from certain axioms in the same way as (say) Group Theory is. The fundamental rôle of the Addition Rule makes Probability part of a subject called Measure Theory which studies length, area, volume, and various generalizations. The 'Independence means Multiply' Rule relates to the 'product-measure' construction which allows us to construct area, volume, etc, from length.

▶▶ **A. Important discussion: What to do about Measure Theory?** I hope that it is already apparent that I believe that intuition is much more important than rigour; but, as I have said earlier, I also believe in clarity and have never believed that fudging adds to clarity. I therefore have a problem here because the first stages of Measure Theory are too difficult for inclusion in a book at this level. But does that mean that we should completely deny ourselves the great advantages of having *some* understanding of that subject? Let us consider an analogous case.

What to do about real numbers in Analysis? Real Analysis makes great claims to be so rigorous: after all, doesn't it have the dreaded ε-δ method?! Real numbers are the foundation of Analysis. Yet, increasingly, students are taught the ε-δ method in considerable detail *without* being shown that the real numbers exist! Students are asked to take *for granted* that the set $\mathbb{R}$ of real numbers has certain properties: that it is a complete ordered field. The question 'How do we know that such a complete ordered field exists?' is ducked. But that approach – as long as it is made honestly – is valid: the construction of the real numbers (either as 'Dedekind cuts' or (much better) as 'equivalence classes of Cauchy sequences of rationals') is too difficult at that stage.

Measure Theory is the foundation of Probability. Borel, Carathéodory, Fubini and (especially) Lebesgue constructed Measure Theory very early last century; and, substantially extending ideas of Borel, Wiener and others, Kolmogorov used it to build the rigorous foundations of modern Probability Theory in 1932. Are we to insult those great mathematicians by ignoring their work completely and regarding Probability as the shambles it was prior to that work? I think not, especially since the *results* of Measure Theory are easy to understand. And I am determined not to tell you things which you would have to unlearn later if you take the subject further. Several references are made to my book [W], [235], for proofs of measure-theoretic results.

It has been traditional in Statistics to say: "Under suitable conditions, the following result holds." Now, it is very silly to scoff at this with a purist remark that it is meaningless. What the statisticians mean is clear: "The result is known to be true in all the cases which we shall meet; but even to *state* the technical conditions under which it holds would complicate things to an unacceptable extent."

By contrast, in Probability, it is possible to state *quite simply* EXACTLY the conditions under which such results as the Strong Law of Large Numbers ('Law of Averages') hold *and* to show what goes wrong when they fail. By taking for granted a few things from Measure Theory, we can get really clear understanding. As mentioned in the Preface, the 'Two-Envelopes Paradox' concerns a simple problem where clear, indeed measure-theoretic, understanding is essential.

So, here's my solution. Everywhere, I shall concentrate on intuition. But I *shall* in passages marked with an Ⓜ symbol provide some measure-theoretic results, which we shall take for granted, so that the whole edifice does stand on secure foundations. You *could* leave the Measure Theory to a second reading if you so wish, but *please* give it a try the first time through.

There is one section of Chapter 2 which contains a reasonable chunk ($3\frac{1}{2}$ pages, counting motivation, magic, discussion, etc) of Measure Theory (in passages marked with the Ⓜ symbol). There is much less Measure Theory in Chapter 3, and hardly any at all in most subsequent chapters.

To continue ... Great fun can be had playing Probability as Pure Maths. One of the surprising developments of the second half of the 20th century was the way in which Probability links deeply with areas of Maths such as Complex Analysis, Potential Theory, Partial Differential Equations, Differential Geometry. It is amazing that in such a well-developed subject as Complex Analysis, the *first* solutions of certain important problems were obtained via the use of Probability.

Using an axiomatic system is great, because the fundamental entities are not defined: they just *are*. An abstract group is a set of elements carrying a 'multiplication' satisfying certain rules: we do not *define* what an individual 'element' is. In Probability, no attempt is made to *define* probability in the real world. Any attempt at such a definition is doomed; and the magic of Maths is that we do not need to – indeed, we must not – make such an attempt. The following discussion will be expanded later.

B. Discussion. The idea of probability as 'long-term relative frequency' is an extremely helpful intuitive one, as we have seen; but it just *cannot* be made into a rigorous definition. Just suppose that I try to define the probability p that this coin in my pocket will fall Heads as the long-term proportion of Heads (the limiting value, assumed to exist, of the Number of Heads divided by the Number of Tosses) if I were to throw the coin infinitely often. Apart from certain impracticalities in this 'definition', it is fundamentally flawed from the point of view of logic. If we assume that the long-term proportion will be p for *every* sequence of tosses which could be named in advance, then this statement would have to be true for the sequence consisting only of those tosses which (during the experiment) produce Heads (and for the corresponding sequence for Tails).

Trying to be 'precise' by making a *definition* out of the 'long-term frequency' idea lands us in real trouble. Measure Theory gets us out of the difficulty in a very subtle way discussed in Chapter 4. As wise and subtle Francis Bacon wrote: *"Histories make men wise; the mathematics, subtile"*.

Remarks. I add some remarks which can serve as pointers to why we need Measure Theory.

In a way, the difficulty concerning the 'long-term relative frequency' idea discussed above is related to a much simpler situation. Suppose that I choose a number between 0 and 1 uniformly at random. (Thus the chance that the chosen number will fall in a subinterval (a, b) of $[0, 1]$ will be $b - a$.) Then, before the experiment, the probability that I will choose any particular number is 0: there are infinitely many numbers all 'equally likely' in some sense; yet the probability that I pick *some* number is 1. Exactly how does this tally?

Contrast the following situation. I *cannot* think of 'choosing a positive integer uniformly at random' in such a way that each positive integer has the same probability p of being chosen as any other. Clearly, p must be 0; and now, there is no way of making things tally. Likewise, I cannot think of choosing a *rational* number in $[0, 1]$ 'uniformly at random'.

1.4 Probability as Applied Maths

Of course, Probability is more than a mathematical game. *Experience shows that mathematical models based on our axiomatic system can be very useful models for everyday things in the real world. Building a model for a particular real-world phenomenon requires careful consideration of what independence properties (or conditional-probability properties) we should insist that our model have.* We shall look at random walks, martingales, Poisson processes, the normal distribution and the Central Limit Theorem. Many of the random processes which we study are in fact Markov chains; but we do not study the general Markov chain because James Norris's recent book [176] covers that theory so well.

It is tempting to argue that our axioms of Probability are so 'self-evident' that the real world must conform to them. However, at its most basic level, our Universe is a universe of elementary particles: electrons and so on. We know that the behaviour of electrons, etc, is governed entirely by probabilistic laws. Yet the Quantum–Probability laws of calculation are very different from those used in the everyday Probability which we study in this book. (To illustrate this, I shall later describe the mind-bending situation relating to Bell's Inequality and the experiments of Alain Aspect and others.) It is somewhat perplexing that the Quantum Probability calculus is profoundly different from the Classical Probability calculus, but there it is. **We cannot dictate how the real world is to behave.**

Our version of Probability *is* as good as we need for the many important applications within Statistics, and for the large number of direct applications within Science and Engineering which do not require Statistics because we know the various parameters either from symmetry or from (say) Physics or from huge data sets.

1.5   First remarks on Statistics

Sadly, it is the case that, as mentioned in the Preface, the world's probabilists and the world's statisticians form two separate communities with relatively little interaction between them. The two subjects do have different ethos; and probabilists' natural inclination towards Pure Maths and/or Physics has tended to lead them further and further from Statistics. But this is crazy: Statistics is based on Probability; and much Probability can be applied to the real world only with the help of Statistics. The extremely important applications of Statistics jolly well ought to interest probabilists and students of Probability. I therefore include in this book some first steps in Statistics (to accompany the first steps the book takes in Probability).

> **True story.**   I learnt the value of Statistics early. In a lecture to students, I had mentioned the result in Subsection 13O that if you have 23 people in a room then there is a probability of more than $\frac{1}{2}$ that two of them have the same birthday. One student came to see me that evening to argue that 23 wasn't anywhere near enough. We went through the proof several times; but he kept insisting that there must be a flaw in the basic theory. After some considerable time, I said, "Look, there must be 23 people still awake in College. Let's ask them their birthdays. To start with, when's yours?" "April 9th", he said. I replied (truthfully!!) "So's mine".

▶▶▶ **A. Probability and Statistics.** Here's an attempt to explain how Probability and Statistics differ and also how they are fundamentally connected.

- *In Probability, we do in effect consider an experiment* **before** *it is performed.* Numbers to be observed or calculated from observations are at that stage Random Variables to be Observed or Calculated from Observations – what I shall call *Pre-Statistics*, rather nebulous mathematical things. We **deduce** *the probability of various outcomes of the experiment in terms of certain basic parameters.*

- *In Statistics, we have to* **infer** *things about the values of the parameters from the observed outcomes of an experiment* **already** *performed.* The Pre-Statistics have now been 'crystallized' into *actual statistics*, the observed numbers or numbers calculated from them. The precise mathematical formulation of the 'crystallization' will be studied later. Pre-Statistics will be denoted by Capital (Upper-Case) Roman Letters, actual statistics by lower-case roman letters.

- *We can decide whether or not operations on actual statistics are sensible only by considering probabilities associated with the Pre-Statistics from which they crystallize. This is the fundamental connection between Probability and Statistics.*

B. General Remarks. The main reason for including the 'Introduction to Probability' in Sections 1.1 and 1.2 was to get far enough into that subject to appreciate that it is interesting before starting on its (initially less interesting) theory. There is less reason for a similar 'Introduction to Statistics' because, once we have the requisite Probability, Statistics is immediately interesting.

It is also immediately controversial! There are no absolutely right answers; and one finds oneself saying; "On the one hand, ...; but on the other, ...; but one can counter that with ...; etc, etc." We don't want too much of that sort of thing in an Introduction. The only statement about Statistics with which everyone would agree is that it is important.

Still, some remarks are in order.

For the simple situations which we consider, I shall not express conclusions about data in terms of Hypothesis Testing – and *that* for Frequentist reasons as much as for Bayesian ones. *I believe that, whenever possible, one should express conclusions about data in terms of appropriate Confidence Intervals (CIs) and Confidence Regions (CRs).*

My criticisms of formal Hypothesis Testing in *simple* situations should not be taken to mean that all Hypothesis Testing is worthless. I believe that it retains value in certain situations: for example, in tests of independence. The idea has an important part to play in exploratory work, where it allows one to test quickly whether certain factors are likely to prove important or not before one proceeds (if there is now any point in so doing) to a more thorough analysis leading if possible to Confidence Intervals or whatever. I do not want to give the idea either that one should ignore books which follow very strictly the Hypothesis Testing route. Indeed, one can acquire a very great deal of statistical wisdom from (for example) Freedman, Pisani and Purves [83].

▶▶ **C. Least squares, regression, ANOVA.** I do want to say a word or two here about the wonderful ideas connected with the Analysis of Variance (ANOVA) method which form a recurring theme. Of course, we shall discuss normal distributions, variance, etc, in full later; but I'm sure that you will already know something about these things. Such details do not really matter in this subsection: it is enough to know that there *are* things called normal distributions, variances, t distributions. *Least-squares ideas tie in perfectly with certain 'likelihood' geometry associated with the normal distribution. I try hard to make this geometry as simple as possible: it can be made quite complicated!*

The idea of the Classical Theory is to use a mathematical model

Observation as Pre-Statistic   =   systematic part + True Error,

True Errors for different Observations being independent normally distributed Random Variables (RVs) with zero mean (because the systematic part has taken up the mean) and unknown common variance σ^2. The simplest case is when we are going to make n measurements $Y_1, Y_2, \ldots, Y_n$ (at the moment Pre-Statistics) of some fixed quantity:

$$Y_k = \mu + \varepsilon_k, \tag{C1}$$

μ being the true value of the quantity and the ε_k Observational Errors. (Since the ε_k can never be observed or calculated from observations, they are Random Variables, but not Pre-Statistics.) After the experiment is performed, we have n actual statistics (actual numbers) $y_1, y_2, \ldots, y_n$, and we mirror (C1) via

$$y_k = m + e_k, \tag{C2}$$

where m is a number, the 'estimated systematic part' and e_k is an unexplained 'residual'. The likelihood geometry tells us that the correct thing is to do a least-squares analysis, choosing m so as to minimize the *residual sum of squares*

$$\text{rss} := \sum e_k^2 = \sum (y_k - m)^2.$$

We find that $m = \bar{y}$, the average of the y_k, and use m as the best estimate of μ. We then use $\text{rss}/(n-1)$ to estimate σ^2, and use this estimate to obtain a (t distribution) CI for μ centred on m. *To justify this, we have to study the Pre-Statistics M and* RSS *which crystallize to the actual statistics m and* rss.

In Linear Regression, we measure (say) the electrical conductivity Y_k of a certain material at temperature x_k, where the 'covariates' $x_1, x_2, \ldots, x_n$ are non-random numbers predetermined by the experimenter in some small temperature range which makes the following linearity assumption reasonable. Before the experiment, we postulate that Nature's model is

$$Y_k = \mu + \beta(x_k - \bar{x}) + \varepsilon_k.$$

Hence, μ is the mean value of Y when $x = \bar{x}$, and β, the slope of the regression line, is a first indication of how conductivity changes with temperature. As usual, we assume that the 'errors' ε_k, a combination of experimental errors and 'random effects' (a term which will later have a different meaning) are independent Random Variables each with the normal distribution of mean 0 and unknown variance σ^2. (There are very good reasons for 'shifting the origin of the x-values to $\bar{x}$' in the way we have done. All computer programs should follow suit! But not all do!) After the experiment, we mirror Nature's model with

$$y_k = m + b(x_k - \bar{x}) + e_k.$$

We choose m and b so as to minimize $\text{rss} = \sum e_k^2$, and then use m and b as best estimates for μ and β. By considering the Pre-Statistics M, B, RSS corresponding to m, b, rss, we obtain CIs for μ and β centred on m and b by using $\text{rss}/(n-2)$ to estimate σ^2 and then using a t-distribution. If we have 97.5% CIs (μ_-, μ_+) for μ and (β_-, β_+) for β, then we can be at least 95% confident that both $\mu \in (\mu_-, \mu_+)$ and $\beta \in (\beta_-, \beta_+)$ are true.

We now consider two-factor ANOVA without interaction. Though agricultural field trials are done in a more sophisticated way than I now describe, the following remarks can still give the idea of ANOVA. We might (for example) apply J fertilizers $1, 2, \ldots, J$ in a sensible way to equal areas in each of K fields $1, 2, \ldots, K$. We assume a model for yields:

$$Y_{jk} = \mu + \alpha_j + \beta_k + \varepsilon_{jk},$$

where μ is an overall mean, α_j is the amount by which fertilizer j increases the yield on average, β_k is associated with the natural fertility of the field k. Because μ is an overall mean, the α_j and β_k will have mean 0. After the experiment, ... (Guess what!) You will know by now that I have no interest in the test of the Null

Hypothesis that all fertilizers are equally good, except as a first exploratory step. Rather, I shall want to quantify differences between fertilizers via Confidence Intervals.

Let us focus on one defect in the experiment just described. (You will be able to think of other defects.) One fertilizer may be better than another on one type of soil, worse on another. So, we upgrade to a model (Two-factor ANOVA with interaction and replication)

$$Y_{jk\ell} = \mu + \alpha_j + \beta_k + \gamma_{jk} + \varepsilon_{jk\ell},$$

where now j refers to fertilizer type, k to soil type, $jk\ell$ to the ℓ-th region of soil type k treated with fertilizer j, and γ_{jk} is an interaction term concerning how well fertilizer j works on soil type k. We need more than one value of ℓ in order to estimate the natural underlying variation described by σ^2. Of course, we arrange the treated plots sensibly. And, certainly, we need to give careful consideration to whether the assumptions of the model tally with common sense and with the data.

And so on; and so on. There is essentially no limit: we can have many factors, all kinds of interactions.

I never did like the standard '$\mathbf{Y} = X\beta + \varepsilon$' method of doing ANOVA. I hope that you will find the way presented here easier *because it better respects the structure* and *allows one to see the geometry* in the basic cases which we consider. Our ANOVA cases possess the special property of having 'orthogonal factors'. They include the examples given above, and are the essential first step to understanding ANOVA, the only cases usually considered in detail in books at this level. But they *are* special, and many cases of experimental design do not fit the 'orthogonal factor' pattern which we study.

Our development of the theory *does* allow non-orthogonal factors, and we look rather carefully at non-orthogonality in connection with Regression. Moreover, we *do* study the '$\mathbf{Y} = X\beta + \varepsilon$' model, partly to show that we can cope with it and understand its geometry.

Watch how considerations of 'sums of squares' develop throughout the book, beginning with the very definition of variance, and guiding the Probability at Lemma 69Ja and elsewhere.

The likelihood geometry of the normal distribution, the most elegant part of Statistics, has been used to great effect, and remains an essential part of Statistics culture. What was a good guide in the past does not lose that property because of advances in computing. But, as we shall see, modern computational techniques mean that we need no longer make the assumption of 'normally distributed errors' to get answers. The new flexibility is exhilarating, but needs to be used wisely.

D. A final remark for the moment. Statistics is, fundamentally, a practical subject justified by its applications to the real world. The subject *proper* starts *after* the mathematical theory (which itself is only begun in this book) is complete. A statistician has to be steeped in the field of application, ideally being involved in deciding how data are to be collected. Possible mathematical models should then be discussed with workers in the field of application. When an appropriate model (or family of models) is agreed, and when the extent of prior knowledge about the values of various parameters is ascertained, analysis of data can begin. Any outliers in the data (results which seem 'out of keeping' with the others) which are shown up by this analysis will need close investigation to see whether they are genuine or the result of experimental error or recording error.

We spend much time examining the analysis of data in simple cases, assuming agreed models and no outliers. We *shall* have some discussion on identifying outliers, and shall discuss (for example) testing of stability of methods via simulation. More importantly, we shall look at how modern computational methods do allow us to utilize more realistic models.

One has to begin somewhere; and if one doesn't know what is in this book and more, then one can do nothing.

1.6 Use of Computers

A. Statistical Packages. It is self-evident that all numerical computations in Statistics should now be done with suitable computer packages. The `Minitab` package is widely used for teaching because it is very easy to use and because it is affordable. I shall include some `Minitab` examples and some examples from the remarkable `WinBUGS` package. Switching packages is not difficult.

Comments on the very important `S-PLUS` and R packages may be found in the Preface.

B. Simulation. The ability of computers to generate *pseudo-random numbers* which are for many purposes reasonable simulations of truly random sequences, has many uses.

In Probability, we approximate the probability of an event by the proportion of times that that event occurs in a long series of simulations of the experiment. Here's an example which you are certainly bright enough to understand in principle (using the comments within /* ... */) even if you have never seen 'C' before.

▶▶ **Ba. Simulation example: A simulation of the average waiting time for HH.**

'C' is not my favourite language, but it is so widely used that it seems sensible to use it here. This simulation uses a Random Number Generator RNG.o which we shall study later.

Only the procedure NextToss() *and the 'function'* main() *matter.* The 'C' compiler always begins at the function main() which here returns the value 0 after the program is completed. The instruction \n within a printf command tells the printer to go to a new line. The scanf is C's highly bizarre way of reading in data. The the first scanf tells the machine to input the value of the long integer Nexpts to be stored in address &Nexpts.

```
/*HH.c : Average Wait for HH for fair coin           DW
 *To compile:   cc HH.c RNG.o -o HH -lm
 *(RNG.o   compiled Random-Number Generator)
 *(-lm to call up mathematics library)
 *To run:    ./HH
 */

#include "RNG.h"  /* Header file -- explained later */
long int Ntosses; /* Total number of tosses */
long int Nexpts;  /* Total number of experiments */
int next;         /* 1 if next toss gives 'H', 0 if 'T'  */
int showall;      /* 1 to show all results, 0 just for average */
double average;

GetParameters(){
  printf("\nInput total number of experiments:    ");
  scanf("%ld", &Nexpts);  /* Read in Nexpts */
  printf("Input  1 to show all results, 0 just for average:    ");
  scanf("%d",&showall);   /* Read in showall */
  setseeds();             /* from RNG */
}

void NextToss(){
  Ntosses++;    /* C parlance for 'Increase Ntosses by 1'  */
  next = WithProb(0.5);
    /* (from RNG) next = 1 with probability 0.5, 0 otherwise  */
  if(showall==1){if (next==1) printf("H");else printf("T");}
}

int main(){
  int last;
  long int k;
```

```
GetParameters();
Ntosses = 0; printf("\n");
for(k = 0; k < Nexpts; k++){ /* do Nexpts times */
  NextToss();
  do{last = next; NextToss();}while (last + next < 2);
  if (showall == 1) printf("\n");
}   /* end for loop */

average = Ntosses/(double)Nexpts;
printf("\nAverage wait for HH was %8.5f\n\n", average);
return 0;
}
```

with results (lines 3–8 relating to setseeds() in a way explained later):

```
Input total number of experiments:   8
Input  1 to show all results, 0 just for average:   1
Which generator?
Input 1 for Wichmann-Hill, 2 for Langlands
1
Input 3 positive-integer seeds
a<30269, b<30307, c<30323
6829 3716 2781
HTTTHH
HH
TTHH
THTTTHTHTHTHTTHTTHTTHTHH
THH
TTTTHTTHTTHH
HH
TTHH
Average wait for HH was  6.87500
-------------------------------
Input total number of experiments:   1000000
Input  1 to show all results, 0 just for average:   0
Which generator?
Input 1 for Wichmann-Hill, 2 for Langlands
1
Input 3 positive-integer seeds
a<30269, b<30307, c<30323
2791 7847 5395
Average wait for HH was  5.99246
-------------------------------
Input total number of experiments:   1000000
Input  1 to show all results, 0 just for average:   0
Which generator?
```

```
Input 1 for Wichmann-Hill, 2 for Langlands
2
Input 3 positive-integer seeds
a,b,c < 65536
3471 7745 58861
Average wait for HH was  5.99899
```

There are many complicated situations where analytic solutions seem impossible and where simulation seems to be the only available method for studying qualitative behaviour. There are other situations, much more complicated still, where one cannot envisage simulation as ever being of any use, and where analysis does work.

C. Simulation in Statistics. The use of **Gibbs sampling** and other 'MCMC' techniques completely transforms Statistics, as we shall see. There are many other uses of simulation in Statistics: for the bootstrap of Simon and Efron, see, for example, Efron and Tibshirani [71], Davison and Hinkley [55].

EVENTS AND PROBABILITIES

We start again 'from scratch', developing Probability in the only order that logic allows.

This chapter is important, but not too exciting (except in discussion of what is perhaps the most brilliant piece of magical trickery in Maths): important in that it covers the Addition Rule; not too exciting in the main because it does not go beyond the Addition Rule to the Multiplication Rules involving conditioning and independence. Without those Multiplication Rules, we can assign actual probabilities to events only in an extremely limited set of cases.

As mentioned at Discussion 25B, we are concerned with constructing a Mathematical Model for a Real-World Experiment: we make no attempt to *define* probability in the real world because such an attempt is doomed. In the mathematical model, *probability* $\mathbb{P}$ is just a function which assigns numbers to *events* in a manner consistent with the Addition Rule (and with conditional-probability considerations) – and nothing more. And events are just (technically, 'measurable') subsets of a certain set, always denoted by Ω. This set is called the *sample space* and it represents the set of all possible outcomes ω of our experiment. In Statistics, the experiment has *been* performed, and we have some information – in many cases, *all* information – about the *actual* outcome ω^{act}, the particular point of Ω actually 'realized'.

The relation between 'sample' in Statistics and our 'sample point' and 'sample space' will become clear as we proceed.

2.1   Possible outcome ω, actual outcome ω^{act}; and Events

▶▶ **A. Sample space Ω of possible outcomes ω.** As was stated earlier, Probability considers an experiment *before* it is performed, and Statistics considers an experiment *after* it has been performed. Probability considers an abstract set Ω, the *sample space*, which represents the set of all possible outcomes of our experiment. A *possible outcome*, or *sample point*, ω is mathematically a point of the 'abstract' set Ω.

For example, for the experiment of tossing a coin three times,

$$\Omega \; = \; \{\texttt{HHH}, \texttt{HHT}, \texttt{HTH}, \texttt{HTT}, \texttt{THH}, \texttt{THT}, \texttt{TTH}, \texttt{TTT}\}, \tag{A1}$$

a finite set.

For the experiment of choosing a point at random between 0 and 1, we naturally take $\Omega = [0, 1]$, an uncountable set. See Appendix A4, p496, for a discussion of countable and uncountable sets.

For the experiment, discussed at Subsection 19B, of tossing a coin until the first time we obtain HH, we can justify taking

$$\Omega \; = \; \{\texttt{HH}, \texttt{THH}, \texttt{HTHH}, \texttt{TTHH}, \dots\},$$

a countable set. However, there *is* the difficulty, which *must* be addressed, that there are uncountably many 'crazy' outcomes in which the pattern HH never occurs. The set of all such crazy outcomes must be *proved* to have probability 0 before we can reduce to the desired Ω.

▶▶ **B. Crystallization of actual outcome; realization.** Our picture is that Tyche, Goddess of Chance, chooses a point ω^{act} of Ω 'at random and *in accordance with the probability law* $\mathbb{P}$' in a sense explained below. This is Tyche's 'experiment', and it determines the *actual outcome* ω^{act} in the real world. The whole sample space Ω is a nebulous, abstract thing. Tyche's choice, as it were, '*crystallizes into existence*' one sample point ω^{act} which becomes *real*. It is important that we regard Tyche's single choice of ω^{act} as determining the entire outcome, the entire *realization*, of the real-world experiment. If the real-world experiment consists of many stages (for example, many coin tosses), then Tyche is in a sense revealing her choice to us in instalments. But she made just the one choice ω^{act}.

▶▶ **C. Event as (measurable) subset of Ω.** For the 'Toss coin three times' experiment, the real-world event '2 Heads in all' occurs if and only if ω^{act} belongs to the subset $\{\texttt{HHT}, \texttt{HTH}, \texttt{THH}\}$ of Ω at (A1) consisting of those possible outcomes

ω which would produce 2 Heads in all. In the mathematical theory, the event '2 Heads in all' is regarded as the subset $\{\text{HHT}, \text{HTH}, \text{THH}\}$ of the sample space Ω. And so for *any* event for *any* experiment.

(*Technical note.* Except in simple cases, essentially those in which Ω is finite or countable, we have to restrict the type of subset of Ω which constitutes an 'event': an event is a 'measurable' subset of Ω.) We therefore have the following table in mind:

Probability Model	Real-world interpretation
Sample space Ω	Set of all outcomes
Point ω of Ω	Possible outcome of experiment
(No counterpart)	Actual outcome ω^{act}
Event F, 'measurable' subset of Ω	The real-world event corresponding to F occurs if and only $\omega^{\text{act}} \in F$
$\mathbb{P}(F)$, a number	Probability that F will occur for an experiment yet to be performed.

We see that Tyche must perform her experiment in such a way that she will choose ω^{act} to be in the set F 'with probability $\mathbb{P}(F)$'. This is what is 'meant'(!) by 'Tyche chooses in accordance with the law $\mathbb{P}$'. Because we are now regarding events F, G, etc, as subsets, we can extend this table:

Event in Maths	Real-world interpretation
Ω, the entire sample space	The certain event 'something happens'
The empty subset $\emptyset$ of Ω	The impossible event 'nothing happens'
The intersection $F \cap G$	'Both F and G occur'
$F_1 \cap F_2 \cap \ldots \cap F_n$	'All of the events $F_1, F_2, \ldots, F_n$ occur simultaneously'
The union $F \cup G$	'At least one of F and G occurs'
$F_1 \cup F_2 \cup \ldots \cup F_n$	'At least one of $F_1, F_2, \ldots, F_n$ occurs'
Complement F^c of F	'F does not occur'
$F \setminus G$	'F occurs, but G does not occur'
$F \subseteq G$	If F occurs, then G must occur

▶▶ **D. Set theory and Probability.** Combining events is exactly the same as combining sets in elementary set theory. I am sure that you know about Venn diagrams. Figure D(i) illustrates parts of the table just considered. The amazing fact (explained

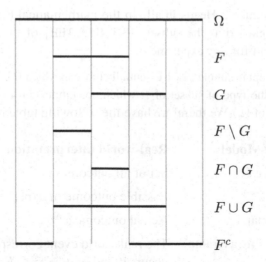

$$\Omega$$

$$F$$

$$G$$

$$F \setminus G$$

$$F \cap G$$

$$F \cup G$$

$$F^c$$

Figure D(i): A simple Venn diagram

later) that the Fundamental Experiment

Choose a number between 0 and 1 uniformly at random

is Universal in that every other experiment is contained within it makes the 1-dimensional pictures particularly appropriate, especially for constructing counterexamples.

You should practise giving probabilistic interpretations of set-theoretic results. For example, one of de Morgan's rules states that

$$(F_1 \cap F_2 \cap \ldots \cap F_n)^c = F_1^c \cup F_2^c \cup \ldots \cup F_n^c. \tag{D1}$$

In real-world terms, with L for the left-hand side and R for the right-hand side of (D1),
L: it is *not* true that *all* of the events $F_1, F_2, \ldots, F_n$ *do* occur;
R: it *is* true that *at least one* of the events $F_1, F_2, \ldots, F_n$ does *not* occur.
Thus (D1) is obvious.

Or again, consider the distributive law:

$$A \cap (B \cup C) = (A \cap B) \cup (A \cap C).$$

Exercise. Use 'event' language (so that the left-hand side signifies 'A definitely occurs and at least one of B and C occurs') to make the distributive law 'obvious'.

[*Note.* Whether an electron regards the distributive law as even being true (even when $C = B^c$) is a moot point. Some people believe that the appropriate logic for quantum theory is 'non-distributive'. Once again, I emphasize that we cannot force Nature to obey our ideas of 'obvious'.]

2.2   Probabilities

Mathematically, probability is just a function $\mathbb{P}$ which assigns to each event F a number $\mathbb{P}(F)$ in $[0, 1]$ such that

$$\mathbb{P}(\Omega) = 1$$

and the so-called Addition Rule holds. One important part of the Addition Rule states the following.

▶ **A. Addition Rule, Part 1 – an Axiom.** *For events F and G,*

$$F \cap G = \emptyset \quad implies \quad \mathbb{P}(F \cup G) = \mathbb{P}(F) + \mathbb{P}(G).$$

If $F \cap G = \emptyset$, then F and G are called *disjoint* or *exclusive*. I emphasize: in the mathematical theory, Rule (A) is an axiom that the probability laws $\mathbb{P}$ which we use must satisfy: and *we do **not** prove this rule*. We have seen its motivation at Subsection 6F.

▶ **B. Lemma.** *(a) For any event F, $\mathbb{P}(F^c) = 1 - \mathbb{P}(F)$.*
(b) For any two events F and G, we have the **Inclusion-Exclusion Principle**

$$\mathbb{P}(F \cup G) = \mathbb{P}(F) + \mathbb{P}(G) - \mathbb{P}(F \cap G).$$

Proof. Part (a) appeared in Subsection 6F.

Part (b) may be seen as follows. Figure 38D(i) may help. The event F may be written the disjoint union of $F \setminus G$ and $F \cap G$. So,

$$\mathbb{P}(F) = \mathbb{P}(F \setminus G) + \mathbb{P}(F \cap G);$$

and, since $F \cup G$ is the disjoint union of $F \setminus G$ and G,

$$\mathbb{P}(F \cup G) = \mathbb{P}(F \setminus G) + \mathbb{P}(G) = \mathbb{P}(F) - \mathbb{P}(F \cap G) + \mathbb{P}(G),$$

the desired result. ☐

C. Lemma. *We have the* extended Addition Rule: *if the events $F_1, F_2, \ldots, F_n$ are disjoint (that is, $F_i \cap F_j = \emptyset$ whenever $i \neq j$),*

$$\mathbb{P}(F_1 \cup F_2 \cup \ldots \cup F_n) = \mathbb{P}(F_1) + \mathbb{P}(F_2) + \cdots + \mathbb{P}(F_n).$$

In shorthand:

$$\mathbb{P}\left(\bigcup_{k=1}^{n} F_k\right) = \sum_{k=1}^{n} \mathbb{P}(F_k).$$

Proof. If A, B and C are disjoint events, the $A \cup B$ is disjoint from C, so that two applications of the Addition Rule for disjoint sets give

$$\mathbb{P}(A \cup B \cup C) = \mathbb{P}((A \cup B) \cup C) = \mathbb{P}(A \cup B) + \mathbb{P}(C) = \mathbb{P}(A) + \mathbb{P}(B) + \mathbb{P}(C).$$

It is not worth dignifying the 'n-event' case with a proof by induction.   □

D. General Inclusion-Exclusion Principle.   *What about the Inclusion-Exclusion Principle for n events?* (Let me be honest and say that one reason for discussing this is so that we can at least do one problem in this chapter!)

Well, for 3 events A, B, C, it would say

$$\mathbb{P}(A \cup B \cup C) = \Sigma_1 - \Sigma_2 + \Sigma_3,$$

where
$$\begin{aligned}
\Sigma_1 &:= \mathbb{P}(A) + \mathbb{P}(B) + \mathbb{P}(C), \\
\Sigma_2 &:= \mathbb{P}(A \cap B) + \mathbb{P}(A \cap C) + \mathbb{P}(B \cap C), \\
\Sigma_3 &:= \mathbb{P}(A \cap B \cap C).
\end{aligned}$$

You can see how this extends Lemma 39B, and that the general case will read as follows.

▶ **Da. Lemma: General Inclusion-Exclusion Principle.** *Let $F_1, F_2, \ldots, F_n$ be any events. Then*

$$\mathbb{P}(F_1 \cup F_2 \cup \ldots \cup F_n) = \Sigma_1 - \Sigma_2 + \Sigma_3 - \cdots + (-1)^{n+1}\Sigma_n,$$

where Σ_r is the sum of terms

$$\mathbb{P}(F_{i_1} \cap F_{i_2} \cap \ldots \cap F_{i_r})$$

over all subsets $\{i_1, i_2, \ldots, i_r\}$ of size r from $\{1, 2, \ldots, n\}$. We shall prove this in the next chapter.

▶ **E. Example: The Hat-Matching Problem.** In this well-known problem, n absent-minded hat-wearing professors attend a meeting. At the end of the meeting, each picks a hat at random. We want to show that the probability at least one of them gets the right hat is

$$1 - \frac{1}{2!} + \frac{1}{3!} - \cdots + (-1)^{n+1}\frac{1}{n!}. \tag{E1}$$

Thus the probability that none of them gets the right hat, namely 1 minus the above expression, is exactly the sum of the first $n + 1$ terms in the expansion of e^{-1}, and so is extremely close to $1/e$ even for moderate n.

Solution. Let F_k be the probability that the kth professor to leave (Prof k) gets the right hat. Now, it is all just as if someone shuffled the whole 'pack' of hats and gave them out randomly to the various professors. So, for *every* k, the chance that the k-th professor to leave gets the right hat is $1/n$. For $i \neq j$, the chance that Prof i and Prof j both get the right hat is $(1/n) \times (1/(n-1))$. One way to see this is by using the idea from the

'National-Lottery' proof 10L of Lemma 9Ka. But since we have not done conditional probabilities, we can argue (more rigorously?!) as follows.

Since there are $n!$ permutations of the n numbers $\{1, 2, \ldots, n\}$, there are $n!$ ways of 'giving out the hats'. We need to know how many ways there are of giving out the n hats such that Prof i and Prof j both get the right hats (others being allowed to get the right hats too). In other words, we need to know how many permutations of $\{1, 2, \ldots, n\}$ keep i and j fixed. But this is just the total number of permutations of the remaining $n - 2$ numbers. Hence for $i \neq j$,

$$\mathbb{P}(F_i \cap F_j) = \frac{(n-2)!}{n!} = \frac{1}{n(n-1)},$$

and, more generally, for any subset $\{i_1, i_2, \ldots, i_r\}$ of $\{1, 2, \ldots, n\}$,

$$\mathbb{P}(F_{i_1} \cap F_{i_2} \cap \ldots \cap F_{i_r}) = \frac{(n-r)!}{n!}.$$

Hence, since by Lemma 9Ka there are $\binom{n}{r}$ subsets of size r from $\{1, 2, \ldots, n\}$,

$$\Sigma_r = \frac{n!}{r!(n-r)!} \times \frac{(n-r)!}{n!} = \frac{1}{r!},$$

and the result follows. □

F. True story – Marx brothers, please note!

I was once the last to leave a restaurant in Swansea, one Friday lunchtime. Just one anorak remained on the hooks outside. It looked exactly like mine, and I took it without thinking. When I put the anorak on that evening, I found it contained house and car keys which were not mine. A lot of detective work identified the true owner (A, say) of that anorak and keys. He lived far away, but his son who lived at Swansea had a spare set of keys to his car. I phoned A, and we agreed to meet at the restaurant on Monday morning to swap coats. I arrived before him, and since I had to go off to give a lecture, left his anorak with the restaurant manager who put it in the restaurant safe. I returned after the lecture. The restaurant manager said that A had been and gone, and that he had left my anorak in the safe. The manager gave me my anorak (it *was* mine!), and I thought that that was the end of the story. However, I later had a furious phone call from A from his home, asking what I was playing at, since the anorak he had collected from the safe (without bothering to check it – for how could it be wrong?) was *not* his! I was puzzled. But it turned out that the submanager of the restaurant was the only other person with a key to the safe – and guess what kind of anorak he had, and where A's anorak was when A collected for the second time the wrong one!

G. Exercise. (a) What is the probability that *exactly* one professor gets the right hat? Express your answer in terms of a suitable $p_{0,m}$, where this latter probability, which we

know, is the probability that if there are m professors in all, then none of them gets the right hat.

(b) What is the probability that exactly r professors get the right hat? Show that, as $n \to \infty$, this probability converges to $e^{-1}/r!$.

2.3   Probability and Measure

I explained my philosophy about Measure Theory in Section 1.3. I think that you *should*

- *have some idea what the Strong Law of Large Numbers, which underpins the whole subject, states;*

- *understand how one gets around the difficulty raised in Discussion 25B;*

- *know what pieces of logic justify things that we* must *do.*

An example of the last point is that we need to be sure that the probability that a population will become extinct at some finite (random) time is the limit as $n \to \infty$ of the probability that the population will be extinct by time n.

In any case, I now get a chance to explain to you one part of the utterly amazing Banach–Tarski Paradox!

Towards the end of the section I discuss the (most) Fundamental Model, that of choosing a number at random between 0 and 1. *Amazingly, as stated earlier, every other model, no matter how complicated the process we are modelling, is contained within this one* (as we shall become convinced later). We can use the Fundamental Model to understand clearly statements of 'Strong Law' type.

Please note that in one brief (and even interesting) page (with surprises!) at Appendix A4, p496, I include everything you need in this book about countable and uncountable sets.

▶ **A.** ⓜ **The Full Addition-Rule Axiom.** *If $F_1, F_2, \ldots$ is an infinite sequence of* disjoint *events, then*

$$\mathbb{P}\left(\bigcup_{k=1}^{\infty} F_k\right) = \sum_{k=1}^{\infty} \mathbb{P}(F_k).$$

It is impossible to prove this by 'letting $n \to \infty$ in Lemma 39C': that is the whole point. With property A one can build a marvellous theory.

But there is a price to pay. Except in very simple cases, we cannot arrange that Property A will hold if we insist that *all* subsets of Ω are events.

▶ **B. Banach–Tarski Paradox (!!!)** These Polish magician-mathematicians brought off one of the most brilliant tricks in Mathematics. They showed – assuming the Axiom of Choice (see Appendix A5, p498) – that one can find a subset A of the unit sphere $S = S^2$ (the surface of the familiar unit ball in $\mathbb{R}^3$) with the most remarkable property. First of all, we can fit together three sets which are rotations of A so that they exactly cover S without overlap. Nothing surprising so far. It seems clear that the area of A is $a/3$, where a is the area of the sphere. However, for any $n = 3, 4, 5, \ldots, \infty$, we can fit together n sets which are rotations of A so that they exactly cover S without overlap. Hence the area of A must be $a/3$ and $a/4$ and $\ldots$ and 0. Conclusion: it is impossible to assign an area to A: A is 'non-measurable'. The 'event' that a point chosen at random on S belongs to A cannot be a true event, for its probability would have to be simultaneously $1/3$, $1/4$, $\ldots$, 0.

See Wagon [233].

▶ **C. ⓜ Full Axiomatization.** In the general theory then, *not all subsets of Ω need be events*. We axiomatize things again. The class $\mathcal{F}$ of all subsets of Ω which *are* events must be what is called a σ-algebra: this means that

- $\Omega \in \mathcal{F}$,
- $F \in \mathcal{F}$ implies that $F^c \in \mathcal{F}$,
- $F_1, F_2, \ldots \in \mathcal{F}$ implies that $\bigcup F_n \in \mathcal{F}$.

Then $\mathbb{P}$ is a map $\mathbb{P} : \mathcal{F} \to [0, 1]$ such that $\mathbb{P}(\Omega) = 1$ and Property 42A holds whenever the F_k are disjoint elements of $\mathcal{F}$. We now have the full axiomatization of the Addition Rule. We say that $\mathbb{P}$ is a *probability measure* on $(\Omega, \mathcal{F})$, and that $(\Omega, \mathcal{F}, \mathbb{P})$ is a *probability triple*.

What makes things work is that

we can always set up an $(\Omega, \mathcal{F}, \mathbb{P})$ triple for the experiment which we wish to model in which $\mathcal{F}$ is large enough to contain every event of which we could ever wish to find the probability.

In the remainder of this section, $(\Omega, \mathcal{F}, \mathbb{P})$ is a probability triple used to model some experiment.

▶ **D. ⓜ Fact: Monotone-Convergence Properties.** *(a) Suppose that we have events $F_1 \subseteq F_2 \subseteq F_3 \subseteq \ldots$ and that $F = \bigcup F_n$. Then $\mathbb{P}(F_n) \uparrow \mathbb{P}(F)$: the sequence $\{\mathbb{P}(F_n)\}$ is non-decreasing with limit $\mathbb{P}(F)$.*

(b) Suppose that we have events $G_1 \supseteq G_2 \supseteq G_3 \supseteq \ldots$ and that $G = \bigcap G_n$. Then $\mathbb{P}(G_n) \downarrow \mathbb{P}(G)$: the sequence $\{\mathbb{P}(G_n)\}$ is non-increasing with limit $\mathbb{P}(G)$. (This uses the fact that $\mathbb{P}(\Omega)$ is finite.)

This is important in that, for example, F_n might be 'population is extinct at time n' and then F is 'population is eventually extinct'.

▶ **E. ⓜ Almost surely.** A statement S about outcomes is said to be *almost surely true* or *to be true with probability* 1 if the truth set T of S, the set of outcomes ω for which S is true, is an element of $\mathcal{F}$, and $\mathbb{P}(T) = 1$. If a statement is *certain*, true for *every* ω, then, of course, its truth set is Ω and, since $\mathbb{P}(\Omega) = 1$, it is almost surely true. (It had better be!) The important point is that *many of the things in which we are most interested are almost surely true without being 'absolutely certain'*. Do note that the probability of an almost sure event is exactly 1, not 99.9% or anything similar.

▶ **F. ⓜ Null set, null event.** An event N is called a *null event* or null set if $N \in \mathcal{F}$ and $\mathbb{P}(N) = 0$. We see that a statement about outcomes is almost surely true if the outcomes for which it is false form a null event.

G. ⓜ Fact. *If $N_1, N_2, \ldots$ is a sequence of null sets, then $\bigcup N_k$ is a null set. If $H_1, H_2, \ldots$ is a sequence of events each of probability 1, then $\bigcap H_k$ has probability 1.*

▶ **H. ⓜ The Fundamental Model.** Consider the experiment 'Choose a (real) number uniformly between 0 and 1. The outcome must be a real number between 0 and 1, so we take Ω to be $[0, 1]$. For each x in $[0, 1]$, we wish the statement 'chosen number is less than or equal to x' to have a probability and for that probability to be x. It is a theorem of Borel and Lebesgue that if $\mathcal{F}$ is the smallest σ-algebra of subsets of $\Omega = [0, 1]$ containing every interval $[0, x]$ where $0 \leq x \leq 1$, then there is a unique probability measure $\mathbb{P}$ on $(\Omega, \mathcal{F})$ such that $\mathbb{P}([0, x]) = x$ whenever $0 \leq x \leq 1$. We therefore have our complete model for choosing a point at random between 0 and 1. Let's call the resulting triple $(\Omega, \mathcal{F}, \mathbb{P})$ the Fundamental Triple.

▶ **I. ⓜ Null sets for the Fundamental Model.** As already mentioned, this most fundamental model contains all other models in a sense to be explained later. It is therefore good that we can understand easily what 'almost sure' or 'with probability 1' means for this model. As explained at F, we need only understand what is a null set for the Fundamental Triple. (Your fears that 'for every $\varepsilon > 0$' might feature are, I'm afraid, justified!) A set N in $\mathcal{F}$ is a null set (it has measure 0) if and only if, for every $\varepsilon > 0$ we can find a sequence of disjoint open subintervals $I_n = (a_n, b_n)$ of $[0, 1]$ (such an open interval is also allowed to have the form $[0, b)$ or $(a, 1]$ to deal with the endpoints) such that

$$N \subseteq \bigcup I_n \quad \text{and} \quad \sum \ell(I_n) < \varepsilon,$$

where $\ell(I_n)$ is the length of I_n. In short, for any $\varepsilon > 0$, you can find an open subset G_ε of $[0, 1]$ containing N and of length at most ε.

J. ⓜ Remark. *Step 1 towards resolving the difficulty at Discussion 25B.* Work with the Fundamental Triple. For $X \in \Omega = [0, 1]$, let $N_x = \{x\}$. Then each N_x is a null event. (For $\varepsilon > 0$, $G_\varepsilon = (x - \frac{1}{3}\varepsilon, x + \frac{1}{3}\varepsilon) \cap [0, 1]$ is an open subset of $[0, 1]$ of length less than ε and containing N_x.) However, $\Omega = \bigcup_{x \in \Omega} N_x$, and Ω is certainly not a null event. The point here is that (as you probably know from Cantor's Diagonal Principle – see Appendix A4, p496) the set $\Omega = [0, 1]$ is not countable: we cannot write

$\Omega = \{x_1, x_2, x_3, \ldots\}$ for some sequence (x_k) of points of Ω; Ω is too large a set to allow us to do this. Otherwise, we would have a contradiction with Lemma G. The set of all subsequences of the sequence of positive integers is also uncountable; and the way out of the difficulty at Discussion 25B begins to emerge.

▶ **K. ⓜ Fact: First Borel–Cantelli Lemma.** *Let* $J_1, J_2, \ldots$ *be a sequence of events. If* $\sum \mathbb{P}(J_k) < \infty$, *then it is almost surely true that only finitely many of the events* J_k *occur.* (Assumed fact)

This First Borel–Cantelli Lemma is a very useful result. In the next chapter, we shall see very clearly why it is true.

▶ **L. ⓜ Borel sets and functions.** The Borel σ-algebra $\mathcal{B}(S)$ on a topological space S is the smallest σ-algebra of subsets of S which contains all open subsets of S (equivalently, the smallest σ-algebra of subsets of S which contains all closed subsets of S). A Borel subset of S is an element of this Borel σ-algebra.

If $S = \mathbb{R}$, then the Borel σ-algebra is the smallest σ-algebra containing every set of the form $(-\infty, x]$, where $x \in \mathbb{R}$. This is just what is needed to study distribution functions in Probability and Statistics. If $S = \mathbb{R}^n$, then the Borel σ-algebra is the smallest σ-algebra containing every subset of the form

$$\{(x_1, x_2, \ldots, x_n) : x_k \le a\}$$

where $a \in \mathbb{R}$ and $k \in \{1, 2, \ldots, n\}$. For the Fundamental Model, $\mathcal{F} = \mathcal{B}([0, 1])$.

Recall that a continuous function $f : S \to \mathbb{R}$ is one such that the inverse image $f^{-1}(B) := \{s \in S : f(s) \in B\}$ of any open subset B of $\mathbb{R}$ is open in S. Analogously, we define a function $f : S \to \mathbb{R}$ to be a Borel function if the inverse image $f^{-1}(B)$ of any Borel subset B of $\mathbb{R}$ is Borel in S: equivalently, if, for every $x \in \mathbb{R}$, $\{s \in S : f(s) \le x\}$ is Borel in S.

Every subset of $\mathbb{R}^n$ **and every function on** $\mathbb{R}^n$ **you are ever likely to meet will be Borel.** Continuous functions are Borel. Limits, limsups, what-have-you of sequences of Borel functions are Borel. You need to be clever to construct explicitly a function which is not Borel; but it can be done.

Technical Remark. After I wrote 'but it can be done' about this last point on a previous occasion, I received many emails demanding 'How?'. So, to avoid a repeat, see Appendix A6, p499.

▶ **M. ⓜ The π-system Lemma.** The last thing I would do is to add unnecessary bits to this discussion of Measure Theory. I must give the idea of the π-system Lemma because it is crucial for proving that certain key things are uniquely defined, for establishing essential properties of independent Random Variables, etc. We shall not actually use the π-system Lemma to do these things, but I shall tell you where it would clinch an argument.

In Subsection 44H, I stated that there exists a unique measure $\mathbb{P}$ on what we now know to be $\mathcal{B}[0, 1]$ such that $\mathbb{P}([0, x]) = x$ for every x in $[0, 1]$. The hard thing is to prove the existence of $\mathbb{P}$. But it matters greatly that $\mathbb{P}$ is unique; and this is where the π-system Lemma comes in. The subsets of $[0, 1]$ of the form $[0, x]$ form a π-system $\mathcal{I}$, which simply means that the intersection of any two elements of $\mathcal{I}$ is again in $\mathcal{I}$. The π-system Lemma states that **two probability measures which agree on a π-system must also agree on the smallest σ-algebra which contains that π-system**, which guarantees the desired uniqueness for the Fundamental Triple.

RANDOM VARIABLES,

MEANS AND VARIANCES

In this chapter, as in the last, we do some fundamental things, but without the benefit of conditioning and independence (the subjects of the next chapter) to enliven things with really good examples and exercises. I do *not* give you (more than about two very easy) exercises in calculating means and variances by summing series or working out integrals, and this for two reasons: firstly, it teaches you nothing about Probability and Statistics; and secondly, we *shall* calculate lots of means and variances later by efficient indirect methods. I do on rare occasions cheat by giving you exercises which use independence properties familiar from Chapter 1, just to keep some interest in the proceedings. That the next chapter is MUCH more interesting is a promise.

3.1   Random Variables

Intuitively, a Random Variable (RV) is 'a number determined by Chance'; but this is hardly adequate for a mathematical theory. The formal definition is as follows.

▶▶ **A. Mathematical formulation of Random Variable (RV).** A Random Variable is defined to be a **function** (Ⓜ strictly, an $\mathcal{F}$-measurable function) **from** Ω **to** $\mathbb{R}$.

Aa. Question:*Why is it appropriate to axiomatize the notion of Random Variable as being a function on our sample space Ω?*

Answer. Suppose that our RV Y is 'total number of Heads' if I toss my coin 3 times. We can make the picture

$\omega:$	HHH	HHT	HTH	HTT	THH	THT	TTH	TTT
$Y(\omega):$	3	2	2	1	2	1	1	0

and this already displays Y as a function on Ω. Any 'intuitive RV X' will assign a value $X(\omega)$ to every possible outcome ω.

Here's another example. Suppose that our experiment consists of throwing a die twice, that X is the score on the first throw, Y that on the second, and Z is the sum of the scores. A typical possible outcome ω has the form (i, j), where i is the first score and y the second. Then $X(\omega) = i$, $Y(\omega) = j$, and $Z(\omega) = i + j$, and we have the picture:

$\Omega:$	$(1,1)$ $(1,2)$ $\cdots$ $(1,6)$			$X:$	1	1	$\cdots$	1	
	$(2,1)$ $(2,2)$ $\cdots$ $(2,6)$				2	2	$\cdots$	2	
	.	.	$\cdots$	.	.	.	$\cdots$	.	
	$(6,1)$ $(6,2)$ $\cdots$ $(6,6)$				6	6	$\cdots$	6	
	1	2	$\cdots$	6		2	3	$\cdots$	7
$Y:$	1	2	$\cdots$	6	$Z:$	3	4	$\cdots$	8
	.	.	$\cdots$	.		.	.	$\cdots$	.
	1	2	$\cdots$	6		7	8	$\cdots$	12

For an RV X, we wish to be able to talk about its *Distribution Function (DF)* $F_X : \mathbb{R} \to [0, 1]$ (of which more, of course, later) defined by

$$F_X(x) := \mathbb{P}(X \le x) \qquad (x \in \mathbb{R}).$$

We therefore require that, for every $x \in \mathbb{R}$, the subset $L_x := \{\omega : X(\omega) \le x\}$ of Ω (the mathematical formulation of the 'event that $X \le x$') be a true mathematical event, that is, an element of the class $\mathcal{F}$ of events; and then, of course, we interpret $\mathbb{P}(X \le x)$ as $\mathbb{P}(L_x)$.

Ab. Question:*Why the restriction to 'measurable' functions in the proper theory?*

Ⓜ Saying that for every x in $\mathbb{R}$, $L_x \in \mathcal{F}$ is *exactly* saying in Measure Theory that X is an $\mathcal{F}$-measurable map from Ω to $\mathbb{R}$. Compare subsection 45L.

Ⓜ **Fact:** Except for that crazy Banach–Tarski context, every function on Ω we meet in this book *is* $\mathcal{F}$-measurable.

The point, already made at 45L, is that sums, products, pointwise limits (if they exist), etc, of measurable functions are measurable. You cannot break out of the world of measurable things without being rather clever. So, we shall ignore measurability questions, except for intuitive discussions on how more subtle types of measurability which are in the background allow us to do better things.

▶▶ **B. Crystallization, Pre-Statistics and actual statistics.** Again let Y be the number of Heads in 3 tosses of my coin. In Probability, we consider the experiment before it is performed, and Y is a function on the nebulous, abstract set Ω. As discussed at 36B, Tyche's choice 'crystallizes into existence' the actual outcome ω^{act}. Suppose that $\omega^{\text{act}} = \{\text{HHT}\}$. Then the observed, or *realized*, value y^{obs} of Y is $y^{\text{obs}} = Y(\omega^{\text{act}}) = 2$. We call (the terminology being my own)

- Y, the RV, a *Pre-Statistic*,

- y^{obs}, the observed value of Y, an *actual statistic*,

and regard the crystallization of ω^{act} as changing Y to $y^{\mathrm{obs}} = Y(\omega^{\mathrm{act}})$.

A Pre-Statistic is a special kind of Random Variable: a Random Variable Y is a Pre-Statistic if the value $Y(\omega^{\mathrm{act}})$ will be known to the observer (it may have to be calculated from things directly observed) after the experiment is performed; and then $Y(\omega^{\mathrm{act}})$ becomes the observed value y^{obs} of Y. Thus,

A Pre-Statistic is an **Observable** Random Variable.

The model may involve RVs the actual values of which cannot be determined after the experiment: these RVs are not Pre-Statistics. We refer to ω^{act} rather than ω^{obs} precisely because full information about ω^{act} may never be known.

▶▶ **C. Indicator function I_F of an event F.** For an event F, we define

$$I_F(\omega) := \begin{cases} 1 & \text{if } \omega \in F, \\ 0 & \text{if } \omega \notin F. \end{cases}$$

Intuitively, $I_F = 1$ if F occurs, 0 if it doesn't. (*Technical Note.* For any x, $\{\omega : I_F(\omega) \leq x\}$ can only be one of three things: $\emptyset$, F^c, Ω, depending on where x lies in relation to 0 and 1. Hence I_F is certainly measurable.)

We use indicator functions to do our counting. The number Y of Heads I get in n tosses of a coin is

$$Y = X_1 + X_2 + \cdots + X_n, \tag{C1}$$

where X_k is the indicator function of the event 'Heads on kth toss': for the sum counts 1 for every Head, 0 for every Tail.

Ca. Exercise. Let F and G be events. Prove that

$$I_{F^c} = 1 - I_F, \qquad I_{F \cap G} = I_F I_G,$$

as function identities ('true for every ω'). Deduce from de Morgan's rule 38(D1) that $I_{F \cup G} = I_F + I_G - I_F I_G$, and explain why this is otherwise obvious.

3.2   DFs, pmfs and pdfs

As a student, I found the study of

DFs: Distribution Functions
        (Idea: $F_X(x) = \mathbb{P}(X \leq x)$)

pmfs: probability mass functions
        (Idea: $p_X(x) = \mathbb{P}(X = x)$, X 'discrete')

pdfs: probability density functions
        (Idea: $f_X(x)\mathrm{d}x = \mathbb{P}(X \in \mathrm{d}x)$, X 'continuous')

a real 'turn-off' – all horrible integrals and that (and I was someone who liked integrals!)
Take heart: crafty methods will allow us almost completely to avoid summations and
integrations, nice or nasty.

▶▶ **A. The Distribution Function (DF) F_X of X.** Let X be an RV. We define
the Distribution Function (DF) F_X of X to be the function $F_X : \mathbb{R} \to [0,1]$ with

$$F_X(x) := \mathbb{P}(X \leq x) \qquad (x \in \mathbb{R}).$$

As already explained, this is the probability of the event $\{X \leq x\}$ interpreted as
$\{\omega : X(\omega) \leq x\}$. Note that for $a \leq b$,

$$\mathbb{P}(a < X \leq b) = F_X(b) - F_X(a).$$

Reason: For $a \leq b$, events $\{X \leq a\}$ and $\{a < X \leq b\}$ are disjoint with union $\{X \leq b\}$
so that

$$\mathbb{P}(X \leq b) = \mathbb{P}(X \leq a) + \mathbb{P}(a < X \leq b). \qquad \square$$

One advantage of the DF is that it is defined for *any* RV X. Most of the RVs one meets
are either 'discrete' RVs or 'continuous' RVs (these are described below) or mixtures of
those types. However, not all RVs are so simple: for example, RVs with values in fractal
sets, studied more and more these days, are much more complicated; but DFs and Measure
Theory can deal with them.

Important note. Statisticians usually refer to the Distribution Function (DF) as the
Cumulative Distribution Function (**CDF**). In Minitab, for example, one calculates the
value of a DF via a `cdf` command. Several examples are given later.

▶ **B. The 'F-inverse' principle for theory and simulation.** If F is the
distribution function of an RV X, then F is certainly non-decreasing, and it
follows from the monotonicity property 43D of measures that

$$\lim_{y\downarrow-\infty} F(y) = 0, \quad \lim_{y\uparrow\infty} F(y) = 1,$$
$$F \text{ is right-continuous: for } x \in \mathbb{R}, \ \lim_{y\downarrow x} F(y) = F(x). \tag{B1}$$

Suppose now that $F : \mathbb{R} \to [0,1]$ has properties (B1). Both for the theory and for
simulation, it is very useful to be able to construct from an RV U with the U$[0,1]$
distribution an RV X with distribution function F.

Ba. Exercise. Suppose in addition that F is continuous and strictly increasing on $\mathbb{R}$. Explain why, for $0 < u < 1$, there exists a unique number $G(u)$ such that $F(G(u)) = u$. In usual notation, $G = F^{-1}$. Prove that if U has the U$[0, 1]$ distribution, so that

$$\mathbb{P}(U \leq u) = u \quad \text{for } 0 \leq u \leq 1,$$

then $X := G(U)$ has distribution function F. Equivalently, if X has a continuous strictly increasing distribution function F, then $F(X)$ has the $U[0, 1]$ distribution.

For general F satisfying properties 50(B1), we define

$$G(u) := \min\{y : F(y) \geq u\},$$

the right-continuity of F guaranteeing that this is a true minimum, not just an infimum. Then, again, $X := G(U)$ has distribution function F. (Assume this.)

Only in special situations (some of them very important) would we use this idea in simulation, because computation of $G(U)$ can be very time-consuming. Some more efficient methods of simulating from some standard distributions will be explained later.

▶ **C. The probability mass function (pmf) p_X of a discrete RV X.** By a discrete Random Variable, we mean one which takes values in the set $\mathbb{Z}$ of all integers.

Let X be a discrete RV. We define the probability mass function (pmf) p_X of X to be the function $p_X : \mathbb{Z} \to [0, 1]$ with

$$p_X(x) := \mathbb{P}(X = x), \qquad (x \in \mathbb{Z}). \tag{C1}$$

Of course, X may take values just in $\mathbb{Z}^+$, in which case $p_X(x) = 0$ for x negative; etc. (We should allow a discrete RV to take values in an arbitrary finite or countable set, but can generally manage without that.)

Important examples of pmfs (illustrated for certain parameters in Figure C(i)) are

$$\text{the } \textbf{Bernoulli}(p) \text{ pmf} \quad p_X(x) = \begin{cases} p & \text{if } x = 1, \\ q := 1 - p & \text{if } x = 0, \\ 0 & \text{otherwise;} \end{cases}$$

$$\tag{C2}$$

$$\text{the } \textbf{binomial}(n, p) \text{ pmf} \quad p_X(x) = \begin{cases} b(n, p; x) & \text{if } x = 0, 1, \dots, n, \\ 0 & \text{otherwise;} \end{cases}$$

$$\tag{C3}$$

$$\text{the } \textbf{Poisson}(\lambda) \text{ pmf} \quad p_X(x) = \begin{cases} e^{-\lambda} \lambda^x / x! & \text{if } x = 0, 1, 2, \dots, \\ 0 & \text{otherwise.} \end{cases} \tag{C4}$$

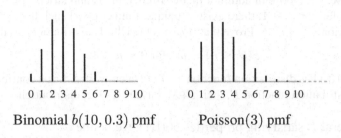

Binomial $b(10, 0.3)$ pmf   Poisson(3) pmf

Figure C(i): Binomial and Poisson pmfs

If X has the Bernoulli(p) pmf, then we say that X has the Bernoulli(p) distribution and write

$$X \sim \text{Bernoulli}(p),$$

with similar remarks for the binomial(n, p) and Poisson(λ) distributions. Of course, the Distribution Function F_X of a discrete RV X with pmf p_X satisfies

$$F_X(x) = \sum_{y \leq x} p_X(y) := \sum_{\mathbb{Z} \ni y \leq x} p_X(y),$$

and, for $a, b \in \mathbb{Z}$ with $a < b$, we have

$$\mathbb{P}(a < X \leq b) = F_X(b) - F_X(a) = \sum_{y=a+1}^{b} p_X(y). \tag{C5}$$

We have already met the binomial pmf at Lemma 10Ma, and we shall be seeing a lot of the Poisson pmf. We have seen in Exercise 41G that, even for moderately large n, the number of professors who get the right hat has almost exactly the Poisson(1) distribution.

D. Exercises. Useful practice!

Da. Exercise. This will be referred to later, so do it! A fair coin will be tossed twice, the number N of Heads will be noted, and then the coin will be tossed N more times. Let X be the *total* number of Heads obtained. Decide on Ω, and make a table with headings

$$\omega \qquad \mathbb{P}(\omega) \qquad X(\omega)$$

Find $p_X(x)$ for $x = 0, 1, 2, 3, 4$.

Db. Exercise. A coin with probability p of Heads is tossed until the first Head is obtained. Let X be the total number of tosses. Find $p_X(x)$ for $x = 1, 2, 3, \ldots$ We say that X has the **geometric**(p) distribution. (The distribution of the total number $X - 1$ of Tails is also called geometric.)

Dc. Exercise. A coin with probability p of Heads is tossed until the first time that a total of n Heads is obtained. Let X be the total number of tosses. Find $p_X(x)$ for $x = n, n + 1, n + 2, \ldots$ We say that X has the **negative-binomial**(n, p) distribution. (The distribution of the total number $X - n$ of Tails is also called negative-binomial.)

▶▶ **E. The probability density function (pdf) f_X of 'continuous' variable X.** We shall call an RV X (with values in $\mathbb{R}$) *'continuous'* if there exists a piecewise-continuous function f_X, called a probability density function (pdf) of X, such that for $a, b \in \mathbb{R}$ with $a < b$,

$$\mathbb{P}(a < X \le b) = F_X(b) - F_X(a) = \int_a^b f_X(x) \, dx. \qquad \text{(E1)}$$

Compare equation 52(C5). See Figure E(i). (*Note.* We refer to *a* pdf rather than *the* pdf because the pdf is only defined modulo sets of measure zero. We'll return to this later.)

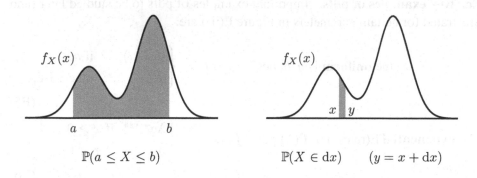

$$\mathbb{P}(a \le X \le b) \qquad\qquad \mathbb{P}(X \in dx) \qquad (y = x + dx)$$

Figure E(i): Probabilities as areas

Of course, it is then true that $\mathbb{P}(X = x) = 0$ for all $x \in \mathbb{R}$ and that

$$\mathbb{P}(a < X < b) = \mathbb{P}(a < X \le b) = \mathbb{P}(a \le X \le b) = \mathbb{P}(a \le X < b).$$

Note that f_X is determined by F_X via

$$f_X(x) = F_X'(x) \text{ except perhaps at finitely many points.} \qquad \text{(E2)}$$

See the examples below for clarification on exceptional points. We must have

$$\int_{-\infty}^{\infty} f_X(x)\, \mathrm{d}x \;=\; \mathbb{P}(-\infty < X < \infty) \;=\; 1. \tag{E3}$$

We often write (E1) in the helpful alternative form (analogous to 51(C1)):

$$\mathbb{P}(X \in \mathrm{d}x) \;=\; f_X(x)\, \mathrm{d}x, \tag{E4}$$

the left-hand side being thought of as '$\mathbb{P}(x < X \le x + \mathrm{d}x)$' and (E4) making rigorous sense of 53(E1) when integrated over $x \in (a, b]$.

Ea. Exercise. A point A is chosen at random in the unit disc $\{(x, y) : x^2 + y^2 \le 1\}$. Let R be the distance from the origin to A. Find $f_R(r)$ and explain the good sense of the answer.

Eb. Note on terminology. Random Variables are functions on Ω. We often need to topologize Ω, and then truly continuous functions on Ω play an important part; and they are not related to the use of 'continuous' which we have been making. I shall always use 'continuous' within '' when meaning 'as in Definition 53E'. (When I speak of a continuous DF, I mean a DF which is continuous, not the DF of a 'continuous' variable.)

Ec. Key examples of pdfs. Important examples of pdfs to be studied later (and illustrated for certain parameters in Figure E(ii)) are:

$$\text{the } \mathbf{uniform}\ \mathrm{U}[a, b] \text{ pdf } \quad f_X(x) \;=\; \begin{cases} (b - a)^{-1} & \text{if } a \le x \le b, \\ 0 & \text{otherwise;} \end{cases} \tag{E5}$$

$$\text{the } \mathbf{exponential}\ \mathrm{E}(\text{rate } \lambda) \text{ or } \mathbf{E}(\lambda) \text{ pdf } \quad f_X(x) \;=\; \begin{cases} \lambda e^{-\lambda x} & \text{if } x \ge 0, \\ 0 & \text{if } x < 0; \end{cases} \tag{E6}$$

$$\text{the } \mathbf{normal}\ \mathrm{N}(\mu, \sigma^2) \text{ pdf } \quad f_X(x) \;=\; \frac{1}{\sigma\sqrt{2\pi}} \exp\left(-\frac{(x - \mu)^2}{2\sigma^2}\right). \tag{E7}$$

If X has the uniform $\mathrm{U}[a, b]$ pmf, then we say that X has the $\mathrm{U}[a, b]$ distribution and write $X \sim U[a, b]$, with similar remarks for the $E(\text{rate } \lambda)$ and $\mathrm{N}(\mu, \sigma^2)$ distributions.

Notes. That the $\mathrm{N}(\mu, \sigma^2)$ pdf does integrate to 1 is verified at Subsection 146B below. How the $\mathrm{U}[a, b]$ pdf is defined at a and b is irrelevant; likewise for the way the $E(\text{rate } \lambda)$ pdf is defined at 0.

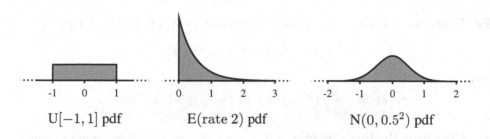

U[−1, 1] pdf   E(rate 2) pdf   N(0, 0.5²) pdf

Figure E(ii): Various pdfs

It will often be convenient to use indicator-function notation in relation to pdfs, so that we write the U[a, b] and E(rate λ) pdfs as

$$(b - a)^{-1} I_{[a,b]}(x), \quad \lambda e^{-\lambda x} I_{[0,\infty)}(x)$$

respectively.

▶▶ **F. A first transformation rule for pdfs.** Suppose that X is a 'continuous' RV with pdf f_X and that ψ is a strictly increasing differentiable function. Write $X = \psi(Z)$. Then

$$F_Z(z) := \mathbb{P}(Z \le z) = \mathbb{P}\big(\psi(Z) \le \psi(z)\big) = \mathbb{P}\big(X \le \psi(z)\big) = F_X\big(\psi(z)\big).$$

Hence, by 53(E2), the pdf of Z is given by

$$f_Z(z) = f_X\big(\psi(z)\big)\psi'(z).$$

We can use heuristics: $x = \psi(z)$, $dx = \psi'(z)dz$, and

$$f_Z(z)dz = f_X(x)dx = f_X(\psi(z))\psi'(z)dz.$$

In many ways, it is better to think in these terms. See the 'Jacobian' treatment of the multi-dimensional case in Chapter 7.

Fa. Exercise. Use this formula to show that if $U \sim$ U[0, 1], then $X = -(\ln U)/\lambda$ has the E(rate λ) distribution. (This is very useful for simulation.) How does this result relate to the F-inverse principle at 50B?

▶ **Fb. Exercise: the standard Cauchy distribution.** A line is drawn through the point $(1, 0)$ in a randomly chosen direction – you formulate this! Let $(0, Y)$ be the point at which it cuts the Y-axis. Show that Y has the *standard Cauchy density*

$$f_Y(y) = \frac{1}{\pi (1 + y^2)} \quad (y \in \mathbb{R}).$$

Prove that $1/Y$ has the same distribution as Y, and explain this geometrically. (*Note.* In this exercise, you can ignore the possibility that the randomly chosen line is parallel to one of the axes, because this has probability 0.)

Fc. Exercise. Let X be a 'continuous' RV and let $Y = X^2$. Prove that, for $y \geq 0$,

$$F_Y(y) = F_X(\sqrt{y}) - F_X(-\sqrt{y}),$$

whence

$$f_Y(y) = \frac{1}{2\sqrt{y}} \{ f_X(\sqrt{y}) + f_X(-\sqrt{y}) \} \text{ when } Y = X^2.$$

G. Exercise: Buffon's needle. This is not really the right place for this example, but it is fun to do now. A large square region of the plane is marked with parallel lines 2 units apart. A needle, 2 unit long is thrown randomly onto the square area. Prove that the probability that the needle crosses one of the lines is $2/\pi$ (if one ignores 'edge effects').

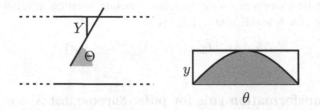

Figure G(i): Hint for the 'Buffon's needle' problem

Hint. Use Figure G(i), in which Y is the distance from the centre of the needle to the nearest line, and Θ the angle between the needle and each of the lines.

3.3   Means in the case when Ω is finite

A. Assumption: Assume in this section that Ω **is finite.** *Assume also that (as will occur in all the cases which concern us) all subsets of Ω are events and all functions on Ω are RVs.* For any event H, we clearly have

$$\mathbb{P}(H) = \sum_{\omega \in H} \mathbb{P}(\omega), \tag{A1}$$

where $\mathbb{P}(\omega) := \mathbb{P}(\{\omega\})$ is the probability of the individual outcome ω.

▶ **B. Definition of $\mathbb{E}(X)$ for finite Ω.** For an RV X, we define

$$\mu_X := \mathbb{E}(X) := \sum_{\omega \in \Omega} X(\omega)\mathbb{P}(\omega). \tag{B1}$$

Thus, $\mathbb{E}(X)$ *is the $\mathbb{P}$-weighted mean of the function X on Ω.*

Ba. Discussion. The 'long-term average' idea. Suppose that $\Omega = \{\omega_1, \omega_2, \ldots, \omega_r\}$. The long-term relative frequency idea would tell us that if the experiment associated with Ω were performed a very large number N of times, then it is likely that we would have outcome ω_k on about $N\mathbb{P}(\omega_k)$ performances. If, therefore, we look at the 'sum of the observed X-values over the performances', then since we will have X-value $X(\omega_k)$ about $N\mathbb{P}(\omega_k)$ times, the sum of the X-values is likely to be about

$$\sum X(\omega_k) \times N\mathbb{P}(\omega_k) = N\mu_X = N\mathbb{E}(X).$$

Hence the average of the X-values should be close to μ_X.

It is very important to stick with the precise 'abstract' definition at 56(B1), and not to try to build in any 'long-term average' idea. We shall eventually *prove* the Strong Law of Large Numbers which formulates the 'long-term average' idea precisely for the mathematical model.

C. Heuristic solution to Puzzle 19A. If I keep throwing a fair die with scores $1, 2, \ldots, 6$ on its faces, then the average gap between successive accumulated totals will be the average of $1, 2, \ldots, 6$, that is, the average of 1 and 6, namely $7/2$. Since the gaps are on average $7/2$, the proportion of 'accumulated scores' amongst all positive integers will be $2/7$. Hence, for any reasonably large number N the probability that at some time my score will equal N would be approximately $2/7$. In particular, this is true for $N = 10000$.

Always be careful about the following though. Suppose instead the scores marked on the die's faces are $2, 4, 6, 8, 10, 12$. What is the average gap between successive accumulated totals now? What is the approximate chance that at some time my score will be 20000? Be careful! $\square$

I emphasize again that for us, 56(B1) is the *definition* of the mean or expectation of X. At the moment, no 'long-term average' ideas feature in the mathematical theory: they will return in correct form in the *proved* 'Strong Law' theorem.

▶ **D. Lemma.** *Suppose that Ω is finite, that X and Y are RVs, and that $c \in \mathbb{R}$. Then*

$$\mathbb{E}(cX) = c\mathbb{E}(X), \qquad \mathbb{E}(X + Y) = \mathbb{E}(X) + \mathbb{E}(Y),$$

so that expectation is a linear functional.

Moreover, for any function $h : \mathbb{R} \to \mathbb{R}$,

$$\mathbb{E}(h(X)) = \sum_x h(x)\,\mathbb{P}(X = x) \tag{D1}$$

summed over the possible values of X. Especially, for an event F,

$$\mathbb{E}(I_F) = \mathbb{P}(F).$$

Note. We need the case $h(x) = x^2$ to find the variance of X, etc, etc.

Proof. By the usual rules for functions,

$$(cX)(\omega) := c\,X(\omega) \quad \text{and} \quad (X+Y)(\omega) := X(\omega)+Y(\omega),$$

and the first part follows.

Summing over ω in Ω can be achieved by first summing over every value x which X can take, and for each such x summing over ω-values for which $X(\omega) = x$. Thus,

$$
\begin{aligned}
\mathbb{E}(h(X)) &= \sum_{\omega \in \Omega} h(X(\omega))\mathbb{P}(\omega) = \sum_{x} \sum_{\{\omega : X(\omega)=x\}} h(X(\omega))\mathbb{P}(\omega) \\
&= \sum_{x} \sum_{\{\omega : X(\omega)=x\}} h(x)\mathbb{P}(\omega) = \sum_{x} h(x) \sum_{\{\omega : X(\omega)=x\}} \mathbb{P}(\omega) \\
&= \sum_{x} h(x)\mathbb{P}(X = x).
\end{aligned}
$$

We have used result 56(A1) for H the event '$X = x$'.

The indicator function I_F takes the value 1 with probability $\mathbb{P}(F)$, else the value 0. So

$$\mathbb{E}(I_F) = 1 \times \mathbb{P}(F) + 0 \times \mathbb{P}(F^c) = \mathbb{P}(F),$$

and that completes the proof.   □

Da. Exercise. Calculate the mean of the binomial(n, p) distribution by using 49(C1).

Db. Exercise. Prove that in the 'hat-matching' problem 40E, if R_n is the number of professors who get the correct hat, then $\mathbb{E}(R_n) = 1$. *Hint:* Again use the idea at 49(C1).

Dc. Exercise. Use intuition to write down without any calculation the value of $\mathbb{E}(X)$ for Exercise 52Da. Then verify that the answer is correct by using that exercise and equation 57(D1).

▶ **E. Application.**   **Proof of the Inclusion-Exclusion Principle.**   Let $F_1, F_2, \ldots, F_n$ be events, let $\varphi_k = I_{F_k}$, and let $U = \bigcup F_k$. By de Morgan's rule 38(D1),

$$U^c = F_1^c \cap F_2^c \cap \ldots \cap F_n^c,$$

whence, by Exercise 49Ca,

$$1 - I_U = (1 - \varphi_1)(1 - \varphi_2)\ldots(1 - \varphi_n).$$

Expanding out the product on the right-hand side (you do this in full for $n = 2$ and $n = 3$ to see what is going on), we obtain

$$1 - I_U = 1 - \sum_i \varphi_i + \sum_i \sum_{j>i} \varphi_i\varphi_j - \cdots + (-1)^n \varphi_1\varphi_2\ldots\varphi_n,$$

whence

$$I_U = \sum \varphi_i - \sum_i \sum_{j>i} \varphi_i \varphi_j + - \cdots + (-1)^{n+1} \varphi_1 \varphi_2 \ldots \varphi_n,$$

$$= \sum I_{F_i} - \sum_i \sum_{j>i} I_{F_i \cap F_j} + - \cdots + (-1)^{n+1} I_{F_1 \cap F_2 \cap \ldots \cap F_n}.$$

Taking expectations, and using the results of Lemma 57D, we obtain

$$\mathbb{P}(U) = \sum \mathbb{P}(F_i) - \sum_i \sum_{j>i} \mathbb{P}(F_i \cap F_j) + \sum_i \sum_{j>i} \sum_{k>j} \mathbb{P}(F_i \cap F_j \cap F_k)$$

$$- \cdots + (-1)^{n+1} \mathbb{P}(F_1 \cap F_2 \cap \ldots \cap F_n);$$

and this is the desired result. □

3.4   Means in general

Now we drop the assumption that Ω is finite. I shall explain/recall the familiar practical rules in terms of pmfs and pdfs at Theorem 62Ha below. But first, I am going to tell you the simple coherent all-embracing version of the theory that Measure Theory provides. Don't worry: just read on – there's no heavy Measure Theory here – I'm putting hardly any 'Warning' symbols!

▶ **A. SF$^+$, the space of simple non-negative RVs.** We call an RV Y simple and non-negative, and write $Y \in \text{SF}^+$, if we can write Y in the form

$$Y = \sum_{k=1}^{r} a_k I_{F_k},$$

where each $a_k \in [0, \infty)$ and each F_k is an event. For this Y, we define (how else?!)

$$\mathbb{E}(Y) := \sum_{k=1}^{r} a_k \mathbb{P}(F_k).$$

This defines $\mathbb{E}$ unambiguously on SF$^+$. (This last statement needs proof, but we skip that. It's messy!)

▶ **B. Mean of an arbitrary non-negative RV.** For two RVs X and Y, we write $Y \leq X$ to mean $Y(\omega) \leq X(\omega)$ for every ω in Ω. For a general non-negative RV $X : \Omega \to [0, \infty]$, we define – with the dreaded 'supremum' revised below –

$$\mathbb{E}(X) := \int_{\omega \in \Omega} X(\omega) \mathbb{P}(d\omega) := \sup\{\mathbb{E}(Y) : Y \in \text{SF}^+, Y \leq X\} \leq \infty.$$

We say that X is **integrable** if $\mathbb{E}(X) < \infty$.

The integral, which is *defined* to be the supremum, is a *Lebesgue integral* which exactly describes the idea of the $\mathbb{P}$-weighted mean of X over Ω. The same remarks about the need to stick with an abstract definition (which will eventually be 'explained' by the Strong Law) apply as in the earlier case of finite Ω. *Allowing infinity.* We allow $\mathbb{E}(X) = \infty$ for *non-negative* X, and even allow $X(\omega)$ to be ∞ for some values of ω, because *it is very convenient to do so*. Of course, if X is non-negative and $\mathbb{P}(X = \infty) > 0$, then, certainly, $\mathbb{E}(X) = \infty$. The point of these remarks is that, as we shall see, *there are a lot of important non-negative RVs for which* $\mathbb{P}(X = \infty) = 0$ *but* $\mathbb{E}(X) = \infty$.

Clarification of the 'sup' bit. Let X be a general non-negative RV. Then, firstly, we have $\mathbb{E}(X) \geq \mathbb{E}(Y)$ whenever $Y \in \mathrm{SF}^+$, and $Y \leq X$ – sensible, yes?! Secondly, if $\mathbb{E}(X) < \infty$, then, for any $\varepsilon > 0$, we can find $Y \in \mathrm{SF}^+$ with $Y \leq X$ and $\mathbb{E}(Y) > \mathbb{E}(X) - \varepsilon$; that is, we can approximate X from below as closely 'as we like in the sense of expectation' by a simple RV. (If $\mathbb{E}(X) = \infty$, then, for any $K > 0$, we can find $Y \in \mathrm{SF}^+$ with $Y \leq X$ and with $\mathbb{E}(Y) > K$.)

C. Facts.   *(a) If X is a non-negative RV, then $\mathbb{E}(X) = 0$ if, and only if, $\mathbb{P}(X = 0) = 1$.*

(b) If X is a non-negative RV and $\mathbb{E}(X) < \infty$, then $\mathbb{P}(X < \infty) = 1$.
These are hardly a surprise! But they are useful. We mentioned result (b) earlier.

▶▶ **D. Fact: Monotone-Convergence Theorem.** *If (X_n) is a sequence of non-negative RVs, and $X_n(\omega) \uparrow X(\omega)$ for each ω, then*

$$\mathbb{E}(X_n) \uparrow \mathbb{E}(X) \leq \infty.$$

Lebesgue's Monotone-Convergence Theorem is the result on which the whole of integration theory rests.

Da. Ⓜ **Application.**   *Proof of the First Borel–Cantelli Lemma 45K.* Let X_k be the indicator of J_k. Let $Y_n := X_1 + \cdots + X_n$. Then, for every n, $\mathbb{E}(Y_n) = \mathbb{P}(J_1) + \cdots + \mathbb{P}(J_n)$. If $\sum \mathbb{P}(J_k) < \infty$, then by the Monotone-Convergence Theorem, $\mathbb{E}(Y_\infty) = \sum \mathbb{P}(J_k) < \infty$, where $Y_\infty := \uparrow \lim Y_n$. Hence Y_∞ is finite with probability 1, and the result follows.
□

▶▶ **E. Means for general RVs.** If X is a general RV, so $X : \Omega \to \mathbb{R}$, then we decompose X into its positive and negative parts:

$$X = X^+ - X^-, \quad X^+(\omega) := \begin{cases} X(\omega) & \text{if } X(\omega) \geq 0, \\ 0 & \text{if } X(\omega) < 0, \end{cases} \quad X^- = (-X)^+.$$

(Of course, $X = X^+ - X^-$ has the usual meaning for functions: we have $X(\omega) = X^+(\omega) - X^-(\omega)$ for every ω.) Then X^+ and X^- are non-negative RVs.

> We say that X is *integrable* if both $\mathbb{E}(X^+)$ and $\mathbb{E}(X^-)$ are finite, and then define
> $$\mathbb{E}(X) = \mathbb{E}(X^+) - \mathbb{E}(X^-).$$

Let's make a standard useful definition.

▶▶ **F. The vector space $\mathcal{L}^1$.** An RV X is said to be in $\mathcal{L}^1$ if it is integrable, that is, if $\mathbb{E}(|X|) < \infty$.

Now, $|X| = X = X^+ + X^-$, so $|X|$ is integrable if and only if both X^+ and X^- are integrable. Moreover, since X^+ and X^- are non-negative,

$$|\mathbb{E}(X)| = |\mathbb{E}(X^+) - \mathbb{E}(X^-)| \leq |\mathbb{E}(X^+)| + |\mathbb{E}(X^-)|$$
$$= \mathbb{E}(X^+) + \mathbb{E}(X^-) = \mathbb{E}(|X|),$$

the familiar result that the modulus of an integral is less than or equal to the integral of the modulus. Note that since $|X + Y| \leq |X| + |Y|$, if $X, Y \in \mathcal{L}^1$, then $X + Y \in \mathcal{L}^1$. We see that

> $\mathcal{L}^1$ is a vector space over $\mathbb{R}$.

▶ **G. A useful result and a useful criterion.** *Let Y be an RV with values in $\mathbb{Z}^+$. Then*

$$\mathbb{E}(Y) = \sum_{n=1}^{\infty} \mathbb{P}(Y \geq n) \leq \infty.$$

For an arbitrary RV X, we have $X \in \mathcal{L}^1$ if and only if

$$\sum_{n \geq 1} \mathbb{P}(|X| \geq n) < \infty.$$

Proof. If $F(n)$ denotes the event $\{\omega : Y(\omega) \geq n\}$, then

$$Y = \sum_{n \geq 1} I_{F(n)}, \quad \text{and} \quad \mathbb{E}(Y) = \sum_{n \geq 1} \mathbb{P}(F(n)) = \sum_{n=1}^{\infty} \mathbb{P}(Y \geq n).$$

Any worries about rigour you might have evaporate when you replace Y by $\min(Y, r)$ ($r \in \mathbb{N}$) and then let $r \uparrow \infty$ using the Monotone-Convergence Theorem.

The criterion is obvious since if Y is the integer part of $|X|$, then $Y \leq |X| < Y + 1$. □

H. Rules for expectations. Now let's look at the practical rules. "Here he goes", you say, "giving us 'practical rules', but with those stupid technical conditions he seems to like. At least he's put them in small type". *My reply.* I wouldn't do it if there wasn't some point. You see, it is simply not always the case that long-term averages converge to a 'mean'. For Cauchy-distributed variables, the 'sample mean of several independent observations' always has the same distribution as a single observation, as we shall see later. Cauchy RVs are not integrable. For some distributions, the long-term average tends even to 'spread out'. Now the whole story is known *exactly*. Why therefore put on blinkers by fudging?

▶▶ **Ha. Theorem: Rules for Expectations.** *Suppose that F is an event, X and Y are integrable RVs (in symbols, $X, Y \in \mathcal{L}^1$), and that $c \in \mathbb{R}$.*

(a) *Then cX and X + Y are integrable, and*

$$\mathbb{E}(I_F) = \mathbb{P}(F), \quad \mathbb{E}(cX) = c\,\mathbb{E}(X), \quad \mathbb{E}(X+Y) = \mathbb{E}(X) + \mathbb{E}(Y).$$

(b) *If X takes integer values, and h is a function defined on the set of integers, then*

$$\mathbb{E}(h(X)) = \sum_{x \in \mathbb{Z}} h(x)\mathbb{P}(X = x),$$

[Ⓜ it being understood that $h(X)$ is integrable if and only if the series converges absolutely.]

(c) *If X is 'continuous' with pdf f_X, then for any piecewise-continuous function $h : \mathbb{R} \to \mathbb{R}$,*

$$\mathbb{E}(h(X)) = \int_{x \in \mathbb{R}} h(x)\mathbb{P}(X \in \mathrm{d}x) = \int_{x \in \mathbb{R}} h(x)f_X(x)\,\mathrm{d}x,$$

[Ⓜ it being understood that $h(X)$ is integrable if and only if the familiar final integral 'converges absolutely', that is, if and only if $\int_{x \in \mathbb{R}} |h(x)|f_X(x)\,\mathrm{d}x < \infty$.]

We are going to take these rules for granted (except for Discussion I). We have seen Rules (a) and (b) in some detail for the case when Ω is finite. Rule (c) is the 'obvious' analogue of Rule (b) for the case when X is 'continuous', so you should believe Rule (c). The real point is that Rule (c) belongs to Measure Theory, where it has a simple and natural proof. See [W] for the details if you really want them now.

I. Ⓜ Discussion. "You are always almost pedantically rigorous", you say, "so give us a clue why $\mathbb{E}(X) = \int x f_X(x)\,\mathrm{d}x$." OK. To keep things simple, let's suppose that X

is bounded and non-negative: $0 \leq X \leq K$ for some integer K. For each positive integer n, define an approximation $X^{(n)}$ to X via

$$X^{(n)}(\omega) := \frac{k}{n} \quad \text{if} \quad \frac{k}{n} \leq X(\omega) < \frac{k+1}{n},$$

so that $X^{(n)}$ is a simple RV

$$X^{(n)} = \sum_{k=0}^{nK} \frac{k}{n} I_{F(n,k)}, \quad \text{where} \quad F(n,k) = \left\{ \omega : \frac{k}{n} \leq X(\omega) < \frac{k+1}{n} \right\}.$$

We have

$$\mathbb{E}\left(X^{(n)}\right) = \sum \frac{k}{n} \mathbb{P}(F(n,k)) = \sum \frac{k}{n} \int_{k/n}^{(k+1)/n} f_X(x) \, \mathrm{d}x$$

$$= \int_0^K s^{(n)}(x) f_X(x) \, \mathrm{d}x,$$

where $s^{(n)}$ is the 'staircase' function with

$$s^{(n)}(x) := \frac{k}{n} \quad \text{if} \quad \frac{k}{n} \leq x < \frac{k+1}{n}, \qquad 0 \leq x \leq K.$$

Draw a picture of $s^{(n)}$, and note that $|s^{(n)}(x) - x| \leq 1/n$ for every x. Note too that $X^{(n)} = s^{(n)}(X)$, so that the above expression for $\mathbb{E} X^{(n)}$ agrees with Rule (c). Since 'the modulus of an integral is less than or equal to the integral of the modulus', we have

$$\left| \mathbb{E}\left(X^{(n)}\right) - \int x f_X(x) \, \mathrm{d}x \right| = \left| \int \left\{ s^{(n)}(x) - x \right\} f_X(x) \, \mathrm{d}x \right|$$

$$\leq \int \left| s^{(n)}(x) - x \right| f_X(x) \, \mathrm{d}x \leq \frac{1}{n} \int f_X(x) \, \mathrm{d}x = \frac{1}{n}.$$

But if $n(r)$ denotes 2^r, then your picture for $s^{(n)}$ shows that $s^{(n(r))}(x) \uparrow x$ for every x, whence $X^{(n(r))} \uparrow X$ and, by the Monotone-Convergence Theorem 60D, $\mathbb{E} X^{(n(r))} \uparrow \mathbb{E} X$; and the result follows.

Rule (a) can be proved by vaguely analogous (but, in Measure Theory, much better) approximation techniques.

▶ **J. Examples.** I emphasize that the following examples do *not* involve any measure-theoretic ideas.

If X has the uniform U$[a,b]$ distribution on $[a,b]$ (see 54(E5)), then

$$\mathbb{E}(X) = \int_{\mathbb{R}} x f_X(x) \, \mathrm{d}x = \int_a^b x \frac{1}{b-a} \, \mathrm{d}x = \frac{1}{b-a} \int_a^b x \, \mathrm{d}x$$

$$= \frac{1}{b-a} \times \frac{b^2 - a^2}{2} = \frac{a+b}{2} \quad \text{of course!}$$

If X has the exponential E(rate λ) distribution (see 54(E6)), then, we could work out

$$\mathbb{E}(X) = \int_{\mathbb{R}} x f_X(x)\, \mathrm{d}x = \int_0^\infty x\,\lambda \mathrm{e}^{-\lambda x}\, \mathrm{d}x = \frac{1}{\lambda}.$$

I have not worked out the integral by integration by parts (but you can if you want): I promised you that we shall see crafty methods which avoid the need to do such things. And it would be silly to work out $\mathbb{E}(Y)$ if Y has the binomial(n, p) distribution via

$$\mathbb{E}(Y) = \sum_{k=0}^n k\mathbb{P}(Y = k) = \sum_{k=0}^n k\binom{n}{k} p^k (1-p)^{n-k} = np;$$

when we do not need to – who wants to do such a sum anyway? (OK, then, just for *you*. The '$k = 0$' term makes no contribution. For $k \geq 1$, the kth term in the sum is np times the probability that $Z = k - 1$ where Z has the binomial $(n-1, p)$ distribution – and it's done!) The right way to find $\mathbb{E}(Y)$ was explained at Exercise 58Da.

I repeat that I am not going to set exercises which just test whether you can work out sums or integrals.

Ja. Exercise. Two points A and C are chosen independently at random in the unit disc $\{(x, y) : x^2 + y^2 \leq 1\}$. Let D be the distance between them. Prove that $\mathbb{E}\left(D^2\right) = 1$. *Hint.* Use Pythagoras's Theorem and exploit symmetry. Later on, you are invited to check that the answer looks right by simulation.

▶▶ **K. Jensen's Inequality.** *Let c be a smooth convex function on a (finite or infinite) interval I in that c'' exists and is continuous on I, and*

$$c''(x) \geq 0 \qquad \textit{for all } x \in I.$$

Let X be an RV in $\mathcal{L}^1$ with values in I, such that $c(X) \in \mathcal{L}^1$. Then

$$\mathbb{E}\, c(X) \geq c(\mu), \qquad \mu := \mathbb{E}(X).$$

If c is strictly convex on I in that $c'' > 0$ on I except perhaps for a finite number of points at which $c'' = 0$, and if $\mathbb{P}(X = \mu) \neq 1$, then $\mathbb{E}\, c(X) > c(\mu)$.

Examples: $c(x) = x^2$ is strictly convex on $\mathbb{R}$; for any real non-zero λ, $c(x) = \mathrm{e}^{\lambda x}$ is strictly convex on $\mathbb{R}$.

Proof. Of course, $\mu \in I$. Since $c'' \geq 0$ on I, the function c' is non-decreasing on I, so that, for $x \geq \mu$ and $x \in I$,

$$c(x) - c(\mu) = \int_\mu^x c'(y)\mathrm{d}y \geq \int_\mu^x c'(\mu)\mathrm{d}y = (x - \mu)c'(\mu).$$

Check that $c(x) \geq c(\mu) + (x - \mu)c'(\mu)$ for all $x \in I$, whence,

$$c(X) \geq c(\mu) + (X - \mu)c'(\mu).$$

Taking expectations yields the result $\mathbb{E}\, c(X) \geq c(\mu)$. The 'strictly convex' version is now easily obtained. □

▶▶ **L. Bounded-Convergence Theorem.** *Suppose that X_n $(n \in \mathbb{N})$ and X are RVs such that $\mathbb{P}(X_n \to X$ as $n \to \infty) = 1$. Suppose further that for some absolute constant K, we have $|X_n(\omega)| \leq K$ for all n and ω. Then,*

$$\mathbb{E}(|X_n - X|) \to 0, \quad \text{and hence } \mathbb{E}(X_n) \to \mathbb{E}(X).$$

Heuristic proof. For a non-negative variable Y and an event F, let us write $\mathbb{E}(Y; F) := \mathbb{E}(Y I_F)$, a notation which will be very useful later. The intuitive idea for the proof of the theorem is that, for fixed $\varepsilon > 0$, we can write

$$\mathbb{E}(|X_n - X|) = \mathbb{E}(|X_n - X|; |X_n - X| \leq \tfrac{1}{2}\varepsilon) + \mathbb{E}(|X_n - X| : |X_n - X| > \tfrac{1}{2}\varepsilon)$$
$$\leq \tfrac{1}{2}\varepsilon + 2K\mathbb{P}(|X_n - X| > \tfrac{1}{2}\varepsilon),$$

and that (this is not obvious) we can choose n_0 such that for $n \geq n_0$ we have $\mathbb{P}(|X_n - X| > \tfrac{1}{2}\varepsilon) < \varepsilon/(4K)$. □

More general convergence theorems are known: in particular, Lebesgue's celebrated Dominated-Convergence Theorem and the 'necessary and sufficient' Uniform-Integrability Theorem. See [W], but remember that the Monotone-Convergence Theorem 60D is the fundamental result.

It is a great pity that Measure Theory courses often give the impression that the convergence theorems are the main things in the subject. The truth is that they contain hardly any of the topic's great subtlety.

3.5 Variances and Covariances

Figure 67E(i) will explain what variance is.

The geometry of the space $\mathcal{L}^2$ now to be introduced is absolutely fundamental to ideas of regression, and Analysis of Variance.

▶▶ **A. The space $\mathcal{L}^2$.** An RV X is said to belong to $\mathcal{L}^2$ if $\mathbb{E}\left(X^2\right) < \infty$; and we then define the $\mathcal{L}^2$ norm of X as

$$\|X\|_2 := \sqrt{\mathbb{E}\left(X^2\right)}.$$

We shall see that the $\mathcal{L}^2$ norm of X agrees with the concept of standard deviation of X if $\mathbb{E}(X) = 0$. (*Note.* Yes, analysts, I know that my 'norm' is not quite a norm, but that's not a point worth quibbling about.)

▶ **B. Lemma.** *Suppose that $X, Y \in \mathcal{L}^2$ and that $c \in \mathbb{R}$. Then*

$$cX \in \mathcal{L}^2, \quad X + Y \in \mathcal{L}^2, \quad XY \in \mathcal{L}^1, \quad X \in \mathcal{L}^1.$$

Proof. It is obvious that $cX \in \mathcal{L}^2$ because $\mathbb{E}\left\{(cX)^2\right\} = c^2\mathbb{E}\left(X^2\right)$. Next, we have

$$0 \leq (X+Y)^2 = 2(X^2+Y^2) - (X-Y)^2 \leq 2(X^2+Y^2),$$

and this makes it clear that $X + Y \in \mathcal{L}^2$. Since

$$0 \leq (|X| - |Y|)^2 = X^2 + Y^2 - 2|X||Y|,$$

we see that

$$\mathbb{E}(|XY|) \leq \tfrac{1}{2}\left\{\mathbb{E}\left(X^2\right) + \mathbb{E}\left(Y^2\right)\right\} < \infty,$$

so that $XY \in \mathcal{L}^1$. Of course, the constant function 1 is in $\mathcal{L}^2$, so that $X = X \times 1$ is in $\mathcal{L}^1$. $\qquad\square$

▶ **Ba. Exercise: Important inequalities for the Strong Law.** Let X and Y be RVs. Prove that

$$(X+Y)^4 \leq 8\left(X^4 + Y^4\right), \tag{B1}$$

and that, for $1 \leq r \leq s$,

$$|X|^r \leq 1 + |X|^s. \tag{B2}$$

Note that, in particular, as also follows from the last part of Lemma 65B,

$$\mathcal{L}^1 \subseteq \mathcal{L}^2.$$

Hints. Of course, it is enough to prove that $(x+y)^4 \leq 8(x^4 + y^4)$ for real numbers x, y (for which read the proof of the above lemma), and that for non-negative x, $x^r \leq 1 + x^s$, for which you say "Either $x \leq 1 \ldots$".

▶▶ **C. The Cauchy–Schwarz Inequality.** *If $X, Y \in \mathcal{L}^2$, then*

$$|\mathbb{E}(XY)| \leq \|X\|_2\|Y\|_2, \quad \text{equivalently,} \quad \{\mathbb{E}(XY)\}^2 \leq \mathbb{E}\left(X^2\right)\mathbb{E}((Y^2), \tag{C1}$$

with equality if and only if $\mathbb{P}(aX + bY = 0) = 1$ for some real numbers a and b, not both zero.

Proof. We may suppose that $\mathbb{E}\left(X^2\right) \neq 0$ (for if $\mathbb{E}\left(X^2\right) = 0$, then we have $\mathbb{P}(X = 0) = 1$ and $\mathbb{E}(XY) = 0$). Let $\beta = \mathbb{E}(XY)/\mathbb{E}(X^2)$. Then (you check!)

$$0 \leq \mathbb{E}\left\{(Y - \beta X)^2\right\} = \mathbb{E}(Y^2) - \{\mathbb{E}(XY)\}^2/\mathbb{E}(X^2),$$

with equality if and only if $\mathbb{P}(Y = \beta X) = 1$. $\qquad\square$

▶▶ **D. Variance, σ_X^2 or $\mathrm{Var}(X)$, of X.** If $X \in \mathcal{L}^2$, then $X \in \mathcal{L}^1$, so μ_X exists, and we can define

$$\sigma_X^2 := \mathrm{Var}(X) := \mathbb{E}\left\{(X - \mu_X)^2\right\}. \tag{D1}$$

Since (remember that μ_X is just a number)

$$\mathbb{E}\left\{(X - \mu_X)^2\right\} = \mathbb{E}\left(X^2 - 2\mu_X X + \mu_X^2\right)$$
$$= \mathbb{E}\left(X^2\right) - 2\mu_X \mathbb{E}(X) + \mu_X^2 = \mathbb{E}\left(X^2\right) - 2\mu_X^2 + \mu_X^2,$$

we also have

$$\sigma_X^2 = \mathrm{Var}(X) = \mathbb{E}\left(X^2\right) - \mu_X^2 = \mathbb{E}\left(X^2\right) - \mathbb{E}(X)^2. \qquad \text{(D2)}$$

This result, often called the **Parallel-Axis result** because of the analogous result on moment of inertia, is the first of a whole series of results on sums of squares which leads up to the Analysis-of-Variance technique. Watch the idea develop. And *do* note that we have

$$\mathrm{Var}(cX) = c^2 \mathrm{Var}(X), \qquad \mathbb{E}\left(X^2\right) = \mathbb{E}(X)^2 + \mathrm{Var}(X), \qquad \text{(D3)}$$

results which we often use.

Da. Example. Suppose that X has the exponential $\mathrm{E}(\lambda)$ distribution, so that the pdf of X is $\lambda e^{-\lambda x} I_{[0,\infty)}(x)$. We already know from Examples 63J that X has mean $\mu = 1/\lambda$, so that, by Rule (c) of Theorem 62Ha, we have (as you can check by integration by parts if you wish)

$$\mathrm{Var}(X) = \mathbb{E}(X^2) - \mu^2 = \int_0^\infty x^2 \lambda e^{-\lambda x}\, dx - \mu^2 = \frac{1}{\lambda^2}.$$

You are reminded that we shall work out such integrals by efficient indirect methods.

▶ **E. Discussion. Variance as a measure of spread.**

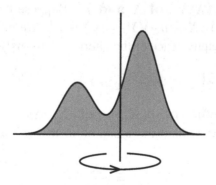

Figure E(i): Variance and moment of inertia

Variance provides a 'quadratic' measure of how widely the distribution of an RV is spread about its mean. The 'moment of inertia' idea makes this precise. Suppose that X is 'continuous'. Imagine that we make a very thin sheet of metal the shape of the area under the graph of f_X, of unit mass per unit area. The variance of X measures how hard it is to spin this metal sheet about a vertical axis through the mean – see Figure E(i). One precise form of this statement is that if the sheet is spinning, at 1 complete revolution per second, then its total kinetic energy is a certain constant ($2\pi^2$) times $\mathrm{Var}(X)$. Experience shows that variance provides a much more useful measure of spread than, for example, $\mathbb{E}(|X - \mu_X|)$.

F. Dull but important exercise.   Label the people in the city of Bath (say in alphabetical order) $1, 2, \ldots, n$. Let h_k be the height of person k. For the experiment of choosing a person in Bath at random, the sample space is $\Omega = \{1, 2, \ldots, n\}$ and $\mathbb{P}(\omega) = 1/n$ for every ω. Let X be the height of a person chosen at random from those in Bath. If $\omega = k$, then $X(\omega) = h_k$. Prove that

$$\mu := \mathbb{E}(X) = \frac{1}{n} \sum_{k=1}^{n} h_k,$$

and that

$$\sigma^2 := \mathrm{Var}(X) = \frac{1}{n} \sum_{k=1}^{n} (h_k - \mu)^2 = \frac{1}{n} \sum_{k=1}^{n} h_k^2 - \mu^2.$$

▶▶ **G. Standard deviation, σ_X or SD(X), of X.** For $X \in \mathcal{L}^2$, we define the standard deviation σ_X or SD(X) of X to be the square root of the variance of X:

$$\sigma_X := \mathrm{SD}(X) := \sqrt{\mathrm{Var}(X)}, \quad \text{so} \quad \mathrm{SD}(cX) = |c| \times \mathrm{SD}(X).$$

(Applied mathematicians call the SD the radius of gyration of our metal sheet.)

▶▶ **H. Covariance $\mathrm{Cov}(X, Y)$ of X and Y.** Suppose that $X, Y \in \mathcal{L}^2$. Then $XY \in \mathcal{L}^1$ and (why?!) $(X - \mu_X)(Y - \mu_Y) \in \mathcal{L}^1$. We may (and do!) therefore define (*you* check the equivalence of the alternative forms!)

$$\mathrm{Cov}(X, Y) := \mathbb{E}\{(X - \mu_X)(Y - \mu_Y)\} = \mathbb{E}(XY) - \mathbb{E}(X)\mathbb{E}(Y).$$

▶▶ **I. Correlation Coefficient $\rho(X, Y)$ of X and Y.**   Under the same assumptions on X and Y (and the assumption that σ_X and σ_Y are non-zero), we define

$$\rho(X, Y) := \frac{\mathrm{Cov}(X, Y)}{\sigma_X \sigma_Y}.$$

That $-1 \le \rho(X, Y) \le 1$ is proved in Lemma Ja below. We say that
    X and Y are *positively correlated* if $\rho(X, Y) > 0$,

equivalently if $\mathrm{Cov}(X, Y) > 0$,
$\quad$ X and Y are *negatively correlated* if $\rho(X, Y) < 0$,
$\quad\quad$ equivalently if $\mathrm{Cov}(X, Y) < 0$,
$\quad$ X and Y are *uncorrelated* if $\rho(X, Y) = 0$,
$\quad\quad$ equivalently if $\mathrm{Cov}(X, Y) = 0$.

What does all this mean? A good answer lies in the next topic which clarifies the rôle of $\rho(X, Y)$ as a 'measure of the degree of linear dependence between X and Y'.

▶ **J. 'Least-squares' ideas; linear regression in Probability.** The important idea here is that we wish to give the 'best predictor' Z, a random variable, of another random variable Y, Z having to take a certain form (as is illustrated in this subsection and in others). The *least-squares-best* predictor Z is that which minimizes the mean-square-error

$$\mathrm{MSE} := \mathbb{E}\{[Y - Z]^2\}.$$

Two important cases are covered by the following lemma.

▶▶ **Ja. Lemma.** *(a) Let Y be an RV in $\mathcal{L}^2$. Suppose that we consider deterministic-constant predictors c of Y, so that*

$$\mathrm{MSE} := \mathbb{E}\{[Y - c]^2\}.$$

Then the best choice of c is

$$c = \mu_Y \text{ and then } MSE = \sigma_Y^2.$$

(b) Let X and Y be RVs in $\mathcal{L}^2$, each with non-zero variance. Let $\rho := \rho(X, Y)$. Suppose that we wish to use a **'regression line'**

$$y = \alpha x + \beta$$

to predict Y from X in the sense that we use $\alpha X + \beta$ as our predictor of Y, so that

$$\mathrm{MSE} := \mathbb{E}\{[Y - (\alpha X + \beta)]^2\}.$$

Then the regression line which leads to the minimum value of MSE *may be rewritten:*

$$\textit{Regression line:} \quad (y - \mu_Y) = \rho\frac{\sigma_Y}{\sigma_X}(x - \mu_X),$$

and then

$$\mathrm{MSE} := \mathbb{E}\{[Y - (\alpha X + \beta)]^2\} = \sigma_Y^2(1 - \rho^2).$$

We therefore have

$$-1 \leq \rho = \rho(X, Y) \leq 1;$$

and if $|\rho| = 1$ *then we have 'almost perfect linearity' in that*

$$(Y - \mu_Y) = \rho \frac{\sigma_Y}{\sigma_X} (X - \mu_X) \quad \text{with probability 1.}$$

Note that the slope of the regression line always has the same sign as ρ, in other words the same sign as $\mathrm{Cov}(X, Y)$.

▶ **Jb. Exercise.** Prove Lemma 69Ja.

Jc. Exercise.   Show that if X has the uniform $\mathrm{U}[-1, 1]$ distribution on $[-1, 1]$, and $Y = X^2$, then $\rho(X, Y) = 0$. *Thus X and Y are uncorrelated even though Y is a deterministic function of X. This emphasizes that ρ can only 'assess' linear dependence.*

Note. Linear regression in Statistics, which has many analogies with what we have just done, is studied in Chapter 8.

▶▶ **K. Lemma.   Variance of a sum.** *Suppose that we have random variables $X_1, X_2, \ldots, X_n \in \mathcal{L}^2$. Then*

$$\mathrm{Var}(X_1 + X_2 + \cdots + X_n) = \sum_i \sum_j \mathrm{Cov}(X_i, X_j) \quad (n^2 \text{ terms})$$

$$= \underbrace{\sum_i \mathrm{Var}(X_i)}_{n \text{ terms}} + \underbrace{\sum_i \sum_{j \neq i} \mathrm{Cov}(X_i, X_j)}_{n(n-1) \text{ terms}}.$$

Proof. Let $\mu_k := \mathbb{E}(X_k)$. Then,

$$(X_1 + X_2 + \cdots + X_n)^2 = (X_1 + X_2 + \cdots + X_n)(X_1 + X_2 + \cdots + X_n)$$

$$= \sum_i \sum_j X_i X_j \quad (n^2 \text{ terms}).$$

and, similarly,

$$(\mu_1 + \mu_2 + \cdots + \mu_n)^2 = \sum_i \sum_j \mu_i \mu_j \quad (n^2 \text{ terms}).$$

If we take expectations of the first of these equations and then subtract the second, we obtain precisely the result we want. (To get the second form, we use the fact that $\mathrm{Cov}(X_i, X_i) = \mathrm{Var}(X_i)$.)   □

Ka. Exercise. Prove that if R_n is the number of professors who get the right hat in Example 40E, then $\mathrm{Var}(R_n) = 1$ whenever $n \geq 2$. *Hint.* Let X_k be the indicator function of F_k in the Solution to Example 40E. Find $\mathbb{E}(X_k)$, $\mathrm{Var}(X_k)$ and $\mathrm{Cov}(X_i, X_j)$ $(i \neq j)$. Why is it not surprising that the covariance is positive?

▶ **Kb. Exercise.** The problem here is to show that if I take an ordered sample of two different people from the n people who live in the city of Bath and record their heights H_1, H_2, then the correlation coefficient $\rho(H_1, H_2)$ is $-1/(n-1)$. Why is it not surprising that the covariance is negative? *Hint.* Suppose that I get the computer to arrange all the people in Bath in random order, all $n!$ permutations being equally likely. Let X_k be the height of the kth person chosen by the computer. Then what is

$$\text{Var}(X_1 + X_2 + \cdots + X_n)?$$

(If I repeated the experiment of finding $X_1 + X_2 + \cdots + X_n$, how much different is the answer I would get?) Now use Lemma 70K to express the common covariance $C := \text{Cov}(X_i, X_j)$ $(i \neq j)$ in terms of the common variance $\sigma^2 = \text{Var}(X_i)$.

The above is a 'trick' solution. A much more illuminating solution of this problem will be given later. But here's a dull solution.

Dull, but instructive, solution. Consider the experiment of choosing 2 different people at random from the population of Bath. We regard a possible ω as an ordered pair (i, j), where $i, j \in \{1, 2, \ldots, n\}$ and $j \neq i$. There are $n(n-1)$ such ordered pairs, and the statement that we choose 'at random' here conveys that every ω has probability $\{n(n-1)\}^{-1}$. For $\omega = (i, j)$, we have $H_1(\omega) = h_i$ and $H_2(\omega) = h_j$. We have

$$
\begin{aligned}
\mathbb{E}\,(H_1 H_2) &= \frac{1}{n(n-1)} \sum_i \sum_{j \neq i} h_i h_j \\
&= \frac{1}{n(n-1)} \sum_i \sum_j h_i h_j - \frac{1}{n(n-1)} \sum h_i^2 \\
&= \frac{1}{n-1} \frac{1}{n} \left(\sum h_i \right)^2 - \frac{1}{n-1} \frac{1}{n} \sum h_i^2 \\
&= \frac{1}{n-1} n\mu^2 - \frac{1}{n-1} (\mu^2 + \sigma^2) = \mu^2 - \frac{\sigma^2}{n-1}.
\end{aligned}
$$

Hence,

$$C := \text{Cov}\,(H_1, H_2) = \mathbb{E}\,(H_1 H_2) - \mathbb{E}\,(H_1)\,\mathbb{E}\,(H_2) = -\frac{\sigma^2}{n-1},$$

and the desired result follows. □

Kc. Exercise. Adapt the trick method of doing the last example to show that if $X_1, X_2, \ldots, X_n$ are any RVs in $\mathcal{L}^2$, such that

$$\text{Var}(X_i) = 1, \qquad \rho(X_i, X_j) = \rho \quad (i \neq j),$$

then $\rho \geq -1/(n-1)$. First though, explain why it is obvious that if $n = 3$, then ρ cannot be -1.

Kd. Exercise. Prove that if X, Y, Z are RVs in $\mathcal{L}^2$ such that

$$\text{Var}(X) = \text{Var}(Y) = \text{Var}(Z) = 1, \qquad \rho(X, Y) = \rho(X, Z) = \rho,$$

then

$$\rho(Y, Z) \geq 2\rho^2 - 1.$$

First though, explain why it is obvious that if $\rho = \pm 1$, then $\rho(Y, Z) = 1$. *Hint for the main part.* Consider $\mathrm{Var}(Y + Z - \alpha X)$. *Note.* The described result will become obvious for other reasons given in Chapter 8.

▶▶ **L. Exercise. Median and the $\mathcal{L}^1$ analogue of Part(a) of Lemma 69Ja.** Suppose that X is a 'continuous' variable with strictly positive pdf f on $\mathbb{R}$. Then (why?) there will exist a unique real number m, called the **median** of X such that

$$\mathbb{P}(X \leq m) = \mathbb{P}(X \geq m) = \tfrac{1}{2}.$$

Suppose that $X \in \mathcal{L}^1$, that is, $\mathbb{E}(|X|) < \infty$. Prove that *the value $c = m$ minimizes the mean absolute error* $\mathbb{E}(|X - c|)$. *Hint.* Draw a graph of the function

$$x \mapsto |x - m| - |x - c|$$

in the cases $m < c, m > c$, and confirm that, for $x \neq m$,

$$|x - m| - |x - c| \leq (m - c)\left\{ I_{(-\infty, m)}(x) - I_{(m, \infty)}(x) \right\}.$$

4

CONDITIONING AND INDEPENDENCE

Hallelujah! At last, things can liven up.

4.1 Conditional probabilities

▶ **A. Definition of conditional probability** $\mathbb{P}(B \mid A)$. Suppose that A and B are events, (measurable) subsets of the sample space Ω for our experiment $\mathcal{E}$. Suppose further that $\mathbb{P}(A) \neq 0$. Then we define the conditional probability $\mathbb{P}(B \mid A)$ of B given A ('that B occurs given that A occurs') via

$$\mathbb{P}(B \mid A) := \frac{\mathbb{P}(A \cap B)}{\mathbb{P}(A)}. \tag{A1}$$

We saw motivation for this at 6G. We also considered briefly there, in relation to an experiment already performed, Bayesian and Frequentist views of the interpretation of conditional probability in the real world.

Definition (A1) exhibits $\mathbb{P}(B \mid A)$ as the $\mathbb{P}$-weighted fraction of A which is (also) covered by B.

Aa. Remark. If Ω is discrete and (as is the case for all examples which then concern us) each one-point set $\{\omega\}$ is an event, then conditioning on A effectively means replacing Ω by A and $\mathbb{P}(\omega)$ by $\mathbb{P}(\omega \mid A) := \mathbb{P}(\{\omega\} \mid A)$ for $\omega \in \mathcal{A}$.

Ab. Example. For the 'Car and Goats' problems at 15P, let's write

$$\omega = (\text{Contestant's original choice, Host's choice}).$$

For an outcome $\omega \in \Omega$, write $X(\omega) = 1$ or 0 according to whether the contestant would win or lose if she sticks to her original choice, $Y(\omega) = 1$ or 0 according to whether the contestant would win or lose if she switches.

ω	$\mathbb{P}(\omega)$	$X(\omega)$	$Y(\omega)$
(C,G_1)	$\frac{1}{6}$	1	0
(C,G_2)	$\frac{1}{6}$	1	0
(G_1,G_2)	$\frac{1}{3}$	0	1
(G_2,G_1)	$\frac{1}{3}$	0	1

Table A(i): Original problem

For the original problem at 15Pa, we have Table A(i) with no conditional probabilities. We see that $\mathbb{E}(X) = \frac{1}{3}, \mathbb{E}(Y) = \frac{2}{3}$.

For the modified problem at 15Pc, we have Table A(ii), where A stands for the event that the host happens to reveal a goat. We see that (with obvious 'conditional expectation' notation) $\mathbb{E}(X \mid A) = \frac{1}{2}, \mathbb{E}(Y \mid A) = \frac{1}{2}$. Conditional expectations are a major topic in Chapter 9.

	ω	$\mathbb{P}(\omega)$	$\mathbb{P}(\omega \mid A)$	$X(\omega)$	$Y(\omega)$
	(C,G_1)	$\frac{1}{6}$	$\frac{1}{4}$	1	0
A	(C,G_2)	$\frac{1}{6}$	$\frac{1}{4}$	1	0
	(G_1,G_2)	$\frac{1}{6}$	$\frac{1}{4}$	0	1
	(G_2,G_1)	$\frac{1}{6}$	$\frac{1}{4}$	0	1
$\Omega \setminus A$	(G_1,C)	$\frac{1}{6}$	0	0	0
	(G_2,C)	$\frac{1}{6}$	0	0	0

Table A(ii): When host doesn't know

Ac. Exercise. Prove that if B and C are disjoint events, then (if $\mathbb{P}(A) \neq 0$)

$$\mathbb{P}(B \cup C \mid A) = \mathbb{P}(B \mid A) + \mathbb{P}(C \mid A).$$

Ad. Exercise. I saw in Lindley's thought-provoking book [153] the following exercise. 'Let A, B and C denote events of positive probability. Say that A *favours* B if $\mathbb{P}(B \mid A) > \mathbb{P}(B)$. Is it generally the case that if A favours B and B favours C, then A favours C?' *Because of the 'Universality of the Fundamental Model' already alluded to, if you wish to produce a counterexample to any such statement, you can always take for the basic experiment $\mathcal{E}$ that of choosing a number between 0 and 1 according to the uniform distribution.* But you don't have to do the question this way.

▶ **B. General Multiplication Rules.** As several examples in Chapter 1 showed, we often use the General Multiplication Rules

$$\begin{aligned} \mathbb{P}(A \cap B) &= \mathbb{P}(A)\mathbb{P}(B \mid A), \\ \mathbb{P}(A \cap B \cap C) &= \mathbb{P}(A)\mathbb{P}(B \mid A)\mathbb{P}(C \mid A \cap B), \quad \text{etc,} \end{aligned} \qquad \text{(B1)}$$

to assign probabilities to events. Have another look at the 'National Lottery' Proof 10L of the binomial-distribution formula, and at the Birthdays Problem 13O from this point of view.

C. Bayes' Theorem. This theorem, due to the Reverend Thomas Bayes (1763), is now an immediate consequence of the definition 73A of conditional probability. However, it systematizes arguments such as that used in the 'Test for disease' problem with which we began Chapter 1. More significantly, it is the basis for the Bayesian approach to Statistics.

▶▶ **Ca. Theorem (Bayes).** *Suppose that* $H_1, H_2, \ldots, H_n$ *are exclusive (disjoint) and exhaustive events:*

$$H_i \cap H_j = \emptyset \ (i \neq j), \qquad \bigcup H_k = \Omega.$$

Thus, precisely one of the H_k *will occur on any performance of our experiment* $\mathcal{E}$. *Assume that* $\mathbb{P}(H_k) > 0$ *for every* k.

Let K *be some event with* $\mathbb{P}(K) > 0$. *Then we have the decomposition result*

$$\mathbb{P}(K) = \sum_{j=1}^{n} \mathbb{P}(H_j \cap K) = \sum_{j=1}^{n} \mathbb{P}(H_j) \mathbb{P}(K \mid H_j). \tag{C1}$$

Now,

$$\mathbb{P}(H_i \mid K) = \frac{\mathbb{P}(H_i) \mathbb{P}(K \mid H_i)}{\mathbb{P}(K)}.$$

We write

$$\mathbb{P}(H_i \mid K) \propto \mathbb{P}(H_i) \mathbb{P}(K \mid H_i), \tag{C2}$$

the constant of proportionality (in fact, $1/\mathbb{P}(K)$*) being determined by the fact (see Exercise 74Ac) that*

$$\sum_i \mathbb{P}(H_i \mid K) = 1 \quad (= \mathbb{P}(\Omega \mid K)).$$

Proof. The way I have worded it, the theorem virtually proves itself. As an exercise, you should work through it, beginning with

$$K = K \cap \Omega = K \cap (H_1 \cup H_2 \cup \ldots \cup H_n)$$

whence, from the distributive law of set theory,

$$K = (K \cap H_1) \cup (K \cap H_2) \cup \ldots \cup (K \cap H_n).$$

This expresses K as a disjoint union. Now use the Addition Rule and the General Multiplication Rule. □

Of course, the rule (C1) generalizes the rule

$$\mathbb{P}(B) = \mathbb{P}(A)\mathbb{P}(B \mid A) + \mathbb{P}(A^c)\mathbb{P}(B \mid A^c), \tag{C3}$$

which we used several times in Chapter 1, where the equation appeared as 8(I1).

▶ **D. Exercise: Pólya's Urn.**   This model is more 'a thing of beauty and a joy for ever' than ever Keats's Grecian urn was! It is very rich mathematically. We shall return to it on several occasions.

At time 0, an urn contains 1 Red and 1 Black ball. Just before each time $1, 2, 3, \ldots$, Pólya chooses a ball at random from the urn, and then replaces it together with a new ball of the same colour. Calculate (for $n = 0, 1, 2, \ldots$ and for $1 \leq r \leq n + 1$)

$$p_{n,r} := \mathbb{P}(\text{at time } n, \text{ the urn contains } r \text{ Red balls}).$$

(*Hint.* If the urn contains r Red balls at time $n > 0$, then it must have contained either r or $r - 1$ Red balls at time $n - 1$. As Pólya himself would tell you (Pólya [185, 186]), you should work out the answer using formula 75(C1) for $n = 1, 2, 3$, guess the answer for the general case, and prove it by induction.)

It is a theorem that the following statement (discussed in Chapter 9) is true with probability 1: as $n \to \infty$, the proportion of Red balls in the urn converges to a limit Θ (a Random Variable). What is

$$\mathbb{P}(\Theta \leq x) \text{ for } 0 \leq x \leq 1?$$

Hint. Apply common sense to your earlier result.

E. Exercise*: A car-parking/dimerization problem.   This problem has been studied by many authors, of whom perhaps the first was E S Page.

On the side of a street, sites

$$1 \quad\quad 2 \quad\quad 3 \quad\quad\quad\quad\quad\quad n$$

are marked, a car's length apart. A car must park so as to occupy a closed interval $[i, i+1]$ for some i with $1 \leq i \leq n-1$ (and then no car may park to occupy $[i-1, i]$ or $[i+1, i+2]$). The first driver to arrive chooses one of the $n - 1$ available parking positions at random. The next to arrive then chooses his parking position at random from those then available; and so on, until only isolated sites (no good for parking) are left. Show that if $n = 5$, then, for the final configuration, we have (with obvious notation):

$$\mathbb{P}(\overline{1 \quad 2} \; \; \overline{3 \quad 4} \; \; 5) = \mathbb{P}(1 \; \; \overline{2 \quad 3} \; \; \overline{4 \quad 5}) = 3/8, \quad \mathbb{P}(\overline{1 \quad 2} \; \; 3 \; \; \overline{4 \quad 5}) = 1/4.$$

[For clarity: Site 5 ends up isolated in the first configuration, Site 1 in the second, Site 3 in the third.] For general n, let p_n be the probability that the right-most site n ends up isolated. Show that

$$(n - 1)p_n = p_1 + p_2 + \cdots + p_{n-2}, \quad\quad p_n - p_{n-1} = -(p_{n-1} - p_{n-2})/(n - 1),$$

and deduce that

$$p_n = \sum_{j=0}^{n-1} \frac{(-1)^j}{j!},$$

so that $p_n \approx e^{-1}$ for large n. Argue convincingly that

$$p_{i,n} := \mathbb{P}[\text{site } i \text{ ends up isolated}] = p_i p_{n-i+1},$$

which, when n is large is very close to e^{-2} for nearly all i. Hence, it is very plausible (and true!) that if μ_n is the mean number of sites which end up isolated when we start with n sites, then

$$\mu_n/n \to e^{-2} \text{ as } n \to \infty.$$

Is it true that

$$\mathbb{P}(\text{interval } [i-1, i] \text{ becomes occupied}) = (1 - p_i)p_{n+1-i}?$$

Notes. Chemists are interested in this type of dimerization problem. Problems of '2 × 1'-dimer coverings of regions on a '1 × 1' lattice are absolutely fascinating and relevant to Statistical Mechanics. See references at the end of Chapter 9.

►► **F. Reformulation of result 75(C2).** We now reformulate result 75(C2) in the *language* of Bayesian Statistics, though what we do here is uncontroversial Probability.

Examples will make clear what this is all about. Think of $H_1, H_2, \ldots, H_n$ as '*hypotheses*' precisely one of which is true. Think of K as representing the *known information* we have about the outcome of the experiment: we are told that *event K occurred*. The *absolute probability* $\mathbb{P}(H_i)$ is called *the prior degree of belief in* H_i. The conditional probability $\mathbb{P}(H_i \mid K)$ of H_i given that K occurred is called the *posterior degree of belief in* H_i. The conditional probability $\mathbb{P}(K \mid H_i)$ of getting the observed result K when H_i is true is called the *likelihood* of K given H_i. Then 75(C2) reads

Posterior degree of belief $\propto$ (Prior degree of belief)× Likelihood.   (F1)

Here's a key 'Weighing the Odds' formula!

Fa. Example. Return again to the 'Test for Disease' Example 3B. Consider the two 'hypotheses':

H_1: the chosen person does *not* have the disease;

H_2: the chosen person *does* have the disease;

We have the following actual information:

K: the test on the chosen person proved positive.

Table F(i) shows a Bayesian table for analyzing the problem:

We work with 'numbers proportional to' ('odds ratios') in all but the last column, where we normalize so as to make the column sum 1. We have of course used formula (F1).

Hypothesis H	Prior $\mathbb{P}(H)$ $\propto$	Likelihood $\mathbb{P}(K\mid H)$ $\propto$	Posterior $\mathbb{P}(H\mid K)$ $\propto$	Simplified ratio $\propto$	Posterior $\mathbb{P}(H\mid K)$ $=$
H_1	99	10	990	11	$11/12$
H_2	1	90	90	1	$1/12$

Table F(i): Bayesian table for test-for-disease model

▶ **G. A step towards Bayesian Statistics.** The Bayesian approach to Statistics regards the whole of Statistics as a set of applications of (F1), but allows a much wider interpretation of 'degree of belief' than is allowed in Probability or in the traditional view of Statistics.

I now ask you to consider a situation which keeps things strictly within Probability. Please do not regard the description in 'light-hearted' language as being a kind of 'send-up' of the Bayesian approach. *I think that Bayesian Statistics has* **very** *great value.* In the following discussion, we treat at an intuitive level conditional pdfs, etc, studied rigorously later.

Suppose that God chose a number Θ between 0 and 1 at random according to the uniform $U[0,1]$ distribution, and made a coin with probability Θ of Heads. Suppose that He then presented the coin to His servant, the Rev'd Thomas Bayes, and explained to Bayes how He had chosen Θ, but did *not* reveal the chosen value of Θ. Suppose that Bayes then tossed the coin n times, and obtained the result

$$K : \text{THHT} \ldots \text{HT}, \quad \text{with } r \text{ Heads in the } n \text{ tosses.} \tag{G1}$$

Think of dividing up $[0,1]$ into lots of little intervals '$d\theta = (\theta, \theta + d\theta]$', and of $H_{d\theta}$ as $\Theta \in d\theta$. The absolute pdf of Θ, denoted in standard Bayesian fashion by $\pi(\cdot)$ rather than $f_\Theta(\cdot)$, is given by

$$\pi(\theta) = 1 \text{ on } [0,1], \quad \textit{uniform prior,}$$

so that

$$\mathbb{P}(H_{d\theta}) := \mathbb{P}(\theta < \Theta \leq \theta + d\theta) = \pi(\theta)d\theta.$$

The *likelihood* $\text{lhd}(\theta; K)$ of getting the actual observation K if $H_{d\theta}$ is true (that is, to all intents and purposes, if $\Theta = \theta$), is, by the usual independence argument

$$\text{lhd}(\theta; K) = (1-\theta)\theta\theta(1-\theta)\ldots\theta(1-\theta) = \theta^r(1-\theta)^{n-r}.$$

By rule 77(F1),

$$\mathbb{P}(H_{d\theta} \mid K) \propto \mathbb{P}(H_{d\theta}) \times \text{lhd}(\theta; K),$$

leading to the formula for the *conditional pdf of Θ given K*:

$$\pi(\theta \mid K) \propto \pi(\theta) \times \text{lhd}(\theta; K) = \theta^r(1-\theta)^{n-r} \quad (0 \leq \theta \leq 1). \tag{G2}$$

We want $\int_0^1 \pi(\theta \mid K) d\theta = 1$. Now, without doing any integration, we shall prove at 107L below that

$$\int_0^1 \theta^r (1 - \theta)^s d\theta = \frac{r!s!}{(r + s + 1)!} \qquad (r, s \in \mathbb{Z}^+). \tag{G3}$$

Hence, Bayes' degree of belief as to the value of Θ based on the result 78(G1) and the uniform prior is summarized by the conditional pdf

$$\pi(\theta \mid K) = (n + 1) \binom{n}{r} \theta^r (1 - \theta)^{n-r} I_{[0,1]}(\theta). \tag{G4}$$

H. A surprising isomorphism. It is no coincidence that I used the symbol Θ for the limiting proportion of Red balls in Pólya urn because *Pólya's urn is mathematically identical (isomorphic) to the Bayesian analysis of coin tossing using a uniform prior for the probability Θ of Heads*. We shall see more of this later, but discuss the key point now.

The absolute probability that Bayes would get outcome $K = \text{THHT} \ldots \text{HT}$, with r Heads and $n - r$ Tails, is

$$\int_0^1 \theta^r (1 - \theta)^{n-r} d\theta = \frac{r!(n - r)!}{(n + 1)!}, \tag{H1}$$

using (G3). *Intuition.* Use decomposition rule 75(C1), thinking of the left-hand side as

$$\mathbb{P}(K) = \sum \mathbb{P}(H_{d\theta}) \mathbb{P}(K \mid H_{d\theta}) = \int (d\theta) \theta^r (1 - \theta)^{n-r}.$$

The absolute probability that Pólya would choose the sequence BRRB...RB of colours corresponding to K, with R for H and B for T, is (see Clarification below)

$$\frac{1}{2} \times \frac{1}{3} \times \frac{2}{4} \times \frac{2}{5} \times \ldots \times \frac{r}{n} \times \frac{n - r}{n + 1} = \frac{r!(n - r)!}{(n + 1)!},$$

so the absolute probability that Bayes would get any result is the same as that that Pólya would get the corresponding result.

Clarification. For the described Pólya result corresponding to K, just before time $n - 1$ (equivalently, at time $n - 2$) there would be r Red balls (the original plus $r - 1$ new ones) out of n balls, and just before time n (equivalently, at time $n - 1$) there would be $n - r$ Black balls out of $n + 1$ balls.

Note. Under the isomorphism as completed via the π-system Lemma (see 45M), the existence of the limit Θ for Pólya's urn is essentially just the Strong Law for coin-tossing.

I. Solution to Eddington's Problem 19C. For the 'Two Liars' version, for which K is 'C affirms that D lied', we make the Bayes table I(i). The 'T/L' under D or C signifies whether or not that person spoke the Truth or Lied. The 'TL' in the first row of the table proper under 'Summary' signifies that 'D spoke the Truth and C lied'. The probability of this is $\frac{1}{3} \times \frac{2}{3} \propto 1 \times 2 = 2$.

Hypothesis	D T/L	C affirms that D lied	C T/L	Summary	Odds ∝
D spoke truth	T	Yes	L	TL	2
D lied	L	No	T	LT	2

Table I(i): Bayes table for 'Two Liars' problem

Hyp	D T/L	(C)	C T/L	(B)	B T/L	(A)	A T/L	Summary	Odds ∝
D spoke truth	T	Yes	L	Yes	L	Yes	T	TLLT	4
	T	Yes	L	No	T	Yes	L	TLTL	4
	T	No	T	Yes	T	Yes	T	TTTT	1
	T	No	T	No	L	Yes	L	TTLL	4
D lied	L	Yes	T	Yes	L	Yes	T	LTLT	4
	L	Yes	T	No	T	Yes	L	LTTL	4
	L	No	L	Yes	T	Yes	T	LLTT	4
	L	No	L	no	L	Yes	L	LLLL	16

Table I(ii): Bayes table for 'Four Liars' problem

The summed odds that D spoke the truth is 2 and the total of the odds is 4. Hence the conditional probability that D spoke the truth is $2/4 = 1/2$.

For the 'Four Liars' problem, we make the Bayes table I(ii) in which we use the shorthand
(C): C affirms that D lied,
(B): B denies that C affirms that D lied,
(A): A affirms that B denies that C affirms that D lied.
The table should be self-explanatory.

The summed odds that D spoke the truth is 13, and the total of the odds is 41. Hence the conditional probability that D spoke the truth is $13/41$.

▶ **J. Hazard functions in discrete time.**   Engineers study failure times of components, and actuaries life-times of people, in terms of *hazard functions* which are *conditional failure rates*. Let's first examine the situation in discrete time.

Let T be an RV taking values in $\mathbb{N} := \{1, 2, 3, \ldots\}$. *Think of T as the time just before which a component fails.* Define the *Reliability Function R for T* via

$$R_n := R(n) := \mathbb{P}(T > n) = 1 - F(n),$$

F being the DF of T. We use whichever of R_n and $R(n)$ is typographically neater at the time.

The hazard function h for T is defined via

$$h_n := \mathbb{P}(T = n \mid T > n - 1) = \mathbb{P}(\text{failure just before } n \mid \text{OK at time } n - 1).$$

(This is really defined only provided that $R_{n-1} > 0$.) Since

$$\{T > n - 1\} \cap \{T = n\} = \{T = n\},$$

we have

$$h_n = \frac{\mathbb{P}(T = n)}{\mathbb{P}(T > n - 1)}, \quad \mathbb{P}(T = n) = R_{n-1}h_n.$$

However,

$$\mathbb{P}(T = n) = R_{n-1} - R_n \quad \text{(you check!)},$$

so that

$$R_n = R_{n-1}(1 - h_n) = (1 - h_1)(1 - h_2)\ldots(1 - h_n),$$
$$p_n = (1 - h_1)(1 - h_2)\ldots(1 - h_{n-1})h_n,$$

where p is the pmf of T: $p_n := \mathbb{P}(T = n)$.

Geometric distribution. If the hazard function is constant, $h_n = c$ for $n \in \mathbb{N}$, then T has the geometric distribution with pmf

$$\mathbb{P}(T = n) = (1 - c)^{n-1}c \quad (n = 1, 2, 3, \ldots).$$

This is the distribution of the time of the first Head if we toss a coin with probability c of Heads repeatedly.

▶ **K. Hazard functions in continuous time.** Now let T be a 'continuous' variable with values in $[0, \infty)$. Define its Reliability Function R as in the discrete case:

$$R(t) := \mathbb{P}(T > t) = 1 - F(t).$$

The intuitive idea is that the hazard function of T is defined via

$$h(t)\mathrm{d}t = \mathbb{P}(t - \mathrm{d}t < T \le t \mid T > t - \mathrm{d}t). \tag{K1}$$

This is the exact analogue of the discrete case, and it is exactly the right way to think in more advanced theory where much more general 'hazard functions' are considered. We interpret (K1) in the obvious way:

$$h(t)\mathrm{d}t = \frac{\mathbb{P}(t - \mathrm{d}t < T \le t)}{\mathbb{P}(T > t - \mathrm{d}t)} = \frac{f(t)\mathrm{d}t}{R(t)},$$

where f is the pdf of T, so that $f(t) = F'(t) = -R'(t)$. Putting things together, we have

$$h(t) = -\frac{R'(t)}{R(t)} = -\frac{\mathrm{d}}{\mathrm{d}t}\ln R(t), \quad R(0) = 1,$$

so that, for $t \ge 0$,

$$R(t) = \exp\left\{-\int_0^t h(s)\,\mathrm{d}s\right\}, \qquad f(t) = h(t)\exp\left\{-\int_0^t h(s)\,\mathrm{d}s\right\}.$$

These are the analogues of the 'discrete' results in the previous subsection.

If h is constant, $h(t) = \lambda$ for $t \geq 0$, then T has the exponential $E(\lambda)$ distribution with pdf $\lambda e^{-\lambda t} I_{[0,\infty)}(t)$.

It is usually assumed for failure of components that the hazard function increases with time. Important cases are the **Weibull**(a, b) distribution in which, for $t \geq 0$,

$$R(t) = \exp\{-at^b\}, \qquad h(t) = abt^{b-1},$$

with $b > 1$, and the **shifted Weibull** for which it is assumed that T is definitely greater than some t_0 and that $T - t_0$ is Weibull(a, b).

▶▶ **L. Lack-of-memory property of the exponential distribution.**   Let T have the exponential $E(\lambda)$ distribution. *We always think of this distribution via its reliability function:*

$$R(t) := \mathbb{P}(T > t) = e^{-\lambda t}.$$

The fact that T has constant hazard function is reflected in the crucial 'lack of memory' property of the exponential distribution which I now explain.

Suppose that $s > 0$ and $t > 0$. Then

$$\mathbb{P}(T > t + s \mid T > s) := \frac{\mathbb{P}(T > s;\ T > s + t)}{\mathbb{P}(T > s)}$$

$$= \frac{\mathbb{P}(T > s + t)}{\mathbb{P}(T > s)} = \frac{e^{-\lambda(t+s)}}{e^{-\lambda s}} = e^{-\lambda t}.$$

Hence,

$$\mathbb{P}(T - s > t \mid T > s) = e^{-\lambda t} = \mathbb{P}(T > t).$$

In other words, *if we are waiting for time T to occur, and we have already waited up to current time s without T's having occurred, then (conditionally on this information) the further time $T - s$ which we have to wait has the same distribution as the time we have to wait from the beginning!* The variable has no memory of the fact that a time s has already elapsed. The 'lack of memory' property is somewhat counter-intuitive – especially in forms which we shall meet later.

The exponential distribution is as important to probabilists as the normal distribution.

▶▶ **M. The recursive nature of conditioning.**   This important topic adds greatly to the power of the conditioning idea. This will be illustrated in the subsection 84O below. We consider Bayes' Theorem in which the known event K has the form

$$K = K_1 \cap K_2 \cap \ldots \cap K_n.$$

Imagine that an experiment is performed in n stages, and that K_r represents information about the rth stage. For an event H regarded as a 'hypothesis', we wish to calculate $\mathbb{P}(H \mid K)$ by calculating recursively:

$$\mathbb{P}_0(H) := \mathbb{P}(H), \quad \mathbb{P}_1(H) := \mathbb{P}(H \mid K_1), \quad \mathbb{P}_2(H) := \mathbb{P}(H \mid K_1 \cap K_2),$$

etc. We know that

$$\mathbb{P}_1(H) = \frac{\mathbb{P}_0(H)\mathbb{P}_0(K_1 \mid H)}{\mathbb{P}_0(K_1)} \propto \mathbb{P}_0(H)\mathbb{P}_0(K_1 \mid H).$$

In the above equation, we think of $\mathbb{P}_0(H)$ as the *prior* probability of H at the first stage, $\mathbb{P}(K_1 \mid H)$ as the likelihood of K_1 given H at the first stage, and $\mathbb{P}(H \mid K_1)$ as the *posterior* probability of H given the information K_1 about the outcome of the first stage.

We now want $\mathbb{P}_1(\cdot)$ to play the rôle of the *prior* probability as we go into the second stage. So, it makes sense to define

$$\mathbb{P}_1(B \mid A) := \frac{\mathbb{P}_1(A \cap B)}{\mathbb{P}_1(A)} = \frac{\mathbb{P}(K_1 \cap A \cap B)}{\mathbb{P}(K_1 \cap A)} = \mathbb{P}(B \mid K_1 \cap A).$$

We have

$$\begin{aligned}
\mathbb{P}(H \mid K_1 \cap K_2) &= \frac{\mathbb{P}(K_1 \cap K_2 \cap H)}{\mathbb{P}(K_1 \cap K_2)} \\
&= \frac{\mathbb{P}(K_1 \cap H \cap K_2)}{\mathbb{P}(K_1 \cap K_2)} = \frac{\mathbb{P}(K_1)\mathbb{P}(H \mid K_1)\mathbb{P}(K_2 \mid H \cap K_1)}{\mathbb{P}(K_1)\mathbb{P}(K_2 \mid K_1)},
\end{aligned}$$

so that we do indeed have

$$\mathbb{P}_2(H) = \frac{\mathbb{P}_1(H)\mathbb{P}_1(K_2 \mid H)}{\mathbb{P}_1(K_2)}.$$

This is exactly the kind of recursion we were looking for, but do note that the 'likelihood' $\mathbb{P}_1(K_2 \mid H)$ at the second stage is

$$\mathbb{P}(K_2 \mid H \cap K_1).$$

Extending this idea, we arrive at the following lemma.

N. Lemma. Our recursion takes the form:

$$\mathbb{P}_1(H) = \frac{\mathbb{P}(H)\mathbb{P}(K_1 \mid H)}{\mathbb{P}(K_1)} \propto \mathbb{P}(H)\mathbb{P}(K_1 \mid H),$$

$$\mathbb{P}_2(H) = \frac{\mathbb{P}_1(H)\mathbb{P}_1(K_2 \mid H)}{\mathbb{P}_1(K_2)} \propto \mathbb{P}_1(H)\mathbb{P}(K_2 \mid H \cap K_1),$$

$$\mathbb{P}_3(H) = \frac{\mathbb{P}_2(H)\mathbb{P}_2(K_3 \mid H)}{\mathbb{P}_2(K_3)} \propto \mathbb{P}_2(H)\mathbb{P}(K_3 \mid H \cap K_1 \cap K_2).$$

Example. Suppose that in the 'Test for disease' problem, a person is chosen at random and tested *twice* for the disease, and that we have

K_1: the first test proves positive,

K_2: the second test proves positive.

Each test, independently of the other and of whether or not the person has the disease, has a 90% probability of being correct. The recursive calculation is described in the Bayes table N(i).

Hypothesis H: Person is	$\mathbb{P}(H)$ $\propto$	$\mathbb{P}(K_1\mid H)$ $\propto$	$\mathbb{P}_1(H)$ $\propto$	$\mathbb{P}_1(K_2\mid H)$ $\propto$	$\mathbb{P}_2(H)$ $\propto$	$\mathbb{P}_2(H)$ $=$
OK	99	10	11	10	110	0.55
Diseased	1	90	1	90	90	0.45

Table N(i): Recursive calculation

Of course, we could have used $K = K_1 \cap K_2$, and used

K: both tests positive

in the single calculation in Table N(ii).

Hypothesis H: Person is	$\mathbb{P}(H)$ $\propto$	$\mathbb{P}(K\mid H)$ $\propto$	$\mathbb{P}(H\mid K)$ $\propto$	$\mathbb{P}(H\mid K)$ $=$
OK	99	0.1^2	99	0.55
Diseased	1	0.9^2	81	0.45

Table N(ii): 'Single' calculation

The whole point of recursion is that the two methods agree.

O. A Bayesian change-point-detection filter.

A machine produces items at a rate of 1 per minute. Suppose that the time T (in minutes) at which a certain crucial component in the machine fails is a discrete RV with constant hazard function c; that before time T each item produced by the machine has a probability p_1 of being defective; but that, from time T on, each item has probability p_2 of being defective. We wish to find recursively

$$\texttt{Phappened[n]} := \mathbb{P}(T \le n \mid \mathcal{F}_n)$$

where $\mathcal{F}_n$ is the information available to us at time n, that is, the sequence

$$\varepsilon_1, \varepsilon_2, \ldots, \varepsilon_n = \texttt{eps[1]}, \texttt{eps[2]}, \ldots, \texttt{eps[n]}$$

where $\varepsilon_k = 1$ if the kth item is defective, 0 otherwise. Because the type of calculation now being discussed can only be done sensibly on the computer, I mix mathematical and programming notation freely.

Actual T-value (vertical line) and observed Random Walk

$p_1 = 0.10,\ p_2 = 0.18$ $\qquad\qquad$ $p_1 = 0.10,\ p_2 = 0.30$

$c = 0.008$

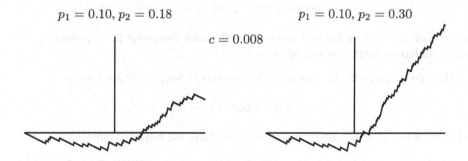

Probability T has happened given current information

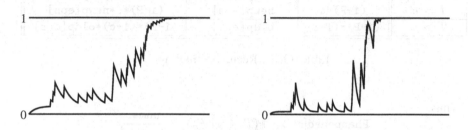

Maximum-height-scaled histogram for T given entire Random Walk

Figure O(i): Change-point detection filter

Write

$$\texttt{oldp}[1] := p_1, \quad \texttt{oldp}[0] := 1 - \texttt{oldp}[1],$$
$$\texttt{newp}[1] := p_2, \quad \texttt{newp}[0] := 1 - \texttt{newp}[1],$$

For $\texttt{eps} = 0$ or 1, write

$$\texttt{mixedp}[\texttt{eps}] := c*\texttt{newp}[\texttt{eps}] + (1 - c)*\texttt{oldp}[\texttt{eps}].$$

We visualize the way in which the available information evolves by plotting the graph of a Random Walk in which

$$\texttt{Walk[n]} := (\text{number of defectives by time } n) - (n \times \texttt{oldp[1]}).$$

Before time T, the walk has no tendency to drift: each increment is on average 0. After time T each increment is on average $p_2 - p_1$.

How do we calculate $\texttt{Phappened[n]}$ recursively? Suppose that we write

$$\text{P} = \texttt{Phappened[n} - 1]$$

and eps for the result regarding the nth item. We have the Bayesian table O(ii).

H:	$\mathbb{P}(H \| \mathcal{F}_{n-1})$ $=$	$\mathbb{P}(\text{eps} \| H, \mathcal{F}_{n-1})$ $=$	$\mathbb{P}(H \| \mathcal{F}_n)$ $\propto$
$T \leq n - 1$	P	newp[eps]	P*newp[eps]
$T = n$	(1-P)*c	newp[eps]	(1-P)*c*newp[eps]
$T > n$	(1-P)*(1-c)	oldp[eps]	(1-P)*(1-c)*oldp[eps]

Table O(ii): Recursive filtering

Thus

$$\texttt{Phappened[n]} := \mathbb{P}(T \leq n \,|\, \mathcal{F}_n) = \frac{\texttt{numer}}{\texttt{denom}},$$

where

$$\texttt{numer} := \text{P} * \text{newp[eps]} + (1 - \text{P}) * \text{c} * \text{newp[eps]},$$
$$\texttt{denom} := \text{numer} + (1 - \text{P}) * (1 - \text{c}) * \text{oldp[eps]}$$

The above logic is built into the procedure $\texttt{Update()}$ in the program which produced Figure 85O(i).

```
void Update(){
  int k, eps;
  double W, P, numer, denom, temp;
  n++;
  if (n < T) eps = WithProb(oldp[1]);
    else eps = WithProb(newp[1]);
  W = Walk[n] = Walk[n-1] + eps - oldp[1];

  P = Phappened[n-1];
  numer = P*newp[eps] + (1.0 - P)*c*newp[eps];
  denom = numer + (1.0 - P)*(1.0 - c)*oldp[eps];
  Phappened[n] = numer/denom;
```

```
for(k=0; k<n; k++) OddsTeq[k] = OddsTeq[k]*newp[eps];
  OddsTeq[n] = OddsTeq[n] * mixedp[eps];
  for (k=n+1; k <= L; k++) OddsTeq[k] = OddsTeq[k]*oldp[eps];

  if (n == L){
    maxOddsAfter = 0.0;
    for (k=0; k<=L; k++){
        temp = OddsAfter[k] = OddsTeq[k];
        if (temp > maxOddsAfter) maxOddsAfter = temp;
    }
  }
}
```

OddsTeq[k] represents odds that $T = k$ given the current information $\mathcal{F}_n$. Check out the logic for that part of the program.

Figure 85O(i) shows two runs, using the same seeds for comparison purposes, the left-hand side being with $p_1 = 0.10$, $p_2 = 0.18$, and the right-hand side with $p_1 = 0.10$, $p_2 = 0.30$. The value of c was 0.008. The value of T was 'cooked' to be 150, right in the middle of each run. The 'Langlands' generator was used with seeds 3561, 2765, 3763. The histogram for T at the bottom plots the OddsAfter function scaled to have constant maximum height when, ideally, it should have constant area under the graph.

The operation of estimation in real time is called *filtering*. Note that the filter is always very eager (too eager!?) to believe that T has already happened once the Random Walk shows any noticeable increase, and that for a long time, it is willing to change its mind very considerably. The fluctuations can well be more marked than in the example shown.

This kind of program can be generalized very considerably and made much more sophisticated (for example, learning more about unknown parameters on each run on real-world data), and it has a huge variety of applications. I have studied some medical ones where one wished to detect as soon as possible growth of bacteria in a laboratory culture using a sample from a patient, but in which there was substantial 'noise' in the observations.

In practice, the tendency of filters to fluctuate rather wildly needs tuning out. In our example, a crude way to do this would be to plot the minimum of Phappened[m] over a window $n - a \leq m \leq n$ which moves with n. Of course, this would delay 'proper' detection. One has to strike a balance.

P. Application of conditional probability to Genetics.
Genetics provides an excellent area in which to practise your skill with conditional probability – as well as being of fundamental importance in the real world. Extensions of the arguments we use here are important in genetic counselling.

We cheat in that we use independence (and even conditional independence) intuitively in this discussion.

We shall have to content ourselves with a very over-simplified picture of things, though one which conveys some of the main ideas. As stated in the Preface, helpful comments from Richard Dawkins and from Richard Tilney-Bassett made me revise substantially an early draft which was misleading in several respects. Any faults in this draft are of course mine.

First, we take a very naive view, skipping most of the verbose terminology typical of Biology (and Chemistry). A little of the reality is mentioned in Subsection 94S.

Every normal cell in one human being has the same genetic makeup. Gametes (sperms or eggs) are the exception. Each normal cell contains 23 *pairs* of 'chromosomes'. One pair of chromosomes is special, and is, amongst other things, the pair which determines the sex of the individual. *In this subsection, we concentrate on the other 22 'autosomal' pairs of chromosomes.* The story for the final pair is in subsection 91Q. Each of the pairs of chromosomes can be thought of as a string of beads, each bead on one chromosome being 'matched' with one on its companion. A matched pair of beads or *sites* such as in the diagram

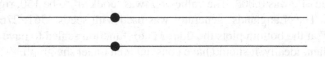

constitutes a *locus*. We focus attention on one particular locus on a particular chromosome pair. On each of the two sites forming this locus sits one of the two types (*alleles*), a or A, of the *gene* associated with that locus. The *pair* of genes at the locus forms one of the 3 *genotypes*

aa, aA (indistinguishable from Aa), AA.

Think of a gamete (sperm or egg) as being formed by randomly picking one chromosome from each pair of chromosomes in a normal cell, so that a gamete has 23 *single* chromosomes. This is, you understand, a mathematical picture of a process which, from any point of view, borders on the miraculous. When a sperm and egg fuse to create the first cell of a child, the 23 single chromosomes in the sperm will pair up correctly with the 23 single chromosomes in the egg to form a normal cell with 23 *pairs* of chromosomes. Each chromosome pair in Baby therefore consists of one chromosome obtained from Mum and one obtained from Dad.

Important Note. A chromosome 'obtained from Dad' need not be identical to one of the chromosomes in Dad's normal cells. See discussion of *crossing-over* in Subsection 94S below. For the one pair of genes at one locus, the focus of our current study, this does not affect probabilities.   □

For the particular locus in which we are interested, we have, with $\{\cdot, \cdot\}$ signifying *unordered pair*, the probabilities in Table P(i) for Baby's genotype given the genotypes of Mum and Dad. The second column in the table will be explained below.

Hypothesis H: {Mum,Dad}	Hardy–Weinberg $\mathbb{P}(H)$	Baby		
		$\mathbb{P}(\mathtt{aa} \mid H)$	$\mathbb{P}(\mathtt{aA} \mid H)$	$\mathbb{P}(\mathtt{AA} \mid H)$
{aa,aa}	p^4	1		
{aa,aA}	$4p^3q$	$\frac{1}{2}$	$\frac{1}{2}$	
{aa,AA}	$2p^2q^2$		1	
{aA,aA}	$4p^2q^2$	$\frac{1}{4}$	$\frac{1}{2}$	$\frac{1}{4}$
{aA,AA}	$4pq^3$		$\frac{1}{2}$	$\frac{1}{2}$
{AA,AA}	q^4			1

Table P(i): $\mathbb{P}(\text{Baby} \mid \{\text{Mum,Dad}\})$

We now assume *random mating* and *equal fertility for genotypes*. Thus, people choose mates entirely at random, in particular there being no preference for someone of the same genotype for our locus for instance. Genotypes for our locus have no effect on fertility. (This is of course very far from true for certain genes.)

Suppose that we have a large population in which at what we consider to be generation 0, the fractions of aa, aA and AA genotypes are

$$u_f, \ 2v_f, \ w_f \text{ amongst females,} \qquad u_m, \ 2v_m, \ w_m \text{ amongst males.}$$

Then the proportion of a genes amongst the female population is $p_f := u_f + v_f$, and of A genes amongst the female population is $q_f = 1 - p_f = v_f + w_f$. We have the obviously analogous results for males. For a couple chosen at random from the 0th generation, the probabilities that their first baby will be of genotypes aa, aA and AA are respectively

$$u := p_f p_m, \quad 2v := p_f q_m + q_f p_m, \quad w := q_f q_m$$

irrespective of whether Baby is male or female. (The baby will be of genotype aa if and only if it inherits an a gene from its Mum (which has probability p_f) and an a type gene from its Dad (which has probability p_m), and random mating makes these events independent; hence the $p_f p_m$. So, $u, 2v, w$ will be the approximate genotype frequencies in generation 1 for both males and females. Hence, for each sex in the 2nd generation, we shall have approximate genotype frequencies

$$p^2, \quad 2pq, \quad q^2 \quad \text{where } p = u + v, \ q = v + w.$$

And you can see that for the 3rd generation, the frequencies will be approximately the same as for the second; and so on. This is the famous *Hardy–Weinberg law*. But be very careful about the 'and so on'. Genotype frequencies *will* fluctuate randomly from their predicted values, and there is *no* restoring force. The Hardy–Weinberg law is a reasonable guide to the state of affairs for a very large population over a few generations.

We now assume that genotype frequencies for our locus are, and have been for a generation or two,

$$p^2, 2pq, q^2 \text{ for } \mathtt{aa}, \mathtt{aA}, \mathtt{AA},$$

for both sexes.

Check out the {Mum,Dad} probabilities in Table 89 P(i). For example, the {aa,aA} possibility is assigned probability $4p^3q$ because two ordered pairs each with probability $p^2 \times 2pq$ contribute to this probability. Note that the sum of the probabilities in the second column of Table 89 P(i) is $(p+q)^4$, namely 1, as it should be. Always try to cross-check your calculations.

Phenotypes, recessive and dominant genes. In some important cases, the a-type gene is *recessive* and aa-people are *'exceptional'* in some good or bad (or neither!) way, while both aA and AA people are *'standard'*: one says that aA and AA people have the same *phenotype* because there is no observable difference. The recessive a gene in an aA person is dominated by the dominant A gene.

[[For certain kinds of flowers, for example, an aa combination results in a white flower, whereas both aA and AA result in purple. (It is often stated that Mendel's work which started Genetics involved this situation with sweet peas, but one doesn't see much mention of this case in Mendel's papers! See Stern and Sherwood [217].) For other kinds of flowers (snapdragons, for example) , the aA combination can produce flowers of 'intermediate' colour: there is incomplete dominance. There are situations in which one has codominance, where an aA individual has characteristics of both the aa and aA individuals (for example, two types of antigen on blood cells). Whatever I could say about a certain human disease being caused by a recessive gene would be an oversimplification.]]

Pa. Exercise.   Consider the 'recessive gene' situation with genotype frequencies p^2, $2pq$, q^2 as just described. John and Mary (who are not related) are standard. They decide to have two children – perhaps against the 'random mating' rule! Find – using the hints below –
(a) the probability that their 1st child is standard (answer: $\frac{1+2p}{(1+p)^2}$),
(b) the conditional probability that their 2nd child is standard given that their 1st child is exceptional (answer obvious),
(c) the conditional probability that their 2nd child is exceptional given that their 1st child is standard (answer: $\frac{3p^2}{4(1+2p)}$).
Hints. Use the notation
   K: John and Mary are standard,
   S_1: their 1st child is standard,
   E_2: their 2nd child is exceptional.
For (c), We wish to find $\mathbb{P}(K \cap S_1 \cap E_2)/\mathbb{P}(K \cap S_1)$. Let H_i $(1 \le i \le 6)$ be one of the 6 'hypotheses' about {John, Mary}. See Table 89 P(i). Then

$$\mathbb{P}(K \cap S_1 \cap E_2) \;=\; \sum_i \mathbb{P}(H_i)\mathbb{P}(K \cap S_1 \cap E_2 \,|\, H_i).$$

For a fixed couple, the division of chromosome pairs for the 2nd child is independent of that for the 1st child. More precisely, S_1 and E_2 are conditionally independent given H_i:

$$\mathbb{P}(S_1 \cap E_2 \,|\, H_i) \;=\; \mathbb{P}(S_1 \,|\, H_i)\,\mathbb{P}(E_2 \,|\, H_i).$$

We can obviously extend this to include K because each $\mathbb{P}(K \mid H_i)$ is either 0 or 1.

Make a table with headings

$$H \quad \mathbb{P}(H) \quad \mathbb{P}(H \cap K) \quad \mathbb{P}(H \cap K \cap S_1) \quad \mathbb{P}(H \cap K \cap S_1 \cap E_2).$$

The column sums of the last four columns are 1, $\mathbb{P}(K)$, $\mathbb{P}(K \cap S_1)$, $\mathbb{P}(K \cap S_1 \cap E_2)$, respectively.

Pb. Exercise. Again consider the recessive-gene situation we have been studying. Suppose we know that Peter and Jane (who are not related) are standard and their four parents are standard.

(a) Show that, with the implicit conditioning here made explicit,

$$\mathbb{P}(\text{Peter is } \mathtt{aA} \mid \text{he and his parents are standard}) = \frac{2p}{1 + 2p}.$$

(b) Show that the chance that the first child of Peter and Jane is exceptional is $p^2/(1+2p)^2$. *Hint.* Use Part (a) in a 'recursive' argument.

Pc. Remark. Blood type in humans *is* controlled by the genotype at a particular locus, but in that case the gene has 3 types A, B and O, not our 2 types a and A. However, AA and AO give rise to the same phenotype (blood group 'A') and BB and BO give rise to the same phenotype (blood group 'B'). There are 4 phenotypes: 'A', 'B', 'AB' and 'O', the last corresponding to genotype OO. But there are much more serious complications than these facts indicate, in a full study of Genetics.

Q. X-linked (or sex-linked) genes.
In females, that 23rd pair of chromosomes is a proper pair with loci just like the 22 other pairs. Each chromosome in the 23rd pair in females is called an X-chromosome. For a gene with two types b and B associated with a locus on her 23rd chromosome pair, a female can be of genotype bb, bB or BB, just as before.

In males, however, the 23rd chromosome pair consists of an X-chromosome and a quite different, much smaller, Y-chromosome. There is now no concept of a locus as a matched pair of sites for genes. The man will have either a b gene or a B gene on the X-chromosome in the 23rd pair, so his genotype is either b or B. We suppose for a gene associated with an X-chromosome that the gene frequencies are

$$p^2, \ 2pq, \ q^2 \text{ for bb, bB, BB in females,}$$

and

$$p, \ q \text{ for b, B in males.}$$

If b is recessive, then a man will be exceptional if he has genotype b, a woman if she has genotype bb.

[[One can show (assuming random mating, etc) that whatever the current genotype frequencies for males and females in a large population, something very close to a $(p^2, 2pq, q^2; p, q)$ genotype frequency situation will evolve in very few generations.]]

A child will be female if it inherits its father's X-chromosome, male if it inherits its father's Y-chromosome. In relation to the X-linked gene we are currently considering, a man cannot pass it to his son, but can to a grandson via a daughter. (This is the situation for the haemophilia gene, for example. The royal families of Europe provide a well-known demonstration. Queen Victoria was a carrier of the haemophilia gene, and indeed all of those royals infected were her descendants.)

Qa. Exercise. (a) Show that the $(p^2, 2pq, q^2; p, q)$ genotype frequencies for our X-linked gene are approximately carried forward to the next generation.

Suppose now that we have the $(p^2, 2pq, q^2; p, q)$ situation, and that our X-linked gene is recessive. A couple have two children, a boy and a girl. Use the result in Part (a) in showing that
(b) the conditional probability that the boy is standard given that the girl is standard is $(1 - \frac{1}{2}p^2)/(1 + p)$,
(c) the conditional probability that the girl is standard given that the boy is standard is $1 - \frac{1}{2}p^2$.

R. Pedigree analysis for rare diseases associated with one locus. We
concentrate on diseases known to be associated with the situation at one locus.

Let's begin with some comments which apply whether or not the disease is rare.

In deciding whether or not the gene which causes the disease is recessive or dominant (even rare diseases can be caused by a dominant gene!), we can use the following principle: *if a dominant gene is responsible for the disease and a child has the disease, then at least one parent must have the disease.*

In deciding whether or not the gene which causes the disease is X-linked, remember the principle: *if a recessive X-linked gene causes the disease, then the daughter of a healthy father must be healthy.*

In pedigree diagrams, a square represents a male, a circle a female. An individual with a black number on a white background represents a healthy individual, one with a white number on a black background a diseased one.

Consider diagram R(i)(a). The principles mentioned above show that the disease is recessive and autosomal (not X-linked). Note that individuals 1 and 2 must both be aA. (We equate aA and Aa.) Given this, it is true that, independently of the fact that individual 5 is aa, individual 6 has absolute probabilities $\frac{1}{4}, \frac{1}{2}, \frac{1}{4}$ of being aa, aA, AA. However, this has to be conditioned by the fact that individual 6 is healthy. Check that, given the information in the diagram, the probability that the first child of 6 and 7 would be diseased is $1/9$.

Rare diseases. Now look at Figure R(i)(b) which relates to a **rare** disease caused by the situation at one locus. The picture is copied from Exercise 25 of Chapter 4 in the superb book by Griffiths *et al* praised earlier. Again it is clear that the bad gene is recessive autosomal.

We wish to know the probability that the first child of individuals 21 and 22 would be diseased. Now, if that child were to be diseased, then both 21 and 22 must be aA. Since a

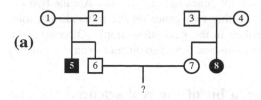

(a)

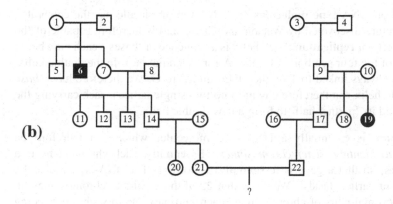

(b)

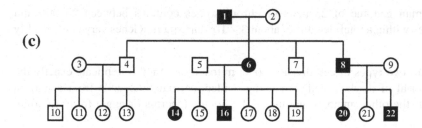

(c)

Figure R(i): Some pedigree diagrams

alleles are rare in the population, we can assume that they have not been brought into the family by any of 8, 15, 16, so, we assume that each of these is **AA**.

Ra. Exercise. Show that (under the above assumptions) the probability that the first child of individuals 21 and 22 would be diseased is 1/72.

Rb. Exercise. Suppose that Figure R(i)(c) shows a pedigree for a rare disease. Do you think that the bad gene is recessive or dominant? Explain your answer.

The examples in Griffiths *et al* are splendid brain-teasers; and of course, things get much more interesting when you look at several genes rather than one. Are the two loci on the same chromosome? How likely is it that two genes on the same chromosome are 'separated' by the crossing-over described in the next subsection? (Of course, this depends on how far apart they are on the chromosome.) And so on; and so on. Fascinating – even in the Human Genome age!

S. Chromosomes, crossovers, etc: a bit of the real science.   Genetics is immensely complicated. 'Essentially' in the following brief discussion will hide a multitude of complications.

Life is based on **DNA-type molecules**. A DNA-type molecule has the celebrated **double-helix** structure discovered by Watson and Crick, and is therefore capable of the fundamental property of **replication**. Each helix is a sequence of 'bases': each base being described by one of the four symbols A,C,G,T (A for 'adenine' etc). In the double helix, A on one helix is always linked to T on the other; and C to G. We therefore have '*base pairs*'. The double helix can therefore unwind into two single helices, each carrying the information to build its 'partner half' to form a new double helix.

A **chromosome** is essentially a DNA-type molecule, which can therefore be replicated into two *identical sister chromosomes*. Essentially, each chromosome is a sequence of **genes**, so that a gene is (essentially) a portion of a DNA-type molecule. These genes are our earlier 'beads'. We know that 22 of these pairs of chromosomes (in a human) are autosomal pairs of chromosomes, each chromosome in such a pair being partnered by a homologue. Each gene on an autosomal chromosome is an allele of the gene at the other half of the locus on its homologous partner.

The human genome of 23 pairs of chromosomes contains between $25,000$ and $40,000$ genes within a total 'length' of about 3×10^9 base pairs. Genes vary considerably in length.

There are two types of cell division: one, mitosis, designed to replicate exactly the genetic material of a whole cell, producing 2 identical daughter cells from a normal 'parent' cell; the other, meiosis, producing 4 gametes (sperms or eggs) from a normal cell.

Mitosis is the more straightforward (but when it goes wrong, it can cause disease ...). Essentially, each chromosome is replicated and the daughter chromosomes migrate, each pairing up correctly with a homologue, to form at cell division the desired two daughter cells.

Meiosis also starts off by duplicating each chromosome in a normal cell. Recall that the object is to produce 4 gametes, so that after each original chromosome is replicated, we have the correct amount of genetic material. However, in later stages of meiosis, it is possible for two non-identical but homologous autosomal chromosomes to **cross over** so that, for example, from the two gene sequences of the original chromosomes

$$\cdots u_1 \cdots u_{r-1} u_r u_{r+1} \cdots u_n \cdots ,$$

$$\cdots v_1 \cdots v_{r-1} v_r v_{r+1} \cdots v_n \cdots ,$$

we obtain the *new* (**recombinant**) chromosomes

$$\cdots u_1 \cdots u_{r-1} u_r v_{r+1} \cdots v_n \cdots ,$$
$$\cdots v_1 \cdots v_{r-1} v_r u_{r+1} \cdots u_n \cdots .$$

Crossovers can be more complex than this simple one involving two chromosomes.

If one ignored the possibility of crossover, Mum could produce 2^{23} different eggs, Dad 2^{23} different sperms. When one takes crossovers into account, these numbers become immeasurably larger. (However, one still sees the 2^{23} in many accounts.)

A subsection of a chromosome is said to be a centimorgan (after the American geneticist T H Morgan) if there is a 1% probability that it will contain a crossover point. A centimorgan contains roughly 1 million base pairs, so there are about 3000 centimorgans in the human genome. The probability of crossovers during the production of a gamete is therefore very high indeed.

T. Thoughts.   Richard Dawkins has explained with unsurpassed clarity Darwin's insight that random events sieved in a non-random way by Natural Selection can explain most 'miracles of creation' such as the human eye. (I am still not sure quite how it explains the much greater miracle of Bach, Mozart and Beethoven!) Do read Dawkins' books *The Selfish Gene, The Blind Watchmaker*, etc, etc. So successful is Evolution that a very important application of Probability mimics it in random algorithms, genetic algorithms, neural nets, etc. (*There* are some words for Web searches!) A (classical) computer operating 'randomly' can solve deterministic problems forever outside the range of deterministic techniques, but a quantum computer could in some situations achieve much more still.

In *River out of Eden* [57], Dawkins describes the crossover phenomenon with his customary vigour: 'A child's chromosomes are an unbelievably scrambled mishmash of its grandparents' chromosomes and so on back to distant ancestors.' His book is a veritable psalm (for atheists?!), full of poetic 'verses' which haunt the mind: 'The universe we observe has precisely the properties we would expect if there is, at bottom, no design, no purpose, no evil and no good, nothing but blind, pitiless indifference. ... DNA neither cares nor knows. DNA just is. And we dance to its music.'

As Dawkins has pointed out, there is a sense in which we are now beginning to break free from the tyranny of Natural Selection: we can 'adjust the musical score', valuing and protecting the weak, seeking to eliminate certain diseases, etc, all without entering in a type of brave new world about which I would have as many qualms as anyone.

Apology. I could not have done justice to Genetics in this small space even if I'd had the expertise. I have not explained what constitutes a gene within a chromosome (or chromosome pair), and I have not explained about alleles, about homologous chromosomes, etc.

When I finish this book, learning more about Genetics will be a high priority, though not as high a one as trying at long last to learn to play Bach, Mozart and Beethoven acceptably well. Those guys did have some very remarkable genomes. Millions of extraordinary 'ordinary' people have very remarkable genomes too.

4.2   Independence

▶ **A. Independence for** 2 **events.**   Two events A and B are called independent if (and, since this is a definition, only if)

$$\mathbb{P}(A \cap B) = \mathbb{P}(A)\mathbb{P}(B). \tag{A1}$$

We have already seen the motivation, recalled here. If $\mathbb{P}(A) \neq 0$ then A and B are independent if, and only if, $\mathbb{P}(B \mid A) = \mathbb{P}(B)$, that is, if and only if knowledge that A has occurred does not affect our degree of belief that B has occurred. (*You* give an explanation which will satisfy a Frequentist.) Of course, the definition is symmetric in A and B.

Equation (A1) is the first of many 'Independence means Multiply' Rules. Note that it may be written

$$\mathbb{E}\left(I_A I_B\right) = \mathbb{E}\left(I_A\right)\mathbb{E}\left(I_B\right),$$

and that (still assuming that A and B are independent) we have, for example,

$$\begin{aligned}
\mathbb{P}(A^c \cap B^c) &= \mathbb{E}\left\{(1 - I_A)(1 - I_B)\right\} = \mathbb{E}\left(1 - I_A - I_B + I_A I_B\right) \\
&= 1 - \mathbb{P}(A) - \mathbb{P}(B) + \mathbb{P}(A)\mathbb{P}(B) = \{1 - \mathbb{P}(A)\}\{1 - \mathbb{P}(B)\} \\
&= \mathbb{P}(A^c)\mathbb{P}(B^c).
\end{aligned}$$

The general moral is that if A and B are independent, then so are A^c and B^c; likewise for A^c and B; and for A and B^c.

▶▶ **B. Independence for a sequence of events.**   Suppose that $A_1, A_2, \ldots$ is a finite or infinite sequence of events. Then $A_1, A_2, \ldots$ are called independent if whenever $i_1, i_2, \ldots, i_r$ are distinct elements of $\mathbb{N}$ such that each A_{i_k} is defined, then

$$\mathbb{P}\left(A_{i_1} \cap A_{i_2} \ldots \cap A_{i_r}\right) = \mathbb{P}\left(A_{i_1}\right)\mathbb{P}\left(A_{i_2}\right) \ldots \mathbb{P}\left(A_{i_r}\right). \tag{B1}$$

By using indicator functions as in the case of 2 events, we can prove that one can put ^{c} complement symbols on any sets on the left-hand side of (B1) provided, of course, that one does the same on the right-hand side. The 'π-system Lemma' from Subsection 45M may be used to do such things much more neatly.

▶▶ **C. Discussion: Using independence to assign probabilities.**   This too we have seen before.

If we toss a coin with probability p of Heads, $q = 1 - p$ of Tails n times, then outcome

$$\omega = \mathtt{HTT}\ldots\mathtt{H}$$

is assigned a probability

$$\mathbb{P}(\omega) = pqq \ldots p$$

because we believe that the results on different tosses are independent: results on one subset of tosses do not influence those on a disjoint subset of tosses.

We take for Ω the Cartesian product $\{H, T\}^n$ of all sequences ω of length n of H's and T's. We assign the appropriate probability based on independence to each ω, thereby defining a 'product measure' on the Cartesian product. Define

$$X_k(\omega) = \begin{cases} 1 & \text{if } \omega_k = H, \\ 0 & \text{if } \omega_k = T. \end{cases}$$

Then $\mathbb{E}(X_k) = 1 \times \mathbb{P}(X_k = 1) + 0 \times \mathbb{P}(X_k = 0) = p$. (We take this as intuitively obvious; strictly speaking, one ought to justify for the sake of consistency that the sum of all $\mathbb{P}(\omega)$ values with $\omega_k = 1$ does equal p, but we skip this.)

If Y is the total number of Heads, then

$$Y = X_1 + X_2 + \cdots + X_n,$$
$$\mathbb{E}(Y) = \mathbb{E}(X_1) + \mathbb{E}(X_2) + \cdots + \mathbb{E}(X_n) = p + p + \cdots + p = np.$$

(You are reminded however that the Addition Rule does not require independence.)

This coin-tossing case typifies the way in which we use independence to assign probabilities – at least in simple situations. **Independence is always closely associated with 'product measure on Cartesian products'.** We do not pursue this here.

Ca. Exercise. Suppose that we throw a fair coin twice. Let

$$A = \text{'H on 1st toss'}, \quad B = \text{'H on 2nd toss'}, \quad C = \text{'HH or TT'}.$$

Prove that A and B are independent, B and C are independent, C and A are independent, but that A, B and C are not independent.

D. Exercise: The 'Five Nations' Problem*. This is a good exercise for you to puzzle out, easy once you spot the way to do it.

Each of five people A, B, C, D, E, plays each of the others (making 10 games in all) in a fair-coin-tossing game: for each game, independently of the others, each of the two players involved has probability $\frac{1}{2}$ of winning. Find the probability that each wins two games.

[[Someone asked me this after it happened one year that each team in the 'Five Nations' Rugby Championship won two games. This was in the period between the great days when Wales were incomparable and the modern era when England and France dominate the Northern Hemisphere. We have now welcomed Italy into the 'Six Nations' Championship.]]

▶ **E. The Second Borel–Cantelli Lemma.** This result helps us understand many things, including what can 'go wrong' for the Strong Law for variables not in $\mathcal{L}^1$. The result says the following.

Lemma. *Suppose that $J_1, J_2, \ldots$ is an infinite sequence of **independent** events such that*

$$\sum \mathbb{P}(J_n) = \infty.$$

Then the probability that infinitely many J_n occur is 1.

Ea. Exercise. Prove that under the assumptions of the Lemma,

$$\mathbb{P}(\text{none of the events } J_1, J_2, \ldots, J_n \text{ occurs}) \;\leq\; \frac{1}{1 + p_1 + p_2 + \cdots + p_n} \qquad \text{(E1)}$$

Hint. Let $p_k := \mathbb{P}(J_k)$. Calculate the probability in (E1) exactly. Then use the facts that for $0 \leq p \leq 1$ and $x, y \geq 0$,

$$1 - p \leq \frac{1}{1 + p}, \quad (1 + x)(1 + y) \geq 1 + x + y.$$

Proof of the Lemma. Inequality (E1) makes it clear that the probability that none of the events J_k occurs is 0. But we can just as easily show for every n that $\mathbb{P}(H_n) = 1$ where H_n is the event that at least one of the events $J_{n+1}, J_{n+2}, \ldots$ occurs. Measure Theory (Lemma 44G) now clinches the fact that the intersection $\bigcap H_n$ of all the events H_n has probability 1; and this is just what we want. □

Eb. Exercise. Give another proof based on the inequality at 14Ob.

Question. Why would the Second Borel–Cantelli Lemma be *obviously* false if the independence assumption were dropped?

▶▶ **F. Independence of Random Variables.** Random Variables $X_1, X_2, \ldots$ in a finite or infinite sequence are called independent if whenever $i_1, i_2, \ldots, i_r$ are distinct numbers such that each X_{i_k} exists, and $x_{i_1}, x_{i_2}, \ldots, x_{i_r} \in \mathbb{R}$, we have

$$\mathbb{P}(X_{i_1} \leq x_{i_1}; \; X_{i_2} \leq x_{i_2}; \ldots X_{i_r} \leq x_{i_r})$$
$$= \mathbb{P}(X_{i_1} \leq x_{i_1}) \mathbb{P}(X_{i_2} \leq x_{i_2}) \ldots \mathbb{P}(X_{i_r} \leq x_{i_r}).$$

If $X_1, X_2, \ldots$ are discrete, this is equivalent (assumed fact) to the statement that whenever $i_1, i_2, \ldots, i_r$ are distinct numbers such that each X_{i_k} exists, and $x_{i_1}, x_{i_2}, \ldots, x_{i_r} \in \mathbb{Z}$, we have

$$\mathbb{P}(X_{i_1} = x_{i_1}; \; X_{i_2} = x_{i_2}; \ldots X_{i_r} = x_{i_r})$$
$$= \mathbb{P}(X_{i_1} = x_{i_1}) \mathbb{P}(X_{i_2} = x_{i_2}) \ldots \mathbb{P}(X_{i_r} = x_{i_r}).$$

One can prove (by the π-system Lemma mentioned earlier) that if $X_1, X_2, \ldots$ are independent, if $1 \leq m \leq n$ and U and V are nice ('Borel measurable') functions of $\mathbb{R}^m$ and $\mathbb{R}^{n-m}$ respectively, then

$$\mathbb{P}\{(X_1, X_2, \ldots, X_m) \in U; \ (X_{m+1}, X_{m+2}, \ldots, X_n) \in V\}$$
$$= \mathbb{P}\{(X_1, X_2, \ldots, X_m) \in U\}\,\mathbb{P}\{(X_{m+1}, X_{m+2}, \ldots, X_n) \in V\},$$

as common sense suggests. The following Fact is a consequence, and we make much use of it.

▶ **G. Fact.** *If $X_1, X_2, \ldots$ are independent RVs and $f_1, f_2, \ldots$ are nice (technically, Borel measurable) functions on $\mathbb{R}$, then $f_1(X_1), f_2(X_2), \ldots$ are independent RVs. Moreover, if f is a nice function on $\mathbb{R}^n$, then $f(X_1, X_2, \ldots, X_n), X_{n+1}, X_{n+2}, \ldots$ are independent.*

This result is assumed.

▶ **H. Sampling with replacement.** Suppose that we take a sample of size n *with* replacement from the N people, numbered $1, 2, \ldots, N$ in their alphabetical order, who live in Bath. 'With replacement' signifies that after choosing our first person at random, we make the choice of the next from the entire population; and so on. We may therefore choose the same person many times.

It is 'obvious' that the heights $X_1, X_2, \ldots, X_n$ of the n people we choose (listed in the order in which we choose them) are independent RVs; for, since we are sampling *with* replacement, no choices can influence any others.

To model the situation, we take for Ω the Cartesian product $\{1, 2, \ldots, N\}^n$ of all N^n ordered n-tuples

$$\omega = (\omega_1, \omega_2, \ldots, \omega_n), \quad \text{each } \omega_k \text{ in } \{1, 2, \ldots, N\}.$$

To each $\omega \in \Omega$ we assign a probability $\mathbb{P}(\omega) = 1/N^n$. Now let $Z_k(\omega) := \omega_k$, so that Z_k represents the kth person chosen. Then, if $\omega = (z_1, z_2, \ldots, z_n)$,

$$\mathbb{P}(Z_1 = z_1; \ Z_2 = z_2; \ \ldots ; Z_n = z_n) = \mathbb{P}(\omega)$$
$$= \frac{1}{N^n} = \mathbb{P}(Z_1 = z_1)\,\mathbb{P}(Z_2 = z_2)\ldots\mathbb{P}(Z_n = z_n).$$

Since, for fixed k, there are N^{n-1} values of ω with $Z_k(\omega) = z_k$, we *do* have the 'obvious' fact that $\mathbb{P}(Z_k = z_k) = \frac{1}{N}$. We see that, as anticipated, $Z_1, Z_2, \ldots, Z_n$ are independent.

If $h(i)$ is the height of person i, then $X_k = h(Z_k)$, and by Lemma G, $X_1, X_2, \ldots, X_n$ are independent.

The independence of $X_1, X_2, \ldots, X_n$ is, as we shall see, a highly desirable theoretical property, so sampling *with* replacement is almost always assumed in theoretical studies. In practice, we sample *without* replacement and rely on the fact that $n \ll N$ to say that the difference is negligible. See Exercise 71Kb for a partial study of sampling without replacement.

►► **I. The 'Boys and Girls' problem.** *This problem makes a start on rather counter-intuitive properties of Poisson distributions and processes which will reappear in our later studies of both Probability and Statistics. It is our first look at* surprising *independence properties.*

Ia. Exercise. The number of births during a day at a hospital is assumed to be Poisson with parameter λ. Each birth is a boy with probability p, a girl with probability $q = 1 - p$, independently of other births (and of the total number of births). Let B be the numbers of boys, G the number of girls, born during the day.

Prove that

$$\mathbb{P}(B = b, \ G = g) \ = \ \frac{e^{-\lambda p}(\lambda p)^b}{b!} \frac{e^{-\lambda q}(\lambda q)^g}{g!},$$

and deduce that

B has the Poisson distribution of parameter λp,
G has the Poisson distribution of parameter λq,
B and G are independent.

[If there are on average 20 births per day and on one day 18 boys are born, you would still expect on average 10 (not 2) girls to be born on that day – a bit weird!]

J. Exercise: A 'number-theoretic' problem. *This is a good problem from which to learn Probability,* so do it even if you are the unique person who doesn't like Number Theory a heck of a lot more.

Recall that a *prime* element of $\mathbb{N}$ is one of $2, 3, 5, 7, \ldots$, my point being that 1 is NOT a prime.

Let $s > 1$, and choose a random number X in $\mathbb{N}$ according to the **Euler**(s) (or Dirichlet(s)) distribution:

$$\mathbb{P}(X = n) \ = \ \frac{n^{-s}}{\zeta(s)} \quad (n \in \mathbb{N}),$$

where ζ is the famous *Riemann zeta-function* defined by

$$\zeta(s) \ := \ \sum_{n=1}^{\infty} n^{-s},$$

the series converging for $s > 1$. [*Remark.* The most important unsolved problem in Maths is that of proving the Riemann Hypothesis about the behaviour of $\zeta(s)$ where s ranges over the set $\mathbb{C}$ of complex numbers. The function 'blows up' (has a 'simple pole') at $s = 1$ but extends to a very nice ('analytic') function on the remainder of $\mathbb{C}$. Amazingly, the function ζ on $\mathbb{C}$ has a strikingly simple probabilistic definition – not given here. The Riemann Hypothesis states that all infinitely many non-trivial zeros of ζ lie on the line $\Re(z) = \frac{1}{2}$. (The 'trivial' zeros are at the negative even integers $-2, -4, -6, \ldots$).]

Let E_m be the event 'X is divisible by m'.
(a) Prove that $\mathbb{P}(E_m) = m^{-s}$ for $m \in \mathbb{N}$.
(b) Prove that the events $(E_p : p$ prime) are independent. *Hint.* If p_1 and p_2 are distinct

primes, then a number is divisible by both p_1 and p_2 if and only if it is divisible by $p_1 p_2$; and similarly for more than two distinct primes.

(c) By considering $\bigcap (E_p^c)$, prove *Euler's formula*

$$\frac{1}{\zeta(s)} = \prod_{p \text{ prime}} \left(1 - \frac{1}{p^s}\right).$$

Don't worry about rigour here: Lemma 43D provides that.

(d)* Let Y, independent of X, be chosen with the Euler(s) distribution, and let H be the highest common factor (greatest common divisor) of X and Y. Let B_p (p prime) be the event that 'both X and Y are divisible by p'. How does the event $\bigcap (B_p^c)$ relate to H? Prove that H has the Euler($2s$) distribution.

▶▶▶ **K. Theorem: 'Independence means Multiply for expectations'.** Let $X_1, X_2, \ldots, X_n$ be independent RVs, each in $\mathcal{L}^1$. Then $X_1 X_2 \ldots X_n \in \mathcal{L}^1$ and

$$\mathbb{E}\left(X_1 X_2 \ldots X_n\right) = \mathbb{E}\left(X_1\right)\mathbb{E}\left(X_2\right)\ldots\mathbb{E}\left(X_n\right).$$

Proof. (This result is really an application of Fubini's Theorem in Measure Theory.) Suppose that X and Y are RVs each taking only finitely many values. Then

$$X = \sum x\, I_{\{X=x\}}, \quad \mathbb{E}(X) = \sum x\, \mathbb{P}(X = x),$$
$$Y = \sum y\, I_{\{Y=y\}}, \quad \mathbb{E}(Y) = \sum y\, \mathbb{P}(Y = y).$$

Then

$$XY = \sum\sum xy\, I_{\{X=x\}}I_{\{Y=y\}} = \sum\sum xy\, I_{\{X=x;\, Y=y\}},$$

and

$$\mathbb{E}(XY) = \sum\sum xy\mathbb{P}(X = x;\ Y = y)$$
$$= \sum\sum xy\mathbb{P}(X = x)\mathbb{P}(Y = y) \quad \text{(by independence)}$$
$$= \left(\sum x\, \mathbb{P}(X = x)\right)\left(\sum y\, \mathbb{P}(Y = y)\right) = \mathbb{E}(XY).$$

The Monotone-Convergence Theorem 60D is now used to extend the result to yield

$$\mathbb{E}(XY) = \mathbb{E}(X)\mathbb{E}(Y) \leq \infty$$

for arbitrary non-negative RVs. The linearity of expectation now yields the desired result for two arbitrary variables in $\mathcal{L}^1$. Induction on the number of variables completes the proof. □

▶▶ **L. Lemma: Variance of a sum of independent RVs.**   *Suppose that* $X_1, X_2, \ldots, X_n$ *are* **independent** *RVs, each in* $\mathcal{L}^2$. *Then*

$$\mathrm{Cov}\,(X_i, X_j) = 0 \quad (i \neq j),$$
$$\mathrm{Var}\,(X_1 + X_2 + \cdots + X_n) = \mathrm{Var}\,(X_1) + \mathrm{Var}\,(X_2) + \cdots + \mathrm{Var}\,(X_n)\,.$$

Proof. This is now trivial. We have, for $i \neq j$, using Theorem 101K,

$$\mathrm{Cov}\,(X_i, X_j) = \mathbb{E}\,(X_i X_j) - \mathbb{E}\,(X_i)\,\mathbb{E}\,(X_j) = \mathbb{E}\,(X_i)\,\mathbb{E}\,(X_j) - \mathbb{E}\,(X_i)\,\mathbb{E}\,(X_j) = 0.$$

The 'Var' result now follows from the general addition rule for variances, Lemma 70K.   □

M. Proving almost-sure results.   We can prove for random sequences results of remarkable precision on how they must behave 'with probability 1', results which come close to defying our notion of randomness. I give here a simple example.

Suppose that $X_1, X_2, \ldots$ are independent RVs each with the exponential $E(1)$ distribution. Thus,

$$\mathbb{P}(X_k > x) = \mathrm{e}^{-x}.$$

We know that the series

$$\sum \frac{1}{n^c}, \quad \sum \frac{1}{n(\ln n)^c}, \quad \sum \frac{1}{n(\ln n)(\ln \ln n)^c}$$

all converge if $c > 1$ and all diverge if $c \leq 1$. Let $h : \mathbb{N} \to (0, \infty)$, and define

$$J_n := \{X_n \geq h(n)\}.$$

By using the two Borel–Cantelli Lemmas 45K and 98E, we see that if

$$h(n) = c \ln(n) \text{ or } h(n) = \ln n + c \ln \ln n \text{ or } h(n) = \ln n + \ln \ln n + c \ln \ln \ln n,$$

then, in each case,

$$\mathbb{P}(\text{infinitely many } J_n \text{ occur}) = \begin{cases} 0 & \text{if } c > 1 \\ 1 & \text{if } c \leq 1. \end{cases}$$

If you know about lim sups, you can see that we can prove for example that each of the following statements holds with probability 1:

$$\limsup \frac{X_n}{\ln n} = 1; \quad \limsup \frac{X_n - \ln n}{\ln \ln n} = 1; \quad \limsup \frac{X_n - \ln n - \ln \ln n}{\ln \ln \ln n} = 1;$$

etc, etc. Moreover the statement that *all of these statements are true* has probability 1. You can see why it's all a bit worrying – and there are still more amazingly precise 'with probability 1' statements.

Of course, results such as those just described cannot be illustrated by simulation. Statisticians might well add that they cannot be seen in the real world either!

4.3 Laws of large numbers

▶▶ **A. IID (Independent, Identically Distributed) RVs.** By a (finite or infinite) sequence of IID Random Variables $X_1, X_2, \ldots$, we mean a sequence of **independent** RVs each with the same distribution function F:

$$\mathbb{P}(X_k \leq x) = F(x) \quad \text{for all } k \text{ and all } x.$$

Then, if one X_k is in $\mathcal{L}^1$, so are they all, and they all have the same mean μ (say); and if one X_k is in $\mathcal{L}^2$, so are they all, and they all have the same variance σ^2 (say).

B. Fact. *Suppose that $X_1, X_2, \ldots$ are IID Random Variables with the same* continuous *distribution function. Then, with probability 1, the X_k's take distinct values.*

This result, useful in many situations (see the later study of Order Statistics, for example) is included at this stage to allow you to do two important exercises (which are not unrelated!).

C. Exercise: Renyi's 'Record Problem'. Suppose that $X_1, X_2, \ldots$ are IID Random Variables with the same *continuous* distribution function. Say that a record occurs at time 1, and write $E_1 := \Omega$. For $n = 2, 3, \ldots$, let E_n be the event that a record occurs at time n in that

$$X_n > X_m \quad \text{for every } m < n.$$

Convince your teacher that the events $E_1, E_2, \ldots$ are independent, E_n having probability $1/n$. You may assume the intuitively obvious fact that if one arranges $X_1, X_2, \ldots X_n$ in increasing order as

$$X_{N(1,n)}, X_{N(2,n)}, \ldots, X_{N(n,n)}$$

then $N(1, n), N(2, n), \ldots, N(n, n)$ will be a random permutation of $1, 2, \ldots, n$, each of the $n!$ permutations being equally likely. Think about how this fact may be proved.

D. Exercise: The 'Car Convoy' Problem. Cars travel down a one-way single-track road. The nth driver would like to drive at speed V_n, where $V_1, V_2, \ldots$ are independent random variables all with the same continuous distribution function. (With probability 1, all the V_k will be different.) Cars will get bunched into 'convoys'. If $V_2 > V_1 > V_3$, then the first convoy will consist of cars 1 and 2, and will be of length 2. Let L be the length of the first convoy. Find $\mathbb{P}(L = n)$, and deduce that $\mathbb{E}(L) = \infty$. How would you explain to the layman that this is not crazy?

Later on, you will be asked to prove that the probability that the nth convoy consists of precisely 1 car is $1/2^n$. Any ideas now?

▶▶▶ **E. Lemma: Mean and variance of Sample Mean.**   Let $X_1, X_2, \ldots, X_n$ be IID Random Variables in $\mathcal{L}^2$, and hence with common mean μ and common variance σ^2. Define (S signifying 'sum', and A 'average')

$$ S_n := X_1 + X_2 + \cdots + X_n, \quad A_n := \frac{X_1 + X_2 + \cdots + X_n}{n} = \frac{S_n}{n}. $$

Then

$$ \mathbb{E}(S_n) = n\mu, \quad \mathrm{Var}(S_n) = n\sigma^2, \quad \mathbb{E}(A_n) = \mu, \quad \mathrm{Var}(A_n) = \frac{\sigma^2}{n}. $$

Proof. By the linearity property 62Ha(a), we have (without assuming independence)

$$ \mathbb{E}(S_n) = \mathbb{E}(X_1 + X_2 + \cdots + X_n) = \mathbb{E}(X_1) + \mathbb{E}(X_2) + \cdots + \mathbb{E}(X_n) $$
$$ = \mu + \mu + \cdots + \mu = n\mu, $$
$$ \mathbb{E}(A_n) = \mathbb{E}\left(\frac{1}{n}S_n\right) = \frac{1}{n}\mathbb{E}(S_n) = \mu. $$

By Lemma 102L, now using independence,

$$ \mathrm{Var}(S_n) = \mathrm{Var}(X_1) + \mathrm{Var}(X_2) + \cdots + \mathrm{Var}(X_n) = n\sigma^2, $$

and, by the first property at 67(D3),

$$ \mathrm{Var}(A_n) = \frac{1}{n^2}\mathrm{Var}(S_n) = \frac{\sigma^2}{n}. $$

□

Statisticians call A_n the Sample Mean of $X_1, X_2, \ldots, X_n$, and usually denote it by $\overline{X}$. (You will see later why I am using capital letters!) This does not show the explicit dependence on n – not a problem is Statistics in which sample size is generally fixed, but certainly a problem in Probability.

Ea. Exercise.   Let Y_n be the total number of Heads in n tosses of a coin with probability p of Heads. Show that $\mathrm{Var}(Y_n) = npq$. Please do this *without* a binomial coefficient in sight! Show that the maximum value of pq is $\frac{1}{4}$.

▶▶ **F. Exercise: Unbiased Estimator of Sample Variance.**   Suppose that

$X_1, X_2, \ldots$ are IID Random Variables each with mean μ and finite variance σ^2. Let

$$\overline{X} := \frac{X_1 + X_2 + \cdots + X_n}{n}, \tag{F1}$$

$$\boxed{\text{RSS} := \sum_{k=1}^{n} (X_k - \overline{X})^2 = \left(\sum_{k=1}^{n} X_k^2\right) - n\overline{X}^2,} \tag{F2}$$

the last equality being a Parallel-Axis result. We use 'RSS' for Residual Sum of Squares' – for reasons which will appear later.

Using $\mathbb{E}\left(X^2\right) = \mathbb{E}(X)^2 + \text{Var}(X)$, prove that

$$\boxed{\mathbb{E}(V) = \sigma^2, \text{ where } V := \frac{\text{RSS}}{n-1}.}$$

Dividing by $n-1$ rather than by n is sometimes called 'Bessel's correction'. Contrast the division by N in the 'population variance' in Exercise 68F.

We use V as a Variance Estimator for σ^2 partly because it is unbiased: as you see, it is right on average. However, there are much better reasons to do with 'degrees of freedom', as we'll see later.

It is not all clear that being right on average is that good a thing when you are dealing with just one sample. This is one of many situations in which criticisms once levelled by Frequentists against Bayesians apply with greater force to things on which they themselves used to concentrate.

G. Discussion. Suppose that $X_1, X_2, \ldots$ is an IID sequence in $\mathcal{L}^2$, each X_k having mean μ and variance σ^2. Then, for large n,

$$A_n \text{ has mean } \mu \text{ and small variance } \sigma^2/n,$$

so that the distribution of A_n is 'concentrated around μ'. We see the possibility of proving various rigorous assertions of 'convergence of A_n to μ' in our mathematical model.

First we make precise the notion that if an RV Z in $\mathcal{L}^2$ has small variance, then there is small probability that Z differs from its mean by a significant amount. The name of the discoverer of the celebrated result which follows has more English transliterations than Tchaikowsky: he features as Cebysev, Tchebychoff, ...

▶▶ **H. Tchebychev's Inequality.** *Let Z be an RV in $\mathcal{L}^2$ with mean μ. Then, for $c > 0$,*

$$c^2 \mathbb{P}(|Z - \mu| \geq c) \leq \text{Var}(Z).$$

Proof. Let F be the event $F := \{\omega : |Z(\omega) - \mu| \geq c\}$. Then, for $c > 0$,

$$(Z - \mu)^2 \geq c^2 I_F, \tag{H1}$$

as I'll now explain. Inequality (H1) says that for *every* ω,

$$\left(Z(\omega) - \mu\right)^2 \geq c^2 I_F(\omega). \tag{H2}$$

Why is this true? Well, if $\omega \in F$, then
    left-hand side $\geq c^2$ by definition of F,
    right-hand side $= c^2$ by definition of I_F.
If $\omega \notin F$, then
    left-hand side ≥ 0 because it is a square,
    right-hand side $= 0$ by definition of I_F.
Hence (H2) is indeed true for every ω.

Taking expectations at (H1), we obtain

$$\text{Var}(Z) = \mathbb{E}\left\{(Z - \mu)^2\right\} \geq c^2 \mathbb{P}(F) = c^2 \mathbb{P}(|Z - \mu| \geq c),$$

as required.   □

Ha. Exercise. (See Exercise 104Ea.) Let Y be the number of heads obtained if we toss a fair coin 100 times. Use Tchebychev's inequality to obtain the *bound*:

$$\mathbb{P}(41 \leq Y \leq 59) = \mathbb{P}(|Y - 50| < 10) \geq 0.75.$$

Notes. The Central Limit Theorem (done later) gives the *approximation*

$$\mathbb{P}(41 \leq X \leq 59) = \mathbb{P}(40.5 < X \leq 59.5) \approx \Phi(1.9) - \Phi(-1.9) = 0.9426.$$

For this problem, the exact value (to 5 places of decimals) is 0.94311. (Even though Tchebychev's inequality is quite crude, it is good enough to allow us to prove the Weak Law of Large Numbers and other important results.)

Prove that if S_n is the number of Heads in n tosses of a coin with probability p of Heads, then

$$\mathbb{P}\left(|n^{-1}S_n - p| \geq \delta\right) \leq \frac{1}{4n\delta^2}.$$

▶ **I. Exercise: Markov's Inequality.** This inequality is more fundamental than Tchebychev's. It is possible to get tight bounds by using it in clever ways. Prove that *if W is a non-negative RV in $\mathcal{L}^1$ and $c > 0$, then*

(for non-negative W and $c > 0$) $c\mathbb{P}(W \geq c) \leq \mathbb{E}(W)$.

[*Hint.* What do you choose for the event F?] Why does Markov's inequality imply Tchebychev's?

▶▶ **J. Theorem: Weak Law of Large Numbers (WLLN).** *Suppose that* $X_1, X_2, \ldots$ *is an infinite sequence of IID Random Variables, each in* $\mathcal{L}^2$, *each with mean* μ *and variance* σ^2. *Let*

$$A_n := \frac{X_1 + X_2 + \cdots + X_n}{n}.$$

Then, for every fixed $\varepsilon > 0$,

$$\mathbb{P}\big(|A_n - \mu| > \varepsilon\big) \to 0 \quad as\ n \to \infty.$$

Proof. By Lemma 104E and Tchebychev's inequality,

$$\varepsilon^2 \mathbb{P}\big(|A_n - \mu| > \varepsilon\big) \le \frac{\sigma^2}{n}, \quad \mathbb{P}\big(|A_n - \mu| > \varepsilon\big) \le \frac{\sigma^2}{n\varepsilon^2}.$$

The desired result is now obvious. □

K. Discussion. The conclusion of the WLLN is expressed by saying that '$A_n \to \mu$ *in probability*': *Sample Mean converges in probability to true mean*. The Strong Law of Large Numbers (SLLN) is in many ways a much more satisfying result for mathematicians: it says that the set of ω for which $A_n(\omega) \to \mu$ in the true sense of convergence you meet in Analysis has probability 1; and it does not need the finiteness of σ^2.

One has to be careful about whether the WLLN or the SLLN or the later Central Limit Theorem (CLT) is really what corresponds to the 'long-term relative frequency' idea in the real world, at least as far as Statistics is concerned. In the real world, we do *not* take infinite samples; and *good estimates* of how far, for example, Sample Mean is likely to be from true mean, or how far the Empirical Distribution is likely to be from the true distribution (see later), is what matters to a statistician. (The business of *getting* such estimates, however, belongs to Probability.)

I want to take you through to a clearer statement of the SLLN in a way which, I hope, you can understand intuitively even if you have not read Section 2.3 on Measure Theory. The proof of the general case of the Strong Law is too difficult for this book, but we shall see why the result is true in a class of situations which covers nearly all those met in practice.

First, however, we see Tchebychev's inequality in action in other important contexts, some important in Statistics, and one relevant to Mathematics.

▶ **L. Order Statistics.** Let r and s be positive integers. Suppose that $U_1, U_2, \ldots, U_{r+s+1}$ are IID Random Variables each with the U[0, 1] distribution. By Lemma 103B, with probability 1, no two of the U's are equal. We may therefore rearrange the U's in ascending order as the *Order Statistics*

$$V_1 < V_2 < \cdots < V_{r+s+1}.$$

We want to find the pdf of V_{r+1}. We have, intuitively,

$$\mathbb{P}(V_{r+1} \in dv) = \sum_{k=1}^{r+s+1} \mathbb{P}(U_k \in dv; A_k),$$

where A_k is the event that precisely r of the $r + s$ values $U_1, \ldots, U_{k-1}, U_{k+1}, \ldots, U_{r+s+1}$ are less than v. Because U_k does not feature in the event A_k, the event $\{U_k \in dv\}$ is independent of the event A_k. Moreover, A_k is the chance that if we throw a coin with probability v of Heads $r + s$ times, then we shall obtain precisely r Heads. Thus,

$$\mathbb{P}(V_{r+1} \in dv) = (r+s+1)\binom{r+s}{r} v^r (1-v)^s dv = \frac{(r+s+1)!}{r!s!} v^r (1-v)^s dv,$$

and since this must integrate to 1, we have proved formula 79(G3). In the terminology of Chapter 6, V_r has the Beta$(r + 1, s + 1)$ distribution.

La. Exercise. Prove that

$$\mathbb{E}(V_{r+1}) = \frac{r+1}{r+s+2}, \quad \text{Var}(V_{r+1}) = \frac{(r+1)(s+1)}{(r+s+2)^2(r+s+3)}.$$

M. Convergence in probability of Sample Median.
Let $U_1, U_2, \ldots, U_{2r+1}$ be IID Variables, each U$[0,1]$. Since

$$\mathbb{P}(U_k \leq \tfrac{1}{2}) = \mathbb{P}(U_k \geq \tfrac{1}{2}) = \tfrac{1}{2},$$

we say (see Subsection 72L) that $\tfrac{1}{2}$ is the true *median* of each U_k. Let $V_1, V_2, \ldots, V_{2r+1}$ be the U's arranged in increasing order, the Order Statistics. Then the 'middle' value $M := V_{r+1}$ is called the *Sample Median* of the sample $U_1, U_2, \ldots, U_{2r+1}$. By Exercise La with $s = r$, we have

$$\mathbb{E}(M) = \tfrac{1}{2}, \quad \text{Var}(M) = \frac{1}{4(2r+3)}.$$

Tchebychev's inequality would now make precise the idea that if r is large, then the Sample Median M is likely to be close to the true median $\tfrac{1}{2}$. The details are now obvious, if you wish to supply them.

Now suppose that $X_1, X_2, \ldots, X_{2r+1}$ are IID Random Variables with strictly positive pdf f on $\mathbb{R}$. Let m now denote the true median of a typical X, so that $F(m) = \tfrac{1}{2}$, where F is the common distribution function of the X's. Now $U_1, U_2, \ldots, U_{2r+1}$, where $U_k = F(X_k)$, are IID Random Variables, each with the U$[0,1]$ distribution. See Subsection 50B and Lemma 99G. Let $Y_1, Y_2, \ldots, Y_{2r+1}$ be the X_k's arranged in increasing order, the Order Statistics for the X's. Then the Order Statistics $V_1, V_2, \ldots, V_{2r+1}$ for the U's are given by $V_k = F(Y_k)$.

Hence, $F(Y_{r+1})$ is likely to be close to $\tfrac{1}{2}$, whence, by continuity of F^{-1}, the Sample Median Y_{r+1} of $X_1, X_2, \ldots, X_{2r+1}$ is likely to be close to m. If you like Analysis, formulating this precisely will cause you no difficulty (the proof of the Weierstrass approximation in Subsection O below will help); if you don't like Analysis, you will be happy to see things intuitively.

See Subsection 165L for a much stronger result.

N. Using the Sample Median in Statistics. A situation in which the results of the previous subsection become useful is the following. One of the infamous Cauchy distributions has pdf

$$f(x) \; = \; \frac{1}{\pi \left(1 + (x - \theta)^2\right)}.$$

Suppose that we are interested in estimating θ from a sample. As mentioned earlier, the Cauchy distribution has *no* mean: a variable with this distribution is not in $\mathcal{L}^1$. The Sample Mean of $2r + 1$ IID observations from this Cauchy distribution does not concentrate: it has the same Cauchy distribution as a single observation. (As already mentioned, we shall see how to prove such things later.) Thus the Sample Mean is here useless for estimating θ. However, θ is the true median of each X, and by the previous subsection, for large r, the Sample Median M is likely to be close to θ. So, we can use M as an Estimator for θ.

See Note 196Ec for how to do this with precision.

O. Probabilistic proof of the Weierstrass approximation theorem. I mentioned earlier that Probability may be used to prove results in other branches of Maths. Here's an example. The Weierstrass approximation theorem, a fundamental result in Analysis, states the following:

> Let f be a continuous function on $[0, 1]$. Let $\varepsilon > 0$ be given. Then there exists a polynomial B (called after Bernstein) such that
>
> $$|f(x) - B(x)| < \varepsilon \text{ for all } x \text{ in } [0, 1].$$

Proof. Let $p \in [0, 1]$ and let S_n be the number of Heads if a coin with probability p of Heads is tossed n times. Then

$$B_n(p) \; := \; \mathbb{E}\left\{ f\left(n^{-1}S_n\right) \right\} \; = \; \sum_{s=0}^{n} f\left(n^{-1}s\right) \mathbb{P}(S_n = s)$$

$$= \; \sum_{s=0}^{n} f\left(n^{-1}s\right) \binom{n}{s} p^s q^{n-s},$$

so that B_n is a polynomial of degree n in p. We have

$$\left| B_n(p) - f(p) \right| \; = \; \left| \mathbb{E}\left\{ f\left(n^{-1}S_n\right) - f(p) \right\} \right|.$$

We need to exploit with precision the fact that for large n, $n^{-1}S_n$ is likely to be close to p; especially, we need our bound on the left-hand side to hold simultaneously for all p.

You can see the possibility of making the proof work now. If you like Analysis, read on; if not, jump to the next subsection.

We can draw two standard conclusions from the fact that f is a continuous function on the compact (in the present context, closed and bounded) interval $[0, 1]$. First, f is bounded: for some $K > 0$,

$$|f(x)| \leq K \text{ for all } x \text{ in } [0, 1].$$

Second, f is uniformly continuous on $[0, 1]$, whence, for our given ε, we can find, and *fix*, $\delta > 0$ such that

$$x, y \in [0, 1] \text{ and } |x - y| < \delta \text{ imply that } |f(x) - f(y)| < \tfrac{1}{2}\varepsilon.$$

Write

$$W_n := \left| f\left(n^{-1}S_n\right) - f(p) \right|, \qquad Z_n := \left| n^{-1}S_n - p \right|.$$

Then $Z_n < \delta$ implies that $W_n < \tfrac{1}{2}\varepsilon$. Let $F_n := \{\omega : Z_n(\omega) < \delta\}$. Then we have just seen that (for every ω) $I_F W_n < \tfrac{1}{2}\delta$ (think!). Since $0 \le W_n \le 2K$ for every ω, we have $I_{F_n^c} W_n \le 2K I_{F_n^c}$ for every ω. Using at the first step the result that 'the modulus of an expectation does not exceed the expectation of the modulus' (see 61F), we have, *for every p in* $[0, 1]$,

$$
\begin{aligned}
\left| B_n(p) - f(p) \right| &\le \mathbb{E}\left(W_n\right) = \mathbb{E}\left(I_{F_n} W_n\right) + \mathbb{E}\left(I_{F_n^c} W_n\right) \\
&\le \tfrac{1}{2}\varepsilon + 2K\mathbb{E}\left(I_{F_n^c}\right) = \tfrac{1}{2}\varepsilon + 2K\mathbb{P}(F_n^c) \\
&= \tfrac{1}{2}\varepsilon + 2K\mathbb{P}(Z_n \ge \delta) \le \tfrac{1}{2}\varepsilon + 2K\frac{1}{4n\delta^2},
\end{aligned}
$$

the last inequality from Exercise 106Ha. Hence, for $n > n_0$, where n_0 is the least integer with $2K/(4n\delta^2) < \tfrac{1}{2}\varepsilon$, we have $\left| B_n(p) - f(p) \right| < \varepsilon$ for all p in $[0, 1]$. $\square$

▶▶ **P. A model for tossing a fair coin infinitely often.**   *Advice.* Read this through the first time skipping the small-font indented sections; then read it all.

The model we use is the Fundamental Model for choosing a point U in $[0, 1]$ at random according to the uniform $\mathrm{U}[0, 1]$ distribution. So we take $\Omega = [0, 1]$, $U(\omega) := \omega$. We define the unique probability measure on nice subsets of Ω which satisfies $\mathbb{P}([0, x]) = x$ for every x.

> [We want each '$U \le x$' where $0 \le x \le 1$ to be an event: in other words we want each subset of Ω of the form $[0, x]$ to be an event. We take the smallest appropriate class ('σ-algebra') of events which contains each such subset. That there is a unique measure on the class of events such that $\mathbb{P}([0, x]) = x$ for every x, is known. Section 2.3 contains more on all this.]

Imagine expanding each ω in Ω in binary:

$$\omega = 0.\omega_1\omega_2\omega_3 \ldots \quad \text{(binary)}$$

and writing H for each 0 and T for each 1.

> [Each dyadic rational of the form $r2^{-n}$ in $[0, 1)$ has two binary expansions: for example,
>
> $$\tfrac{1}{2} = 0.10000\ldots = 0.01111\ldots \quad \text{(binary)}.$$
>
> However, the set $\mathbb{D}$ of such dyadic rationals satisfies $\mathbb{P}(\mathbb{D}) = 0$: the chance that Tyche would choose a point in $\mathbb{D}$ is exactly 0. The awkward points in $\mathbb{D}$ are not relevant. We can pretend that they do not exist – or even force them not to exist by working with $\Omega \setminus \mathbb{D}$.]

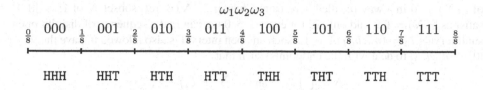

Figure P(i): Coin tossing and the Fundamental Model

Figure P(i) shows what happens if we divide $\Omega = [0, 1]$ into 8 equal intervals. Between 0 and $\frac{1}{8}$, the binary expansion of ω begins 0.000, leading to the sequence HHH. Thus each of the 8 outcomes of the first 3 tosses is assigned a probability $\frac{1}{8}$ exactly as required by independence. The Fundamental Model contains every '*finite* fair-coin-tossing' model; and, really because of the uniqueness of $\mathbb{P}$ discussed earlier (really because of the π-system Lemma at 45M), it forms the essentially unique model for tossing a fair coin *infinitely* often.

For
$$\omega = 0.\omega_1\omega_2\omega_3\ldots \quad \text{(binary)},$$

define
$$X_k(\omega) := \omega_k, \quad A_n(\omega) := \frac{X_1(\omega) + \cdots + X_n(\omega)}{n}.$$

Thus $A_n(\omega)$ is the proportion of Tails in the first n tosses for the realization ω.

▶▶ **Q. Theorem: Borel's Strong Law for Coin Tossing.** *Let G be the set of* '*good' outcomes ω for which*
$$A_n(\omega) \to \tfrac{1}{2}$$

in the usual sense of convergence in Analysis (explained in Clarification below). Thus, G is the set of outcomes for which the 'long-term relative frequency' (LTRF) idea works. Then G is an event, and $\mathbb{P}(G) = 1$.

For proof, see Subsection 113T.

▶▶ **Clarification.** *We put ω in G if and only if for every $\varepsilon > 0$ there exists an integer $n_0(\omega)$,* **which is allowed to depend on** *ω, such that for $n \geq n_0(\omega)$, we have*

$$|A_n(\omega) - \tfrac{1}{2}| < \varepsilon.$$

Another way of expressing Borel's result is that the set N of bad outcomes ω for which $A_n(\omega)$ does not converge to $\frac{1}{2}$ (the sequence may not converge at all, or converge to the wrong value) satisfies $\mathbb{P}(N) = 0$. We can understand the geometrical significance of $\mathbb{P}(N) = 0$ in a way recalled here from Chapter 2. A (Borel) subset N of $\Omega = [0,1]$ satisfies $\mathbb{P}(N) = 0$ if and only if, for every $\varepsilon > 0$ we can find a sequence of disjoint open subintervals $I_n = (a_n, b_n)$ of $[0,1]$ (such an open interval is also allowed to have the form $[0, b)$ or $(a, 1]$ to deal with the endpoints) such that

$$N \subseteq \bigcup I_n \quad \text{and} \quad \sum \ell(I_n) < \varepsilon,$$

where $\ell(I_n)$ is the length of I_n. In short, for every $\varepsilon > 0$, you can find an open subset G_ε of $[0,1]$ containing N and of length at most ε.

I hope that that helps explain the statement of Borel's Strong Law. The statement of the general Kolmogorov Strong Law is now no more difficult.

▶▶ **R. Universality of the Fundamental Model.** As mentioned earlier, it is an amazing fact that *every experiment, no matter how complicated (perhaps involving the evolution in time of several interacting populations, for example) can be reduced to the single experiment of choosing a number between 0 and 1 according to the uniform distribution for which we have our Fundamental Model.* You will start to believe this very shortly.

Again, let

$$\omega = 0.\omega_1\omega_2\omega_3 \ldots \quad \text{(binary)},$$

in the Fundamental Model. Now define

$$\begin{aligned}
U_1(\omega) &:= 0.\omega_1\omega_3\omega_6 \ldots \quad \text{(binary)}, \\
U_2(\omega) &:= 0.\omega_2\omega_5\omega_9 \ldots \quad \text{(binary)}, \\
U_3(\omega) &:= 0.\omega_4\omega_8\omega_{13} \ldots \quad \text{(binary)},
\end{aligned} \tag{R1}$$

and so on, the suffices being arranged in a series of '45°' lines.

Now *use your intuition.* Because each U_k corresponds to a fair-coin-tossing sequence, each U_k has the $U[0,1]$ distribution. Because the coin-tossing sequences for the different U_k's involve different subsets of tosses, the U_k's are independent. So,

$U_1, U_2, \ldots$ is an IID sequence, each with the $U[0,1]$ distribution.

Thus our Fundamental Model contains infinitely many independent copies of itself!!

Now, by the F-inverse Principle at 51Ba, we can produce from the U sequence a sequence of independent RVs with arbitrary distributions. If we want an IID sequence, then we simply take $X_k = G(U_k)$, with G as explained at Exercise 51Ba and the discussion which followed it.

We still have the same geometrical interpretation of a null subset of Ω, so we can understand:

▶▶▶ S. Fact (Kolmogorov): Strong Law of Large Numbers (SLLN).
Suppose that $X_1, X_2, \ldots$ are IID Random Variables. Define

$$A_n := \frac{X_1 + \cdots + X_n}{n}.$$

(a) If each X_k is in $\mathcal{L}^1$, and μ denotes the common mean, then

$$\mathbb{P}(A_n \to \mu) = 1,$$

the set of ω such that $A_n(\omega) \to \mu$ being an event.
(b) If each X_k is not in $\mathcal{L}^1$, then the set of ω for which $\lim A_n(\omega)$ exists (the limit being allowed to depend on ω) is an event of probability 0.

Here then is the ultimate expression in the mathematical theory of the long-term-average idea. *Please re-read the Clarification which follows Theorem 111Q and which obviously applies here too.*

What I have been saying is that the geometric picture of null sets for the Fundamental Triple allows us to understand the SLLN provided that we use the Fundamental Model with $X_k = G(U_k)$ with the U_k as at 112(R1). That the theorem follows for any probability triple $(\Omega, \mathcal{F}, \mathbb{P})$ may be deduced from its truth in the 'Fundamental Representation' via an Isomorphism Theorem which hinges on the π-system at Lemma 45M. This is all about interpretation: proving the SLLN for an arbitrary triple is no more difficult (and no less difficult) than for the Fundamental Representation.

Kolmogorov's theorem is difficult to prove in full generality, but it is possible to give a simple proof of Part (a) assuming an additional condition which holds in most practical situations. It is also possible to prove Part (b) quite easily. I now do these two things.

▶ T. Theorem: A special case of the Strong Law. *Suppose that $X_1, X_2, \ldots$ are IID Random Variables and that the common value K of $\mathbb{E}\left(X_k^4\right)$ is finite. Then*

$$\mathbb{P}\left(A_n \to \mu\right) = 1,$$

where μ is the common mean of the X_k's and A_n is as usual.

Proof. We can replace X_k by $X_k - \mu$ and thereby reduce the problem to the case when, as we now assume,

$$\mu = 0.$$

For justification of this step and later steps, inequality 66(B2) shows that $\mathbb{E}\left(|X|\right)$, $\mathbb{E}\left(X^2\right)$ and $\mathbb{E}\left(|X|^3\right)$ are finite, and inequality 66(B1) shows that

$$\mathbb{E}\left[(X_k - \mu)^4\right] \leq 8(K + \mu^4),$$

and we now redefine K to be the expression on the right-hand side.

Let $S_n := X_1 + \cdots + X_n$, as usual. Then

$$S_n^4 = (X_1 + \cdots + X_n)(X_1 + \cdots + X_n)(X_1 + \cdots + X_n)(X_1 + \cdots + X_n). \quad \text{(T1)}$$

Consider calculating $\mathbb{E}\left(S_n^4\right)$ by using this expansion. Let i, j, k, ℓ be distinct numbers. Then, since $\mathbb{E}\left(X_i\right) = 0$, the 'Independence means Multiply' rule shows that

$$\mathbb{E}\left(X_i X_j^3\right) = 0, \quad \mathbb{E}\left(X_i X_j X_k^2\right) = 0, \quad \mathbb{E}(X_i X_j X_k X_\ell) = 0.$$

So,

$$\mathbb{E}\left(S_n^4\right) = \sum_{k=1}^{n} \mathbb{E}\left(X_k^4\right) + 6 \sum_i \sum_{j>i} \mathbb{E}\left(X_i^2 X_j^2\right).$$

The 6 is $\binom{4}{2}$: think of the ways in which $X_1^2 X_2^2$ can appear in the right-hand side of (T1). But, by the proof of Lemma 65B,

$$\mathbb{E}\left(X_i^2 X_j^2\right) \leq \tfrac{1}{2}\left\{\mathbb{E}\left(X_i^4\right) + \mathbb{E}\left(X_j^4\right)\right\} \leq K.$$

Hence,

$$\mathbb{E}\left(A_n^4\right) \leq \frac{1}{n^4}\left[nK + 3n(n-1)K\right] \leq \frac{3K}{n^2},$$

and $\mathbb{E}\sum A_n^4 < \infty$. Thus, with probability 1, $\sum A_n^4 < \infty$, and so $A_n \to 0$. Rigour for these last steps mirrors exactly that used in the proof 60Da of the First Borel–Cantelli Lemma.

The very keen reader will have noticed that I skipped proving that the set of ω for which $A_n(\omega) \to \mu$ is an event. See [W] if you are that interested. $\quad\square$

U. Proof of Part (b) of Theorem 113S. If ω is such that $A_n(\omega) \to L(\omega)$, then

$$\frac{X_n(\omega)}{n} = A_n(\omega) - \frac{n-1}{n} A_{n-1}(\omega) \to L(\omega) - L(\omega) = 0.$$

Now, if X is an RV with the same distribution as each X_k and if $\mathbb{E}\left(|X|\right) = \infty$, then, by result 61G,

$$\sum \mathbb{P}\left(|X_n| \geq n\right) = \sum \mathbb{P}\left(|X| \geq n\right) = \infty,$$

and, by independence and the Second Borel–Cantelli Lemma 98E, the set of ω for which $|X_n(\omega)| \geq n$ for infinitely many n has probability 1. The desired result follows. $\quad\square$

▶▶ **V. Probability and long-term relative frequency.** Let's stick to the case of the mathematical model for coin tossing, for a coin with probability p of Heads. So suppose that $X_1, X_2, \ldots$ are IID Random Variables with (as in our discussion of Borel's Theorem) $\mathbb{P}(X = 0) = p$, $\mathbb{P}(X = 1) = q$. Then, the number of Tails in the first n tosses is

$$S_n := X_1 + X_2 + \cdots + X_n.$$

Manifestly, $\mathbb{E}\left(X_n^4\right)$ is finite: it is q. We therefore know that, **with probability** 1, the proportion $A_n := S_n/n$ of Tails in the first n tosses will converge to q. For any fixed subsequence $n(1) < n(2) < \ldots$,

$$\frac{1}{r}\left(X_{n(1)} + X_{n(2)} + \cdots + X_{n(r)}\right) \to q$$

with probability 1. But the set of ω such that

$$\frac{1}{r}\left(X_{n(1)}(\omega) + X_{n(2)}(\omega) + \cdots + X_{n(r)}(\omega)\right) \to q$$

simultaneously for *all* subsequences $n(\cdot)$ is clearly empty. Convince yourself of this. But there is no contradiction. The set of all increasing subsequences of $\mathbb{N}$ is uncountable (see Appendix A4, p496); but Measure Theory restricts the second part of result 44G to a *countable* collection of H's. So the logical flaw in the naive use of the 'long-term relative frequency' idea mentioned at Discussion 25B is completely resolved. Philosophers and others, please note!

W. Convergence in probability (to a constant). *The conclusion of the Weak Law does not prevent there being rare occasions on which A_n differs significantly from μ.* Let me try to give you some understanding of this.

We saw in subsection 102M that if $X_1, X_2, \ldots$ are IID Random Variables each with the exponential E(1) distribution, then it is not true that $X_n/(\ln n) \to 0$ with probability 1. However, it is obvious that for $\varepsilon > 0$, we have (for $n > 1$),

$$\mathbb{P}\left(\frac{X_n}{\ln n} > \varepsilon\right) = \mathbb{P}(X_n > \varepsilon \ln n) = e^{-\varepsilon \ln n} = n^{-\varepsilon},$$

so that $\mathbb{P}\left(\frac{X_n}{\ln n} > \varepsilon\right) \to 0$ as $n \to \infty$, and $X_n/(\ln n)$ converges to 0 in probability.

Suppose that $X_1, X_2, \ldots$ are IID Random Variables with $\mathbb{P}(X_k = \pm 1) = \frac{1}{2}$. Tchebychev's inequality shows that for every $\varepsilon > 0$,

$$\mathbb{P}(Y_n > \varepsilon) \to 0, \text{ where } Y_n := \frac{S_n}{\sqrt{2n \ln \ln n}},$$

so that $Y_n \to 0$ in probability. However, the celebrated **Law of the Iterated Logarithm (LIL)** states that, with probability 1,

$$\limsup Y_n = 1.$$

There are astoundingly precise extensions of the LIL, Strassen's Law being the most remarkable. See, for example, Deuschel and Stroock [61].

▶ **X. Empirical Distribution Function.**   Let $\mathbf{Y} = (Y_1, Y_2, Y_3, \ldots)$ be a sequence of IID Random Variables with common Distribution Function F. The Sample $(Y_1, Y_2, \ldots, Y_n)$ determines the so-called **Empirical Distribution Function (EDF)** $F_n(\,\cdot\,; \mathbf{Y})$, where $F_n(x; \mathbf{Y})$ is the Random Variable

$$F_n(x; \mathbf{Y}) := \frac{1}{n} \sharp\{k : k \leq n; \; Y_k \leq x\}, \tag{X1}$$

the proportion of $k \leq n$ such that $Y_k \leq x$. By the SLLN, we have for a fixed x the result that with probability 1, $F_n(x; \mathbf{Y}) \to F(x)$.

▶▶ **Y. The 'Fundamental Theorem of Statistics' (Glivenko–Cantelli).**
Make the assumptions and use the notation of the previous subsection. Then, with probability 1, $F_n(\,\cdot\,; \mathbf{Y})$ converges to $F(\cdot)$ uniformly over $\mathbb{R}$.

Clarification. We say that $\omega \in G$ if and only if it is true that for every $\varepsilon > 0$, there exists an integer $n_0(\omega)$, *which is allowed to depend on* ω, such that for $n \geq n_0(\omega)$,

$$|F_n(x; \mathbf{Y}(\omega)) - F(x)| < \varepsilon \quad \text{for every } x \text{ in } \mathbb{R},$$

where, $F_n(x; \mathbf{Y}(\omega)) := n^{-1}\sharp\{k : k \leq n; \; Y_k(\omega) \leq x\}$. Then G is an event and $\mathbb{P}(G) = 1$.

     This fact can be seen as providing part of the justification for (amongst many other things) the bootstrap technique of Simon and Efron – see Efron and Tibshirani [71], Davison and Hinkley [55].

▶ **Ya. Very instructive (but tricky) Exercise*.** Prove the Glivenko–Cantelli Theorem for the case when F is continuous and strictly increasing.

4.4   Random Walks: a first look

Later, we shall see other methods of doing several of the problems in this section.

▶▶ **A. Random Walk; $\mathbb{P}_a$; Simple Random Walk SRW(p).** *For a general Random Walk $(W_n : n \geq 0)$ on the set $\mathbb{Z}$ of all integers, W_n represents the position of a particle at time n, and*

$$W_n = a + X_1 + X_2 + \cdots + X_n, \quad so \; W_0 = a,$$

where $a \in \mathbb{Z}$ is the starting position, and the 'jumps' $X_1, X_2, \ldots$ are IID $\mathbb{Z}$-valued RVs. Probabilities associated with the random walk will depend on a and on the distribution function of the X_k variables. It will be very convenient to be able to indicate the dependence on a.

We write $\mathbb{P}_a$ for the probability law for the random walk when it starts at a and $\mathbb{E}_a$ for the corresponding expectation.

Simple Random Walk SRW(p), where $0 < p < 1$, is the case when each X_k takes one of the values $+1, -1$, and

$$\mathbb{P}(X_k = +1) = p, \quad \mathbb{P}(X_k = -1) = q := 1 - p.$$

Then $\mathbb{P}_a(X_k = 1) = p$ for all a and k.

In this case, W_n is the fortune at (just after) time n of a gambler who at each of the times $1, 2, 3, \ldots$ plays a game in which, independently of previous results, she wins \$1 with probability p or loses \$1 with probability q.

▶ **B. Gambler's Ruin: probabilities.** We use SRW(p) started at $a \geq 0$ as our model for the gambler's fortune. She plays until her fortune reaches either b (where $b > a$) or 0. Let G be the event that her fortune does reach b. We wish to find

$$x(a) := \mathbb{P}_a(G) = \mathbb{P}_a(W \text{ visits } b \text{ before } 0).$$

We can do this by describing the function x as the solution of the difference relation

$$x(a) = px(a+1) + qx(a-1) \quad (0 < a < b); \qquad x(0) = 0; \; x(b) = 1. \quad \text{(B1)}$$

The boundary conditions are obvious, and the 'recurrence relation' is just decomposing $x(a)$ according to the value of X_1: in detail, for $0 < a < b$,

$$\mathbb{P}_a(G) = \mathbb{P}_a(X_1 = 1)\mathbb{P}_a(G \,|\, X_1 = 1) + \mathbb{P}_a(X_1 = -1)\mathbb{P}_a(G \,|\, X_1 = -1).$$

However, for $0 < a < b$, we have $\mathbb{P}_a(G \,|\, X_1 = +1) = \mathbb{P}_{a+1}(G)$, because the $X_1 = 1$ has taken the Random Walk up to $a + 1$ and since $X_2, X_3, \ldots$ are independent of X_1, the whole system 'starts afresh' at time 1. For a little more on the rigour here, wait for Discussion 387Ca. Solving equation (B1) is just algebra. For the record, here's a refresher course on difference equations.

C. Solving difference equations, 1.   You will remember that to solve the *differential* equation

$$ax''(t) + bx'(t) + cx(t) = 0, \qquad (a \neq 0) \tag{C1}$$

you introduce the auxiliary equation

$$am^2 + bm + c = 0,$$

and if this has different roots α and β, then the general solution of (C1) is $x(t) = Ae^{\alpha t} + Be^{\beta t}$ for some constants A and B, but that if $\alpha = \beta$ then the solution is $x(t) = (A + Bt)e^{\alpha t}$. The story for difference equations is very similar.

To solve the *difference* equation

$$ax(k + 2) + bx(k + 1) + cx(k) = 0. \tag{C2}$$

we solve the auxiliary equation

$$am^2 + bm + c = 0$$

and if this has different roots α and β, then the general solution of (C1) is $x(k) = A\alpha^k + B\beta^k$ for some constants A and B, but that if $\alpha = \beta$ then the solution is $x(k) = (A + Bk)\alpha^k$.

For our SRW example, we have, with $k = a - 1$,

$$px(k + 2) - x(k + 1) + qx(k) = 0,$$
$$pm^2 - m + q = (pm - q)(m - 1) = 0, \quad \alpha = q/p, \ \beta = 1.$$

(Note: 1 will be a solution of virtually every equation we shall see!) Hence, absorbing an α into A and a β into B,

$$x(a) = A\alpha^a + B, \quad 1 = A\alpha^b + B, \quad 0 = A + B, \quad \text{if } p \neq q,$$

and

$$x(a) = (A + Bk), \quad 1 = A + Bb, \quad 0 = A, \quad \text{if } p = q = \tfrac{1}{2}.$$

One can now find A and B from the boundary conditions.

▶ D. Hitting and return probabilities.   Consider SRW(p). Let H denote the event that $W_n = 0$ for some strictly positive time n:

$$H := \{W_n = 0 \text{ for some } n \geq 1\}.$$

> (This restriction to strictly positive times is standard in the advanced theory, partly because it leads to perfect tie-up with electrostatics!! All concepts in electrostatics (equilibrium potential, equilibrium charge, capacity, etc) have very illuminating probabilistic interpretations. Indeed, several of the subsections in this present section relate to electrostatics, but there isn't room to reveal why.)

Let $x(a) := \mathbb{P}_a(H)$. If $a \neq 0$, then $x(a)$ is the obvious probability that the walk started at a will hit 0; but $x(0)$ is the probability that the walk started at 0 will hit 0 at some *positive* time, that is, that it will *return* to 0.

By the arguments of Subsection 117B, we have

$$x(1) \; = \; px(2) + (q \times 1) \; = \; px(2) + q.$$

It would not be correct to have $q \times x(0)$ here: if our walk starts at 1 and jumps down at time 1, then it *definitely* hits 0 at a positive time, whereas the 'return probability' $x(0)$ need not be 1.

We now use intuition for which the rigour will be provided in the next section. We have

$$x(2) = x(1)^2, \tag{D1}$$

because to go from 2 down to 0, the walk has to get from 2 down to 1, and then from 1 down to 0. It is obvious that the probability of getting from 2 down to 1 is the same as that of getting from 1 down to 0. It is *intuitively* clear that if the walk gets down from 2 to 1, then after first hitting 1, it starts afresh, so, using independence,

$$x(2) \; = \; \mathbb{P}_2(\text{hit } 0) \; = \; \mathbb{P}_2(\text{hit } 1)\mathbb{P}_2(\text{hit } 0 \mid \text{hit } 1)$$
$$= \; \mathbb{P}_2(\text{hit } 1)\mathbb{P}_1(\text{hit } 0) \; = \; x(1)^2. \tag{D2}$$

We now have

$$x(1) \; = \; px(1)^2 + q, \quad [px(1) - q][x(1) - 1] \; = \; 0,$$

whence $x(1) = q/p$ or $x(1) = 1$. If $q \geq p$, then since $x(1)$ cannot be greater than 1, we must have $x(1) = 1$.

What happens if $q < p$? Let's again use our intuition. Since $\mathbb{E}(X_k) = p - q$, then, by the SLLN, the walk will tend to $+\infty$, so there will be a positive probability that if started at 1 it will drift to $+\infty$ without hitting 0. Hence we must have $x(1) = q/p$, not $x(1) = 1$.

Da. Bringing-in time. But can we prove this without appealing to the SLLN? Yes, we can. We bring the time into the story (thus switching from electrostatics to the theory of heat flow!) So, let

$$H_n \; := \; \{W_m = 0 \text{ for some } m \text{ satisfying } 1 \leq m \leq n\}.$$

Let $x(a, n) := \mathbb{P}_a(H_n)$. Our intuition, confirmed by property 43D(a), is that $x(a)$ is the 'monotonically increasing limit'

$$x(a) \; = \; \uparrow \lim_{n \to \infty} x(a, n). \tag{D3}$$

For $n \geq 1$, we have

$$x(1, n) \ = \ px(2, n-1) + q$$

– Yes?! Also, we have

$$x(2, n) \ \leq \ x(1, n-1)x(1, n-1).$$

Think about this. If the particle gets from 2 to 0 in n steps, it must get from 2 to 1 in at most $n-1$ steps and from 1 to 0 in at most $n-1$ steps.

Hence, $x(1, n) \leq px(1, n-1)^2 + q$, and since $x(1, 1) = q \leq q/p$, we have $x(1, 2) \leq p(q/p)^2 + q = q/p$. By induction $x(1, n) \leq q/p$ and now, by 119(D3), $x(1) \leq q/p$.

You can now check that in all cases,

$$1 - x(0) \ = \ \mathbb{P}_0(\text{no return to } 0) \ = \ |p - q|,$$

a nice fact if we think of $|p - q|$ as the bias in a coin and of the random walk as accumulated 'Number of Heads minus Number of Tails'.

▶ **E. Expected number of visits to initial state.** Consider any Random Walk on $\mathbb{Z}$. Let r be the probability of a return to 0 if, as we now assume, the process starts at 0. Let N be the total number of visits to 0 including that at time 0. Using the intuitive idea that the Random Walk starts afresh at the time of first return to 0 (if this is finite), we have, for the expectation of N when the starting point is 0,

$$\mathbb{E}_0(N) \ = \ 1 + r\mathbb{E}_0(N) + (1 - r) \times 0, \tag{E1}$$

the '1' representing the visit at time 0. We saw such things at subsection 21E. Hence,

$$\mathbb{E}_0(N) \ = \ \frac{1}{1 - r} \quad (= \infty \text{ if } r = 1). \tag{E2}$$

However, if $\xi_n := I_{\{W_n = 0\}}$, then

$$N \ = \ \xi_0 + \xi_1 + \xi_2 + \cdots, \tag{E3}$$

the sum counting 1 for every visit to 0. Thus

$$\mathbb{E}_0(N) \ = \ \sum_{n=0}^{\infty} \mathbb{E}_0(\xi_n) \ = \ \sum_{n=0}^{\infty} \mathbb{P}_0(W_n = 0), \tag{E4}$$

and this raises the possibility of finding the return probability r by summing the series on the right-hand side of (E4) and then using (E2). An important example is given in the next subsection.

The intuition about 'starting afresh' which we have used here is justified in Section 4.5, and the type of decomposition used at (E1) is studied in detail in Chapter 9.

▶ **F. Recurrence of Symmetric RW on $\mathbb{Z}^2$.** Consider the integer lattice $\mathbb{Z}^2$ in the plane, the typical point of which is (u, v), where $u, v \in \mathbb{Z}$. The lattice is illustrated on the left-hand side of Figure F(i). Each point of $\mathbb{Z}^2$ has 4 neighbours; and in Symmetric Random Walk on $\mathbb{Z}^2$, a particle jumps from its current position to a randomly chosen one of its 4 neighbours with probability $\frac{1}{4}$ each, independently of its past history. We are interested in the return probability to $O = (0, 0)$.

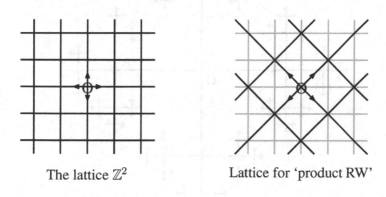

The lattice $\mathbb{Z}^2$ Lattice for 'product RW'

Figure F(i): To help derive recurrence of RW on $\mathbb{Z}^2$

Suppose that $U = (U_n : n = 0, 1, 2, \ldots)$ and V are independent SRW($\frac{1}{2}$) processes on $\mathbb{Z}$, with $U_0 = V_0 = 0$. Consider the 'product' process $\tilde{\mathbf{W}}$ with $\tilde{\mathbf{W}}_n = (U_n, V_n)$. A little thought shows that $\tilde{\mathbf{W}}$ is a Symmetric Random Walk on the 45° lattice on the right of Figure F(i). However,

$$\mathbb{P}\left(\tilde{\mathbf{W}}_{2n} = (0, 0)\right) = \mathbb{P}(U_{2n} = 0)\,\mathbb{P}(V_{2n} = 0) = b(2n, \tfrac{1}{2}; n)^2 \sim \frac{1}{\pi n},$$

by result 11(N2). Hence, $\sum \mathbb{P}\left(\tilde{\mathbf{W}}_{2n} = (0, 0)\right) = \infty$, and, by the argument of the previous subsection, *the return probability to $(0, 0)$ for Symmetric Random Walk on $\mathbb{Z}^2$ is* 1: we say that the Random Walk is *recurrent*.

For Symmetric Random Walk $\mathbf{W}$ on $\mathbb{Z}^3$ (or on $\mathbb{Z}^d$ for $d > 3$), the return probability is strictly less than 1: we say that the Random Walk is *transient*. We cannot use the same type of argument because each site in $\mathbb{Z}^3$ has 6 neighbours whereas a site for the 'product' of 3 SRW processes on $\mathbb{Z}$ would have 8 neighbours. The intuitive point is that, for $d \geq 3$, $\mathbb{Z}^d$ is a 'big' space; and the Symmetric Random Walk on it will (with probability 1) drift to infinity. However, (with probability 1) it will not drift to infinity in any particular direction: the set of values $\mathbf{W}_n / \|\mathbf{W}_n\|$ will be dense in the unit sphere.

It is not easy to see intuitively what is a 'big' space: for example, one can find a Random Walk on the set $\mathbb{Q}$ of rationals which will with probability 1 visit every rational infinitely often. There is no similar recurrent Random Walk on the 'free group fractal' of the next exercise.

G. Random Walk on the 'fractal' free group on 2 generators.   Random

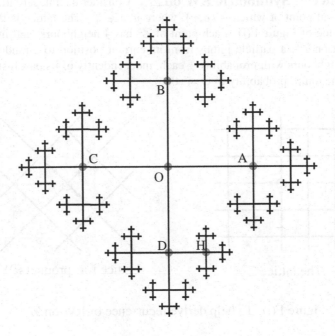

Figure G(i): Free group as fractal

walks (or rather, their continuous analogues, Brownian motions) on non-Abelian groups such as rotation groups, are very important. But here's a simple example which is reinterpretable via a picture. You should enjoy the Exercise!

The free group $\mathbb{G}$ on 2 generators a and b consists of the empty word (identity or unit element) e and all 'words' obtained by stringing together symbols a, a^{-1}, b, b^{-1}. Thus,

$$a^{-1}bb^{-1}abba^{-1}b^{-1}ba^{-1}$$

is a word which we can cancel down to the reduced word b^2a^{-2} of length 4 (in that as far as length is concerned, it is regarded as $bba^{-1}a^{-1}$). Henceforth, we think of $\mathbb{G}$ as consisting of *reduced* words. The group is 'free' in that two reduced words are regarded as equal if and only if they are identical. There are no relations such as $ab = ba^2$ between the generators.

We can build a random walk on $\mathbb{G}$ by starting with the element e and then multiplying it on the right by a randomly chosen one of a, a^{-1}, b, b^{-1}, each having probability $\frac{1}{4}$ of being chosen, independently of what has happened previously. At each jump, the reduced word either increases by one in length, or is reduced by one in length.

We can picture Γ as the fractal sketched in Figure G(i). Of course, the fractal should go on for ever, not stop at reduced words of length up to 5. We have (for one possible convention) $O = e$, $A = a$, $B = b$, $H = b^{-1}a$, etc. The 'freeness' of the group means that there are no loops in the picture. (Physicists might call the picture a Bethe lattice.) From any point, the particle is equally likely to jump to any of its 4 'neighbours'. The shrinking in size as we move away from O is a misleading feature of the picture. Still, the picture *is* helpful, as you should learn from the following problems. It must be understood that 'hit A before B' means 'the particle does hit A and does so before it hits B (if it does)'.

Ga. Exercise*. Prove that

$$\mathbb{P}_O(\text{hit A}) = \tfrac{1}{3}, \qquad \mathbb{P}_O(\text{hit A before B}) = \tfrac{3}{10},$$
$$\mathbb{P}_O(\text{hit both A and B}) = \tfrac{1}{15},$$
$$\mathbb{P}_O(\text{hit H before either A or B}) = \tfrac{9}{101}.$$

With $\mathbb{P}_O$ probability 1, $n^{-1}(\text{length of reduced word})$ will tend to a deterministic limit c. What is the value of c?

H. Fractal random variables. Consider the full geometric picture of which part is sketched in Figure 122G(i). We are now thinking of the points of the free group as the nodes of the graph in Figure 122G(i) embedded in the plane, with the distances as shown. The Random Walk (W, say) on the vertices of the fractal will, with probability 1, converge to a point W_∞ which belongs to the set E (say) of 'endpoints' of the fractal. This set E is a true fractal, a set of fractional dimension. The RV W_∞ is not discrete because $\mathbb{P}(W = x) = 0$ for every point x in the plane; but neither is it 'continuous': it has no pdf in the plane. However, such RVs present no problem to Measure Theory.

The set E is a much prettier relative of the classical Cantor subset C of $[0, 1]$. We can think of Cantor measure on the Cantor set as the law of the RV $W_\infty := \lim W_n$ where

$$W_n := \frac{1}{2} + \sum_{k=1}^{n} \frac{X_k}{3^k},$$

where $X_1, X_2, \ldots$ are IID RVs with $\mathbb{P}(X_k = \pm 1) = \tfrac{1}{2}$. Everything is coin tossing! Again, W_∞ is neither discrete nor 'continuous'.

▶▶ **I. Sharpening our intuition.** Consider a Random Walk W on $\mathbb{Z}$:

$$W_n := a + X_1 + \cdots + X_n,$$

where $X_1, X_2, \ldots$ are IID with

$$\mathbb{P}(X_n = -1) = \tfrac{2}{7}, \quad \mathbb{P}(X_n = 1) = \tfrac{2}{7}, \quad \mathbb{P}(X_n = 2) = \tfrac{3}{7}.$$

Since the Random Walk W can only go down by 1 unit at any step, we have, for $k \geq 1$,

$$\mathbb{P}_k(\text{hit 0}) = \alpha^k$$

for some α. Compare subsection 118D. We have

$$\mathbb{P}_1(\text{hit } 0) = \tfrac{2}{7} + \tfrac{2}{7}\mathbb{P}_2(\text{hit } 0) + \tfrac{3}{7}\mathbb{P}_3(\text{hit } 0),$$

so that

$$\alpha = \tfrac{2}{7} + \tfrac{2}{7}\alpha^2 + \tfrac{3}{7}\alpha^3, \quad (\alpha - 1)(3\alpha - 1)(\alpha + 2) = 0.$$

Since $\mathbb{E}(X_n) = \tfrac{6}{7} > 0$, the Random Walk will drift to $+\infty$, so we must have $\alpha < 1$. Hence,

$$\alpha = \tfrac{1}{3}.$$

For $k \geq -1$, noting the definition carefully, let

$$y_k := \mathbb{P}_0\left(W_n = k \text{ for some } n \geq 0\right).$$

Then $y_{-1} = \alpha = \tfrac{1}{3}$, $y_0 = 1$ and, for $k \geq 1$ (but *not*, note, for $k = 0$),

$$y_k = \tfrac{2}{7}y_{k+1} + \tfrac{2}{7}y_{k-1} + \tfrac{3}{7}y_{k-2}.$$

The auxiliary equation (see subsection 118C)

$$\tfrac{2}{7}m^3 - m^2 + \tfrac{2}{7}m + \tfrac{3}{7} = 0$$

has roots $1, -\tfrac{1}{2}, 3$ (the inverses of those for the α equation), so we have

$$y_k = A + B\left(-\tfrac{1}{2}\right)^k + C\left(3\right)^k, \quad k \geq -1.$$

Since $0 \leq y_k \leq 1$ for every k, C must be 0. Next,

$$1 = y_0 = A + B, \quad \tfrac{1}{3} = y_{-1}1 = A - 2B, \quad A = \tfrac{7}{9}, \quad B = \tfrac{2}{9}.$$

The return probability r to 0 satisfies

$$r = \tfrac{2}{7}y_1 + \tfrac{2}{7}\alpha + \tfrac{3}{7}\alpha^2 = \tfrac{1}{3}.$$

Hence, if the process starts at 0, *as we now assume it does*, then the total number of visits to 0 is $1/(1 - r) = 3/2$.

If k is very large, then the probability y_k that k is hit is very close to $A = \tfrac{7}{9}$; and if k is hit, then the process spends on average a total time $3/2$ at k. Hence, for large K,

$$\mathbb{E}_0\left[\text{total time spent in states } 0, 1, 2, \ldots, K\right] \sim K \times \tfrac{7}{9} \times \tfrac{3}{2} = \tfrac{7}{6}K.$$

This tallies exactly with the fact that since $\mathbb{E}(X_n) = 6/7$, the process drifts to infinity at rate $6/7$, so will hit K at about time $7K/6$. The total time spent in states less than 0 and the period between the first hit on K and the last visit to $0, 1, 2, \ldots, K$ are negligible compared with the total time in $0, 1, 2, \ldots, K$.

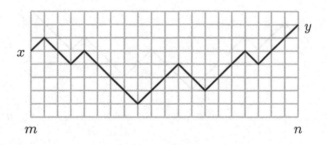

Figure J(i): A simple path from (m, x) to (n, y)

J. Counting simple paths. We are now going to look at some counting arguments for Random Walks. Many of the things we do in the remainder of this section via counting can be done efficiently by other arguments. However, not all can: for some problems, we *need* the counting arguments.

Figure J(i) shows a 'simple path' joining (m, x) to (n, y), where m, n, x, y are integers with $n > m$. Each piece of such a simple path on the 'integer grid' has slope 1 or -1. We think of the horizontal axis as the time-axis.

▶ **Ja. Lemma.** *Suppose that m, n, x, y are integers with $n > m$. Then the number of simple paths from (m, x) to (n, y) is*

$$\binom{n - m}{u}, \quad \text{where } u = \frac{n - m + y - x}{2}.$$

If u is not an integer in the range $\{0, 1, 2, \ldots, n - m\}$, then the binomial coefficient is 0: of course, if there is a simple path from (m, x) to (n, y), then $n - m$ must have the same parity as $y - x$.

Proof. To get from (m, x) to (n, y) requires u 'ups' and d 'downs', where

$$u + d = n - m, \quad u - d = y - x, \quad \text{whence } u = \frac{n - m + y - x}{2}.$$

The u 'ups' can occupy any subset of size u of the $n - m$ available time-intervals. □

▶ **K. The Reflection Principle.** *Suppose that $x > 0$, $y > 0$ and $n > m$. Then the number of simple paths from (m, x) to (n, y) which touch or cross the time-axis ('$x = 0$') equals the total number of simple paths from (m, x) to $(n, -y)$, equivalently, the total number of paths from $(m, -x)$ to (n, y).*

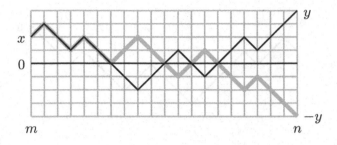

Figure K(i): Reflection Principle illustrated

Proof. For any path from (m, x) to (n, y) which touches or crosses the time-axis, we can reflect the path in the time-axis from the first time it hits the time-axis on. In the figure, the original path is a thin black line, the transformed path the wide grey line, the two paths agreeing until the time-axis is hit. You can see that this 'reflection' sets up a one-one correspondence between the two sets of paths which we want to prove have the same cardinality.   □

▶ **L. The Ballot Theorem.** *Consider an election between two candidates A and B in which A gets a votes and B gets b votes, where $a > b$. Suppose that the ballot slips on which people register their votes are arranged randomly before the count takes place. Then the probability that A is strictly in the lead from the first ballot slip counted right through the count is $(a - b)/(a + b)$.*

Please note that this is an important result, not in any way confined to its 'election' context. Clearly one feels that such a simple answer must have a simple explanation. I shall provide one in Chapter 9. In the meantime, ...

La. Exercise. Prove the Ballot Theorem. [*Hints.* Think of a simple path as going up [down] for every ballot slip in favour of A [B]. What is the total number of paths from $(0, 0)$ to $(a + b, a - b)$? How many paths go from $(1, 1)$ to $(a + b, a - b)$? How many such paths do so without touching or crossing the axis?]

Also find the probability that B is never in the lead throughout the count. (Paths may now touch the time axis.)

Lb. Exercise: Application to return times. Consider SRW(p) started at 0. Let T be the first time that the particle returns to 0 with (standard convention) $T = \infty$ if the particle never returns there. Prove that

$$\mathbb{P}_0(T_0 = 2n) = \frac{2}{2n - 1}\binom{2n - 1}{n}p^n q^n.$$

Hints. If $T = 2n$, then W_{2n-1} is either 1 or -1. Use the Ballot Theorem to calculate the conditional probability that $W_k \neq 0$ for $1 \leq k \leq 2n - 1$ given that $W_{2n-1} = 1$.

Lc. Exercise: Maximum of SRW. Let W be SRW(p), and define $M_n := \max_{k \leq n} W_k$. Prove that, for $m \geq w$ and $m \geq 0$,

$$\mathbb{P}_0\left(M_n \geq m; W_n = w\right) = \binom{n}{u} p^v q^{n-v}$$

where $u := m + \frac{1}{2}(n - w)$ and $v = \frac{1}{2}(w + n)$.

Deduce that if $p = q = \frac{1}{2}$ and $w \leq m$, then

$$\mathbb{P}_0\left(M_n = m; W_n = w\right) = \mathbb{P}_0\left(W_n = 2m - w\right) - \mathbb{P}_0\left(W_n = 2m + 2 - w\right).$$

and then that

$$\mathbb{P}(M_n = m) = \mathbb{P}_0(W_n = m) + \mathbb{P}_0(W_n = m + 1).$$

4.5   A simple 'strong Markov principle'

ORIENTATION. You might wish to skip this section on a first reading.

We now take a closer look at equation 119(D2), where we claimed that for SRW(p),

$$\mathbb{P}_2(\text{hit } 0) = \mathbb{P}_2(\text{hit } 1)\mathbb{P}_1(\text{hit } 0).$$

We claimed therefore that the process starts afresh at the first time T that it hits 1. But T is a *random* time, and our claim, however intuitively appealing, demands proof. (A further complication is that T might be infinite: in other words, the particle may never hit 1 when started from 2.)

To make sense of starting afresh needs a little thought, but **it involves fundamental ideas which will reappear later – as fundamental ideas do.** This is why this topic is having a whole section to itself. Perhaps the first use of stopping times was in Statistics – in Wald's *sequential sampling* (described at 416H). We need the ideas in the next section for very practical simulation issues.

For what kind of random time will the process start afresh? Suppose that we consider coin tossing, the coin being tossed at time $1, 2, 3 \ldots$. Let T be $U - 1$, where U is the time of the first Head after time 1. The process certainly does not start afresh at this time T, because the next toss will definitely produce a Head. Here T is 'cheating' because it looks forwards in time.

We need to formulate the idea of a *stopping time* T, that is, a random time which 'depends only on history up to the present'. We then need to describe the class $\mathcal{F}_T$ of events which describe history up to a stopping time T, and then we want to prove that things start afresh at such a stopping time. The *intuitive* ideas of stopping times and of $\mathcal{F}_T$ are easy, and are what I want you to understand. I have put them in bold type.

OK, let's take it in stages.

▶▶ **A. Stopping time T; history $\mathcal{F}_n$ up to time n.** The reason for the name 'stopping time' will become clear in Chapter 9.

The situation with which we are concerned is the following. Suppose that we have a stochastic process **X**, that is, a sequence $X_1, X_2, X_3, \ldots$ of RVs. Suppose that T is an RV taking values in $\{1, 2, 3, \ldots; \infty\}$. **Think of the information available at time n as being knowledge of the values $X_1, X_2, \ldots, X_n$.**

Intuitively, T is a stopping time if whether or not $T \leq n$ can be calculated from the values $X_1, X_2, \ldots, X_n$. In our discrete-time setting, it is easier to say that T is a stopping time if whether or not $T = n$ can be calculated from the values $X_1, X_2, \ldots, X_n$. Thus,

$$T := \min\{n : X_n > a\},$$
$$T := \min\{n : S_n > a\}, \quad \text{where } S_n := X_1 + X_2 + \cdots + X_n,$$
$$T := \min\{n : X_{n-1} > a\}$$

are stopping times. **Generally, a stopping time is the first time that something observable by that time happens.**

Here is an example which is *not* a stopping time:

$$T := \min\{n : X_{n+1} > a\}.$$

We don't know X_{n+1} at time n.

We write $\mathcal{F}_n$ for all the collection of all events which depend only on the values $X_1, X_2, \ldots, X_n$. Each such event is of the form

$$\{\omega : (X_1(\omega), X_2(\omega), \ldots, X_n(\omega)) \in U\}$$

where U is a nice subset of $\mathbb{R}^{n+1}$; and here, 'nice' means precisely 'Borel' – see 45L. The class $\mathcal{F}_n$ is then a σ-algebra. We think of $\mathcal{F}_n$ as the mathematical formulation of the history of X up to time n.

Formally then, T is defined to be a stopping time relative to the evolution described by $\{\mathcal{F}_n : n \in \mathbb{N}\}$ if for every n, the event $\{T \leq n\}$ is in $\mathcal{F}_n$, equivalently if, for every n, the event $\{T = n\}$ is in $\mathcal{F}_n$.

Ⓜ If $T := \min\{n : X_n > a\}$, then

$$\{\omega : T(\omega) = n\} = \{\omega : (X_1(\omega), \ldots, X_n(\omega)) \in U\}$$

where

$$U := \{(x_0, \ldots, x_n) \in \mathbb{R}^{n+1} : x_k \leq a \, (1 \leq k < n); \, x_n > a\}.$$

The set U is Borel. Indeed,

$$U = \left(\bigcap_{1 \leq k < n} \{x_k \leq a\}\right) \cap \{x_n > a\},$$

and the sets $\{x_k \leq a\}$ are closed (hence Borel) and $\{x_n > a\}$ is open (hence Borel). As a finite intersection of Borel sets, U is Borel.

▶ **B. History $\mathcal{F}_T$ up to time T.** We know that $\mathcal{F}_n$ represents the history up to fixed time n. But if we want to formulate and prove the idea that things start afresh at time T, we have first to formulate the history $\mathcal{F}_T$ up to time T. So when will an event H belong to $\mathcal{F}_T$? The answer is simple (once you get used to it): *we say that*

$$H \in \mathcal{F}_T$$

if and only if, for each n,

$$H \cap \{T \le n\} \in \mathcal{F}_n;$$

equivalently, if and only if, for each n,

$$H \cap \{T = n\} \in \mathcal{F}_n.$$

So, $H \in \mathcal{F}_T$ if, for every n, for those outcomes for which $T = n$, you will know by time n whether or not H has occurred.

If $T := \min\{n : X_n > a\}$, and $b < a$, then

$$H := \{X_n < b \text{ for some } n < T\}$$

is in $\mathcal{F}_T$.

▶ **C. A simple 'Strong Markov Principle'.** *Let the process $X = (X_1, X_2, \ldots)$ be an IID sequence, each X_k having distribution function F. Let T be a stopping time for X, and let $H \in \mathcal{F}_T$. If $T(\omega)$ is finite, then we define*

$$X_{T+r}(\omega) := X_{T(\omega)+r}(\omega).$$

(a) Suppose first that $T(\omega)$ is finite with probability 1. Then the variables

$$I_H, X_{T+1}, X_{T+2}, \ldots$$

are independent; and each X_{T+k} has DF F. To summarize:

$$X_{T+1}, X_{T+2}, \ldots \text{ is an IID sequence with common DF } F$$

and is independent of any event in $\mathcal{F}_T$. *For any k, for $x_1, x_2, \ldots, x_k \in \mathbb{R}$, and for $H \in \mathcal{F}_T$, we have*

$$\mathbb{P}\left(H; X_{T+1} \le x_1; X_{T+2} \le x_2; \ldots; X_{T+k} \le x_k \right)$$
$$= \mathbb{P}(H)F(x_1)F(x_2)\ldots F(x_k). \tag{C1}$$

(b) For general T for which $\mathbb{P}(T = \infty)$ might be positive, equation (C1) holds when H is replaced by $H \cap \{T < \infty\}$, and then there is no problem about the meaning of X_{T+1}, etc, because we restrict attention to the event that T is finite.

Proof. The short proof is easier than the statement to which we have built up.

We have

$$\mathbb{P}\left(H \cap \{T < \infty\}; X_{T+1} \leq x_1; \ldots; X_{T+k} \leq x_k \right)$$

$$= \sum_{1 \leq n < \infty} \mathbb{P}(H \cap \{T = n\}; X_{T+1} \leq x_1; \ldots; X_{T+k} \leq x_k) \qquad \text{(C2)}$$

$$= \sum_{1 \leq n < \infty} \mathbb{P}(H \cap \{T = n\}; X_{n+1} \leq x_1; \ldots; X_{n+k} \leq x_k)$$

However, *for fixed n*, $H \cap \{T = n\} \in \mathcal{F}_n$ by definition of $\mathcal{F}_T$: it therefore depends only on the RVs $X_1, X_2, \ldots, X_n$, and so is independent of $X_{n+1}, X_{n+2}, \ldots, X_{n+k}$. Thus the expression at (C2) is

$$\sum_{1 \leq n < \infty} \mathbb{P}(H \cap \{T = n\}) \, \mathbb{P}(X_{n+1} \leq x_1) \ldots \mathbb{P}(X_{n+k} \leq x_k)$$

$$= F(x_1)F(x_2)\ldots F(x_k) \sum_{1 \leq n < \infty} \mathbb{P}(H \cap \{T = n\})$$

$$= F(x_1)F(x_2)\ldots F(x_k)\mathbb{P}(H \cap \{T < \infty\}).$$

The proof is complete. $\qquad\qquad\qquad\qquad\qquad\qquad\qquad\qquad\qquad\qquad\qquad\quad\square$

D. The Random Walk example. We had

$$W_n = 2 + X_1 + X_2 + \cdots + X_n,$$

where $X_1, X_2, \ldots$ are IID, with $\mathbb{P}(X_k = 1) = p$ and $\mathbb{P}(X_k = -1) = q$. We had $T = \min\{n : W_n = 1\}$, and this is a stopping time for the X-process. Conditionally on $\{T < \infty\}$, which is trivially in $\mathcal{F}_T$, the variables $X_{T+1}, X_{T+2}, \ldots$ are IID each with the same distribution as X_1, and the process truly starts afresh at time T, leading to

$$\mathbb{P}_2 \,(\text{hit } 0 \,|\, \text{hit } 1) = \mathbb{P}_1 \,(\text{hit } 0)$$

as required.

The intuition about 'starting afresh' used in subsections 21E and 120E are also justified by our Strong Markov Principle.

4.6   Simulation of IID sequences

▶ **A. 'Random-number generators'.** The fundamental problem is to get the computer to produce a sequence $u_1, u_2, \ldots$ of numbers between 0 and 1 which one can regard as a simulation of a realization of values chosen independently each from the uniform $U[0, 1]$ distribution. What the computer actually does is generate a sequence of integers in a range $\{0, 1, 2, \ldots, M-1\}$ and then take $u_k = x_k/M$. The initial value x_1 is a '*seed*' which is either input by the user or calculated from the readings on the computer clock. The computer is programmed with a function which I'll call N (for 'next'), and it calculates the x-sequence recursively via $x_k = N(x_{k-1})$.

We shall discuss three types of generator. In the discussion, it will be convenient to use Euler's φ-function defined on positive integers as follows:

$$\varphi(n) := \#(C(n)), \quad \text{where } C(n) := \{m : 1 \le m < n; \ \mathrm{hcf}(m, n) = 1\}$$

where 'hcf' denotes highest common factor (or greatest common divisor). Thus, $C(n)$ is the set of positive integers less than n which are '*coprime*' to n, and $\varphi(n)$ is the number of elements in $C(n)$. For a prime p, we clearly have $\varphi(p) = p - 1$.

B. Multiplicative generators. The idea here is to take

$$x_k = a x_{k-1} \quad \mathrm{mod}\ M.$$

One of the cases in which this makes good sense is when M is a prime p and a is a 'primitive root' for p. (One of Fermat's theorems says that for $1 \le a < p - 1$, we have $a^{p-1} = 1 \mod p$. We say that a is a primitive root if $a^d \ne 1 \mod p$ for $1 \le d < p-1$.) Then whatever number x_1 is taken in $\{1, 2, \ldots, p - 1\}$, the set $\{x_1, x_2, \ldots, x_{p-1}\}$ is the set $\{1, 2, \ldots, p - 1\}$ arranged in some order: we have 'full period'.

For example, if $p = 3$ and $a = 2$, then x will cycle through the values (1 2), and if $p = 5$ and $a = 3$, then x will cycle through (1 3 4 2). By this, I mean in the latter case, for example, that if $x_1 = 3$, then $x_2 = 4$, $x_3 = 2$, $x_4 = 1$ – and then we had better stop. Of course in practice, the computer will use a huge prime p.

It is known from Number Theory that *a prime p has $\varphi(p - 1)$ primitive roots.*

C. 'Mixed' multiplicative generators (MMGs). No random-number generator is perfect. It has been suggested that defects can be 'averaged out' if one mixes a number of different random-number generators as follows.

Let the rth generator $(1 \le r \le s)$ generate a sequence $x_{1,r}, x_{2,r}, \ldots$ via the multiplicative generator

$$x_{k,r} = a_r x_{k-1,r} \quad \mathrm{mod}\ p_r,$$

where $p_1, p_2, \ldots, p_s$ are distinct primes, and for each r, a_r is a primitive root for p_r. For each r, we can think of $u_{1,r}, u_{2,r}, \ldots$, where $u_{k,r} = x_{k,r}/p_r$, as a perhaps crude random 'uniform' sequence. One then mixes these sequences by defining

$$u_k := (u_{k,1} + u_{k,2} + \cdots + u_{k,s}) \quad \mathrm{mod}\ 1,$$

the 'mod 1' signifying that we take the fractional part: $2.67 \mod 1 = 0.67$, for example.

Ca. Exercise. Prove that *if U has the uniform $U[0, 1]$ distribution and V is an RV independent of U with any distribution at all, then $(U + V) \mod 1$ has the $U[0, 1]$ distribution.*

So, there is good sense in the MMG idea, and it seems to work well in practice.

Even so, mathematicians will feel some unease because there is a sense in which mixing does not really achieve what it was intended to do. Note that

$$u_k = \frac{x_k}{M} \quad \text{where } M := p_1 p_2 \ldots p_s,$$

and

$$x_k = (p_2 p_3 \ldots p_s)\, x_{k,1} + \cdots + (p_1 p_2 \ldots p_{s-1})\, x_{k,s} \quad \bmod M.$$

It follows already that $x \in \mathcal{C}(M)$, and it is well known that

$$\varphi(M) = (p_1 - 1)(p_2 - 1) \ldots (p_s - 1).$$

According to the Chinese Remainder Theorem, there is a unique number a in $\{1, 2, \ldots, M - 1\}$ such that

$$a = a_r \quad \bmod p_r \quad \text{for } 1 \leq r \leq s.$$

We find that

$$x_k = a x_{k-1} \quad \bmod M,$$

so the new 'mixed generator' is still a multiplicative generator, but one *not* of full period, by which I mean not even of period $\varphi(M)$. The period, Period, of the generator is the lowest common multiple of $p_1 - 1$, $p_2 - 1$, $\ldots$, $p_s - 1$.

Consider mixing the two trivial examples we looked at earlier:

$$x_{k,1} = 2x_{k-1,1} \quad \bmod 3, \qquad x_{k,2} = 3x_{k-1,2} \quad \bmod 5,$$
$$x_k = 5x_{k,1} + 3x_{k,2}, \qquad M = 15.$$

We spot that $a = 8$ in the Chinese Remainder Theorem, so

$$x_k = 8x_{k-1} \quad \bmod 15,$$

and x will cycle either through (1 8 4 2) or (14 7 11 13) depending on which cycle contains the seed. This is, of course, an extreme example.

In general, $\mathcal{C}(m)$ will be partitioned into $\varphi(M)/\text{Period}$ sets each of Period elements and the x values will cycle through one of these sets.

D. Congruential generators.

The idea here is to use a recurrence

$$x_k = a x_{k-1} + b \quad \bmod M.$$

Usually, M is taken to be a power of 2 for ease of computer calculation.

Lemma. *If $M = 2^r$, where $r \geq 2$ and*

$$b \text{ is odd}, \qquad\qquad 0 < b < 2^r,$$
$$a = 1 \quad \bmod 4, \qquad 0 < a < 2^r,$$

and we set $M = 2^r$,

$$x_k := ax_{k-1} + b \mod M,$$

then, for any $x \in \{0, 1, 2, \ldots, M - 1\}$, the numbers $x_1, x_2, \ldots, x_M$ are the numbers $0, 1, 2, \ldots, M - 1$ in some order: we have full period.

For proof, see Ripley [196].

E. Discussion. *The period is just one of many factors to be considered in assessing the appropriateness of a random-number generator.* For example, if we take $a = 1$ and $b = 1$ in the Lemma, we obtain full period but a sequence which we can hardly think of as random!

Generators must be subjected to lots of tests: see Ripley [196]. Not only must $(u_k : k \in N)$ be uniform, but also

$$\{(u_k, u_{k+1}, \ldots, u_{k+d}) : k \in \mathbb{N}\}$$

must be uniform in the d dimensional cube $[0, 1]^d$. It is important to check this for values of d up to (say) 6; some early generators failed this test. Testing out the generator on known results is of course useful. We tested two generators on the 'Waiting for HH' example.

Warning. Ripley's book alerts us to the possibility of very disturbing 'lattice structures' in multiplicative and congruential generators and 'spiral structures' in the most common way of generating normally distributed RVs.

▶ **F. Generators used in this book.** In each simulation done in 'C' in this book, I use one of two random-number generators, both of which are programmed into RNG.o as described below.

One generator, due to Wichmann and Hill, is the mixed multiplicative generator with

$$s = 3; \; a_1 = 171, \; p_1 = 30269; \; a_2 = 172, \; p_2 = 30307;$$
$$a_3 = 170, \; p_3 = 30323.$$

My thanks to the Stats group at Bath for telling me about this one. Any worries one has (I had!) because of the 'mathematical' reasons discussed above are completely dispelled by the fact that very sensible people are happy with the generator (and have tested it in several ways) and by the sheer 'scale' of the thing. As a single multiplicative generator, it reads

$$x_k = ax_{k-1} \mod M, \text{ where } a = 16555425264690,$$
$$M = 27817185604309, \quad \text{period} = 6953607871644.$$

The set $\mathcal{C}(M)$ is partitioned into 4 disjoint cycles in the sense explained earlier.

The other generator I use is a congruential generator which I saw in a paper by the great number-theorist, Langlands, and co-workers, testing by simulation absolutely astonishing predictions by the theoretical physicist Cardy on Percolation, one of the most challenging branches of Probability. See Langlands, Pouliot and Saint-Aubin [145], and the book [102] by Grimmett. The generator reads

$$x_k = ax_{k-1} + b \mod m,$$

where

$$a = 142412240584757, \quad b = 11, \quad m = 2^{48} = 281474976710656.$$

By our Lemma, it has full period m. The percolation problems are very subtle, and I was impressed by the very close agreement between the simulation results and Cardy's predictions. The random-number generator is getting right some extremely complex probabilities.

I simulated 'Waiting for HHH' for a coin with probability 0.25 of Heads by Wichmann-Hill and Langlands, performing the experiment 1000000 times. Results:

    `Wichmann-Hill(2561, 7539, 23307)`: mean 83.93570;

    `Langlands(9267, 47521, 37551)`: mean 84.03192.

By Exercise 22Ga, the true answer is 84. Moreover, the (true) standard deviation of the average of a million waiting times is 0.081756 (to 6 places). Each of the two generators is here very good on something which could catch generators out.

But it is always a good idea to have two or more generators at one's disposal, especially in the light of the Warning.

▶ **G. C code for a random-number generator.**   Please note that there is an explanation of 'static' variables in 'C' at Appendix A7, p499. Part of the C code for the random-number generator I use is contained in a header file which begins

```
/*  RNG.h                           DW  */
#if defined RNG_h
#else
#define RNG_h

#define E 2.718281828459
/* #define PI 3.141592653590 necessary on some machines */
int WhichGen;   /*decides which generator*/
void setseeds();
void ShowOldSeeds();
double Unif(); /*Sample from U[0,1]*/
int WithProb(double p); /*Gives 1 with probability p, 0 otherwise*/
```

The corresponding beginning of the program RNG.c – some of which was written by Bill Browne and some by me – reads:

```
/* RNG.c - William Browne and David Williams
   Contains some random number generation routines
```

```
  cc -c RNG.c -o RNG.o                    to compile
*/

#include <stdio.h>
#include <math.h>
#include "RNG.h"
/* Following used by 'Langlands' generator */
static const C2 = 33157;
static const C1 = 61187;
static const C0 = 53;
static const M = 65536.0;

/* x0,...,x3 current 'seeds'; oldx0, etc, original seeds */
static unsigned long int x0, x1, x2, x3, oldx0, oldx1, oldx2, oldx3;
static double Msq;

void setseeds(){
  printf("Which generator?");
  printf("\nInput 1 for Wichmann-Hill, 2 for Langlands\n");
  scanf("%d", &WhichGen);
  printf("\n\nInput 3 positive-integer seeds\n");
  if (WhichGen==1){
    printf("a<30269, b<30307, c<30323\n\n");
    scanf("%d%d%d", &oldx1, &oldx2, &oldx3);
    x1 = oldx1;   x2 = oldx2;  x3 = oldx3;
  }
  else{
    printf("a,b,c < 65536\n");
    scanf("%d%d%d", &oldx0, &oldx1, &oldx2);
    x0 = oldx0;   x1 = oldx1;  x2 = oldx2;
    Msq = (double) M * M;
  }
}

void ShowOldSeeds(){
  printf("\n\nGenerator used was ");
  if(WhichGen==1)
   printf("Wichmann-Hill(%7d, %7d,%7d)", oldx1, oldx2, oldx3);
  else printf("Langlands(%7d,%7d,%7d)", oldx0, oldx1, oldx2);
}

double WHunif(){
   x1 = (171 * x1)%30269; x2 = (172 * x2)%30307;
   x3 = (170 * x3)%30323;
   return fmod(x1/30269.0 + x2/30307.0 + x3/30323.0, 1.0);
}
```

```
double Lunif(){
  unsigned long y0, y1, y2, t0, t1, t2, u1, u2, w2;
  double temp;
      t0 = C0*x0 + 11;   y0 = t0 % M;   u1 = t0/M;
      t1 = C1*x0 + C0*x1 + u1; y1 = t1 % M; u2 = t1/M;
      w2 = (C1 * x1)%M;
      t2 = C2*x0 + w2 + C0*x2 + u2; y2 = t2 % M;
      x0= y0; x1 = y1; x2 = y2;

  temp = (double) x2 * M + x1;
  return((temp+0.5)/Msq);
}


double Unif(){
  if (WhichGen == 1) return WHunif();
  else return Lunif();
}


int WithProb(double p){
   if (Unif() <= p) return 1; else return 0;
}
```

From now on, I quote parts of RNG.c only, the corresponding parts of RNG.h being obvious.

▶ **H. Simulating from exponential distributions.**   Compare Exercise 55Fa. Let $\lambda > 0$. If U has the uniform U$[0,1]$ distribution, then $V := -(\ln U)/\lambda$ has the exponential E(rate λ) distribution because, for $x \geq 0$,

$$\mathbb{P}(V > x) = \mathbb{P}(\ln U < -\lambda x) = \mathbb{P}(U < e^{-\lambda x}) = e^{-\lambda x}.$$

This is applying the F-inverse principle at 50B but to the uniform $1 - U$ instead of U. We therefore have, for generating E(rate λ) variables:

```
double Expl(double lambda){
  return (-log(Unif())/lambda);
}
```

The two-sided exponential distribution E2S(rate λ).   The exponential distribution E(rateλ) has pdf $\lambda e^{-\lambda x}$ on $[0, \infty)$. The two-sided exponential distribution E2S(rate λ) has pdf $\frac{1}{2}\lambda e^{-\lambda|x|}$ on $\mathbb{R}$. The idea is explained in

```
double E2S(double lambda){
  double oneside = Expl(lambda);
  if (WithProb(0.5)==1) return oneside; else return (-oneside);
}
```

▶ **I. Rejection Sampling.** As remarked earlier, the F-inverse principle is not that efficient for sampling from most distributions. Rejection sampling is a useful general technique. Lemma 139J provides the theoretical basis. Here's the idea.

We wish to obtain a sample corresponding to pdf f. Suppose that we have a means of simulating from a distribution with pdf g where

$$f(x) \le Kg(x) \quad \text{for all } x. \tag{I1}$$

Every time the computer provides us with a value z corresponding to pdf g, we accept z with probability $v(z) := f(z)/(Kg(z))$; otherwise, we reject z. Then the accepted values $x_1, x_2, \ldots$ (say) form a sequence corresponding to IID values from the distribution with pdf f. On average, it takes K z-values to get each x-value, so we want K to be small.

Example. Suppose that f is the pdf of the standard normal distribution (or Gauss distribution) N$(0, 1)$, so that

$$f(x) = \frac{1}{\sqrt{2\pi}} \exp\left(-\tfrac{1}{2}x^2\right),$$

and take g to be the pdf of the E2S(λ) distribution: $g(x) = \tfrac{1}{2}\lambda \exp(-\lambda|x|)$. Then

$$\sup_{x \in \mathbb{R}} \frac{f(x)}{g(x)} = \sup_{x > 0} \frac{2}{\lambda\sqrt{2\pi}} \exp\left(-\tfrac{1}{2}x^2 + \lambda x\right) = \frac{2}{\lambda\sqrt{2\pi}} \exp\left(\tfrac{1}{2}\lambda^2\right)/\lambda,$$

the supremum being achieved when $x = \lambda$. To obtain the smallest K, we need to minimize the last expression, equivalently, to minimize its logarithm, and we find that

$$\frac{\mathrm{d}}{\mathrm{d}\lambda}\left(\tfrac{1}{2}\lambda^2 - \ln\lambda\right) = \lambda - \frac{1}{\lambda} = 0,$$

so $\lambda = 1$. Thus, pick z according to pdf $\tfrac{1}{2}e^{-|x|}$, and accept z with probability $v = 2\exp\left(-\tfrac{1}{2}z^2 + z\right)/\sqrt{2\pi}$. This gives us a realization of IID N$(0, 1)$ variables $X_1, X_2, \ldots$. For each X, we have to try on average $\sqrt{2e/\pi} = 1.315$ values of z. Rearranging so as to avoid the use of $|\cdot|$, we can use

```
double aGauss(){    /* simulates N(0,1) variable */
  double z,v,w;
  do{z = Expl(1.0);
     w = z - 1.0;
     v = exp(-w*w/2.0);
  }while (WithProb(v)==0);
  if (WithProb(0.5) == 1) return z; else return (-z);
}
```

Do check that the above rearrangement works.

Results obtained in conjunction with Figure I(i):

```
Generator used was Wichmann-Hill(3718,6734,23115)
Number of z's per x = 1.3154
Empirical P(RV<=1.960)=0.9750
Empirical P(RV<=2.576)=0.9951
```

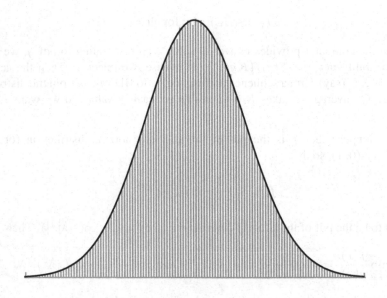

Figure I(i): Histogram from aGauss()

Figure I(i) shows a 128-bin histogram obtained from a sample of size 1 million from aGauss() with the normal curve superimposed. The little spikes at the ends correspond to tail probabilities. For an N(0, 1) variable N, $\mathbb{P}(N \leq 1.96) = 0.975$. The proportion of simulated X-values with $X \leq 1.96$ was, as stated, 0.9750. From this and the corresponding result for 0.995, we see that aGauss() is getting the small tails correct. The number of z-values used in the simulation shows spectacular agreement with the theoretical expectation.

The computer 'rejects z with probability v' by choosing a U[0, 1] representative u and rejecting z if $u \leq v$. Lemma J provides full theoretical justification for the method.

Ia. Speeding up Rejection Sampling. Again suppose that we have $f(\cdot) \leq Kg(\cdot)$. One way to speed up Rejection Sampling is to pick 'lower' and 'upper' functions $a(\cdot)$ and $b(\cdot)$ with

$$a(z) \leq v(z) := \frac{f(z)}{Kg(z)} \leq b(z)$$

where $a(\cdot)$ and $b(\cdot)$ are 'close' and where they are simple functions which may be evaluated very rapidly. We can then adopt the strategy:

Repeat{
    Choose z according to pdf g,
    Choose u according to U$[0, 1]$;
    If $u > b(z)$ then reject z;
    Else if $u < a(z)$ accept z;
    Else if $u < v(z)$ accept z;
} until z is accepted.

This type of idea is utilized in Chapter 7.

Ib. Adaptive Rejection Sampling. This is a much cleverer idea (due to Gilks and Wild, and in more general form to Gilks) which works for log-concave functions f, that is for functions f for which $\ln(f)$ is concave, that is, $-\ln(f)$ is convex.

The idea is to keep refining the pdf g (in our original version of Rejection Sampling) to bring it closer and closer to f, utilizing numbers already calculated to update the piecewise-linear function $\ln(g)$. I wish I had space to explain further. See Gilks [95] and the references therein.

Ic. Exercise. Check the result of Exercise 64Ja by simulation. (A solution is given in Chapter 11.)

J. Lemma. (The statement of the lemma looks much more complicated than it is!) *Suppose that f and g are pdfs, and that 137(I1) holds for some K. Let $Z_1, Z_2, \ldots$ be IID RVs with pdf g and $U_1, U_2, \ldots$ be an IID U$[0, 1]$ sequence independent of the (Z_k) sequence. Let $V_n := f(Z_n)/Kg(Z_n)$. Define $T_0 := 0$ and for $k = 1, 2, 3, \ldots$, define*

$$T_k := \min\{n : n > T_{k-1}; \ U_n \leq V_n\}, \quad X_k := Z_{T_k}.$$

Then $X_1, X_2, \ldots$ is a IID sequence of RVs with common pdf f. **Addendum.** Define the event $A_n := \{U_n \leq V_n\}$ that Z_n is accepted. Then the events $A_1, A_2, \ldots$ are independent, each with probability $p := 1/K$. The gaps $T_1 - T_0, \ T_2 - T_1, \ T_3 - T_2, \ldots$ between acceptance times are independent each with the geometric distribution of mean K:

$$\mathbb{P}(T_k - T_{k-1} = m) = q^{m-1}p \quad (m = 1, 2, 3 \ldots),$$

where $q := 1 - p$. The IID sequence $X_1, X_2, \ldots$ is independent of the IID sequence $T_1 - T_0, \ T_2 - T_1, \ T_3 - T_2, \ldots$.

Partial proof. We have, with '$X \in dx$' thought of as '$x < X \leq x + dx$' as usual,

$$\mathbb{P}(T_1 = 1; X_1 \in dx) = \mathbb{P}(Z_1 \in dx; \ U_1 \leq f(x)/Kg(x))$$

$$= g(x)dx\frac{f(x)}{Kg(x)} = pf(x)dx \text{ where } p := \frac{1}{K}.$$

Hence, the probability that Z_1 is accepted is

$$\mathbb{P}(T_1 = 1) = \int_x \mathbb{P}(T_1 = 1; X_1 \in dx) = p\int f(x)dx = p.$$

Next, we have, using 'starting afresh at the fixed time 1' which does not need the Strong Markov Principle,

$$\mathbb{P}(T_1 = 2; X_1 \in dx) = \mathbb{P}(Z_1 \text{ not accepted})\,\mathbb{P}(T_1 = 1; X_1 \in dx)$$
$$= (1-p)pf(x)dx = qpf(x)dx \quad (q := 1-p),$$

and $\mathbb{P}(T_1 = 2) = qp$. And so on. We therefore have

$$\mathbb{P}(X_1 \in dx) = \sum_{1 \le m < \infty} \mathbb{P}(T_1 = n; X_1 \in dx) = \sum q^{n-1}pf(x)dx = f(x)dx.$$

We see that T_1 and X_1 are independent. All the other independence properties follow from the Strong Markov Principle since if we regard the information $\mathcal{F}_n$ available at time n as being knowledge of the values of

$$Z_1, U_1, Z_2, U_2, \ldots, Z_n, U_n,$$

then each T_k is a stopping time relative to the evolution determined by the $\mathcal{F}_n$. $\qquad\square$

Ja. Exercise. This is a place to have another think, if necessary, about the last part of the car-convoy problem at Exercise 103D.

GENERATING FUNCTIONS;

AND THE CENTRAL LIMIT THEOREM

ORIENTATION. Sums of IID RVs are of central importance in both Probability and Statistics. Pmfs and pdfs are not that easy to use: for example it is already rather tricky to find the pdf of the sum of n IID RVs each with the uniform $U[0, 1]$ distribution. To handle sums of IID RVs effectively, we need new methods; and **generating functions** provide these.

The main result in this chapter is the miraculous **Central Limit Theorem (CLT)** which says that the sum of a large number of IID RVs each with ANY distribution of finite variance has approximately a *normal* distribution. The **normal distribution** is therefore very special in this sense: we discover why. (We shall see later that the normal distribution is very special in several other ways, too.) There are at least three nice methods of proving the CLT, the first by generating functions, the second by using the fact that one can find the sequence of sums of IID Variables within a 'Brownian-motion' process, and the third (not quite so general) by using the maximum-entropy property of the normal distribution. We take a heuristic look at the first of these. Section I.8 of Volume 1 of Rogers & Williams [199] describes the second method: it requires much more background material. For hints on the maximum-entropy method, see 198H below. *Historical note:* The first case of the CLT was proved by de Moivre in 1706!

Some results in this chapter are too difficult to *prove* in full at this level; but *precise* statements (and, sometimes, sketched proofs) of the results are given.

The chapter contains many important exercises, but they are all rather routine, I'm afraid – not much challenge here. But *do* them, nevertheless.

5.1   General comments on the
use of Generating Functions (GFs)

In this book, we mainly use three types of Generating Functions:
▶ **the probability generating function (pgf)**

$$g_X(\alpha) := \mathbb{E}\left(\alpha^X\right) \quad (0 \leq \alpha \leq 1)$$

for $\mathbb{Z}^+$-valued X;
▶ **the Moment Generating Function (MGF)**

$$M_X(\alpha) := \mathbb{E}\left(e^{\alpha X}\right)$$

for suitable X and suitable α;
▶ **the Characteristic Function (CF)**

$$\varphi_X(\alpha) := \mathbb{E}\left(e^{i\alpha X}\right)$$

valid for all X and all real α.

As usual, F_X will denote the Distribution Function of X.

For any generating function $G_X(\alpha)$ of X, we need the following.

▶ **1. A Uniqueness Theorem**: each of the functions F_X and G_X (on the appropriate domains) determines the other. Of course, F_X always determines G_X. The result in the other direction follows from the Convergence Theorem described below.

▶ **2. An effective method of calculating from** G_X **the moments** $\mu_r(X) :=$ $\mathbb{E}\left(X^r\right)$ $(r = 1, 2, 3, \ldots)$ of X which exist.

▶ **3. An 'Independence means Multiply' Theorem**: if X and Y are independent, then
$$G_{X+Y}(\alpha) = G_X(\alpha)G_Y(\alpha). \tag{1}$$

Obviously, by combining results 1 and 3, we can in principle find the distribution of the sum of two independent RVs.

▶ **4. A Convergence Theorem**: we have $G_{X_n}(\alpha) \to G_X(\alpha)$ for all relevant α if and only if $F_{X_n}(x) \to F_X(x)$ at every point x at which $F_X(x)$ is continuous. Such continuity points of F are dense in $\mathbb{R}$, so this is good enough. (If for every ω, $X_n(\omega) = 1/n$ and $X(\omega) = 0$, then $F_{X_n}(0) = 0$ for every n, but $F_X(0) = 1$. This explains why 'convergence at continuity points of F_X' is the right condition.)

Remark. We shall *prove* the relevant properties only for the 'pgf' case. It requires too much Analysis to deal with the other cases.

We shall use two other types of generating function: the **Laplace Transform (LT)**

$$L_X(\alpha) := \mathbb{E}\left(e^{-\alpha X}\right) \quad (\alpha \geq 0)$$

for RVs with values in $[0, \infty)$ and the **Cumulant Generating Function (CGF)**

$$C_X(\alpha) := \ln M_X(\alpha).$$

Of course, we have 'Independence means *Add*' for CGFs.

5.2 Probability generating functions (pgfs)

▶▶ **A. Probability generating function (pgf) $g_X(\cdot)$ of X.** Let X be a $\mathbb{Z}^+$ RV taking values in $\mathbb{Z}^+ := \{0, 1, 2, \ldots\}$. We define the probability generating function (pgf) $g_X(\cdot)$ of X on $[0, 1]$ via

$$g_X(\alpha) := \mathbb{E}\left(\alpha^X\right) = \sum_{k=0}^{\infty} \alpha^k \mathbb{P}(X = k), \quad 0 \leq \alpha \leq 1.$$

Note that

$$g_X(1) = 1, \quad g_X(0) = \mathbb{P}(X = 0).$$

In this section, it is understood that we deal only with $\mathbb{Z}^+$ RVs.

▶ **Aa. Example: pgf of Poisson(λ).** If $\lambda > 0$ and X has the Poisson(λ) distribution, then

$$\mathbb{P}(X = k) = e^{-\lambda}\frac{\lambda^k}{k!} \quad (k \in \mathbb{Z}^+),$$

so that

$$g_X(\alpha) = e^{-\lambda}\sum_{k=0}^{\infty}\frac{(\alpha\lambda)^k}{k!} = e^{-\lambda}e^{\alpha\lambda}$$

and

$$g_X(\alpha) = e^{\lambda(\alpha-1)}.$$

a result of considerable importance.

▶ **B. The Uniqueness Theorem for pgfs.** *If we know the values $g_X(\alpha)$, $\alpha \in [0, 1]$, then we can find $\mathbb{P}(X = k)$ for $k \in \mathbb{Z}^+$. This follows from the Convergence Theorem E below.*

▶ **C. Finding moments.** We have – with discussion of rigour below –

$$g_X(\alpha) = \mathbb{E}\left(\alpha^X\right) = \sum \alpha^k \mathbb{P}(X = k),$$

$$g_X'(\alpha) = \mathbb{E}\left(X\alpha^{X-1}\right) = \sum k\alpha^{k-1}\mathbb{P}(X = k),$$

$$g_X'(1) = \mathbb{E}(X) = \sum k\mathbb{P}(X = k) \leq \infty,$$

$$g_X''(\alpha) = \mathbb{E}\left(X(X-1)\alpha^{X-2}\right),$$

$$g_X''(1) = \mathbb{E}\left(X(X-1)\right) \leq \infty,$$

so we can find $\mathbb{E}(X)$, $\mathbb{E}\left(X^2\right)$, and hence, if these are finite, $\text{Var}(X)$, from g_X. Rigour is provided by standard results on differentiation of power series together with monotonicity properties arising from the fact that the coefficients in the power series determining a pgf are non-negative. Of course, the pgf power series has radius of convergence at least 1. In a moment, we do some serious Analysis on the Convergence Theorem; and that's enough for this section.

Ca. Exercise: Poisson mean and variance.   Use g_X to show that if X has the Poisson(λ) distribution, then $\mathbb{E}(X) = \lambda$ and $\text{Var}(X) = \lambda$.

▶ **D. 'Independence means Multiply' Lemma.** *If X and Y are independent ($\mathbb{Z}^+$ valued) RVs, then*

$$g_{X+Y}(\alpha) = g_X(\alpha)g_Y(\alpha).$$

Proof. Using Lemma 99G and Theorem 101K, we have

$$g_{X+Y}(\alpha) = \mathbb{E}\left(\alpha^{X+Y}\right) = \mathbb{E}\left(\alpha^X\alpha^Y\right) = \left(\mathbb{E}\ \alpha^X\right)\left(\mathbb{E}\ \alpha^Y\right) = g_X(\alpha)g_Y(\alpha). \quad \square$$

Da. Exercise.   (a) Use Lemma D to show that if Y is the number of Heads in n tosses of a coin with probability p of Heads, then $g_Y(\alpha) = (q + p\alpha)^n$. This ties in the Binomial Theorem with the binomial distribution.

▶ (b) **Summing independent Poisson variables.**   Use Lemma D and Example 143Aa to show that if X and Y are independent, $X \sim$ Poisson(λ), $Y \sim$ Poisson(μ), then $X + Y \sim$ Poisson($\lambda + \mu$).

E. The Convergence Theorem for pgfs.   *Suppose that X and X_n (where $n = 1, 2, 3, \ldots$) are $\mathbb{Z}^+$ valued RVs, and that*

$$g_{X_n}(\alpha) \to g_X(\alpha) \quad for\ 0 \leq \alpha \leq 1.$$

Then

$$\mathbb{P}(X_n = k) \to \mathbb{P}(X = k) \qquad (k \in \mathbb{Z}^+).$$

We need the following lemma.

Lemma. *Suppose that $a_{k,n}$ and a_k ($k \in \mathbb{Z}^+$, $n \in \mathbb{N}$) are non-negative constants with $\sum_k a_{k,n} \leq 1$ for every n and $\sum a_k \leq 1$. Suppose that for $0 < \alpha \leq 1$, we have*

$$\hat{a}_n(\alpha) := \sum_{k=0}^{\infty} a_{k,n}\alpha^k \to \hat{a}(\alpha) := \sum_{k=0}^{\infty} a_k\alpha^k.$$

Then $a_{0,n} \to a_0$.

Proof of Lemma. Note that, for $0 < \alpha \leq 1$,

$$\hat{a}_n(\alpha) = a_{0,n} + A_n(\alpha), \quad \hat{a}(\alpha) = a_0 + A(\alpha),$$

where

$$0 \leq A_n(\alpha) := \sum_{k=1}^{\infty} a_{k,n}\alpha^k = \alpha \sum_{k=1}^{\infty} a_{k,n}\alpha^{k-1} \leq \alpha \sum a_{k,n} \leq \alpha,$$

and similarly $0 \leq A(\alpha) \leq \alpha$.

Let ε be given with $0 < \varepsilon \leq 3$. Choose N so that for $n \geq N$,

$$\left| \hat{a}_n \left(\tfrac{1}{3} \right) - \hat{a} \left(\tfrac{1}{3} \right) \right| < \tfrac{1}{3}\varepsilon.$$

Then, for $n \geq N$,

$$\begin{aligned}
|a_{0,n} - a_0| &= \left| \left\{ \hat{a}_n \left(\tfrac{1}{3}\varepsilon \right) - A_n \left(\tfrac{1}{3}\varepsilon \right) \right\} - \left\{ \hat{a} \left(\tfrac{1}{3}\varepsilon \right) - A \left(\tfrac{1}{3}\varepsilon \right) \right\} \right| \\
&\leq \left| \hat{a}_n \left(\tfrac{1}{3}\varepsilon \right) - \hat{a} \left(\tfrac{1}{3}\varepsilon \right) \right| + \left| A_n \left(\tfrac{1}{3}\varepsilon \right) \right| + \left| A \left(\tfrac{1}{3}\varepsilon \right) \right| \\
&\leq \tfrac{1}{3}\varepsilon + \tfrac{1}{3}\varepsilon + \tfrac{1}{3}\varepsilon = \varepsilon.
\end{aligned}$$

$\square$

Proof of Theorem. Immediately from the Lemma,

$$\mathbb{P}(X_n = 0) \to \mathbb{P}(X = 0), \qquad g_{X_n}(0) \to g_X(0).$$

Next, for $0 < \alpha \leq 1$,

$$\frac{g_{X_n}(\alpha) - \mathbb{P}(X_n = 0)}{\alpha} \to \frac{g_X(\alpha) - \mathbb{P}(X = 0)}{\alpha}$$

as $n \to \infty$. Thus,

$$\sum_{k=1}^{\infty} \mathbb{P}(X_n = k)\alpha^{k-1} \to \sum_{k=1}^{\infty} \mathbb{P}(X = k)\alpha^{k-1} \quad \text{for } 0 < \alpha \leq 1.$$

By the Lemma, $\mathbb{P}(X_n = 1) \to \mathbb{P}(X = 1)$. You can see that the whole result now follows by induction. $\square$

If X and Y are RVs with the same pgf, then we can take $X_n = Y$ for all n, whence the DF of X_n converges to the DF of X: in other words, $F_Y = F_X$. This proves the Uniqueness Theorem 143B.

Ea. Exercise: 'Poisson approximation to Binomial'. Suppose that $\lambda > 0$ and that X_n has the binomial $B(n, p_n)$ distribution, where $\mathbb{E}(X_n) = np_n \to \lambda$ as $n \to \infty$. Prove from the Convergence Theorem and result 144Da(a) that

$$\mathbb{P}(X_n = k) \to p_\lambda(k) := e^{-\lambda}\frac{\lambda^k}{k!}, \quad (k = 0, 1, 2, \ldots).$$

You may assume initially that if $nx_n \to y$ then $(1 + x_n)^n \to e^y$. How do you prove this by taking logarithms and using l'Hôpital's Rule?

Compare $b(20, 0.15, k)$ with $p_3(k)$ for $k = 0, 1, 2, 3$. Figure 52C(i) compares the Binomial$(10, 0.3)$ and Poisson(3) pmfs.

We shall see several uses of pgfs in Chapter 9.

5.3   Moment Generating Functions (MGFs)

▶▶ **A. Moment Generating Function (MGF) M_X of X.**   Let X be an RV with values in $\mathbb{R}$. We say that X is *MGF good* if $\mathbb{E}(e^{\alpha X}) < \infty$ for all α in some non-empty open interval $(-\delta, \delta)$ $(\delta > 0)$ containing the origin. For definiteness, we define $\delta(X)$ to be the supremum of all such δ, so that $\delta(X)$ may now be ∞. We then define

$$M_X(\alpha) := \mathbb{E}(e^{\alpha X}) \quad \text{for } -\delta(X) < \alpha < \delta(X).$$

If X is 'continuous' with pdf f_X, then, of course,

$$M_X(\alpha) = \int_{\mathbb{R}} e^{\alpha x} f_X(x)\,dx, \quad -\delta(X) < \alpha < \delta(X).$$

▶ **Aa. Exercise.** (a) Suppose that X has the **exponential** E(rate λ) distribution. Check that

$$\delta(X) = \lambda, \quad M_X(\alpha) = \frac{\lambda}{\lambda - \alpha} \quad (-\lambda < \alpha < \lambda).$$

(b) Suppose that X has the **Poisson**(λ) distribution. Check that

$$\delta(X) = \infty, \quad M_X(\alpha) = \exp\{\lambda(e^\alpha - 1)\}.$$

B. The 'normal integral'.   The standard normal N$(0, 1)$ distribution has pdf

$$\varphi(x) := \frac{1}{\sqrt{2\pi}}\,e^{-\frac{1}{2}x^2}. \tag{B1}$$

To verify that

$$I := \int_{\mathbb{R}} \varphi(x) \, \mathrm{d}x = 1, \tag{B2}$$

we use a change to polar coordinates ('$\mathrm{d}x\mathrm{d}y = r\mathrm{d}r\mathrm{d}\theta$') (see the 'Jacobian Theorem' 246C) and a substitution $u = \frac{1}{2}r^2$ within the calculation:

$$I^2 = \int\int_{\mathbb{R}^2} \varphi(x)\varphi(y) \, \mathrm{d}x\mathrm{d}y = \frac{1}{2\pi} \int\int \exp\left\{-\tfrac{1}{2}(x^2 + y^2)\right\} \, \mathrm{d}x\mathrm{d}y$$

$$= \frac{1}{2\pi} \int_0^\infty \exp(-\tfrac{1}{2}r^2)r \, \mathrm{d}r \int_0^{2\pi} \mathrm{d}\theta = \int_0^\infty e^{-u} \, \mathrm{d}u = 1.$$

Suppose that a Random Variable G is distributed $N(0,1)$. Let $X = \mu + \sigma G$, where $\mu \in \mathbb{R}$ and $\sigma > 0$. Then

$$F_X(x) := \mathbb{P}(X \le x) = \mathbb{P}\left(G \le \frac{x-\mu}{\sigma}\right) = F_G\left(\frac{x-\mu}{\sigma}\right)$$

so that, since $F_G'(y) = \varphi(y)$,

$$f_X(x) = F_X'(x) = \frac{1}{\sigma} \varphi\left(\frac{x-\mu}{\sigma}\right) = \frac{1}{\sigma\sqrt{2\pi}} \exp\left(-\frac{(x-\mu)^2}{2\sigma^2}\right),$$

as given at equation 54(E7).

▶ **C. MGFs for normal variables.** Suppose that G has the standard normal $N(0,1)$ distribution. Then, since $\varphi(x)$ tends to 0 as $x \to \infty$ faster than any exponential, we have $\delta(G) = \infty$. For $\alpha \in \mathbb{R}$, we have

$$M_G(\alpha) = \int_{\mathbb{R}} e^{\alpha x} \frac{1}{\sqrt{2\pi}} e^{-\frac{1}{2}x^2} \, \mathrm{d}x = e^{\frac{1}{2}\alpha^2} \int \frac{1}{\sqrt{2\pi}} e^{-\frac{1}{2}(x-\alpha)^2} \, \mathrm{d}x = e^{\frac{1}{2}\alpha^2}$$

because the second integral is that of the pdf of the $N(\alpha, 1)$ pdf and so must be 1.

Important advice: the 'Change pdf' trick. You will find that it is often possible to work out an integral in this way – by expressing it as a multiple of the integral of a 'new' pdf. □

So

$$\text{if } G \sim N(0,1), \text{ then } M_G(\alpha) = e^{\frac{1}{2}\alpha^2} \quad (\alpha \in \mathbb{R}). \tag{C1}$$

Of course, there is something very special about this MGF.

If $X \sim N(\mu, \sigma^2)$, then we can write $X = \mu + \sigma G$ (use the result at 55F), so we have $e^{\alpha X} = e^{\alpha\mu}e^{\alpha\sigma G}$, and

$$\text{if } X \sim N(\mu, \sigma^2), \text{ then } M_X(\alpha) = e^{\alpha\mu + \frac{1}{2}\alpha^2\sigma^2} \quad (\alpha \in \mathbb{R}). \tag{C2}$$

▶ **Ca. The log-normal distribution.** This distribution is often used in Statistics to model the distribution of a non-negative Random Variable Y. One assumes that

$$\ln(Y) \sim N(\mu, \sigma^2). \tag{C3}$$

It is easy to calculate the mean and variance of Y from (C2); but it is much preferable to think of $\ln(Y)$ as the 'Observation', an important case of *transforming variables* to get appropriate models.

▶▶ **D. The Gamma function $\Gamma(\cdot)$ on $(0, \infty)$.** We define this important function via

$$\Gamma(K) := \int_0^\infty x^{K-1} e^{-x}\, dx \qquad (K > 0).$$

You can easily prove by integration by parts that,

$$\text{for } K > 0, \ \Gamma(K+1) = K\Gamma(K), \text{ so that } \Gamma(n) = (n-1)! \text{ for } n \in \mathbb{N}. \tag{D1}$$

One method of numerical calculation of $\ln\Gamma(K)$ for $K > 0$. We can use the asymptotic expansion (see Note below) which generalizes Stirling's formula:

$$\ln\Gamma(K) = (K - \tfrac{1}{2})\ln K - K + \tfrac{1}{2}\ln(2\pi)$$
$$+ \frac{1}{12K} - \frac{1}{360K^3} + \frac{1}{1260K^5} - \frac{1}{1680K^7} + O(K^{-9}), \tag{D2}$$

in a C program:

```
double loggam(double K){
  int i;
  double y, corr;
  if (K>= 10){
    return ((K - 0.5)*log(K) - K + 0.5 * log(2*PI)
    + 1.0/(12.0 * K) - 1.0/(360.0 * pow(K,3))
    + 1/(1260.0*pow(K,5)) - 1/(1680.0*pow(K,7)));
  }
  else{
    corr = 0.0; y = K;
    do{corr = corr - log(y); y=y+1;}while (y<=10);
    return (corr + loggam(y));
  }
}
```

Notes on asymptotic expansions and on the above program segment. Formula (D2) is an *asymptotic* expansion. The 'series', in which the coefficients are calculated from the so-called Bernoulli numbers, does not converge for any K. As we know from Appendix

A1, p495, the term $O(K^{-9})$ remains bounded in modulus by some constant multiple of K^{-9} as $K \to \infty$. Note that in the program we use the relation $\Gamma(K) = \Gamma(K+1)/K$ to end up with a value of K at least 10, for which the asymptotic expansion gives small percentage error. You can try it out for $\Gamma(1) = 1$ and for $\Gamma(\frac{1}{2}) = \sqrt{\pi}$.

Da. Exercise. Deduce from equation 147(B2) that $\Gamma(\frac{1}{2}) = \sqrt{\pi}$.

▶▶ **E. The Gamma$(K, \textbf{rate } \lambda)$ distribution on $(0, \infty)$.** The pdf of this distribution (of which, for example, the celebrated χ^2 distributions – of which much more later – are a special case) is defined to be

$$f_X(x) = \frac{\lambda^K x^{K-1} e^{-\lambda x}}{\Gamma(K)} I_{[0,\infty)}(x).$$

Note that *the exponential $E(\textrm{rate } \lambda)$ distribution is the same as the Gamma$(1, \textrm{rate } \lambda)$ distribution.* You will see pictures of ('standardized') Gamma$(K, 1)$ for $K = 1, 2, 4, 10$ in Figure 157B(i). Sketch the density for $K = \frac{1}{2}$.

Warning. Especially in the Statistics literature and in statistical packages, the parameters used are K and $\mu = 1/\lambda$. We then refer to the Gamma$(K, \textrm{mean } \mu)$ distribution, of which the mean is $K\mu(!)$. If $X \sim$ Gamma$(K, \textrm{mean } \mu)$, then $X/\mu \sim$ Gamma$(K, 1)$.

Packages and Tables. On Minitab, one would find $\mathbb{P}(X < 67.50)$, if $X \sim$ Gamma$(25, \textrm{mean } 2.0)$, via

```
cdf 67.50;
gamma 25 2.0.
```

giving 0.975. This tallies with the table, p517, for the chi-squared distributions studied in Subsection 152J because, *if $2K$ is an integer, then*

the Gamma$(K, \textrm{mean } 2)$ distribution is exactly the χ^2_{2K} distribution,   (E1)

so that if H has the χ^2_{50} distribution, then

$$\mathbb{P}(X \le 67.5) = 1 - \mathbb{P}(H \ge 67.5) = 1 - 0.025 = 0.975.$$

Ea. Programming DFs of Gamma distributions. Let $K > 0$, and define

$$h(x) := \sum_{n=0}^{\infty} \frac{x^n}{\Gamma(K+n+1)},$$

a rapidly convergent series, this time. Then using the 'recurrence relation' 148(D1), we find that (*Exercise!*)

$$xh'(x) = \frac{1}{\Gamma(K)} + (x - K)h(x),$$

whence

$$(x^K e^{-x} h)' = \frac{1}{\Gamma(K)} x^{K-1} e^{-x},$$

and the DF of the Gamma$(K, 1)$ distribution is given by

$$F(x) := \int_0^x \frac{y^{K-1} e^{-y}}{\Gamma(K)} \, dy = x^K e^{-x} h(x).$$

It is therefore easy to calculate $F(x)$ on the computer. The `FGrate` function gives the DF for the Gamma$(K, \text{rate } \lambda)$ distribution and the `FGmean` function gives the DF for the Gamma$(K, \text{mean } \mu)$ distribution. The `htol` is a parameter less than 'half tolerance' (usually set for the purposes of this book to 10^{-5} to guarantee an answer correct in the 4th decimal place.)

```
double Fgamma(double K, double x){
  int n;
  double sum, f, g, ratio, term;

  g = loggam(K + 1);

  term = sum = exp(-g)*pow(x,K)*exp(-x);  n=1;
  do{ratio = x/(K+n);  term = term * ratio;
    sum = sum + term; n++;
  }while ((fabs(term) > htol) || (ratio > 0.5));
  return sum;
}
double FGrate(double x){return Fgamma(K, lm * x);}
double FGmean(double x){return Fgamma(K, x/mu);}
```

▶ **Eb. Exercise.** By making the substitution $\lambda x = y$, show that the pdf of Gamma$(K, \text{rate } \lambda)$ does integrate to 1. By using the 'Change pdf' trick, or otherwise, show that if $X \sim$ Gamma$(K, \text{rate } \lambda)$ then, for $-\lambda < \alpha < \lambda$,

$$M_X(\alpha) = \left(\frac{\lambda}{\lambda - \alpha} \right)^K.$$

▶ **F. Simulating Gamma variables.** It is important, particularly for Bayesian Statistics, to be able to simulate RVs with Gamma distributions efficiently; and it is not at all obvious how to do this. Very many methods are available – see Ripley's book for those used in 1986. We look at two taken from that book (but solving the exercises there by providing sketched proofs).

Fa. Gamma$(K, 1)$ simulation for $K < 1$ (Ahrens and Dieter). Here is the 'two-type rejection sampling' method:

Repeat
        Choose U with the U$[0, 1]$ distribution;

if $U > e/(K + e)$, let $X = -\ln\{(K + e)(1 - U)/(Ke)\}$ so that $X > 1$
  and accept X with probability X^{K-1};
if $U \le e/(K + e)$, let $X = \{(K + e)U/e\}^{1/K}$ so that $X < 1$,
  and accept X with probability e^{-X};

until X is accepted.

Sketched proof that this works. For $x > 1$, we have

$$du = \frac{Ke}{K + e} e^{-x} dx$$

and when we allow for acceptance probability, the probability that U will lead to an accepted value of X in dx ($x > 1$) is

$$\frac{Ke}{K + e} e^{-x} x^{K-1} dx.$$

For $x < 1$,

$$du = \frac{Ke}{K + e} x^{K-1} dx,$$

and after allowing for acceptance probability, we see that the probability that U will lead to an accepted value of X in dx ($x \le 1$) is also

$$\frac{Ke}{K + e} e^{-x} x^{K-1} dx.$$

Hence X is truly Gamma$(K, 1)$ and

$$\text{acceptance probability} = \frac{Ke}{K + e} \Gamma(K).$$

As $K \downarrow 0$, this tends to 1; as $K \uparrow 1$, it tends to $e/(e + 1)$; and when $K = \frac{1}{2}$ it equals $e\sqrt{\pi}/(1 + 2e)$; so the method is good.

'C' code for the above Ahrens–Dieter method and for simulating Gamma variables for which $K > 1$ is given in Subsection 256I.

▶ **G. Finding moments: the name 'MGF'.** For 'MGF good' X and for $|\alpha| < \delta(X)$, we have – at least formally, but by easily justified rigour –

$$M_X(\alpha) = \mathbb{E}\, e^{\alpha X} = \mathbb{E}\left(1 + \alpha X + \frac{\alpha^2 X^2}{2!} + \cdots\right) = 1 + \mu_1 \alpha + \frac{1}{2!}\mu_2 \alpha^2 + \cdots,$$

where μ_r is the *r*th *moment* of X:

$$\mu_r := \mu_r(X) := \mathbb{E}\,(X^r) \quad (r = 1, 2, 3, \ldots).$$

We may therefore read off the moments μ_r as coefficients in this power series.

▶ **Ga. Exercise: Moments of normal and Gamma distributions.** (a) Prove that if G has the $N(0,1)$ distribution, then $\mathbb{E}(G) = 0$ and $\text{Var}(G) = 1$. Deduce that

$$\text{if } X \sim N(\mu, \sigma^2) \text{ then } \mathbb{E}(X) = \mu, \ \text{Var}(X) = \sigma^2.$$

(b) Use its MGF to prove that an RV X with the $E(\text{rate } \lambda)$ distribution has kth moment $\mathbb{E}\left(X^k\right) = k!/\lambda^k$. Hence prove 148(D1). Note that

$$\text{if } X \sim E(\text{rate } \lambda), \text{ then } \mathbb{E}(X) = \frac{1}{\lambda}, \ \text{Var}(X) = \frac{1}{\lambda^2}.$$

(c) Prove more generally that

$$\text{if } X \sim \text{Gamma}(K, \text{rate } \lambda), \text{ then } \mathbb{E}(X) = \frac{K}{\lambda}, \ \text{Var}(X) = \frac{K}{\lambda^2}.$$

▶ **H. Fact: The Uniqueness Theorem for MGFs.** *Suppose that X and Y are 'MGF good' RVs, and that for some $\delta > 0$,*

$$M_X(\alpha) = M_Y(\alpha) \text{ for } -\delta < \alpha < \delta.$$

Then $F_X = F_Y$.

As already stated, this is too difficult to prove at this level.

▶ **I. The 'Independence means Multiply' Lemma for MGFs.** *Suppose that X and Y are independent 'MGF good' RVs. Then for $|\alpha| < \min(\delta(X), \delta(Y))$, we have*

$$M_{X+Y}(\alpha) = M_X(\alpha)M_Y(\alpha).$$

Ia. Exercise. (a) Prove Lemma I. *Hint.* See the proof of 144D.
▶▶ (b) **Sums of independent 'normals' are 'normal'.** Prove that if X and Y are independent, $X \sim N\left(\mu_1, \sigma_1^2\right)$, $Y \sim N\left(\mu_2, \sigma_2^2\right)$, then $X + Y \sim N\left(\mu_1 + \mu_2, \sigma_1^2 + \sigma_2^2\right)$.
▶ (c) **Sums of independent Gamma variables.** Prove that if X and Y are independent, and $X \sim \text{Gamma}(K_1, \text{rate } \lambda)$, $Y \sim \text{Gamma}(K_2, \text{rate } \lambda)$, then $X + Y \sim \text{Gamma}(K_1 + K_2, \text{rate } \lambda)$. Note that it follows that the sum of n IID RVs each with the $E(\text{rate } \lambda)$ distribution has the $\text{Gamma}(n, \text{rate } \lambda)$ distribution, a result of considerable importance.

▶▶ **J. Chi-squared distributions.** These distributions play a fundamental rôle in Statistics. For $\nu = 1, 2, 3, \ldots$, the χ^2 distribution with ν degrees of freedom is defined to be the distribution of

$$G_1^2 + G_2^2 + \cdots + G_\nu^2,$$

where $G_1, G_2, \ldots, G_\mu$ are IID each $N(0,1)$. You have been told at 149(E1) that

the χ_ν^2 distribution is the $\text{Gamma}(\frac{1}{2}\nu, \text{rate } \frac{1}{2})$ distribution.   (J1)

Proof. That the χ_1^2 distribution is the Gamma($\frac{1}{2}$, rate $\frac{1}{2}$) distribution is immediate from Exercise 56Fc. The case of general ν is now obvious from Exercise Ia(c). □

▶ **Ja. The non-central $\chi^2(\rho^2)$ distribution.** This is another important distribution for Statistics. Again let $G_1, G_2, \ldots, G_\nu$ be IID, each N(0, 1). Let $a_1, a_2, \ldots, a_\nu$ be real numbers. Then, as we shall see in Chapter 8, the distribution of

$$W := (a_1 + G_1)^2 + (a_2 + G_2)^2 + \cdots + (a_\nu + G_\nu)^2$$

depends only on ν and $\rho^2 = a_1^2 + a_2^2 + \cdots a_\nu^2$ ($\rho \geq 0$); it is called the non-central $\chi_\nu^2(\rho^2)$ distribution, and ρ^2 is called the *non-centrality parameter*. We write $W \sim$ non-central $\chi_\nu^2(\rho^2)$.

Jb. Exercise. Prove that in the case when $\nu = 1$,

$$F_W(w) = \Phi\left(\sqrt{w} - \rho\right) - \Phi\left(-\sqrt{w} - \rho\right) \quad (w \geq 0),$$
$$f_W(w) = (2\pi w)^{-\frac{1}{2}} e^{-\frac{1}{2}w - \frac{1}{2}\rho^2} \cosh\left(\rho\sqrt{w}\right) \quad (w > 0).$$

▶ **K. Fact: The Convergence Theorem for MGFs.** *Suppose that X and X_n ($n = 1, 2, 3\ldots$) are RVs such that for some $\delta > 0$, we have $\delta(X_n) > \delta$ for each n and $\delta(X) > \delta$. Suppose also that*

$$M_{X_n}(\alpha) \to M_X(\alpha) \quad \text{for } |\alpha| < \delta.$$

Then $F_{X_n}(x) \to F_X(x)$ at every point x at which $F_X(x)$ is continuous.

This is another result too difficult to prove here.

▶ **L. Cumulant Generating Function C_X; cumulants $\kappa_r(X)$.** If X is MGF good, then we define the Cumulant Generating Function C_X of X via

$$C_X(\alpha) := \ln M_X(\alpha) \quad \text{for } |\alpha| < \delta(X).$$

Using the Taylor series

$$\ln(1 + y) = \int_0^y \frac{1}{1+t}\, dt = \int_0^y \left(1 - t + t^2 - t^3 + \cdots\right) dt$$
$$= y - \frac{1}{2}y^2 + \frac{1}{3}y^3 - \frac{1}{4}y^4 + \cdots,$$

we see intuitively that, provided $|\alpha|$ is small,

$$C_X(\alpha) = \ln\left(1 + \mu_1\alpha + \frac{1}{2!}\mu_2\alpha^2 + \cdots\right)$$
$$= \left(\mu_1\alpha + \frac{1}{2!}\mu_2\alpha^2 + \cdots\right) - \frac{1}{2}(\cdot)^2 + \frac{1}{3}(\cdot)^3 - \cdots$$
$$= \kappa_1\alpha + \frac{1}{2!}\kappa_2\alpha^2 + \frac{1}{3!}\kappa_3\alpha^3 + \cdots,$$

for some constants

$$\kappa_r = \kappa_r(X) \quad (r = 1, 2, 3, \ldots).$$

The constant $\kappa_r(X)$ is called the rth *cumulant* of X.

▶ **M. Lemma.** *Let X be MGF good, and let c be a constant. Write κ_r for $\kappa_r(X)$. Then the following results are true.*
(a) $\kappa_1(X + c) = \kappa_1 + c, \quad \kappa_r(X + c) = \kappa_r \ (r \geq 2)$.
(b) $\kappa_r(cX) = c^r \kappa_r$.
(c) $\kappa_1 = \mu_1 = \mu := \mathbb{E}(X)$.
(d) $\kappa_2 = \mathbb{E}\left\{(X - \mu)^2\right\} = \sigma_X^2 = \mathrm{Var}(X)$.
(e) $\kappa_3 = \mathbb{E}\left\{(X - \mu)^3\right\}$.
(f) $\kappa_4 = \mathbb{E}\left\{(X - \mu)^4\right\} - 3\mathbb{E}\left\{(X - \mu)^2\right\}^2$.

It is clear that if all the moments $\mu_k := \mathbb{E}(X^k)$ of X exist, then we can define the cumulants $\kappa_r(X)$ as polynomials in the μ_k even if X is not MGF good.

We have (with statisticians understanding 'skewness' and 'kurtosis')

$$\kappa_1 = \text{mean}, \quad \kappa_2 = \text{variance}, \quad \frac{\kappa_3}{\kappa_2^{3/2}} =: \text{'skewness'}, \quad \frac{\kappa_4}{\kappa_2^2} =: \text{'kurtosis'}.$$

See Subsection 163J below for 'correcting the CLT by allowing for skewness, kurtosis, etc'.

Ma. Exercise. Prove Lemma M. *Hint.* Use Part (a) and the fact that $X = (X - \mu) + \mu$ in doing Parts (c)–(f). For $r \geq 2$, $\kappa_r(X) = \kappa_r(X - \mu)$, and '$\mu_1$ for $X - \mu$ is 0'.

▶ **N. Lemma.** *The following results hold and (**Fact**) the described properties characterize Normal and Poisson distributions amongst distributions with finite moments.*
(a) If $X \sim \mathrm{N}(0, 1)$, then $\kappa_1 = 0$, $\kappa_2 = 1$, $\kappa_r = 0 \ (r \geq 3)$.
(b) If $X \sim \mathrm{Poisson}(\lambda)$, then $\kappa_r = \lambda \ (r \geq 1)$.

Na. Exercise. Prove results (a) and (b) of Lemma N. Accept the 'characterization' part for now. There are notes on it at (Pb) below.

▶ **O. Lemma.** *If X and Y are independent MGF good RVs, then with $\delta := \min(\delta(X), \delta(Y))$,*

$$\begin{aligned} C_{X+Y}(\alpha) &= C_X(\alpha) + C_Y(\alpha) \ (|\alpha| < \delta), \\ \kappa_r(X + Y) &= \kappa_r(X) + \kappa_r(Y) \ (r \geq 1). \end{aligned}$$

The second equality is true for a fixed r provided only that $\mathbb{E}(|X|^r)$ and $\mathbb{E}(|Y|^r)$ exist.

Oa. Exercise. Prove the above lemma, assuming throughout that X and Y are MGF good.

P. A cumulant guarantee of convergence to the normal distribution.
This is the way that statisticians see things – and for good practical reasons. Mathematicians prefer Theorem 168E below.

▶ **Pa. Fact.** *If V_n is a sequence of RVs each possessing moments of all orders, and if*

$$\kappa_2\left(V_n\right) \to 1, \quad \kappa_r\left(V_n\right) \to 0 \ (r \neq 2), \tag{P1}$$

then, with $N \sim \mathrm{N}(0,1)$, we have, for every x in $\mathbb{R}$,

$$\mathbb{P}\left(V_n \leq x\right) \to \mathbb{P}(N \leq x) = \Phi(x) := \int_{-\infty}^{x} \varphi(y)\,\mathrm{d}y.$$

A *heuristic* idea is that under suitable extra conditions, (P1) implies that, for small $|\alpha|$,

$$M_{V_n}(\alpha) = \exp\left(C_{V_n}(\alpha)\right) \to \exp\left(\tfrac{1}{2}\alpha^2\right) = M_N(\alpha),$$

and now we use Theorem 153K. However, this heuristic idea is dangerous ... .

▶ **Pb. Warning.** It is possible to find two RVs X and Y with $\mathbb{E}\left(|X|^r\right) < \infty$ and $\mathbb{E}\left(|Y|^r\right) < \infty$ for $r \in \mathbb{N}$, such that $\mathbb{E}(X^r) = \mathbb{E}(Y^r)$ for every $r \in \mathbb{N}$ but X and Y have *different* distributions. Note that X and Y have the same cumulants. If $X_n := X$ for $n \in \mathbb{N}$, then $\kappa_r(X_n) \to \kappa_r(Y)$ for every $r \in \mathbb{N}$, but the distribution of X_n does not tend to the distribution of Y. See Appendix A8, p500, for how to construct the pair (X, Y).

Regard proving the positive results stated as Facts within (154N) and at Pa (in each of which *normal* distributions and their cumulants play a key part) as easy exercises after you have read [W]. [[*Technical Notes* for such exercises. At 154N, prove that the characteristic function of X must be that of the normal distribution. In regard to Fact Pa, it follows already from the fact that $\kappa_1(V_n) \to 0$ and $\kappa_2(V_n) \to 1$ that the distributions of the V_n form a 'tight' family. Of course, Result 154N is needed to complete the proof of Fact Pa.]]

▶ **Q. The Laplace transform $L_X(\alpha)$.** If X is a non-negative Variable, then

$$L_X(\alpha) := \mathbb{E}\left(\mathrm{e}^{-\alpha X}\right)$$

exists for all $\alpha \geq 0$. The distribution of X is determined by the function $L_X(\cdot)$ on $[0, \infty)$. We have $-L'_X(0) = \mathbb{E}(X) \leq \infty$, etc. For independent non-negative X and Y, we have $L_{X+Y} = L_X L_Y$. Finally, for non-negative X_n and X, we have $L_{X_n}(\alpha) \to L_X(\alpha)$ as $n \to \infty$ for all $\alpha \geq 0$ if and only if $F_{X_n}(x) \to F_X(x)$ at every point x of continuity of F_X.

5.4   The Central Limit Theorem (CLT)

▶ **A. Standardized form Y^* of a Random Variable Y.** Let Y be an RV in $\mathcal{L}^2$, so that $\mathbb{E}(Y)$ and $\text{Var}(Y)$ exist. Let $\text{SD}(Y)$ denote the standard deviation $\sqrt{\text{Var}(Y)}$ as usual. We suppose that $\text{SD}(Y) > 0$, so that Y involves some randomness.

We define the *standardized form Y^* of Y* via

$$Y^* = \frac{Y - \mathbb{E}(Y)}{\text{SD}(Y)}.$$

The RV Y^* is standardized in the sense that $\mathbb{E}(Y^*) = 0$, $\text{SD}(Y^*) = 1$. We have $Y \sim \text{N}(\mu, \sigma^2)$ for some μ and $\sigma^2 > 0$ if and only if $Y^* \sim \text{N}(0,1)$.

▶▶▶ **B. Fact: The Central Limit Theorem (CLT) (de Moivre 1706, Laplace 1812, Lindeberg 1922, ...).** Let $X_1, X_2, \ldots$ be IID RVs with ANY common distribution of finite mean μ and variance $\sigma^2 > 0$. Let

$$S_n := X_1 + X_2 + \cdots + X_n,$$

as usual. We know from Lemma 104E that

$$\mathbb{E}(S_n) = n\mu, \quad \text{Var}(S_n) = n\sigma^2, \quad \text{SD}(S_n) = \sigma\sqrt{n}.$$

(a) To formulate precisely the idea that "S_n is approximately $\text{N}(n\mu, n\sigma^2)$", we let S_n^* be the standardized form of S_n. Then, as $n \to \infty$, for every x in $\mathbb{R}$,

$$\mathbb{P}(S_n^* \leq x) \to \Phi(x) := \int_{-\infty}^x \varphi(y)\, dy = \int_{-\infty}^x \frac{1}{\sqrt{2\pi}} e^{-\frac{1}{2}y^2}\, dy.$$

(b) If A_n denotes the average $A_n := S_n/n$, then $\mathbb{E}(A_n) = \mu$, $\text{Var}(A_n) = \sigma^2/n$, $\text{SD}(A_n) = \frac{\sigma}{\sqrt{n}}$. Then $A_n^* = S_n^*$ and we formulate precisely the idea that A_n is approximately $\text{N}(\mu, \sigma^2/n)$ via

$$\mathbb{P}(A_n^* \leq x) = \mathbb{P}(S_n^* \leq x) \to \Phi(x).$$

(c) If each X_k has the $\text{N}(\mu, \sigma^2)$ distribution, then we have, exactly, $S_n \sim \text{N}(n\mu, n\sigma^2)$ and $A_n \sim \text{N}(\mu, \sigma^2/n)$.

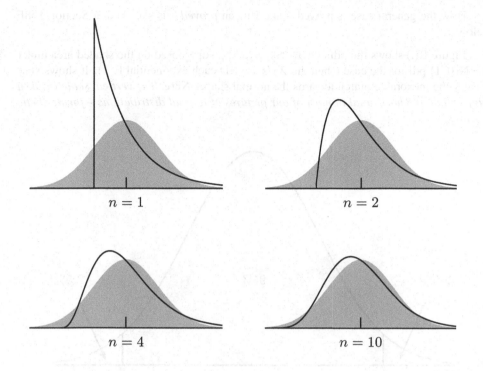

Figure B(i): pdf of standardized sum of n E(1) variables

Sketched proof under 'finite moments' assumption. By replacing each X_k by *its* standardized version, we can – and do – reduce the problem to that in which

$$\mu = 0, \quad \sigma^2 = 1, \quad S_n^* = \frac{S_n}{\sqrt{n}}.$$

Let us assume that if X is an RV with the same distribution as each X_k, then $\mathbb{E}\left(|X|^k\right) < \infty$ for every positive integer k. Then

$$\kappa_1(X) = \mathbb{E}(X) = 0, \quad \kappa_2(X) = \sigma_X^2 = 1.$$

By Part (b) of Lemma 154M and Lemma 154O, we have

$$\kappa_r\left(S_n^*\right) = \left(\frac{1}{\sqrt{n}}\right)^r \kappa_r\left(S_n\right) = \left(\frac{1}{\sqrt{n}}\right)^r n\kappa_r(X),$$

so that as $n \to \infty$,

$$\kappa_1\left(S_n^*\right) = 0, \quad \kappa_2\left(S_n^*\right) = 1, \quad \kappa_r\left(S_n^*\right) \to 0 \ (r \geq 3).$$

By Fact 155Pa, the result follows. $\qquad\qquad\qquad\qquad\qquad\qquad\qquad\qquad\square$

How the general case is proved – and I mean *proved* – is sketched at Section 168F below.

Figure B(i) shows the pdfs of S_1^*, S_2^*, S_4^*, S_{10}^*, superposed on the shaded area under the N(0, 1) pdf for the case when the X_k's are IID each exponential E(1). It shows very clearly the inexorable march towards the normal shape. **Note**. *The vertical scale is taken bigger than the horizontal in most of our pictures of normal distributions – for aesthetic reasons.*

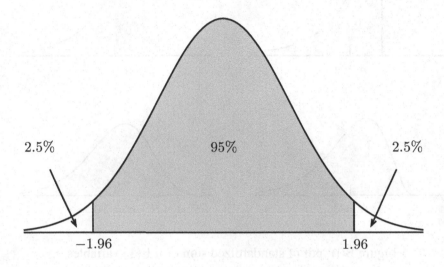

2.5%   95%   2.5%

-1.96 1.96

Figure B(ii): Some N(0, 1) probabilities

In practice, we commonly use the facts (illustrated in Figure B(ii), where the graph is that of φ, and discussed in the next subsection) that if $G \sim \mathrm{N}(0, 1)$, then, to sufficient accuracy,

$$\begin{aligned} \mathbb{P}(G \le 1.96) &= 0.975, & \mathbb{P}(|G| \le 1.96) &= 0.95, \\ \mathbb{P}(G \le 2.58) &= 0.995, & \mathbb{P}(|G| \le 2.58) &= 0.99. \end{aligned} \tag{B1}$$

Hence, we can say that, if n is large, then

$$\begin{aligned} &\mathbb{P}\big(\big|S_n - \mathbb{E}(S_n)\big| \le 1.96\,\mathrm{SD}(S_n)\big) \\ &= \mathbb{P}(|S_n - n\mu| \le 1.96\sigma\sqrt{n}) = \mathbb{P}(|A_n - \mu| \le 1.96\sigma/\sqrt{n}) \approx 0.95. \end{aligned} \tag{B2}$$

C. Working with normal distributions. The N(0, 1) Distribution Function Φ:

$$\Phi(x) = \mathbb{P}(G \le x) = \int_{-\infty}^{x} \frac{1}{\sqrt{2\pi}} \, e^{-\frac{1}{2}y^2} \, dy$$

is tabulated and available on all Statistical Packages. A simple method of calculating Φ in 'C' is given later in this subsection. One problem with the tabulations is that many different ways of tabulating Distribution Functions are used. The tables at the end of this book follow the most commonly used convention of tabulating Φ in the obvious way, but basing other tabulations on tail probabilities (focusing on $1 - F(x)$ instead of $F(x)$). Do be careful whether a table in another book gives (for whatever distribution) $\mathbb{P}(Y \leq x)$ or $\mathbb{P}(Y > x)$ or $\mathbb{P}(0 \leq Y \leq x)$ or whatever. It is easy to switch from one to the other. Because the famous bell-shaped graph of the pdf φ is symmetric about 0, we have, for example (as illustrated in Figure B(ii) for the case when $x = 1.96$ for which $\Phi(x) = 0.975$),

$$\Phi(-x) = 1 - \Phi(x) \quad (x \in \mathbb{R}).$$

If $G \sim \mathrm{N}(0, 1)$, then

$$\mathbb{P}(a \leq G \leq b) = \Phi(b) - \Phi(a).$$

Check from the table, p515, that if $G \sim \mathrm{N}(0, 1)$, then

$$\mathbb{P}(-1.234 \leq G \leq 0.567) = 0.7147 - 0.1086 = 0.606.$$

Minitab uses cdf for (Cumulative) Distribution Function, and, **with** \\ **denoting 'new line'** (not of course typed in Minitab),

```
minitab \\ cdf -1.234; \\ normal 0 1.
```

gives $\Phi(-1.234) = 0.1086$.

If $X \sim \mathrm{N}\left(\mu, \sigma^2\right)$, then $X = \mu + \sigma G$ and so

$$\mathbb{P}(a \leq X \leq b) = \mathbb{P}\left(\frac{a - \mu}{\sigma} \leq \frac{X - \mu}{\sigma} \leq \frac{b - \mu}{\sigma}\right)$$

$$= \mathbb{P}\left(\frac{a - \mu}{\sigma} \leq G \leq \frac{b - \mu}{\sigma}\right)$$

$$= \Phi\left(\frac{b - \mu}{\sigma}\right) - \Phi\left(\frac{a - \mu}{\sigma}\right).$$

Check that if $X \sim N(1.2, 1.69)$, then

$$\mathbb{P}(0.1 \leq X \leq 1.7) = 0.451.$$

Warning. Most computer packages use the parameters μ and σ (the standard deviation, not the variance) for the $\mathrm{N}\left(\mu, \sigma^2\right)$ distribution. So, for example, on Minitab ,

```
\\ cdf 1.7 K2; \\ normal 1.2 1.3.
\\ cdf 0.1 K1; \\ normal 1.2 1.3.
\\ print K2-K1
```

would be needed.

Calculation of Φ in C. There is a huge literature on clever ways to compute Φ. Nowadays, one doesn't have to be clever. One method is as follows.

Let $f(x) = \exp\left(\frac{1}{2}x^2\right) \int_0^x \exp\left(-\frac{1}{2}y^2\right) \, dy$. Then

$$f(0) := 0, \qquad f'(x) = 1 + xf(x), \qquad f(x) := \int_0^x \{1 + yf(y)\} \, dy.$$

Picard's Theorem on Ordinary Differential Equations solves this via

$$f_0(x) = 0 \ (\forall x), \qquad f_{n+1}(x) = \int_0^x \{1 + yf_n(y)\} \, dy,$$

and here guarantees that $f_n(x) \to f(x)$ as $n \to \infty$. Hence, we see that

$$f(x) = \frac{x}{1} + \frac{x^3}{1.3} + \frac{x^5}{1.3.5} + \frac{x^7}{1.3.5.7} + \cdots.$$

Of course,

$$\Phi(x) = \frac{1}{2} + \frac{1}{\sqrt{2\pi}} e^{-\frac{1}{2}x^2} f(x).$$

This may be implemented as follows:

```
double Phi(double x){
  int j;
  double sum, term, ratio;

  term = x*exp(- x*x/2.0)/ sqrt(2.0 * PI);
  sum = term; j = 2;
  do{
    ratio = x*x/(2*j - 1);
    term = term * ratio;
    sum = sum + term;
    j++;
  }while ((fabs(term) > htol) || (ratio > 0.5));
  return(0.5 + sum);
}
```

Ca. Simulating from a general normal distribution.   For simulating from general
Normal distributions, we use (with precision `prec` the inverse of variance):

```
double SDnormal(double mean, double sd)
{
    double random;
    random = mean + sd * aGauss();
    return random;
}
double PRECnormal(double mean, double prec)
{
    double sd = 1/sqrt(prec);
    return SDnormal(mean,sd);
}
```

▶ **D. 'Integer' or 'continuity' correction.** Suppose that each X_k in the CLT takes only integer values. Then we have for integers a and b,

$$\mathbb{P}\left(a \leq S_n \leq b\right) = \mathbb{P}\left(a - \tfrac{1}{2} \leq S_n \leq b + \tfrac{1}{2}\right)$$

by obvious logic. A picture would suggest (*correctly*) that it is better to use the right-hand side in applying the CLT:

$$\mathbb{P}\left(a \leq S_n \leq b\right) \approx \Phi\left(\frac{b + \tfrac{1}{2} - n\mu}{\sigma\sqrt{n}}\right) - \Phi\left(\frac{a - \tfrac{1}{2} - n\mu}{\sigma\sqrt{n}}\right).$$

This is obviously better when $a = b(!)$, and you can now explain result 11(N2) that $b(2n, \tfrac{1}{2}; n) \sim (\pi n)^{-\tfrac{1}{2}}$.

Check that the CLT gives the estimate

$$\mathbb{P}\left(41 \leq Y \leq 59\right) \approx 0.9426,$$

mentioned at 106Ha. As stated there, the exact answer to 5 places is 0.94311.

▶ **E. Important note on convergence in the CLT.** In the CLT, we have

$$\mathbb{P}\left(S_n^* \leq x\right) \to \Phi(x) = \mathbb{P}(G \leq x),$$

where G is an N(0, 1) variable. Thus, the statement that

$$\mathbb{P}\left(S_n^* \in B\right) \to \mathbb{P}(G \in B) \tag{E1}$$

is true if B has the form $(-\infty, x]$. However, (E1) need not be true for all nice sets B.

Suppose for example that $X_1, X_2, \ldots$ are IID, each with $\mathbb{P}(X_k = \pm 1) = \tfrac{1}{2}$. Then every S_n^* must take values in the *countable* set C of numbers of the form $r/\sqrt{s}$, where r is an integer and s is a positive integer. We have

$$1 = \mathbb{P}\left(S_n^* \in C\right) \nrightarrow \mathbb{P}(G \in C) = 0.$$

Here is a positive result, which depends on the famous Scheffé's Lemma in Measure Theory.

▶ **Ea. Fact.** *If Z_n has pdf f_n and Z has pdf f, and $f_n(x) \to f(x)$ for every x, then*

$$\mathbb{P}\left(Z_n \in B\right) \to \mathbb{P}(Z \in B)$$

for every nice (measurable) set B.

Sometimes we can prove cases of the CLT directly (and with stronger conclusion) by showing that the hypotheses of Fact Ea hold where $Z_n = S_n^*$ and $Z = G$. This will be illustrated by examples.

F. Application to exponential RVs. We discuss the case illustrated at Figure 157B(i). Suppose that each X_k has the E(rate λ) distribution for some $\lambda > 0$. We know that the E(rate λ) distribution has mean $1/\lambda$ and standard deviation $1/\lambda$ – see Exercise 152Ga. We know – see Part (c) of Exercise 152Ia — that S_n is distributed Gamma(n, rate λ) (with mean n/λ and variance n/λ^2, of course), so that for $y \in \mathbb{R}$, we have, as $n \to \infty$,

$$\text{if } y \in \mathbb{R} \text{ and } z_n := \frac{n}{\lambda} + y\frac{\sqrt{n}}{\lambda}, \text{ then}$$

$$\mathbb{P}(S_n \le z_n) = \int_0^{z_n} \frac{\lambda^n x^{n-1} e^{-\lambda x}}{(n-1)!}\, dx = \mathbb{P}(S_n^* \le y) \to \Phi(y). \tag{F1}$$

Especially, for large n (and n does need to be large when dealing with exponential variables),

$$\mathbb{P}\left(|A_n - \mu| \le 1.645\frac{\mu}{\sqrt{n}}\right) \approx 90\%, \tag{F2}$$

where $\mu = 1/\lambda$ is the mean of one of the X_k's.

We can prove the CLT for this situation directly. By Subsection 55F, we have

$$f_{S_n^*}(x) = \frac{\sqrt{n}}{\lambda} f_{S_n}\left(\frac{n}{\lambda} + x\frac{\sqrt{n}}{\lambda}\right),$$

and, by using Stirling's formula 11(N1), and the facts that

$$\left(\frac{n}{n-1}\right)^{n-1} \to e, \quad \ln(1+y) = y - \tfrac{1}{2}y^2 + O\left(y^3\right) \text{ for small } y$$

(see 496(ApA 2.4)) we find that (*Exercise!*)

$$f_{S_n^*}(x) \to \varphi(x) \text{ for all } x.$$

By Fact 161Ea, this is a stronger conclusion than that of the CLT.

G. Exercise. Show that if each X_k in the CLT has the Poisson(λ) distribution, then, for large n,

$$\mathbb{P}\left(|A_n - \lambda| \le 1.96\sqrt{\lambda/n}\right) \approx 95\%.$$

Show that if $Y_n \sim$ Poisson(n), so that Y has the same distribution as the sum of n IID Poisson(1) variables, then for $y \in \mathbb{R}$, as $n \to \infty$,

$$\mathbb{P}\left(Y_n \le n + y\sqrt{n}\right) = \sum_{0 \le k \le n + y\sqrt{n}} e^{-n}\frac{n^k}{k!} \to \Phi(y).$$

H. Exercise: Numerical Examples. Use the Integer Correction when appropriate.
(a) Let S be the number of sixes and T the total score if I throw a fair die 60 times. What are $\mathbb{E}(S)$ and Var(S)? Show that $\mathbb{E}(T) = 210$ and Var(T) = 175. Show that the CLT leads to the estimates:

(i) $\mathbb{P}(10 \le S \le 13) \approx 0.456$,   (ii) $\mathbb{P}(200 \le T \le 230) \approx 0.726$.

Note. The exact answer to (i) (to 4 places) is 0.4384.

(b) An RV X has the Poisson distribution with parameter 25. Use the CLT to show that

$$\mathbb{P}(22 \leq X \leq 28) \approx 0.5160.$$

Note. The exact answer (to 5 places) is 0.51610.

(c) Show that if Y is the sum of 10 IID RVs each with the exponential E(1) distribution, a case pictured in Figure 157B(i), then 162(F2) would lead us to estimate the values y such that $\mathbb{P}(Y > y) = 0.025$ and 0.01 respectively as 16.2 and 17.4, whereas the correct values from the table, p517, are 17.1 and 18.8. See 149(E1). Better methods of approximating Gamma distributions by normal ones are given in the Subsections K and J below.

I. Accuracy of the CLT.
We have already seen numerical instances of how accurate the CLT is. Here now is an amazing theoretical result on the accuracy, a uniform bound over x.

▶ **Fact: The Berry–Esseen Theorem.** *In the CLT, we have*

$$\sup_{x \in \mathbb{R}} \left| \mathbb{P}\left(S_n^* \leq x\right) - \Phi(x) \right| \leq 10 \frac{\mathbb{E}\left(|X - \mu|^3\right)}{\sigma^3 \sqrt{n}}. \tag{I1}$$

Suppose that we consider approximating the Binomial$(2n, \frac{1}{2})$ distribution using the CLT. We know from Exercise 106Ha that the jump $b(2n, \frac{1}{2}; n)$ in $\mathbb{P}\left(S_n^* \leq x\right)$ at 0 is asymptotic to $1/\sqrt{\pi n}$. Since Φ is continuous, the supremum in (I1) must be at least equal to this jump. The right-hand side in this case is $10/\sqrt{2n}$. The order of magnitude in the Berry–Esseen Theorem cannot be improved.

For proof of the Berry–Esseen Theorem, see Stroock [221].

▶ **J. Improvements of the CLT: Edgeworth expansions; saddle-point method.** If in the CLT each X_k has mean 0 and variance 1 with higher cumulants $\kappa_3, \kappa_4, \ldots$, then, as our heuristic proof of the CLT showed,

$$M_{S_n^*}(\alpha) = \exp\left\{C_{S_n^*}(\alpha)\right\} = \exp\left\{\frac{\alpha^2}{2!} + \frac{\kappa_3 \alpha^3}{3!\sqrt{n}} + \frac{\kappa_4 \alpha^4}{4!n} + \cdots\right\}$$

$$= e^{\frac{1}{2}\alpha^2} \exp\left\{\frac{\kappa_3 \alpha^3}{3!\sqrt{n}} + O\left(\frac{1}{n}\right)\right\} = e^{\frac{1}{2}\alpha^2}\left\{1 + \frac{\kappa_3 \alpha^3}{3!\sqrt{n}} + O\left(\frac{1}{n}\right)\right\}.$$

But, by differentiating $\int_{\mathbb{R}} e^{\alpha x} \varphi(x) \mathrm{d}x = e^{\frac{1}{2}\alpha^2}$ with respect to α three times and combining the results obtained, we find that

$$\int_{\mathbb{R}} e^{\alpha x}\left(x^3 - 3x\right)\varphi(x)\mathrm{d}x = \alpha^3 e^{\frac{1}{2}\alpha^2},$$

suggesting that a better approximation than the CLT would be (under suitable conditions)

$$\mathbb{P}\left(S_n^* \leq x\right) \approx \Phi(x) + \frac{\kappa_3}{3!\sqrt{n}} \int_{-\infty}^{x} \left(y^3 - 3y\right)\varphi(y)\mathrm{d}y$$

$$= \Phi(x) - \frac{\kappa_3}{3!\sqrt{n}}(x^2 - 1)\varphi(x).$$

(You check that $(\mathrm{d}/\mathrm{d}x)\{(1-x^2)\varphi(x)\} = (x^3 - 3x)\varphi(x)$.) One can of course expand in similar fashion to include the higher cumulants.

We saw at (c) of Exercise 162H that the CLT does not approximate Gamma distributions particularly well. Using the example there, we have if $Y \sim \text{Gamma}(10, 1)$, then

$$\mathbb{P}(Y \leq 17.08) = 0.975.$$

With $x = (17.08 - 10)/\sqrt{10}$, we have

$$\Phi(x) = 0.988, \qquad \kappa_3 = 2, \qquad \Phi(x) - \frac{\kappa_3}{3!\sqrt{n}}(x^2 - 1)\varphi(x) = 0.974,$$

showing that correcting for skewness here leads to significant improvement.

There are more powerful ways of improving on the CLT. Of particular importance is the saddle-point method applied to the inversion formula for obtaining Distribution Functions from Characteristic Functions. This was much developed in the statistical context in papers by Daniels. See, for example, Daniels [53], Reid [192], Barndorff-Nielsen and Cox [8, 9].

K. Another method of approximating Gamma distributions. This topic is optional, but is referred to in most sets of tables.

For large K, it is the case that

$$\text{if } V \sim \text{Gamma}(K, 1), \text{ then } \sqrt{V} \text{ is approximately } \mathrm{N}\left(\sqrt{K - \tfrac{1}{4}}, \left(\tfrac{1}{2}\right)^2\right). \qquad \text{(K1)}$$

Ka. Note. Because of 149(E1), this is equivalent to the statement usually mentioned in χ^2 tables, that, *for large ν,*

$$\text{if } H \sim \chi^2_\nu \text{ then } \sqrt{2H} \text{ is approximately } \mathrm{N}\left(\sqrt{2\nu - 1}, 1\right).$$

For $\mu = 100$, we have

$$\left(\sqrt{199} + 1.96\right)^2 / 2 = 129.07, \qquad \left(\sqrt{199} - 1.96\right)^2 / 2 = 73.77,$$

agreeing quite closely with the values 129.56 and 74.22 in the last row of the table on p517.

Kb. Exercise. Show that if Y is the sum of 10 IID RVs each with the exponential E(1) distribution, a case pictured in Figure 157B(i), then (K1) would lead us to estimate the values y such that $\mathbb{P}(Y > y) = 0.025$ and 0.01 respectively as 16.8 and 18.4 – significantly better than the CLT approximations at part (c) of Exercise 162H.

Proof (and rigorous formulation) of (K1). Let $a \in \mathbb{R}$. We now use 55(F) to obtain the distribution of Z if X has the Gamma$(K, 1)$ distribution and $\sqrt{X} = Z + \sqrt{K - a}$. We have

$$f_X(x) = \frac{x^{K-1}e^{-x}}{\Gamma(K)}, \qquad x = \psi(z) = \left(z + \sqrt{K - a}\right)^2.$$

Thus,

$$f_Z(z) = f_X(\psi(z))\psi'(z) = \frac{2(z + \sqrt{K-a})^{2K-1} e^{-(z+\sqrt{K-a})^2}}{\Gamma(K)},$$

whence

$$\ln f_Z(z) = \ln 2 + (2K-1)\ln\left(\sqrt{K-a} + z\right) - \left(z + \sqrt{K-a}\right)^2 - \ln\Gamma(K)$$

$$= \ln 2 + (K - \tfrac{1}{2})\ln(K-a) + (2K-1)\ln\left(1 + \frac{z}{\sqrt{K-a}}\right)$$

$$- \left(z + \sqrt{K-a}\right)^2 - \ln\Gamma(K).$$

Fix z. After writing (see Appendix A1, p495 and Appendix A3, p496 for 'O' and 'o' notation)

$$\ln\left(1 + \frac{z}{\sqrt{K-a}}\right) = \frac{z}{\sqrt{K-a}} - \frac{1}{2}\frac{z^2}{K-a} + O\left(K^{-3/2}\right),$$

$$\frac{z}{\sqrt{K-a}} = \frac{z}{\sqrt{K}}\left(1 + \tfrac{1}{2}\frac{a}{K} + O\left(K^{-2}\right)\right),$$

$$\ln\Gamma(K) = (K - \tfrac{1}{2})\ln K - K + \tfrac{1}{2}\ln(2\pi) + o(1),$$

the first being 496(ApA 2.4), the second the result of applying 496(ApA 2.2) to the expression $(1 - a/K)^{-\frac{1}{2}}$, and the last Stirling's formula 11(N1), we find that, whatever the value of a, as $K \to \infty$,

$$f_Z(z) \to \frac{2}{\sqrt{2\pi}}e^{-2z^2},$$

the pdf of the normal distribution with mean 0 and SD $\tfrac{1}{2}$.

Now note that if $W \sim N\left(\sqrt{K-a}, \tfrac{1}{2}^2\right)$, then

$$\mathbb{E}(W^2) = \mathbb{E}(W)^2 + \mathrm{Var}(W) = K - a + \tfrac{1}{4},$$

so it makes sense to choose $a = \tfrac{1}{4}$.

L. 'Central Limit Theorem for medians'. The following result leads to a simple effective method of obtaining Confidence Intervals in some situations.

La. Theorem. *Suppose that $Y_1, Y_2, \ldots, Y_n$ are IID, each 'continuous' with strictly positive pdf f on $\mathbb{R}$ (or on some subinterval of $\mathbb{R}$). Let m_0 be the median of each Y_k, so that $F(m_0) = \tfrac{1}{2}$, where F is the Distribution Function of each Y_k. Then the Sample Median $M = M_n$ of $(Y_1, Y_2, \ldots, Y_n)$ has approximately the*

$$N\left(m_0, \frac{1}{4nf(m_0)^2}\right)$$

distribution in the precise sense that, as $n \to \infty$,

$$\mathbb{P}\left(2n^{\frac{1}{2}}f(m_0)(M_n - m_0) \le x\right) \to \Phi(x) \quad (x \in \mathbb{R}).$$

Recall that the Sample Median of $(Y_1, Y_2, \ldots, Y_n)$ is defined to be $Y_{(\frac{1}{2}(n+1))}$ if n is odd, and $\frac{1}{2}\left\{Y_{(\frac{1}{2}n)} + Y_{(\frac{1}{2}n+1)}\right\}$ if n is even, where

$$Y_{(1)} < Y_{(2)} < \cdots < Y_{(n)}$$

are the Y_k's arranged in increasing order. (With probability 1, no two of the Y_k's are equal.)

Proof when each Y_k has the U[0, 1] *distribution.* We discuss only the case when n is odd: $n = 2R + 1$. From Subsection 107L, we know that

$$\mathbb{P}(M \in dm) = (2R + 1)\binom{2R}{R} m^R (1 - m)^R dm. \tag{L1}$$

If we put

$$m = \frac{1}{2} + \frac{1}{2}\frac{w}{\sqrt{2R}}, \qquad dm = \frac{dw}{2\sqrt{2R}}$$

and use

$$\binom{2R}{R}\frac{1}{2^{2R}} \sim \frac{1}{\sqrt{\pi R}} \quad \text{(see 11(N2))}, \qquad \left(1 - \frac{w^2}{2R}\right)^R \to e^{-\frac{1}{2}w^2},$$

we find that, as $n \to \infty$, the right-hand side of (L1) converges to $\varphi(w)dw$, where φ is the pdf of N(0, 1). Hence, for large n, we have approximately

$$M = \frac{1}{2} + \frac{W}{2\sqrt{2R}} \quad \text{where} \quad W \sim \text{N}(0, 1).$$

This proves the result – even rigorously because of Fact 161Ea.   □

Proof of the general case for odd n. Each $F(Y_k)$ is U[0, 1], by Exercise 51Ba. But $F(M)$ is the Sample Median of the $F(Y_k)$, and so, with obvious notation,

$$\frac{1}{2} + \frac{W}{2\sqrt{2R}} = F(M) \approx F(m_0) + (M - m_0) f(m_0)$$

and, since $F(m_0) = \frac{1}{2}$, we have $(M - m_0) \approx W/\{2f(m_0)\sqrt{n}\}$. The result follows, and it is not difficult to make this step fully rigorous.

5.5   Characteristic Functions (CFs)

▶▶ **A. The Characteristic Function φ_X of X.** (Apologies for the fact that φ's denotes CFs and the pdf of the normal – both standard notations.) The CF φ_X of X is the map $\varphi_X : \mathbb{R} \to \mathbb{C}$ defined by

$$\varphi_X(\alpha) := \mathbb{E}\left(e^{i\alpha X}\right) = \mathbb{E}(\cos \alpha X) + i\,\mathbb{E}(\sin \alpha X), \tag{A1}$$

the integral existing for ALL α in $\mathbb{R}$.

One obvious advantage of the CF is that it *is* defined for *all* α. Another advantage is that $|\varphi_X(\alpha)| \leq 1$ for all α, and that φ_X has other nice boundedness properties. A huge additional advantage is that it is possible to calculate φ_X exactly in numerous important cases via Cauchy's miraculous Calculus of Residues which you study in courses on Complex Analysis. Moreover, there is an explicit inversion formula giving the Distribution Function in terms of the Characteristic Function – a strong version of the required Uniqueness Theorem. In particular, if $\int_{\mathbb{R}} |\varphi_X(\alpha)| \, d\alpha < \infty$, then X has pdf f_X given by

$$f_X(x) = \frac{1}{2\pi} \int_{-\infty}^{\infty} e^{-i\alpha x} \varphi_X(\alpha) \, d\alpha. \tag{A2}$$

It is all part of Fourier-transform theory.

We can replace α by $i\alpha$ in the MGF formulae we found for binomial, Poisson, normal, Gamma distributions (provided that we use the correct branch of the logarithm in the last one). Especially, we have

$$\varphi_G(\alpha) = e^{-\frac{1}{2}\alpha^2} \quad (\alpha \in \mathbb{R}) \tag{A3}$$

for $G \sim N(0, 1)$.

B. CF for the standard Cauchy distribution. The standard Cauchy distribution has pdf $\{\pi(1 + x^2)\}^{-1}$ on $\mathbb{R}$. A variable X with the standard Cauchy distribution has no moment μ_k for any $k = 1, 2, 3, \ldots$. But, of course, it has a CF, which is found to be

$$\varphi_X(\alpha) = e^{-|\alpha|}$$

by using Residue Calculus. Note how this checks with (A2).

▶ **C. Fact: Uniqueness Theorem for CFs.** *If* $\varphi_X(\alpha) = \varphi_Y(\alpha)$ *for ALL* $\alpha \in \mathbb{R}$, *then* $F_X(x) = F_Y(x)$ *for all* $x \in \mathbb{R}$.

Notes. It is *not* enough here to have $\varphi_X(\alpha) = \varphi_Y(\alpha)$ for all α in some open interval containing 0: it has to be on the whole of $\mathbb{R}$.

I should also mention that $\varphi_X'(0)$ can exist even for X not in $\mathcal{L}^1$.

▶ **D. The 'Independence means Multiply' Lemma for CFs.** *Suppose that X and Y are independent RVs. Then, for all* $\alpha \in \mathbb{R}$,

$$\varphi_{X+Y}(\alpha) = \varphi_X(\alpha)\varphi_Y(\alpha). \quad \text{(Exercise!)}$$

Da. Exercise: sums of Cauchy RVs. Show that if $X_1, X_2, \ldots, X_n$ are IID RVs each with the standard Cauchy distribution, then the sample mean $\overline{X}$ also has the standard Cauchy distribution. This fact that the mean of Cauchy variables is just as spread as a single variable was discussed at 109N, where it was seen that the median *does* concentrate. For more on the Cauchy median, see 196Ec.

▶▶ **E. Fact: Lévy's Convergence Theorem for CFs.** *Let X_n and X be RVs. We have $\varphi_{X_n}(\alpha) \to \varphi_X(\alpha)$ for all $\alpha \in \mathbb{R}$ if and only if $F_{X_n}(x) \to F_X(x)$ at every point x at which F is continuous.*

This is the perfect mathematical result which can be used to prove convergence to (say) Cauchy distributions as much as to normal ones. (For a wide class of variables without finite variance, there are other forms of the CLT involving different normalizations from $1/\sqrt{n}$ and non-normal limits.)

▶ **F. Note on 'CF' proof of the CLT.** For the details of rigour, see, for example, [W]. The heuristics are obvious. Take $\mu = 0$ and $\sigma = 1$ in the CLT. Then

$$\varphi_X(\alpha) = 1 - \tfrac{1}{2}\alpha^2 + o(\alpha^2) \quad \text{as } \alpha \to 0.$$

Hence,

$$\varphi_{S_n^*}(\alpha) = \varphi_X\left(\frac{\alpha}{\sqrt{n}}\right)^n = \left(1 - \frac{\alpha^2}{2n} + o\left(\frac{\alpha^2}{n}\right)\right)^n \to e^{-\frac{1}{2}\alpha^2} = \varphi_G(\alpha),$$

and the result follows.

The miracle is not lost: it has been shifted from the result to the proof.

6

CONFIDENCE INTERVALS FOR ONE-PARAMETER MODELS

6.1 Introduction

▶▶ **A. Overview.** This chapter presents a thorough treatment of both Frequentist and Bayesian theories (both of which you should learn) for the simplest, namely one-parameter, situation. The principles studied here extend to the multi-parameter case, and we shall see several of the extensions in later chapters.

We have in mind three notations:

$\mathbf{Y} = (Y_1, Y_2, \ldots, Y_n)$ for Observations as PreStatistics;
$\mathbf{y}^{\text{obs}} = \left(y_1^{\text{obs}}, y_2^{\text{obs}}, \ldots, y_n^{\text{obs}}\right)$ for the corresponding *actual* statistics after the experiment is performed, so that $y_k^{\text{obs}} = Y_k(\omega^{\text{act}})$: $\mathbf{y}^{\text{obs}}$ is the **realization** of $\mathbf{Y}$;
$\mathbf{y} = (y_1, y_2, \ldots, y_n)$ for *possible* values taken by $\mathbf{Y}$.

For the sample mean of $\mathbf{y}^{\text{obs}}$ we shall write $\bar{y}_{\text{obs}}$ (rather than $\overline{y^{\text{obs}}}$ or $\bar{y}^{\text{obs}}$) for typographical reasons.

As far as possible, I stick to the above conventions when it helps clarity; but there are times when sticking to them too ruthlessly becomes so cumbersome as to be unhelpful. So, sometimes, whether Tyche (or Zeus – see later!) is choosing Y_k, y_k^{obs} or y_k is essentially irrelevant: you will understand when you read on.

▶▶ **The probabilistic model.** What exactly are ω and Ω in this context, and how are the Random Variables $Y_1, Y_2, \ldots, Y_n$ functions on Ω? Well, a typical sample point ω is a typical observation or sample $\mathbf{y} = (y_1, y_2, \ldots, y_n)$ and then $Y_k(\omega) = y_k$, so

$$\omega = \mathbf{y} = (y_1, y_2, \ldots, y_n), \quad Y_k(\omega) = y_k, \quad \mathbf{Y}(\omega) = \mathbf{y}, \quad \mathbf{Y}(\omega^{\text{act}}) = \mathbf{y}^{\text{obs}}.$$

Of course, the sample space Ω is the set of all possible samples ω. The probability measure on (measurable) subsets of Ω will depend on a parameter θ, and we shall write $\mathbb{P}(B \mid \theta)$ for the probability of the event B when the parameter is θ. Exactly how to define this probability for the context in which we shall be working, will be seen at 181(A2) below. The notation $\mathbb{P}(B \mid \theta)$ has no 'conditional probability' significance in Frequentist theory: it is just a convenient notation. In Bayesian theory, $\mathbb{P}(B \mid \theta)$ does have a 'conditional probability' significance. I everywhere try to bring the notation and terminology (and results!) of the two approaches into line, thereby undoubtedly offending both schools. The notation $\mathbb{P}(B; \theta)$ would better reflect the Frequentist view.

▶▶ **Notational Nightmare.** The standard notations of Probability and of Statistics are often in conflict. Thus, for example, Ω always denotes the *sample* space in Probability, but often denotes the *parameter* space in Statistics. More importantly for us, probabilists always write S_n for *sum*, writing

$$S \;=\; S_n \;=\; Y_1 + Y_2 + \cdots + Y_n, \qquad \textbf{(We use)}$$

which takes the forms

$$s_{\text{obs}} \;=\; y_1^{\text{obs}} + y_2^{\text{obs}} + \cdots + y_n^{\text{obs}}, \qquad \textbf{(We use)}$$
$$s \;=\; y_1 + y_2 + \cdots + y_n, \qquad \textbf{(We use)}$$

for observed and possible values. However, statisticians use

$$s_n^2 \;=\; \tfrac{1}{n} \sum (y_k - \overline{y})^2, \qquad \text{(We do \textbf{NOT} use)}$$
$$s_{n-1}^2 \;=\; \tfrac{1}{n-1} \sum (y_k - \overline{y})^2, \qquad \text{(We do \textbf{NOT} use)}$$

for variance estimates, with s_n and s_{n-1}, their square roots, as standard-deviation estimates. We use (in later chapters)

$$\hat{\sigma}_{n-1}(\mathbf{Y}) \;=\; \sqrt{\frac{1}{n-1} \sum \left(Y_k - \overline{Y}\right)^2}, \qquad \textbf{(We use)}$$

even though $\hat{\sigma}_{n-1}(\mathbf{Y})$ is not the Maximum-Likelihood Estimator. This is an occasion where in strict accordance with our 'upper-case versus lower-case' conventions, we should use $\hat{\Sigma}_{n-1}(\mathbf{Y})$, but the Σ sign suggests a sum rather than an Estimated Standard Deviation.

If my use of s_{obs} makes statisticians weep, so be it.   □

The one-parameter model which we assume throughout the Frequentist parts of this chapter is that

$Y_1, Y_2, \ldots, Y_n$ **are IID each with pdf or pmf** $f(y \,|\, \theta)$ **depending on an unknown parameter** θ **which lies in a subinterval of** $\mathbb{R}$.

Thus for example $f(y \,|\, \theta)$ could be the pdf of $N(\theta, 1)$ (with $\theta \in \mathbb{R}$) or of $E(\text{mean } \theta)$ (with $\theta \in (0, \infty)$) or the pmf of $Poisson(\theta)$ (with $\theta \in (0, \infty)$) or of $Bernoulli(\theta)$ (with $\theta \in [0, 1]$).

We are interested in obtaining Confidence Intervals (CIs) or Confidence Regions for θ. In the last section of the chapter, I discuss how to test hypotheses on θ if you must.

The **Likelihood Function**

$$\text{lhd}(\theta; \mathbf{y}) = \text{lhd}(\theta; y_1, y_2, \ldots, y_n) = f(y_1|\theta)f(y_2|\theta)\ldots f(y_n|\theta). \quad \text{(A1)}$$

is the second most important thing in (both Frequentist and Bayesian) Statistics, the first being common sense, of course. Here θ denotes the unknown parameter and $\mathbf{y} = (y_1, y_2, \ldots, y_n)$ a possible vector of observations. As a function of $\mathbf{y}$ for fixed θ, $\text{lhd}(\theta; \mathbf{y})$ gives the likelihood ('joint' pmf or pdf) of obtaining the observations $\mathbf{y}$ if the true parameter value is θ. There will be a study of joint pmfs and pdfs in the general context in the next chapter. The independence assumptions made in the current chapter simplify matters greatly. In applications, the validity of such assumptions needs to be examined carefully.

In Statistics, we concentrate on the function $\text{lhd}(\theta; \mathbf{y}^{\text{obs}})$ as a function of θ for the actual observations $\mathbf{y}^{\text{obs}}$.

This is reflected in the fact that θ appears before $\mathbf{y}^{\text{obs}}$ in the notation. A value of θ for which $\text{lhd}(\theta; \mathbf{y}^{\text{obs}})$ is 'large' for the observed data $\mathbf{y}^{\text{obs}} = \mathbf{Y}(\omega^{\text{act}})$ is one which tallies well with the data.

We begin the chapter with commonsense ways of obtaining Frequentist Confidence Intervals for θ for familiar distributions. Then we start to develop a general theory.

Fisher's brilliant concept of **sufficient statistic** justifies our 'collapsing' of the full information $y_1^{\text{obs}}, y_2^{\text{obs}}, \ldots, y_n^{\text{obs}}$ to information just about a lower-dimensional statistic such as $\overline{y}_{\text{obs}}$ in certain important cases. **Ancillary statistics** represent a step in the opposite direction, representing information which we must take fully into account. (This last sentence *will* be clarified later!)

We look briefly at **Point Estimation**, where one wants to give a single 'best estimate' of an unknown parameter, but do so largely because key ideas such as **Fisher information** help clarify the fundamentally important theorem on **Asymptotic normality of Maximum-Likelihood Estimators** (Frequentist) **or of posterior distributions** (Bayesian).

We then study the **Bayesian theory of CIs**. Apart from other major advantages (and some disadvantages), this *greatly* clarifies the theories of sufficient statistics and of ancillary statistics, and copes effortlessly with the important matter of predicting a new observation, a task over which Frequentist theory can stumble. However, Bayesian *practice* is a more messy matter than the elegant Bayesian theory.

In the last section of the chapter, we look briefly at **Hypothesis Testing** from both Frequentist and Bayesian viewpoints. I believe that (even within the terms of Frequentist theory) the fundamental Likelihood-Ratio Test needs some modification. I shall disagree strongly with claims that Bayesian and Frequentist theories are sometimes in serious conflict for large sample sizes.

In Subsection 236J we take a first look at the fundamental problem of **choosing a model**, with particular attention to Akaike's Information Criterion (AIC).

Note on terminology. In fact, Bayesians speak of **Credible Intervals** (credo: I believe). They *never* use the term Confidence Interval, perhaps because Frequentists *do*! However, Confidence Intervals provide one of the more sensible pieces of Frequentist theory, and it seems to me entirely appropriate to transfer the term 'Confidence Interval' to describe the strictly analogous concept on Bayesian theory. A Frequentist CI is then in some important cases a Bayesian CI associated with a particular prior (the 'Frequentist prior'); and Frequentist theory then makes much more sense. (One of the many pieces of heresy in this book is that I *shall* – for certain special situations only – refer to a **Frequentist prior** ($\pi(\theta) = 1$ for a location parameter θ, $\pi(\gamma) = \gamma^{-1}$ on $(0, \infty)$ for a scale parameter γ.) We shall see however that, in spite of heroic efforts, attempts at a general theory of prior densities which represent 'no prior information' are far from being entirely satisfactory.

▶ **B. Regular pdfs.** There are lots of occasions when we need to differentiate an integral with respect to a parameter. For example, we might wish to show that

$$\int_0^\infty \theta e^{-\theta y} dy = 1 \text{ implies that } \int_0^\infty (1 - \theta y) e^{-\theta y} dy = 0$$

by differentiating the first integral with respect to θ. Note the effortless way that this shows that the expectation of an exponential variable with rate θ is $1/\theta$. *The key requirement in differentiating through an integral is that the range of integration does not depend upon θ.* I am not going to fuss over the rigour of things in this book. A rigorous result on when differentiation through an integral is justified is given in, for example, [W], Chapter A16. The idea there is to transform things into theorems on integration in Measure Theory.

By a *regular pdf*, we mean one with which one can safely differentiate through various integrals.

We shall *see* that things work in many examples; and that is good enough for now.

6.2 Some commonsense Frequentist CIs

▶ **ORIENTATION.** 'Commonsense' in the title of this section is meant to convey that the main purpose of the section is to present important situations in which you may easily calculate *sensible* (sometimes rough-and-ready) Frequentist Confidence Intervals (CIs) without yet having any theoretical justification that these CIs are in any sense best.

For the time being, we stick to the 'Frequency School' or 'Frequentist' concept of a Confidence Interval. It is appropriate to use Frequentist Confidence Intervals when

- *either* one has no reliable prior information about the values of the unknown parameter in which one is interested,

- *or* one is using a sample so large that one's conclusions are insensitive to prior information (something clarified when we study the Bayesian approach!).

▶▶ **A. The Frequentist Definition of a Confidence Interval.** To keep things simple, suppose that the probabilistic structure of observations $Y_1, Y_2, \ldots, Y_n$ (Pre-Statistics) to be made in an experiment yet to be performed, is completely determined by one unknown parameter $\theta \in \mathbb{R}$ for which we seek a $C\%$ CI. We write

$$\mathbb{P}(E \mid \theta)$$

for the probability of an event E if the true value of the parameter is θ. In the Frequentist approach, the notation has no 'conditional probability' significance.

Examples. (a) The RVs $Y_1, Y_2, \ldots, Y_n$ might have the form

$$Y_k = \theta + \varepsilon_k, \qquad \varepsilon_1, \varepsilon_2, \ldots, \varepsilon_n \text{ IID, each } N(0, 1).$$

For example, θ might be the true value of a quantity to be measured, and ε_k the error in the kth measurement (all in suitable units). It is being assumed that long experience has convinced us that this model is reasonable.

(b) We might be interested in estimating the fraction θ of the adult population who will vote for some particular political party. We take a large sample of size n, and let $Y_k = 1$ if the kth chosen person says that he/she will vote for that party, $Y_k = 0$ if not.

Aa. Definition. Suppose that we can find functions τ_- and τ_+ mapping $\mathbb{R}^n$ to $\mathbb{R}$ such that, for every θ,

$$\mathbb{P}\left(\tau_-(\mathbf{Y}) \leq \theta \leq \tau_+(\mathbf{Y}) \,\middle|\, \theta\right)$$

$$= \mathbb{P}\left(\tau_-(Y_1, Y_2, \ldots, Y_n) \leq \theta \leq \tau_+(Y_1, Y_2, \ldots, Y_n) \,\middle|\, \theta\right) = C\%.$$

(A1)

Thus, *whatever the true value of* θ, the probability that the **random** interval $[\tau_-(\mathbf{Y}), \tau_+(\mathbf{Y})]$ contains θ is $C\%$. *After* the performance of the experiment, the Pre-Statistics $Y_1, Y_2, \ldots, Y_n$ have been crystallized into actual statistics $y_1^{\text{obs}}, y_2^{\text{obs}}, \ldots, y_n^{\text{obs}}$ where $y_k^{\text{obs}} = Y_k(\omega^{\text{act}})$, the point ω^{act} being Tyche's choice of ω from Ω. See 36B and 48B. *We say that*

$$\left[\tau_-(\mathbf{y}^{\text{obs}}), \, \tau_+(\mathbf{y}^{\text{obs}})\right]$$

$$:= \left[\tau_-\left(y_1^{\text{obs}}, y_2^{\text{obs}}, \ldots, y_n^{\text{obs}}\right), \, \tau_+\left(y_1^{\text{obs}}, y_2^{\text{obs}}, \ldots, y_n^{\text{obs}}\right)\right]$$

is a $C\%$ *CI for* θ.

Frequentists, as their name implies, interpret this in LTRF (Long-Term Relative Frequency) terms. We have sorted out the Strong Law, so will no longer be fussy about precise wording of the LTRF idea.

Ab. Interpretation.   A Frequentist will say that *if he were to repeat the experiment lots of times, each performance being independent of the others, and announce after each performance that, for that performance,*

$$\theta \in [\tau_-(observed\ \mathbf{y}), \, \tau_+(observed\ \mathbf{y})],$$

then he would be right about $C\%$ *of the time*, and would use THAT as his interpretation of the statement that on the *one* occasion that the experiment *is* performed with result $\mathbf{y}^{\text{obs}}$, the interval $[\tau_-(\mathbf{y}^{\text{obs}}), \tau_+(\mathbf{y}^{\text{obs}})]$ is a $C\%$ CI for θ.

Once   the   experiment   has   been   performed, $[\tau_-(\mathbf{y}^{\text{obs}}), \tau_+(\mathbf{y}^{\text{obs}})]$ is a *deterministic* interval with actual numbers as endpoints. From the Frequentist standpoint, either $\theta \in [\tau_-(\mathbf{y}^{\mathrm{obs}}), \tau_+(\mathbf{y}^{\mathrm{obs}})]$ in which case, for any concept 'Prob' of 'probability', $\mathrm{Prob}\big(\theta \in [\tau_-(\mathbf{y}^{\mathrm{obs}}), \tau_+(\mathbf{y}^{\mathrm{obs}})]\big) = 1$ or $\theta \notin [\tau_-(\mathbf{y}^{\mathrm{obs}}), \tau_+(\mathbf{y}^{\mathrm{obs}})]$ in which case $\mathrm{Prob}\big(\theta \in [\tau_-(\mathbf{y}^{\mathrm{obs}}), \tau_+(\mathbf{y}^{\mathrm{obs}})]\big) = 0$. A Frequentist *cannot* say that

$$\mathrm{Prob}\big(\theta \in [\tau_-(\mathbf{y}^{\mathrm{obs}}), \tau_+(\mathbf{y}^{\mathrm{obs}})]\big) = C\%. \tag{A2}$$

There are many echoes here of the 'Homer Simpson' situation at 4C.

▶ **B. CIs for an IID $N(\theta, 1)$ sample.** Now assume that

$$Y_k = \theta + \varepsilon_k, \quad \varepsilon_1, \varepsilon_2, .., \varepsilon_n \text{ IID, each } N(0, 1).$$

We know that $\overline{Y} - \theta$ has the $N(0, 1/n)$ distribution. In that its distribution does not depend on the unknown parameter θ, it is called a **Pivotal Random Variable**. Hence if we choose a with $\Phi(a) - \Phi(-a) = C\%$, then

$$\mathbb{P}\left(-an^{-\frac{1}{2}} \leq \overline{Y} - \theta \leq an^{-\frac{1}{2}} \,\Big|\, \theta\right) = C\%,$$

whence (logic!)

$$\mathbb{P}\left(\overline{Y} - an^{-\frac{1}{2}} \leq \theta \leq \overline{Y} + an^{-\frac{1}{2}} \,\Big|\, \theta\right) = C\%$$

and, after the experiment is performed with observed sample mean $\overline{y}_{\mathrm{obs}}$, we say that

$$\left[\overline{y}_{\mathrm{obs}} - an^{-\frac{1}{2}}, \overline{y}_{\mathrm{obs}} + an^{-\frac{1}{2}}\right] \tag{B1}$$

is a $C\%$ CI for θ. There are lots of other $C\%$ CIs for θ, but this is the best (shortest) two-sided one.

Ba. Near-heresy, 1. Numerically – but in the Frequentist approach definitely NOT philosophically – we obtain a $C\%$ CI $[\tau_-, \tau_+]$ for θ by pretending that θ is a Random Variable Θ with the $N(\overline{y}_{\mathrm{obs}}, \frac{1}{n})$ distribution, and so requiring that $\mathbb{P}(\Theta \in [\tau_-, \tau_+]) = C\%$. Thus if $\pi(\theta \,|\, \mathbf{y}^{\mathrm{obs}})$ (the '*posterior density function*') denotes the pdf of the $N(\overline{y}_{\mathrm{obs}}, \frac{1}{n})$ distribution, then (you check!) an interval I is a $C\%$ CI for θ if and only if

$$\int_I \pi(\theta \,|\, \mathbf{y}^{\mathrm{obs}}) \mathrm{d}\theta = C\%.$$

This is sound Frequentist stuff (well, more-or-less), with Bayesian notation. The 'best' (shortest) $C\%$ CI will clearly take the form

$$\{\theta : \pi(\theta \,|\, \mathbf{y}^{\mathrm{obs}}) \geq c\},$$

and it will be that at (B1).

▶▶ **Bb. Determining sample size in advance.** If, as is being assumed, we really know from past experience that errors are IID each $N(0, 1)$, and we want (say) a 95% CI for θ of length at most 0.2, then we require that $2an^{-\frac{1}{2}} \leq 0.2$, where $a = 1.96$, so that *we know in advance that we shall need at least* 385 *observations*.

Bc. One-sided CIs. It is sometimes appropriate to decide in advance of an experiment to obtain a CI for θ of the form $\{\theta : \theta \leq B\}$ for a constant B. Now, for Pre-Statistics,

$$\mathbb{P}\left(\overline{Y} > \theta - bn^{-\frac{1}{2}} \,|\, \theta\right) = \mathbb{P}\left(\theta < \overline{Y} + bn^{-\frac{1}{2}} \,|\, \theta\right) = C\%,$$

where $1 - \Phi(-b) = C\%$, or, using the symmetry of φ, $\Phi(b) = C\%$. Thus

$$\left(-\infty, \overline{y}_{\text{obs}} + bn^{-\frac{1}{2}}\right]$$

is a $C\%$ CI.

One really must decide *before* the experiment whether to obtain one-sided or two-sided CIs. Analyzing data in ways which depends on the data can be dangerous, but there are many occasions where it is unavoidable.

From now on, I generally stick to two-sided CIs, leaving you to develop one-sided ones.

C. CIs for samples from an exponential distribution.
We shall see later that certain distributions must be exponential to a high degree of accuracy.

Let $Y_1, Y_2, \ldots, Y_n$ be IID RVs, each exponential with MEAN θ (and hence rate-parameter $\lambda = 1/\theta$). Then, each Y_k/θ has the exponential distribution of mean 1, whence $\sum Y_k/\theta = n\overline{Y}/\theta$ has the Gamma$(n, 1)$ distribution. We see that $n\overline{Y}/\theta$ is a Pivotal RV. We can therefore find b, c such that

$$\mathbb{P}\left(b \leq \frac{n\overline{Y}}{\theta} \leq c \,\bigg|\, \theta\right) = C\% = \mathbb{P}\left(\frac{n\overline{Y}}{c} \leq \theta \leq \frac{n\overline{Y}}{b} \,\bigg|\, \theta\right),$$

giving an *exact* $C\%$ CI $[n\overline{y}_{\text{obs}}/c, n\overline{y}_{\text{obs}}/b]$ after the experiment.

Ca. More near-heresy. Let $s_{\text{obs}} = \sum y_k^{\text{obs}}$. Numerically – but in the Frequentist approach definitely NOT philosophically – we can pretend that the rate parameter $\lambda = 1/\theta$ is a Random Variable Λ with the Gamma$(n, \text{rate } s_{\text{obs}})$ distribution in that an interval I will be a $C\%$ CI for λ if and only if $\mathbb{P}(\Lambda \in I) = C\%$; it's as if Λs_{obs} is Gamma$(n, 1)$. You check this. We could find the shortest $C\%$ CI on the computer, but there is not much point in doing so.

Cb. CLT approximation. If n is large, then the CLT used via 162(F2) yields an *approximate* 90% CI

$$\left[\frac{\overline{y}_{\text{obs}}}{1 + a/\sqrt{n}}, \frac{\overline{y}_{\text{obs}}}{1 - a/\sqrt{n}}\right]$$

for θ, where $a = 1.645$.

Suppose, for example, that $n = 25$ and $\overline{y}_{\text{obs}} = 2.3$, and we want a 90% CI for the mean θ. From 149(E1), we may take b to be half of the entry for χ^2_{50} in the table on p517 with 95% tail probability, namely $b = \frac{1}{2} \times 37.46$. Similarly $c = \frac{1}{2} \times 67.50$, and our *exact* 90% CI is $[1.70, 3.31]$. The CLT method gives the *approximate* 90% CI $[1.73, 3.43]$ – quite good.

▶ **D. Difficulties with discrete variables.** If the Random Variables $Y_1, Y_2, \ldots, Y_n$ are discrete, then we are not going to be able to arrange that 174(A1) holds exactly for every θ, since, if for example, each Y_k can take only finitely many values, then for each θ, the probability $\mathbb{P}\big(\tau_-(\mathbf{Y}) \le \theta \le \tau_+(\mathbf{Y}) \mid \theta\big)$ can take only finitely many values.

The sensible thing to do is to require that 174(A1) hold approximately for each θ of real interest, giving an *approximate* $C\%$ CI for θ. Alternatively, we can insist that

$$\mathbb{P}\Big(\tau_-(\mathbf{Y}) \le \theta \le \tau_+(\mathbf{Y}) \mid \theta \Big) \ge C\%$$

for every θ, leading to a **'conservative'** $C\%$ **CI** for θ: we are 'at least $C\%$ confident' that our CI contains the true value.

Da. Exercise. Suppose that n is large and $Y_1, Y_2, \ldots, Y_n$ are IID, each Poisson(λ). Use Exercise 162G to show how to find an approximate 95% CI for λ.

Important Note. There are great difficulties in obtaining exact CIs when we have 'continuous' pdfs outside the exponential family defined in Subsection 184E. *Hence, we are often forced to rely on approximate CIs in the 'continuous' context.* We shall see later how Maximum-Likelihood Estimation can allow us to achieve reasonable approximations on the computer.

▶ **E. CIs for proportions.** Consider the situation mentioned earlier, in which we might be interested in estimating the fraction θ of the adult population who will vote for some particular political party. We take a large sample of size n, and let $Y_k = 1$ if the chosen person says that he/she will vote for that party, $Y_k = 0$ if not.

Because our sample size, though large, will be much smaller than the total population size, we can ignore the difference between sampling without replacement and sampling with replacement, and assume that $Y_1, Y_2, \ldots, Y_n$ are IID each Bernoulli(θ) so that

$$\mathbb{P}(Y_k = 1) = \theta = 1 - \mathbb{P}(Y_k = 0) \qquad \text{for each } k.$$

Then

$$\mathbb{E}\,(Y_k) = \theta, \quad \text{Var}\,(Y_k) = \theta(1 - \theta), \quad \mathbb{E}\,(\overline{Y}) = \theta, \quad \text{SD}\,(\overline{Y}) = \sqrt{\frac{\theta(1 - \theta)}{n}},$$

and the CLT gives, for large n, the approximate result

$$\mathbb{P}\left(\left|\overline{Y} - \theta\right| \le 1.96\sqrt{\frac{\theta(1 - \theta)}{n}} \right) \approx 95\%.$$

If we wish our Estimator $\overline{Y}$ of θ to lie within 0.01 of the true value of θ with 95% probability, we must ensure that

$$1.96\sqrt{\theta(1-\theta)/n} \leq 0.01.$$

Now, $\sqrt{\theta(1-\theta)}$ is always at most 0.5, and if θ lies anywhere close to 0.5 (say, even between 0.3 and 0.7), then $\sqrt{\theta(1-\theta)}$ is acceptably close to 0.5. Since 1.96×0.5 is 1 (as nearly as makes no difference), then, in most cases of interest to politicians,

$$\mathbb{P}\left(|\overline{Y}-\theta| \leq 1/\sqrt{n}\right) \approx 95\%,$$

so that to get the desired accuracy, we need to take a sample of size at least 10000.

If $\overline{Y}$ differs significantly from 0.5 (or even if it doesn't!), we can estimate $\sqrt{\theta(1-\theta)}$ by $\sqrt{\overline{Y}(1-\overline{Y})}$ to get an approximate 95% CI for θ after the experiment is performed of the form

$$\left[\overline{y}_{\text{obs}} - 2\sqrt{\overline{y}_{\text{obs}}(1-\overline{y}_{\text{obs}})/n},\ \overline{y}_{\text{obs}} + 2\sqrt{\overline{y}_{\text{obs}}(1-\overline{y}_{\text{obs}})/n}\right]. \tag{E1}$$

Ea. Notes.   In practice, no-one takes such a big sample. Samples of size 1000 are common, leading of course to a CI about $\sqrt{10} = 3$ times as wide. The inevitability of bias in sampling procedures makes taking larger samples just not worth the effort.

One often finds in practice an unwillingness of administrators either to believe, or (if they do) to act upon, the 'square root' phenomenon which means that very large sample sizes are necessary for reasonably good accuracy.

Eb. Instructive Exercise.   (This is a *two*-parameter example, but who cares?!) Suppose that one takes a large sample of size n from a population. Of these S_1 say that they will vote for Party 1, S_2 for Party 2. Let p_1 and p_2 be the fractions of the whole population who will vote for Party 1 and Party 2 respectively. Show that (assuming the independence associated with sampling with replacement) we have $\mathbb{E}\left(S_2 - S_1\right) = n\left(p_1 - p_2\right)$ and

$$\text{Var}\left(S_2 - S_1\right) = n\left[p_1 + p_2 - (p_2 - p_1)^2\right].$$

How do you find an approximate 95% CI for $p_2 - p_1$ after the experiment is performed?

F. An Important Example.   (Another two-parameter one!) Suppose that one wants to compare the effectiveness of two Treatments A and B in curing some new disease, and that a randomly chosen 1000 diseased people are given Treatment A and 1000 Treatment B with the results in Table F(i). Let p_A [respectively, p_B] be the probability that a person treated with A [B] will recover.

Fa. Important Exercise.   Practitioners in the field use the *difference in log-odds-ratios*:

$$\ln\frac{p_B}{1-p_B} - \ln\frac{p_A}{1-p_A}$$

to quantify difference in effectiveness, but we can here safely use $p_B - p_A$ which is more comprehensible to the layman and me. In regard to PreStatistics, for the kth person

Treatment	Recover	Die	Total
A	870	130	1000
B	890	110	1000
Total	1760	240	2000

Table F(i): Treatment results

treated with A [respectively, B], let Y_k [respectively, Z_k] be 1 if the person recovers, 0 if not. Assume that $Y_1, \ldots, Y_n, Z_1, \ldots, Z_n$ are independent, where $n = 1000$. Show that $\mathbb{E}\left(\overline{Z} - \overline{Y}\right) = p_B - p_A$ and

$$\text{Var}\left(\overline{Z} - \overline{Y}\right) = \left\{ p_A \left(1 - p_A\right) + p_B \left(1 - p_B\right)\right\}/n.$$

Obtain the rough-and-ready '2 estimated SDs of $\overline{Z} - \overline{Y}$ to either side' 95% CI $[-0.009, 0.049]$ for $p_B - p_A$. The way in which this relates to the χ^2 test (which is not appropriate here) will be explained later.

▶ **G. Confounding.** Confounding is one of statisticians' worst nightmares. Here are two instances. (We shall later meet confounding again in connection with regression.)

Ga. Smoking and lung cancer: 'confusion worse confounded'. The percentage of smokers who develop lung cancer is without question greater than the percentage of non-smokers who develop it. That in itself shows only that smoking and lung cancer are *associated*: it does not prove that smoking *causes* lung cancer. Sir Ronald Fisher, greatest of statisticians and one of the greatest geneticists, thought that there is a *confounding factor*, something in the genetic makeup of some people which makes them both (a) more susceptible to lung cancer and (b) more inclined to become smokers: thus, smoking and lung cancer would be associated because in many of the cases where they occur simultaneously, they have a common genetic cause. Later studies showed conclusively that Fisher was wrong in thinking that such confounding could be a *complete* explanation: we now know that smoking does *cause* lung cancer. But the case shows that one always has to be on the lookout for confounding factors.

Gb. Simpson's Paradox; the perverse behaviour of ratios. (The 'Simpson' here refers to real statistician E H Simpson, not to Homer, Marge and family. The paradox had long been known in connection with bowling averages in cricket.) I present the paradox in an extreme form. Table G(i) summarizes the situation which I now describe.

Suppose that 6M (6 million) men and 6M women were infected with a certain disease, making a total of 12M people. First we are told that 6M of these people were given a certain treatment, of whom 2M recovered and 4M died; and that of the 6M who were not treated, 4M recovered and 2M died. If this is all the information we have, then, clearly, the treatment should not be given to people.

	People		Men		Women	
	Recover	Die	Recover	Die	Recover	Die
Treated	2M	4M	1M	0	1M	4M
Untreated	4M	2M	4M	1M	0	1M

Table G(i): 'Simpson Paradox' table

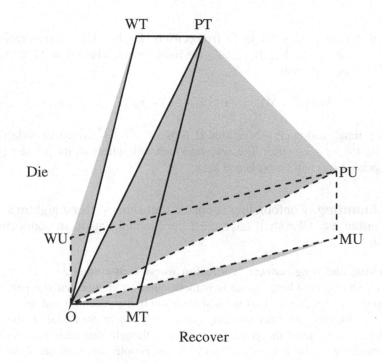

Figure G(ii): Simpson's Paradox: a picture

However, we are then told that amongst men, only 1M were treated, *all* of whom recovered, and that of the 5M men not treated, 4M recovered. We deduce that amongst women, 5M were treated of whom 1M recovered, and 1M were not treated and *all* of those died. Given all this information, it is certainly best to treat a man, who is thereby guaranteed a cure; and it is certainly best to treat a woman, for otherwise she will certainly die.

The People table *confounds* the two disparate populations of men and women, and leads to a false conclusion. It seems hardly fair that Statistics, already having to face difficult philosophical problems, should have to deal in addition with the fact that ratios can behave in a perverse way totally counter to one's first intuition. A picture (Figure G(ii)) might help by providing a way of 'seeing' our example.

In the figure, 'MT' signifies 'Men Treated', etc. The shaded areas reflect the 2×2 tables. The MU vector is steeper than the MT, and the WU vector is steeper than the WT, but the sum PU is less steep than the PT. The lengths of the lines are as important as their directions.

Of course, one is now inclined to worry, given *any* 2×2 table as one's only information, that it might split into two *natural* tables which would lead one to a different conclusion. Nothing in the table itself can indicate whether or not it is 'confounded'.

6.3 Likelihood; sufficiency; exponential family

First, we define the likelihood function.

▶▶ **A. Likelihood Function** $\mathrm{lhd}(\theta; \mathbf{y})$. (This definition will be illustrated by many examples in Subsection C.)

Again, suppose that $Y_1, Y_2, \ldots, Y_n$ are IID RVs each with pdf (or pmf, if each Y_k is discrete) $f(y \mid \theta)$, θ being an unknown parameter. For a possible outcome $y_1, y_2, \ldots, y_n$, we define

$$\mathrm{lhd}(\theta; \mathbf{y}) = \mathrm{lhd}(\theta; y_1, y_2, \ldots, y_n) = f(y_1 \mid \theta) f(y_2 \mid \theta) \ldots f(y_n \mid \theta). \tag{A1}$$

If each Y_k is discrete, then, by independence,

$$\mathbb{P}(Y_1 = y_1; Y_2 = y_2 \ldots; Y_n = y_n \mid \theta)$$
$$= \mathbb{P}(Y_1 = y_1 \mid \theta) \mathbb{P}(Y_2 = y_2 \mid \theta) \ldots \mathbb{P}(Y_n = y_n \mid \theta)$$
$$= f(y_1 \mid \theta) f(y_2 \mid \theta) \ldots f(y_n \mid \theta) = \mathrm{lhd}(\theta; \mathbf{y}),$$

so that $\mathrm{lhd}(\theta; \mathbf{y})$ is the probability of outcome $\mathbf{y}$: thus $\mathrm{lhd}(\theta; \mathbf{y})$ acts as a 'joint pmf' for $(Y_1, Y_2, \ldots, Y_n)$.

If each Y_k is 'continuous' with pdf $f(\cdot \mid \theta)$, then $\mathrm{lhd}(\theta; \mathbf{y})$ acts as a 'joint pdf' for $(Y_1, Y_2, \ldots, Y_n)$ in the sense that (as is proved heuristically below) for any nice (Borel) subset B of $\mathbb{R}^n$,

$$\mathbb{P}(\mathbf{Y} \in B \mid \theta) = \int_B \mathrm{lhd}(\theta; \mathbf{y}) \mathrm{d}\mathbf{y} \tag{A2}$$

$$:= \int \int \ldots \int I_B(\mathbf{y}) \mathrm{lhd}(\theta; y_1, y_2, \ldots, y_n) \mathrm{d}y_1 \mathrm{d}y_2 \ldots \mathrm{d}y_n.$$

If we think of Ω as $\mathbb{R}^n$ the set of all possible samples, then B is the event $\mathbf{Y} \in B$, so the probability measure $\mathbb{P}(B \mid \theta)$ is defined to be either of the integrals at (A2).

Heuristic proof of (A2). Equation (A2) is true if B is a 'hypercube' of the form $\{\mathbf{y} : y_k \in [a_k, b_k] \ (1 \le k \le n)\}$ because then the integral turns into a product

$$\prod_{k=1}^{n} \left(\int_{a_k}^{b_k} f\left(y_k \mid \theta\right) \mathrm{d}y_k \right) = \prod_{k=1}^{n} \mathbb{P}\left(Y_k \in [a_k, b_k] \,\middle|\, \theta \right),$$

tallying with the 'Independence means Multiply' rule

$$\mathbb{P}\left(Y_1 \in [a_1, b_1]; \ldots; Y_n \in [a_n, b_n] \,\middle|\, \theta \right) = \prod_{k=1}^{n} \mathbb{P}\left(Y_k \in [a_k, b_k] \,\middle|\, \theta \right).$$

By standard measure theory, result 181(A2) then holds for all Borel sets B and hence for any set B we need ever consider. It is obvious that 181(A2) holds for sets which are unions of disjoint hypercubes, and intuitively obvious that any nice set B may be approximated by such unions. □

▶▶ **B. Log-likelihood $\ell(\theta; \mathbf{y})$.** It will be very convenient to make the definitions:

$$\ell(\theta; \mathbf{y}) := \ln \mathrm{lhd}(\theta; \mathbf{y}), \qquad f^*\left(y \mid \theta\right) := \ln f(\theta; \mathbf{y}).$$

Then, of course,

$$\ell(\theta; \mathbf{y}) = \sum_{k=1}^{n} f^*\left(y_k \mid \theta\right).$$

▶▶ **C. Important Examples.** The formulae here will be much used later.

Ca. Normal $\mathbf{N}(\theta, 1)$. If each Y_k is $N(\theta, 1)$, then for $y \in \mathbb{R}$,

$$f\left(y \mid \theta\right) = (2\pi)^{-\frac{1}{2}} e^{-\frac{1}{2}(y-\theta)^2}, \qquad f^*\left(y \mid \theta\right) = -\tfrac{1}{2}\ln(2\pi) - \tfrac{1}{2}(y - \theta)^2,$$

so that

$$\ell(\theta; \mathbf{y}) = -\tfrac{1}{2}n\ln(2\pi) - \tfrac{1}{2}\sum_{k=1}^{n}(y_k - \theta)^2$$

$$= -\tfrac{1}{2}n\ln(2\pi) - \tfrac{1}{2}\sum_{k=1}^{n}(y_k - \overline{y})^2 - \tfrac{1}{2}n(\overline{y} - \theta)^2,$$

by the 'Parallel-Axis Theorem'.

Cb. Exponential $\mathbf{E}(\text{mean } \theta)$. If each Y_k is exponential MEAN θ, then, for $y \in [0, \infty)$,

$$f\left(y \mid \theta\right) = \frac{1}{\theta} \exp\left(-\frac{y}{\theta}\right), \qquad f^*\left(y \mid \theta\right) = -\ln\theta - \frac{y}{\theta},$$

so that

$$\ell(\theta; \mathbf{y}) = -n\ln\theta - \frac{1}{\theta}\sum_{k=1}^{n} y_k = -n\ln\theta - \frac{n}{\theta}\overline{y}.$$

Cc. Poisson(θ). If each Y_k is Poisson(θ), then, for $y \in \mathbb{Z}^+$,

$$f(y \mid \theta) = e^{-\theta} \frac{\theta^y}{y!}, \qquad f^*(y \mid \theta) = -\theta + y \ln \theta - \ln(y!),$$

so that

$$\ell(\theta; \mathbf{y}) = -n\theta + n\bar{y} \ln \theta - \sum_{k=1}^{n} \ln(y_k!).$$

Cd. Bernoulli(θ). If each Y_k is Bernoulli(θ), then (you check the convenient expressions of the formulae!) for $y \in \{0, 1\}$,

$$f(y \mid \theta) = \theta^y (1 - \theta)^{1-y}, \qquad f^*(y \mid \theta) = y \ln \theta + (1 - y) \ln(1 - \theta).$$

so that

$$\ell(\theta; \mathbf{y}) = n\{\bar{y} \ln \theta + (1 - \bar{y}) \ln(1 - \theta)\}.$$

▶▶ **D. Sufficient statistics, 1.** It has to be said that Bayesian theory gives a much more satisfying definition of a sufficient statistic. Here we give the first of two Frequentist definitions; but we call it a 'criterion' because the second, given later in the book, is more satisfactory as a definition.

Da. First criterion for sufficiency. Suppose that $Y_1, Y_2, \ldots, Y_n$ are IID, each with pdf (or pmf) $f(\cdot \mid \theta)$. Define

$$\mathrm{lhd}(\theta; \mathbf{y}) = \mathrm{lhd}(\theta; y_1, y_2, \ldots, y_n) = f(y_1 \mid \theta) f(y_2 \mid \theta) \ldots f(y_n \mid \theta),$$

as usual. A statistic

$$t = \tau(\mathbf{y}) = \tau(y_1, y_2, \ldots, y_n)$$

(where τ is some nice function on $\mathbb{R}^n$) is called a sufficient statistic for θ based on the actual sample $y_1, y_2, \ldots, y_n$ if there is factorization

$$\mathrm{lhd}(\theta; \mathbf{y}) = g(\theta; t) h(\mathbf{y}) = g(\theta; \tau(\mathbf{y})) h(\mathbf{y}) \tag{D1}$$

for some functions g and h with the obvious domains. Do note that $h(y)$ **does not involve** θ.

Note that *if t is a sufficient statistic, then so is any one-one function of t.*

The Frequentist interpretation of sufficient statistics. Numerous illustrations of what follows are to be found in this book. Subsection 185F already contains several.

Db. Fact. *Let t be a sufficient statistic, and let T be the PreStatistic which crystallizes to t:*

$$T = \tau(Y_1, Y_2, \ldots, Y_n).$$

Let $f_T(\cdot \mid \theta)$ be the pdf of T when the true value of the parameter is θ.
Then Zeus (father of the Gods) and Tyche can produce IID RVs $Y_1, Y_2, \ldots, Y_n$ each with pdf $f(\cdot \mid \theta)$ as follows.
Stage 1. *Zeus picks T according to the pdf $f_T(\cdot \mid \theta)$. He reports the chosen value of T, but **NOT** the value of θ, to Tyche.*
Stage 2. *Having been told the value of T, but not knowing the value of θ, Tyche performs an experiment which produces the required $Y_1, Y_2, \ldots, Y_n$.*

Since Tyche did not even know the value of θ when she performed Stage 2, the only thing which can matter to making inferences about θ is therefore the value of T chosen by Zeus. In practice, then, given the actual observations $\mathbf{y}^{\text{obs}}$, we must base our inference about θ solely on $t^{\text{obs}} = T(\omega^{\text{act}})$.

I hasten to add that, normally, Tyche knows the value of an unknown parameter. But, in regard to sufficient or ancillary statistics, Zeus sometimes uses Tyche to do a large part of the experiment without her knowing θ; and in these cases, the value of θ is known only to him.

I must stress that while we have been concentrating here on a 1-dimensional sufficient statistic, we generally have to allow our sufficient statistic to be multidimensional. For example, the vector $\mathbf{y}$ of all observations is always a sufficient statistic. (*Note.* The study of Minimal Sufficient Statistics is deferred until Subsection 262D.)

▶▶ **E. The exponential family of pdfs and pmfs.** A pdf (or pmf) $f(y \mid \theta)$ is said to belong to the *exponential family* if it has the form

$$f(y \mid \theta) = \exp\{a(\theta)b(y) + c(\theta) + d(y)\}. \tag{E1}$$

If $Y_1, Y_2, \ldots, Y_n$ are IID each with pdf (or pmf) $f(y \mid \theta)$, then, clearly,

$$t = \sum b(y_k) \text{ is a sufficient statistic for } \theta \text{ based on } y_1, y_2, \ldots, y_n.$$

Amongst regular pdfs, the converse is true: only pdfs (or pmfs) in the exponential family lead to 1-dimensional sufficient statistics. We skip the proof.

Note that the pdfs (or pmfs) of the $N(\theta, 1)$, E(mean θ), Poisson(θ), and Bernoulli(θ) distributions belong to the exponential family. Indeed it is worth making Table E(i), in which we do not include $d(y)$, because it will not be relevant in future discussions (except implicitly). In the Gamma(K_0, mean θ) entry, K_0 is assumed known. Recall that in $N(0, \theta)$, θ signifies variance (not SD).

	$a(\theta)$	$b(y)$	$c(\theta)$
Normal N$(\theta, 1)$	θ	y	$-\frac{1}{2}\theta^2$
Exponential E(mean θ)	$-1/\theta$	y	$-\ln\theta$
Poisson(θ)	$\ln\theta$	y	$-\theta$
Bernoulli(θ)	$\ln\{\theta/(1-\theta)\}$	y	$\ln(1-\theta)$
Gamma$(K_0,$ mean $\theta)$	$-K_0/\theta$	y/K_0	$-K_0\ln\theta$
Normal N$(0, \theta)$	$-1/(2\theta)$	y^2	$-\frac{1}{2}\ln\theta$

Table E(i): Some members of the exponential family

F. Illustrations of Fact Db. The first two examples will be discussed later. It would be too cumbersome to discuss them with our current technology.

Fa. Normal N$(\theta, 1)$. Zeus and Tyche decide to produce IID vectors $Y_1, Y_2, \ldots, Y_n$ each with the N$(\theta, 1)$ distribution. In Stage 1, Zeus chooses $T = \overline{Y}$ with the N$(\theta, 1/n)$ distribution, and reports the chosen value of T to Tyche. In Stage 2, Tyche then chooses the vector $\mathbf{Y}$ in $\mathbb{R}^n$ so that

$$\left(Y_1 - \overline{Y}, Y_2 - \overline{Y}, \ldots, Y_n - \overline{Y}\right)$$

has the standard normal distribution in the hyperplane through $(T, T, \ldots, T)$ perpendicular to $(1, 1, \ldots, 1)$. Then $\mathbf{Y}$ has the desired property.

Fb. Exponential E(mean θ). Here, Zeus and Tyche decide to produce IID vectors $Y_1, Y_2, \ldots, Y_n$ each with the E(mean θ) distribution. Zeus chooses S (which will become $\sum Y_k$) according to the Gamma$(n,$ mean $\theta)$ distribution, and reports the value of S to Tyche. Tyche then picks $n - 1$ points uniformly at random in $[0, S]$, and she labels these points in ascending order as

$$Y_1, Y_1 + Y_2, \ldots, Y_1 + Y_2 + \cdots + Y_{n-1}.$$

Finally, she sets $Y_n = S - (Y_1 + Y_2 + \cdots + Y_{n-1})$.

Fc. Poisson(θ). Now, Zeus and Tyche decide to produce IID vectors $Y_1, Y_2, \ldots, Y_n$ each with the Poisson(θ) distribution. Zeus chooses S according to the Poisson$(n\theta)$ distribution, and reports the value of S to Tyche. She then makes S independent choices each uniformly from the discrete set $\{1, 2, \ldots n\}$, and lets Y_k be the number of times that k is chosen.

Fd. Bernoulli(θ). Finally, Zeus and Tyche decide to produce IID vectors $Y_1, Y_2, \ldots, Y_n$ each with the Bernoulli(θ) distribution. Zeus chooses S according to the Binomial(n, θ) distribution, and reports the value of S to Tyche. She now chooses a random subset K of size S from $\{1, 2, \ldots, n\}$, and sets $Y_k = 1$ if $k \in K$, 0 otherwise.

Fe. Discussion. Consider the **Poisson** case. Let $y_1, y_2, \ldots, y_n \in \mathbb{Z}^+$ and set $s := y_1 + y_2 + \cdots + y_n$. If we make s independent choices of numbers each uniformly within $\{1, 2, \ldots, n\}$, and if η_k is the number of times that k is chosen, then by extending the 'multinomial' arguments we used in Chapter 1, we see that

$$\text{Prob}\,(\eta_1 = y_1; \ldots; \eta_n = y_n) = \frac{s!}{y_1! y_2! \ldots y_n!} \times \frac{1}{n^s},$$

so that

$$\mathbb{P}_{\text{ZT}}\,(Y_1 = y_1; \ldots; Y_n = y_n)$$

$$= \frac{e^{-n\theta}(n\theta)^s}{s!} \times \frac{s!}{y_1! y_2! \ldots y_n!} \times \frac{1}{n^s}$$

$$= \prod_{k=1}^{n} \frac{e^{-\theta}(\theta)^{y_k}}{y_k!} = \mathbb{P}\,(Y_1 = y_1; \ldots; Y_n = y_n),$$

the last probability relating to the case when $Y_1, Y_2, \ldots, Y_n$ are IID each Poisson(θ).

Ff. Exercise. Discuss the **Bernoulli** case.

Fg. Exercise. If Zeus chooses $\sum Y_k^2$ in the $N(0, \theta)$ case, how do you think Tyche then chooses the vector $\mathbf{Y}$? Just give the intuitive answer without worrying about precise formulation.

G. A 'non-regular' example: the U$[0, \theta]$ distribution.
I have no interest in producing silly counterexamples. This example matters in practice.

Suppose that $Y_1, Y_2, \ldots, Y_n$ are IID each with the uniform U$[0, \theta]$ distribution on $[0, \theta]$. Then,

$$f\,(y \mid \theta) = \frac{1}{\theta} I_{[0,\theta]}(y), \qquad \text{lhd}(\theta; y_1, y_2, \ldots, y_n) = \frac{1}{\theta^n} I_{[m,\infty)}(\theta),$$

where $m := \max\,(y_1, y_2, \ldots, y_n)$. Thus, m is a sufficient statistic for θ based on $y_1, y_2, \ldots, y_n$.

Ga. Exercise. Describe intuitively the Zeus–Tyche construction of $(Y_1, Y_2, \ldots, Y_n)$, with Zeus first choosing $M := \max\,(Y_1, Y_2, \ldots, Y_n)$.

▶ ## H. Ancillary statistics; location parameters.
The theory of Ancillary Statistics is an important topic, but a tricky one; and I defer discussion until Subsection 263E.

However, we now examine the implications of the theory for the important case of location parameters, implications studied at 264Eb.

▶ **Ha. Location parameter.** Suppose that $f\,(y \mid \theta) = h(y - \theta)$, where h is a pdf on $\mathbb{R}$. (For example, $f\,(y \mid \theta)$ might be the pdf of the $N(\theta, 1)$ distribution, in which case $h = \varphi$.) The final Frequentist story (see 264Eb) says that CIs may be calculated from $\pi(\theta \mid \mathbf{y}^{\mathrm{obs}})$, where

$$\pi(\theta \mid \mathbf{y}^{\mathrm{obs}}) \propto \mathrm{lhd}(\theta; \mathbf{y}^{\mathrm{obs}}), \tag{H1}$$

the 'constant of proportionality' (depending on $\mathbf{y}^{\mathrm{obs}}$) being chosen so that $\int \pi(\theta \mid \mathbf{y}^{\mathrm{obs}}) d\theta = 1$. (Calculation of this normalizing constant – even on the computer – can prove a real headache, but MCMC methods make its calculation unnecessary.) Thus, for this 1-parameter situation, Frequentist theory agrees exactly with 'Bayesian theory with a uniform prior density', as we shall see.

Hb. Example. Suppose that $Y_1, Y_2, \ldots, Y_n$ are IID each with the uniform distribution on $[\theta - \frac{1}{2}, \theta + \frac{1}{2}]$. Then $\pi(\theta \mid \mathbf{y}^{\mathrm{obs}})$ is the pdf of the uniform distribution on

$$\left[y_{(n)}^{\mathrm{obs}} - \tfrac{1}{2}, y_{(1)}^{\mathrm{obs}} + \tfrac{1}{2} \right],$$

an interval which we know by simple logic MUST contain θ. If we ignore the theory of ancillary statistics, that is, if we concentrate on absolute probabilities consistent with Definition 174Aa, we could miss out on this logic.

Note. This example has special features to do with minimal sufficiency, as we shall see later. But I shall argue there that if Statistics is to have any coherence, then we *must* obtain the correct answer from the general location-parameter result.

6.4 Brief notes on Point Estimation

As explained earlier, this topic is done not so much for its own sake as for the light it throws on Confidence Intervals obtained via Maximum-Likelihood Estimation.

▶ **A. Estimators and estimates.** Consider the usual situation where $Y_1, Y_2, \ldots, Y_n$ are IID, each with pdf (or pmf) $f(y \mid \theta)$, and where, therefore,

$$\mathrm{lhd}(\theta; \mathbf{y}) = \mathrm{lhd}(\theta; y_1, y_2, \ldots, y_n) = f(y_1 \mid \theta) f(y_2 \mid \theta) \ldots f(y_n \mid \theta).$$

An *Estimator T* of θ is just the same as a PreStatistic, and here therefore a function

$$T = \tau(\mathbf{Y}) = \tau(Y_1, Y_2, \ldots, Y_n)$$

of the vector $\mathbf{Y}$ of Observations considered before the experiment is performed. After the experiment, T leads to the *estimate*

$$t^{\mathrm{obs}} = T(\omega^{\mathrm{act}}) = \tau(\mathbf{y}^{\mathrm{obs}}) = \tau\left(y_1^{\mathrm{obs}}, y_2^{\mathrm{obs}}, \ldots, y_n^{\mathrm{obs}} \right),$$

where $y_k^{\mathrm{obs}} = Y_k(\omega^{\mathrm{act}})$ is the value to which Y_k crystallizes.

An Estimator is therefore a rule for getting estimates.

▶ **B. Unbiased Estimators.**   An Estimator T is called *Unbiased for θ if, for every θ*, it is correct 'on average' in that we have

$$\mathbb{E}\left(T \mid \theta\right) = \theta,$$

that is, in full,

$$\int \tau(\mathbf{y})\,\mathrm{lhd}(\theta; \mathbf{y})\mathrm{d}\mathbf{y}$$

$$= \int \int \cdots \int \tau\left(y_1, y_2, \ldots, y_n\right)\mathrm{lhd}(\theta; y_1, y_2, \ldots, y_n)\mathrm{d}y_1\mathrm{d}y_2 \ldots \mathrm{d}y_n = \theta.$$

The integral becomes a sum in the discrete case. We have extended the rule

$$\mathbb{E}\,h(X) \;=\; \int h(x)f_X(x)\mathrm{d}x$$

in the obvious way.

With our definitions, it is meaningless to speak of an unbiased esti*mate*: the estimate is just a number, and could be considered unbiased only if it is exactly equal to the unknown parameter θ.

If each Y_k has mean θ, as is the case for the $N(\theta, 1)$, E(mean θ), Poisson(θ) and Bernoulli(θ) cases we have been studying, then, of course, $\overline{Y}$ is an Unbiased Estimator of θ. We studied an Unbiased Estimator for variance in Exercise 104F.

Ba. Exercise.   Show that if each Y_k is E(rate λ), and $n \geq 2$, then

$$\mathbb{E}\left(\frac{1}{S}\right) = \int_0^\infty \frac{1}{s}\frac{\lambda^n s^{n-1}e^{-\lambda s}}{(n-1)!}\mathrm{d}s = \frac{\lambda}{n-1}$$

where $S = \sum Y_k$, whence $(n-1)/S$ is an Unbiased Estimator of λ.

Bb. Exercise.   Prove that if $Y_1, Y_2, \ldots, Y_n$ are IID each with the $U[0, \theta]$ distribution, then $(n+1)M/n$ is unbiased for θ where $M := \max\left(Y_1, Y_2, \ldots, Y_n\right)$.

Bc. Exercise.   Lehmann showed that sometimes there is only one Unbiased Estimator and it is absurd. I am not claiming that the following example really matters; but it does matter that you can do this exercise! Suppose that Y has the truncated Poisson distribution with pmf $(e^\theta - 1)^{-1}\theta^y/y!$ for $y = 1, 2, 3, \ldots$. Show that the only Unbiased Estimator T of $(1 - e^{-\theta})$ based on Y is obtained by taking $T = 0$ if Y is odd, $T = 2$ if Y is even. Since $1 - e^{-\theta} \in [0, 1)$, this Unbiased Estimator is *really* absurd.

An Unbiased Estimator with small variance is likely to be close to the true value of θ.   There is however an explicitly known lower bound to the value of $\mathrm{Var}(T)$ if $\mathbb{E}\left(T\right) = \theta$. In order to establish this, we need the concept of *Fisher information*, and for that we need the notation and identities in the next subsection.

Terminology: efficiency.   If one Unbiased Estimator has smaller variance than another, it is said to be more efficient.

▶ **C. Useful notation and identities.** Recall that

$$\ell(\theta; \mathbf{y}) = \ln \mathrm{lhd}(\theta; \mathbf{y}) = \sum f^*(y_k \mid \theta).$$

Write

$$\partial_\theta = \frac{\partial}{\partial \theta}, \qquad \partial_\theta^2 = \frac{\partial^2}{\partial \theta^2}. \tag{C1}$$

Note that

$$\partial_\theta \ell(\theta; \mathbf{y}) = \frac{\partial_\theta \mathrm{lhd}(\theta; \mathbf{y})}{\mathrm{lhd}(\theta; \mathbf{y})}. \tag{C2}$$

Terminology: score. I should tell you that the expression $\partial_\theta \ell(\theta; \mathbf{y})$ is called the *score statistic* at parameter value θ.

For regular pdfs or pmfs (see 172B), we have (with sums for integrals in the discrete case)

$$\int_{\mathbb{R}^n} \mathrm{lhd}(\theta; \mathbf{y}) d\mathbf{y} = \mathbb{P}(\mathbf{Y} \in \mathbb{R}^n) = 1,$$

and

$$\mathbb{E}\left(\partial_\theta \ell(\theta; \mathbf{Y}) \,\middle|\, \theta\right) = \int \{\partial_\theta \ell(\theta; \mathbf{y})\} \, \mathrm{lhd}(\theta; \mathbf{y}) d\mathbf{y}$$

$$= \int \frac{\partial_\theta \, \mathrm{lhd}(\theta; \mathbf{y})}{\mathrm{lhd}(\theta; \mathbf{y})} \, \mathrm{lhd}(\theta; \mathbf{y}) d\mathbf{y} = \int \partial_\theta \, \mathrm{lhd}(\theta; \mathbf{y}) d\mathbf{y}$$

$$= \partial_\theta \int \mathrm{lhd}(\theta; \mathbf{y}) d\mathbf{y} = \partial_\theta 1 = 0. \tag{C3}$$

We therefore see that the Score Pre-Statistic at θ has zero $\mathbb{E}(\cdot \mid \theta)$ mean.

Next, by differentiating

$$\int \partial_\theta \{\ell(\theta; \mathbf{y})\} \, \mathrm{lhd}(\theta; \mathbf{y}) d\mathbf{y} = 0,$$

with respect to θ, we obtain (since $\partial_\theta \mathrm{lhd}(\theta; \mathbf{y}) = \{\partial_\theta \ell(\theta; \mathbf{y})\} \mathrm{lhd}(\theta; \mathbf{y})$)

$$0 = \int \{\partial_\theta^2 \ell(\theta; \mathbf{y})\} \, \mathrm{lhd}(\theta; \mathbf{y}) d\mathbf{y} + \int \{\{\partial_\theta \ell(\theta; \mathbf{y})\}^2\} \, \mathrm{lhd}(\theta; \mathbf{y}) d\mathbf{y}$$

$$= \mathbb{E}\left(\partial_\theta^2 \ell(\theta; \mathbf{Y}) \mid \theta\right) + \mathbb{E}\left(\{\partial_\theta \ell(\theta; \mathbf{Y})\}^2 \mid \theta\right). \tag{C4}$$

▶▶ **D. Fisher information.** We now define

$$I_\ell(\theta) := \mathbb{E}\left(\{\partial_\theta \ell(\theta; \mathbf{Y})\}^2 \mid \theta\right) = \mathrm{Var}\left(\partial_\theta \ell(\theta; \mathbf{Y}) \mid \theta\right), \tag{D1}$$

the equality following from (C3). Thus $I_\ell(\theta)$ is the $\mathbb{P}(\cdot\,|\,\theta)$ variance of the Score Pre-Statistic at parameter value θ. Equation (C4) enables us to rewrite (D1) as

$$I_\ell(\theta) \;=\; -\mathbb{E}\left(\partial_\theta^2 \ell(\theta;\mathbf{Y})\,|\,\theta\right) \;=\; -n\mathbb{E}\left(\partial_\theta^2 f^*(Y\,|\,\theta)\right) \;=\; nI_{f^*}(\theta). \qquad \text{(D2)}$$

The expression $I_\ell(\theta)$ is a measure of the amount of information a sample of size n contains about θ. This measure is proportional to sample size.

Convince yourself that the definition at 189(D1) seems sensible. For a good explanation of Fisher information, wait until Exercise 198Fc.

If, for example, each Y_k has the $N(\theta,\sigma_0^2)$ distribution, where the variance σ_0^2 is known, then (you check!) $I_\ell(\theta) = n/\sigma_0^2$. If σ_0^2 is very small, then we have a lot of information in that we can predict θ with high accuracy. (Bayesians refer to the inverse of the variance of a normal variable as the *precision* – on which topic more later.)

▶▶ **E. The Cramér–Rao Minimum-Variance Bound.** *We work with regular pdfs/pmfs. Suppose that $T = \tau(\mathbf{Y})$ is an* **Unbiased Estimator** *of θ based on $Y_1, Y_2, \ldots, Y_n$, where, as usual, $Y_1, Y_2, \ldots, Y_n$ are IID each with pdf (or pmf) $f(y\,|\,\theta)$. Then*

$$\text{Var}(T\,|\,\theta) \;\geq\; \frac{1}{I_\ell(\theta)}, \qquad\qquad \text{(E1)}$$

with equality if and only if

$$T - \theta = \tau(\mathbf{Y}) - \theta = \{\partial_\theta \ell(\theta;\mathbf{Y})\} \times \text{function}(\theta). \qquad \text{(E2)}$$

The right-hand side of (E1) is called the Minimum-Variance Bound for an Unbiased Estimator. When T is an Unbiased Estimator of θ for which (E1) holds with equality, T is called an **MVB** *Unbiased Estimator of θ. Note that then T is a Sufficient Statistic for θ based on $\mathbf{Y}$.*

Proof. Suppose that $T = \tau(\mathbf{Y})$ is an Unbiased Estimator of θ, so that

$$\int \tau(\mathbf{y})\text{lhd}(\theta;\mathbf{y})d\mathbf{y} \;=\; \theta.$$

Differentiate with respect to θ to obtain

$$\int \tau(\mathbf{y})\partial_\theta\text{lhd}(\theta;\mathbf{y})d\mathbf{y} \;=\; 1 \;=\; \int \tau(\mathbf{y})\{\partial_\theta\ell(\theta;\mathbf{y})\}\,\text{lhd}(\theta;\mathbf{y})d\mathbf{y}.$$

Hence, using 189(C3),

$$1 = \mathbb{E}\left\{T\partial_\theta\ell(\theta;\mathbf{Y})\,\Big|\,\theta\right\} = \mathbb{E}\left\{(T-\theta)\partial_\theta\ell(\theta;\mathbf{Y})\,\Big|\,\theta\right\},$$

whence, by the Cauchy–Schwarz Inequality 66C,

$$1 = 1^2 \leq \mathbb{E}\left\{(T-\theta)^2 \,\middle|\, \theta\right\} \mathbb{E}\left\{\partial_\theta \ell(\theta; \mathbf{Y})^2 \,\middle|\, \theta\right\} = \mathrm{Var}(T \,|\, \theta) I_\ell(\theta),$$

with equality if and only if

$$T - \theta = \{\partial_\theta \ell(\theta; \mathbf{Y})\} \times \mathrm{function}(\theta).$$

$\square$

▶ **F. A special exponential family.** If $f(y \,|\, \theta)$ has the special exponential form with $b(y) = y$:

$$f(y \,|\, \theta) = \exp\{a(\theta)y + c(\theta) + d(y)\}, \tag{F1}$$

so that

$$f^*(Y \,|\, \theta) = a(\theta)Y + c(\theta) + d(Y),$$

then, taking $n = 1$ in 189(C3),

$$0 = \mathbb{E}\left(\partial_\theta f^*(Y \,|\, \theta) \,\middle|\, \theta\right) = a'(\theta)\mathbb{E}(Y \,|\, \theta) + c'(\theta),$$

so that (we assume that $a'(\theta) \neq 0$)

$$\mathbb{E}\left(\overline{Y} \,|\, \theta\right) = \mathbb{E}(Y \,|\, \theta) = -\frac{c'(\theta)}{a'(\theta)}, \tag{F2}$$

and $\overline{Y}$ is an Unbiased Estimator of θ if and only if

$$\theta a'(\theta) + c'(\theta) = 0. \tag{F3}$$

Assume now that $\overline{Y}$ is unbiased, equivalently that (F3) holds. Then

$$\partial_\theta \ell(\theta; \mathbf{y}) = n\{a'(\theta)\overline{y} + c'(\theta)\} = na'(\theta)\{\overline{y} - \theta\}, \tag{F4}$$

so that condition 190(E2) holds and $\overline{Y}$ *is an MVB Unbiased Estimator of* θ. In particular,

$$\mathrm{Var}\left(\overline{Y}\right) = \frac{1}{nI_{f^*}(\theta)} = \frac{1}{na'(\theta)}, \tag{F5}$$

because

$$I_{f^*}(\theta) = -\mathbb{E}\{a''(\theta)Y + c''(\theta)\} = -a''(\theta)\theta - c''(\theta)$$
$$= -\frac{\mathrm{d}}{\mathrm{d}\theta}\{\theta a'(\theta) + c'(\theta)\} + a'(\theta) = a'(\theta).$$

Of course, the pdfs/pmfs of the $N(\theta, 1)$, $E(\text{mean } \theta)$, $\mathrm{Poisson}(\theta)$ and $\mathrm{Bernoulli}(\theta)$ distributions are in the special exponential family for which (F3) holds.

Exercise. Consider the $N(0, \theta)$ and $\mathrm{Gamma}(K_0, \text{mean } \theta)$ cases.

▶ **Fa. Exercise and Warning.** If $Y \sim E(\text{mean } \theta)$, then Y is an MVB Unbiased Estimator of θ. But for what value a is the mean-square-error $\mathbb{E}\{(aY - \theta)^2\}$ minimized?

6.5   Maximum-Likelihood Estimators (MLEs) and associated CIs

This is a particularly important topic, crucial to Frequentist theory and, in its large-sample results, to Bayesian theory too. The setup is the familiar one in which $Y_1, Y_2, \ldots, Y_n$ are IID each with pdf/pmf $f(y \mid \theta)$.

▶▶ **A. Maximum-Likelihood Estimators (MLEs); mles.** We have discussed the idea that a value of θ for which $\mathrm{lhd}(\theta; \mathbf{y}^{\mathrm{obs}})$ is large is one that tallies well with the data. Thus the *maximum-likelihood estimate (mle)* of θ, the value (assumed unique) $\hat{\theta}_{\mathrm{obs}}$ at which $\theta \mapsto \mathrm{lhd}(\theta; \mathbf{y}^{\mathrm{obs}})$ is a maximum, has some claims to be the 'best' estimate of θ. We have written the 'obs' in $\hat{\theta}_{\mathrm{obs}}$ because $\hat{\theta}_{\mathrm{obs}}$ may be calculated from the data. We clearly have the *likelihood equation*

$$\partial_\theta \mathrm{lhd}(\theta; \mathbf{y}^{\mathrm{obs}}) = 0, \text{ equivalently, } \partial_\theta \ell(\theta; \mathbf{y}^{\mathrm{obs}}) = 0, \text{ when } \theta = \hat{\theta}_{\mathrm{obs}}. \quad \text{(A1)}$$

If there is a sufficient statistic t, then, as we see from the Factorization Criterion 183(D1), $\hat{\theta}_{\mathrm{obs}}$ must maximize $\theta \mapsto g(\theta; t^{\mathrm{obs}})$ whence $\hat{\theta}_{\mathrm{obs}}$ is a function of t^{obs}.

By our usual conventions, the *Maximum-Likelihood Estimator (MLE)* $\widehat{\Theta}$ of θ is the PreStatistic which crystallizes to $\hat{\theta}_{\mathrm{obs}}$, so $\widehat{\Theta}$ maximizes $\theta \mapsto \mathrm{lhd}(\theta; \mathbf{Y})$. It is worth mentioning now that we sometimes 'correct' MLEs so as to make them Unbiased. We now meet an important case where no such correction is necessary.

If $f(y \mid \theta)$ has the special exponential form at 191(F2):

$$f(y \mid \theta) = \exp\{a(\theta)y + c(\theta) + d(y)\},$$

where $a'(\theta)\theta + c'(\theta) = 0$, we already know that $\overline{Y}$ is an MVB Unbiased Estimator of θ with variance therefore equal to $\{I_\ell(\theta)\}^{-1}$. Since, from 191(F4),

$$\partial_\theta \ell(\theta; \mathbf{Y}) = na'(\theta)\{\overline{Y} - \theta\},$$

it is clear that $\overline{Y}$ is also the MLE of θ. For large n, the CLT shows that, since $\overline{Y}$ is a sample mean of IID variables,

$$\overline{Y} \text{ is approximately N}\left(\theta, \{I_\ell(\theta)\}^{-1}\right).$$

Fact B, one of the most important results in Statistics, establishes an analogous result for all sufficiently regular pdfs/pmfs.

▶▶▶ **B. Fact (Fisher, Wilks, Wald): Asymptotic Normality of MLEs.**
Suppose that $f(y \mid \theta)$ satisfies certain regularity conditions satisfied in all the examples which we shall meet. Then, for large n, the MLE $\widehat{\Theta}$ of θ based on $Y_1, Y_2, \ldots, Y_n$ is both **approximately MVB Unbiased** and **approximately normal**. Thus, for large n,

$$\widehat{\Theta} \text{ is approximately normal } N\left(\theta, \frac{1}{n I_{f^*}(\theta)}\right) = N\left(\theta, \frac{1}{I_\ell(\theta)}\right).$$

We use this to obtain CIs for θ by pretending that θ has posterior density from the

$$N\left(\hat{\theta}_{\text{obs}}, \frac{1}{I_\ell(\hat{\theta}_{\text{obs}})}\right)$$

distribution. We have seen enough instances of this kind of thing.

We shall examine some heuristic reasons why the Fisher–Wilks–Wald result is true, and make computer studies of an interesting case. First, we have to convince ourselves that, for large n, $\widehat{\Theta}$ is likely to be close to the true value θ.

▶ **C. Fact: Consistency of the MLE.** *Again make appropriate regularity assumptions. Assume that $Y_1, Y_2, \ldots$ are IID each with pdf/pmf $f(y \mid \theta)$. Let $\widehat{\Theta}_n$ be the MLE of θ based on $Y_1, Y_2, \ldots, Y_n$. Then, with $\mathbb{P}(\cdot \mid \theta)$ probability 1,*

$$\widehat{\Theta}_n \to \theta \text{ as } n \to \infty.$$

Heuristic idea for proof. If θ is the true value of the parameter and φ is some other value of the parameter, then with $\mathbb{P}(\cdot \mid \theta)$ probability 1,

$$\frac{\text{lhd}(\varphi; Y_1, Y_2, \ldots, Y_n)}{\text{lhd}(\theta; Y_1, Y_2, \ldots, Y_n)} \to 0 \text{ as } n \to \infty. \tag{C1}$$

To prove this (really *prove* it!), note that

$$\mathbb{E}\left(\frac{f(Y \mid \varphi)}{f(Y \mid \theta)} \,\Big|\, \theta\right) = \int \frac{f(y \mid \varphi)}{f(y \mid \theta)} f(y \mid \theta) \, dy = \int f(y \mid \varphi) dy = 1,$$

whence, by Jensen's Inequality 64K,

$$\mathbb{E}\left(\ln \frac{f(Y \mid \varphi)}{f(Y \mid \theta)} \,\Big|\, \theta\right) < 0.$$

The strict inequality in this application of Jensen's inequality is because $(-\ln)$ is strictly convex on $(0, \infty)$ and because we may and do ignore the situation where we have

$f(Y \mid \varphi) = f(Y \mid \theta)$ with $\mathbb{P}(\cdot \mid \theta)$ probability 1 (in which case the $\mathbb{P}(\cdot \mid \varphi)$ and $\mathbb{P}(\cdot \mid \theta)$ probabilities would be equal). The Strong Law now tells us that, with $\mathbb{P}(\cdot \mid \theta)$ probability 1,

$$\ln \frac{\mathrm{lhd}(\varphi; Y_1, Y_2, \ldots, Y_n)}{\mathrm{lhd}(\theta; Y_1, Y_2, \ldots, Y_n)} = \sum_{k=1}^{n} \ln \frac{f(Y_k \mid \varphi)}{f(Y_k \mid \theta)} \to -\infty$$

as $n \to \infty$, and hence 193(C1) is true.

It is intuitively clear that 193(C1) is a big step towards proving Fact 193C, and clear that, for full rigour, it needs some uniformity assertions which are significantly more technical.

▶ **D. A heuristic Frequentist view of Fact 193B.** (For the heuristic Bayesian view, see Subsection 204F.) Make the assumptions of the theorem. Then $\widehat{\Theta}$ satisfies

$$\partial_\theta \ell(\theta; \mathbf{Y})\big|_{\theta = \widehat{\Theta}} = 0,$$

whence by Taylor expansion (since $\widehat{\Theta}$ and θ are likely to be close)

$$\partial_\theta \ell(\theta; \mathbf{Y}) + \left(\widehat{\Theta} - \theta\right) \partial_\theta^2 \ell(\theta; \mathbf{Y}) \approx 0 \text{ 'at the point } \theta\text{'}$$

so that (a kind of Newton–Raphson statement)

$$\widehat{\Theta} - \theta \approx \frac{\partial_\theta \ell(\theta; \mathbf{Y})}{\{-\partial_\theta^2 \ell(\theta; \mathbf{Y})\}}. \tag{D1}$$

Now, the denominator

$$-\partial_\theta^2 \ell(\theta; \mathbf{Y}) = \sum \{-\partial_\theta^2 f^*(Y_k \mid \theta)\}$$

is, by the Strong Law (or Weak Law), likely to be close ('proportionately') to its expectation $n I_{f^*}(\theta)$. Next, the $\partial_\theta f^*(Y_k \mid \theta)$ are independent each with zero mean and variance $I_{f^*}(\theta)$. By the CLT, the numerator

$$\partial_\theta \ell(\theta; \mathbf{Y}) = \sum_{k=1}^{n} \partial_\theta f^*(Y_k \mid \theta)$$

in (D1) is therefore approximately $\mathrm{N}\big(0, \, n I_{f^*}(\theta)\big)$. The result follows.

E. Example: The Cauchy distribution.   We already know that Fact 193B works well for $\mathrm{N}(\theta, 1)$, $\mathrm{N}(0, \theta)$, $\mathrm{E}(\text{mean } \theta)$, $\mathrm{Poisson}(\theta)$, $\mathrm{Bernoulli}(\theta)$.

Let's try it out on the Cauchy density

$$f(y \mid \theta) = \frac{1}{\pi \{1 + (y - \theta)^2\}}.$$

Though this density is notorious for having no mean (hence no variance), that does not influence its behaviour in regard to Fact 193B.

Fact 193B suggests that, for large n,

$$\widehat{\Theta} \approx N\left(\theta, \frac{2}{n}\right). \tag{E1}$$

Ea. Exercise. By differentiating with respect to a for some of the steps, check that

$$\int \frac{1}{\pi(a^2 + y^2)}\, dy = \frac{1}{a}, \qquad \int \frac{2a}{\pi(a^2 + y^2)^2}\, dy = \frac{1}{a^2},$$

$$\int \frac{1}{\pi(a^2 + y^2)^2}\, dy = \frac{1}{2a^3}, \qquad \int \frac{4a}{\pi(a^2 + y^2)^3}\, dy = \frac{3}{2a^4},$$

$$\int \frac{a^2}{\pi(a^2 + y^2)^3}\, dy = \frac{3}{8a^3},$$

$$\int \frac{y^2}{\pi(a^2 + y^2)^3}\, dy = \left(\frac{1}{2} - \frac{3}{8}\right)\frac{1}{a^3} = \frac{1}{8a^3},$$

$$I_{f^*}(\theta) = \frac{1}{2} \text{ for all } \theta.$$

(Alternatively, prove the last result directly using the Calculus of Residues.)   □

We know that, since θ is a location parameter, 'ancillarity' tells us that in the Frequentist approach, we should obtain CIs for θ by pretending that θ is a random variable Θ with pdf $\pi(\theta \mid y)$ proportional as a function of θ to

$$\text{lhd}(\theta; y_1, y_2, \ldots, y_n) = \prod f(y_k \mid \theta);$$

in other words, by 'taking the Bayesian posterior pdf corresponding to a constant prior density'. We therefore have

$$\pi(\theta \mid \mathbf{y}) = I^{-1} \prod f(y_k \mid \theta), \text{ where } I = \int_{-\infty}^{+\infty} \prod f(y_k \mid \varphi)\, d\varphi. \qquad \text{(E2)}$$

The calculation of such integrals is a problem, especially in higher-dimensional situations. We shall see later how Markov Chain Monte Carlo methods avoid the difficulty.

How does this compare with 194(E1)? We can get an idea about this by simulation. A sample of size 40 was taken from the Cauchy distribution with $\theta = 0$. Figure E(i) compares the 'Bayesian posterior' density with the normal curve obtained from 194(E1), based on the first n observations for $n = 3, 15, 40$ successively. The first three y-values on this simulation were 5.01, 0.40 and -8.75 – a bit spread out, but the Cauchy distribution does have large tails. We see that, for this sample, for large n, the posterior density is more concentrated around the mean than the normal, but the normal must eventually tail off more quickly, of course.

Eb. Exercise: Simulating from the Cauchy distribution.   Show that if U has the U$[0, 1]$ distribution, then

$$Y = \tan \pi(U - \tfrac{1}{2})$$

has the Cauchy density with $\theta = 0$.

Here's an alternative method. Show that if we repeat choosing IID U$[-1, 1]$ variables V and W until $V^2 + W^2 \leq 1$, then $Y = V/W$ has the Cauchy density with $\theta = 0$.

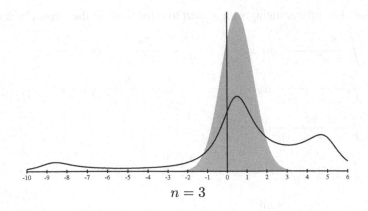

$$n = 3$$

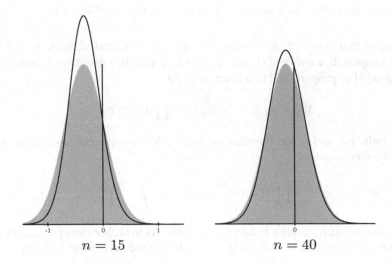

$$n = 15 \qquad\qquad\qquad n = 40$$

Figure E(i): Parts of posterior pdfs for Cauchy sample.

The normal curve is for $N(\hat{\theta}_{\mathrm{obs}}, 2/n)$.

▶ **Ec. Important note on use of median.** We know from Theorem 165La that, for large n, the median of $Y_1, Y_2, \ldots, Y_n$ has approximately the $N(\theta, \frac{\pi^2}{4n})$ distribution, and this gives an easy way of obtaining approximate CIs for θ. One would have a 95% CI

$$(m - r, m + r), \quad \text{where } r = 1.96 * \pi/(2\sqrt{n}).$$

For the '$n = 40$' sample illustrated in Figure E(i), the true degree of confidence of the interval arrived at in this way is 97.3%, not 95%. So things are only very rough guides.

▶▶ **F. 'Randomness', entropy, approximation of pdfs, etc.**   The proof of the Consistency of the MLE in Subsection 193C utilized ideas associated with *entropy*. I'll just say a brief word about this concept.

We know that if $Y_1, Y_2, \ldots, Y_n$ are IID each with pdf/pmf f, then the Strong Law implies that the associated likelihood satisfies

$$\frac{1}{n} \ln \mathrm{lhd}(f; Y_1, Y_2, \ldots, Y_n) := \frac{1}{n} \sum \ln f(Y_k)$$
$$\to \mathbb{E}_f \ln f(Y) = \int f(y) \ln f(y) \, dy,$$

the subscript f in $\mathbb{E}_f$ signifying that the expectation is associated with the pdf/pmf f. We define the *entropy associated with the pdf f* as

$$\mathrm{Ent}(f) := -\int f(y) \ln f(y) \, dy. \tag{F1}$$

We see that, very roughly speaking, the likelihood of the observed outcome is likely to be roughly of order of magnitude $e^{-n\,\mathrm{Ent}(f)}$. The larger $\mathrm{Ent}(f)$, the less the likelihood of the typical outcome, and thus the greater the 'randomness'.

Be careful about sign conventions and normalization conventions in regard to entropy; in Shannon's information theory, $\log_2$ is used rather than $\ln$ (because information is thought of in terms of bits, bytes, etc). However, entropy was, of course, studied much earlier by physicists, especially Clausius and Gibbs, in connection with the *Second Law of Thermodynamics*: *'entropy increases'*.

If we write $Y_k = h(Z_k)$, where h is a smooth strictly increasing function, then $\mathrm{Ent}(f)$ and $\mathrm{Ent}(f_Z)$ are different: the concept of entropy does not have the desired invariance property. We have $f_Z(z) = f(h(z))h'(z)$ (see 55(F)) from which it follows (check!) that

$$\mathrm{Ent}(f_Z) = \mathrm{Ent}(f) + \int f(h(z))h'(z) \ln h'(z) \, dz.$$

In spite of this lack of invariance, entropy is rightly a widely used concept.

The key step in the proof of the Consistency of the MLE was that if $Y_1, Y_2, \ldots, Y_n$ are IID with pdf f, and g is another pdf, then with $\mathbb{P}_f$ probability 1,

$$\frac{1}{n} \ln \frac{\mathrm{lhd}(g; Y_1, Y_2, \ldots, Y_n)}{\mathrm{lhd}(f; Y_1, Y_2, \ldots, Y_n)} \to \mathbb{E}_f \ln \frac{g(Y)}{f(Y)} \le 0,$$

with equality if and only if f and g determine the same distribution. We can clearly use

$$\mathrm{App}(f \leftarrow g) := \mathbb{E}_f \ln \frac{f(Y)}{g(Y)} = \int \ln\left(\frac{f(y)}{g(y)}\right) f(y) \, dy \ge 0 \tag{F2}$$

as a measure of how badly f is approximated by g. This is the *Kullback–Leibler 'relative entropy'*.

Fa. Exercise.   Check that $\mathrm{App}(f \leftarrow g)$ *is invariant under a one-one transformation* $Y = h(Z)$.

Fb. Exercise.   Check that if f is the pdf of $\mathrm{N}(\mu_1, 1)$ and g that of $\mathrm{N}(\mu_2, 1)$, then

$$\mathrm{App}(f \leftarrow g) \;=\; \tfrac{1}{2}(\mu_1 - \mu_2)^2.$$

By extending the above exercise, we can get a really good understanding of Fisher information.

▶▶ **Fc. Exercise.**   Consider the usual situation where $Y_1, Y_2, \ldots, Y_n$ are IID, each with pdf/pmf $f(y \mid \theta)$. Prove (for nice situations) that, as $\varphi \to \theta$,

$$\mathrm{App}\big(f(\cdot \mid \theta) \leftarrow f(\cdot \mid \varphi)\big) \;=\; \tfrac{1}{2}(\varphi - \theta)^2 I_{f*}(\theta) + \mathrm{O}\big((\varphi - \theta)^3\big).$$

Now explain the good sense of Fisher information.

▶ **G. Lemma: Maximum-entropy distributions.**
(a) Amongst all pdfs on $\mathbb{R}$ *with given mean* μ *and variance* σ^2, *that of the* $\mathrm{N}(\mu, \sigma^2)$
distribution has greatest entropy.
(b) Amongst all pdfs on $[0, \infty)$ *with given mean* μ, *that of the* $\mathrm{E}(\mathrm{mean}\ \mu)$
distribution has greatest entropy.
(c) The value of p *which maximizes the entropy*

$$-p \ln p - (1 - p) \ln(1 - p)$$

of the $\mathrm{Bernoulli}(p)$ *pmf is* $\tfrac{1}{2}$.

Proof of Part (a). All three parts are proved similarly. I do Part (a), and leave Parts (b) and (c) as exercises for you.

Let f be a pdf on $\mathbb{R}$ with mean μ and variance σ^2. Let g be the pdf of the $\mathrm{N}(\mu, \sigma^2)$ distribution. Then

$$0 \leq \int \ln\left(\frac{f(y)}{g(y)}\right) f(y)\,\mathrm{d}y \;=\; \int \ln(f(y)) f(y)\mathrm{d}y - \int \ln(g(y)) f(y)\mathrm{d}y$$

$$= -\mathrm{Ent}(f) + \int \ln\left(\sigma\sqrt{2\pi}\right) f(y)\,\mathrm{d}y + \int \frac{(y - \mu)^2}{2\sigma^2} f(y)\,\mathrm{d}y$$

$$= -\mathrm{Ent}(f) + \ln\left(\sigma\sqrt{2\pi}\right) + \tfrac{1}{2}$$

with equality if and only if $f = g$.   □

Ga. Exercise.   Prove Parts (b) and (c) similarly.

▶▶ **H. Entropy and Fisher information for location parameters.**   A
natural question now arises: *Can the maximum-entropy property of the normal distribution explain its rôle in the Central Limit Theorem?* It is a question which

had been considered by many authors prior to its definitive affirmative solution by Barron [11].

Let Y be an RV with location-parameter density $f(y \mid \theta) = h(y - \theta)$ where h is a pdf on $\mathbb{R}$ with associated mean 0 and variance 1. Since $\partial_\theta f^*(y \mid \theta) = -h'(y - \theta)$, we have

$$I_{f^*}(\theta) = \mathbb{E}\left(\{\partial_\theta f^*(y \mid \theta)\}^2\right) = \mathbb{E}_0\left(\left\{\frac{h'(Y)}{h(Y)}\right\}^2\right) =: I(Y) \quad \text{(say)}$$

$I(Y)$ being independent of θ, and $\mathbb{E}_0$ referring to the '$\theta = 0$' situation. Consider also (see Exercise Ha)

$$J(Y) := \mathbb{E}_0\left(\left\{\frac{h'(Y)}{h(Y)} - \frac{\varphi'(Y)}{\varphi(Y)}\right\}^2\right) = I(Y) - 1, \tag{H1}$$

where φ is the pdf of $N(0, 1)$.

Barron's remarkable identity is that if X is a 'continuous' RV with positive pdf on $(0, \infty)$, then

$$\text{App}(f \leftarrow \varphi) = \int_0^{\frac{1}{2}\pi} J(X \cos \theta + Z \sin \theta) \tan \theta \, \mathrm{d}t,$$

Z being independent of X with $Z \sim N(0, 1)$. Barron uses this in proving a strong version of the CLT (with 'Scheffé' convergence of densities) if the typical summand X has continuous pdf. The method will not deal with discrete RVs or with 'fractal' ones, but it is very illuminating even so.

Ha. Exercise. Prove equation (H1), and show that if $J(Y) = 0$, then $Y \sim N(0, 1)$. You may assume that $yh(y) \to 0$ as $|y| \to \infty$. Since $J(Y) \geq 0$, it is clear that $I(Y) \geq 1$. Show that this is the Cramér–Rao MVB result for the location-parameter situation and that $\theta \mapsto N(\theta, 1)$ is the only location-parameter family with mean θ and variance 1 and with an MVB Unbiased Estimator.

6.6   Bayesian Confidence Intervals

I should reiterate my thanks to David Draper here. This section is much improved due to his wise advice.

See towards the end of Subsection 169A for notes on my deliberate misuse of orthodox Bayesian terminology (of which misuse David does not approve).

Let's see how the Bayesian machine works before getting involved with the philosophy.

Suppose that I give you a newly-minted coin, that you throw it 5 times and get 1 Head and 4 tails. The Frequentist 'best estimate' of the probability p that the coin falls Heads would be $1/5$, an absurd estimate because we *know* that the true value of p will be close to $1/2$. (But, of course, the Frequentist Confidence Interval would be very wide.)

How should we build in prior information? We can only think sensibly about this after we have our machine running.

▶▶ **A. THE formula again.** From one point of view, Bayesian Statistics consists of just one formula already discussed in subsections 77F–78G

$$\text{Posterior density} \propto \text{Prior density} \times \text{Likelihood,} \qquad (A1)$$

or, in symbols,

$$\pi\left(\theta \,|\, \mathbf{y}^{\text{obs}}\right) \propto \pi(\theta)\text{lhd}\left(\theta; \mathbf{y}^{\text{obs}}\right) \qquad (A2)$$

with the 'constant of proportionality', a function of $\mathbf{y}^{\text{obs}}$, ensuring that $\pi\left(\theta \,|\, \mathbf{y}^{\text{obs}}\right)$ is a proper density – a true pdf – in that $\int_{\mathbb{R}} \pi\left(\theta \,|\, \mathbf{y}^{\text{obs}}\right) d\theta = 1$. Because of the proportionality at (A2), we need only specify the prior $\pi(\theta)$ 'modulo constant multiples', and we can also absorb functions of $\mathbf{y}^{\text{obs}}$ alone into the $\propto$ sign, being free to omit them from $\text{lhd}(\theta; \mathbf{y}^{\text{obs}})$.

The **posterior density** $\pi\left(\theta \,|\, \mathbf{y}^{\text{obs}}\right)$ is meant to indicate 'degree of belief' in the whereabouts of θ after the experiment is performed with result $\mathbf{y}^{\text{obs}}$. Thus a $C\%$ **Bayesian Confidence Interval** for θ after the experiment is performed is an interval I such that

$$\int_I \pi\left(\theta \,|\, \mathbf{y}^{\text{obs}}\right) d\theta = C\%. \qquad (A3)$$

To a Bayesian,

the posterior density itself provides the correct way of describing inference about θ, not the less informative $C\%$ Confidence Interval derived from it.

Of course, knowledge of the posterior density is equivalent to knowledge of all $C\%$ CIs for every C.

In strict Bayesian theory, there is the obvious analogous interpretation to (A3) of the **prior density** $\pi(\theta)$ as expressing degree of belief in the whereabouts of θ before the experiment is performed. As is discussed later, Bayesian degrees of belief are **subjective**: each person is allowed to have his/her own prior density, though he/she needs to justify its choice in some way; and that person's prior density will result in his/her posterior density. Somewhat controversially, Bayesians treat 'degree of belief' densities in exactly the same way as they do pdfs.

For the purposes of Your calculation, therefore, it is *as if*, as we did in the Rev'd Thomas Bayes story in Subsection 78G, You regard θ as the value of a Random Variable Θ chosen by God according to the pdf $\pi(\cdot)$, where $\pi(\cdot)$ is Your prior density. Your posterior density $\pi\left(\cdot \mid \mathbf{y}^{\text{obs}}\right)$ is then the conditional pdf of Θ given that $\mathbf{Y} = \mathbf{y}^{\text{obs}}$ in a sense already intuitively clear from Subsection 78G and clarified further in our discussion of conditional pdfs in the next chapter. But I also discuss it now, so that you will feel secure.

The fundamental postulate is that

> **conditionally on $\Theta = \theta$,**
> **the Variables $Y_1, Y_2, \ldots, Y_n$ are IID each with pdf $f(\cdot \mid \theta)$.**

So, heuristically,

$$
\begin{aligned}
\mathbb{P}(\mathbf{Y} \in d\mathbf{y} \mid \Theta \in d\theta) &= \mathbb{P}(Y_1 \in dy_1; \ldots; Y_n \in dy_n \mid \Theta \in d\theta) \\
&= \mathbb{P}(Y_1 \in dy_1 \mid \Theta \in d\theta) \ldots \mathbb{P}(Y_n \in dy_n \mid \Theta \in d\theta) \\
&= f(y_1 \mid \theta) dy_1 \ldots f(y_n \mid \theta) dy_n \\
&= \text{lhd}(\theta; y_1, \ldots y_n) dy_1 \ldots dy_n = \text{lhd}(\theta; \mathbf{y}) d\mathbf{y},
\end{aligned}
$$

and so

$$
\begin{aligned}
\mathbb{P}(\Theta \in d\theta; \mathbf{Y} \in d\mathbf{y}) \\
&= \mathbb{P}(\Theta \in d\theta)\mathbb{P}(\mathbf{Y} \in d\mathbf{y} \mid \Theta \in d\theta) \\
&= \pi(\theta)d\theta\, \text{lhd}(\theta; \mathbf{y})d\mathbf{y}.
\end{aligned}
$$

Hence

$$
\mathbb{P}(\mathbf{Y} \in d\mathbf{y}) = \left\{ \int \pi(\theta)\, \text{lhd}(\theta; \mathbf{y})d\theta \right\} d\mathbf{y}, \tag{A4}
$$

and

$$
\begin{aligned}
\mathbb{P}(\Theta \in d\theta \mid \mathbf{Y} \in d\mathbf{y}) &= \frac{\mathbb{P}(\Theta \in d\theta; \mathbf{Y} \in d\mathbf{y})}{\mathbb{P}(\mathbf{Y} \in d\mathbf{y})} \\
&= \frac{\pi(\theta)\, \text{lhd}(\theta; \mathbf{y})d\theta}{\int \pi(\varphi)\, \text{lhd}(\varphi; \mathbf{y})d\varphi},
\end{aligned}
$$

or

$$
\pi(\theta \mid \mathbf{y}) \propto \pi(\theta)\, \text{lhd}(\theta; \mathbf{y}).
$$

Note that $Y_1, Y_2, \ldots, Y_n$, though conditionally independent given θ, will generally not be absolutely independent because the integral at (A4) mixes up the multiplicative property which would hold if Θ were (as in the Frequentist approach) fixed at a deterministic value.

(*Historical Note.* The 'God and the Rev'd Bayes' story does not represent (as Statistics, I mean, not as Theology) the way that Bayes thought of things.)

Especially to express the idea of *no prior information*, or of *vague prior information*, about θ, *improper priors* such as

$$\pi(\theta) = 1 \quad (\theta \in \mathbb{R}) \quad \text{or} \quad \pi(\theta) = 1/\theta \quad (\theta \in (0, \infty))$$

are often used. These priors are *improper* in that they satisfy $\int \pi(\theta) \, d\theta = \infty$, so they cannot be normalized to a proper pdf by dividing by a constant. The only sense we can make of a uniform prior $\pi(\cdot) = 1$ (for example) in terms of God's experiment is that, conditional on the fact that the number chosen by God lies in the interval $[a, b]$, it has the U$[a, b]$ distribution. *Warning.* Using 'conditional probabilities' in settings where the underlying measure has infinite total mass can be very dangerous. It really is best to use proper priors.

For the time being, I shall speak of $\pi(\cdot)$ as 'our' prior, pretending that You and I have agreed on it.

▶▶ **B. Recursive Bayesian updating.** We have already seen in Subsections 82M – 84O how the posterior pdf after one experiment can become the prior for a second experiment; and so on. This is a particularly nice feature of Bayesian theory.

▶ **Ba. Pre-priors and Pre-experiments.** We have to see this topic in practice (as we shall do later) to understand it properly.

It can sometimes help in achieving a sensible balance between prior information and information from our experiment to imagine that Zeus first chooses θ according to a 'Pre-prior' $\pi_Z(\cdot)$, then tells Tyche the function $\pi_Z(\cdot)$ but not the value of θ, that Tyche performs a first experiment (a *pre-experiment* to ours) to determine θ, and that Tyche's posterior density $\pi_T(\cdot)$ is then revealed to us and becomes our prior $\pi(\cdot)$. Note that (heresy, but one with some appeal to Fisher and to K. Pearson) it suggests a way of building prior information into Frequentist theory.

It is helpful to think of Zeus as using a 'vague prior', widely spread and therefore giving little information about the whereabouts of θ. The number of observations made by Tyche is then the **effective prior sample size**, a good indicator of how heavily we are weighting prior information. In the next subsection, however, Tyche does not do an experiment.

▶ **C. 'Vague prior' Bayesian theory for a location parameter.** Suppose that $Y_1, Y_2, \ldots, Y_n$ are IID each with pdf/pmf of the form $f(y \mid \theta) = h(y - \theta)$. If we have no prior information about the whereabouts of θ, it is natural that our prior should be invariant under any shift $\theta \mapsto \theta + c$ where c is a constant. This implies that $\pi(\theta) = \pi(\theta + c)$, so that $\pi(\cdot) = $ constant, and the constant may as well be 1: we say that we use the **uniform prior**. [It is impossible to conceive of a situation where we have *no* prior information: we may for instance know that $|\theta| < 10^6$; but ignore this for now. Considerations of robustness are important here.]

As already explained, the posterior density $\pi \left(\cdot \,|\, \mathbf{y}^{\text{obs}} \right)$ then agrees with the Frequentist 'posterior density' obtained by using ancillary statistics. Or perhaps one should be cynical and say that ancillarity was used to cook the Frequentist answer to agree with the Bayesian with a uniform prior. You can see why I refer to the uniform prior for location parameters as the Frequentist prior.

If each Y_k is $N(0, 1)$, then, since $\pi(\theta) = 1$, we find from 182Ca that

$$\pi \left(\theta \,|\, \mathbf{y}^{\text{obs}} \right) \; \propto \; \text{lhd}(\theta; \mathbf{y}^{\text{obs}}) \; \propto \; \exp \left\{ -\tfrac{1}{2} n (\theta - \overline{y}_{\text{obs}})^2 \right\},$$

functions of $\mathbf{y}^{\text{obs}}$ *being absorbed in the* $\propto$ *signs.* Thus, $\pi \left(\cdot \,|\, \mathbf{y}^{\text{obs}} \right)$ must be the pdf of $N(\overline{y}_{\text{obs}}, 1/n)$, in agreement with the discussion at 175Ba.

D. The improper 'vague prior' for Exponential E(rate λ).

Suppose that each Y_k has the $E(\text{rate } \lambda)$ distribution. Write $Y_k = \exp(W_k)$ and $\lambda = \exp(-a)$. Knowledge of $\mathbf{Y}$ is the same as that of $\mathbf{W}$. The density of W_k in terms of α is given by $f(w \,|\, \alpha) = h(w - \alpha)$, where $h(r) := \exp \left\{ r - e^r \right\}$. Thus $\alpha = -\ln \lambda$ behaves as a location parameter for $W_1, W_2, \ldots, W_n$; and the uniform prior for α translates into a prior $\pi(\lambda) = 1/\lambda$ for λ. Naively, $|d\alpha| = |d\lambda|/\lambda$.

Then,

$$\pi(\lambda \,|\, \mathbf{y}^{\text{obs}}) \; \propto \; \frac{1}{\lambda} \lambda^n e^{-\lambda s_{\text{obs}}} \; \propto \; \lambda^{n-1} e^{-\lambda s_{\text{obs}}},$$

so that $\pi(\lambda \,|\, \mathbf{y}^{\text{obs}})$ is the pdf of the $\text{Gamma}(n, \text{rate } s_{\text{obs}})$ distribution, in agreement with the discussion at 176Ca. This is my justification for calling $\pi(\lambda) = \lambda^{-1}$ prior the Frequentist prior here.

►► E. Sufficiency and ancillarity in Bayesian theory.

If $t = \tau(\mathbf{y})$ is a sufficient statistic in that the factorization 183(D1)

$$\text{lhd}(\theta; \mathbf{y}^{\text{obs}}) \; = \; g(\theta; t) h(\mathbf{y}^{\text{obs}})$$

holds, then, for any prior $\pi(\cdot)$, we have

$$\pi \left(\theta \,|\, \mathbf{y}^{\text{obs}} \right) \; \propto \; \pi(\theta) g(\theta; t^{\text{obs}}) h(\mathbf{y}^{\text{obs}}) \; \propto \; \pi(\theta) g(\theta; t^{\text{obs}})$$

so that $\pi \left(\theta \,|\, \mathbf{y}^{\text{obs}} \right)$ **depends on** $\mathbf{y}^{\text{obs}}$ **only via** t^{obs}:

$$\pi \left(\theta \,|\, \mathbf{y}^{\text{obs}} \right) \; = \; \frac{\pi(\theta) g(\theta; t^{\text{obs}})}{\int \pi(\varphi) g(\varphi; t^{\text{obs}}) \, d\varphi}.$$

Though we have not studied ancillarity, I remark that since we condition on the whole information in $\mathbf{y}^{\text{obs}}$ in Bayesian theory, we automatically condition on ancillary statistics, so there is no more to say.

This is indeed all much simpler than the Frequentist version.

▶▶ **F. Asymptotic normality in Bayesian theory.** Imagine for a moment that we work with a uniform prior $\pi(\theta) = 1$. Then

$$\pi(\theta \mid \mathbf{y}^{\text{obs}}) \propto \text{lhd}(\theta; \mathbf{y}^{\text{obs}}).$$

We take a Taylor expansion of $\ell(\theta; \mathbf{y}^{\text{obs}})$ around $\hat{\theta}_{\text{obs}}$ (at which point, $\partial_\theta \ell(\theta; \mathbf{y}^{\text{obs}}) = 0$) to obtain

$$\ell\left(\theta; \mathbf{y}^{\text{obs}}\right) \approx \ell\left(\hat{\theta}_{\text{obs}}; \mathbf{y}^{\text{obs}}\right) + \tfrac{1}{2}\left(\theta - \hat{\theta}_{\text{obs}}\right)^2 \partial_\theta^2 \ell(\theta; \mathbf{y}^{\text{obs}})\big|_{\theta = \hat{\theta}_{\text{obs}}}.$$

However, the Strong (or Weak) Law shows that

$$\partial_\theta^2 \ell(\theta; \mathbf{y}^{\text{obs}})\big|_{\theta = \hat{\theta}_{\text{obs}}} = \sum \partial_\theta^2 f^*(y_k^{\text{obs}} \mid \theta)\big|_{\theta = \hat{\theta}_{\text{obs}}}$$

is likely to be close to $n\mathbb{E}\,\partial_\theta^2 f^*(Y \mid \theta) = -nI_{f^*}(\theta)$, whence, as a function of θ,

$$\pi(\theta \mid \mathbf{y}^{\text{obs}}) \propto \text{lhd}(\theta; \mathbf{y}^{\text{obs}}) \approx \text{function}(\mathbf{y}^{\text{obs}}) \exp\left\{-\tfrac{1}{2}(\theta - \hat{\theta}_{\text{obs}})^2 I_\ell(\hat{\theta}_{\text{obs}})\right\},$$

whence, $\pi(\theta \mid \mathbf{y}^{\text{obs}})$ must be close to the density of $\text{N}\left(\hat{\theta}_{\text{obs}},\, 1/I_\ell(\hat{\theta}_{\text{obs}})\right)$.

If some other prior density $\pi(\theta)$ is used, then $\pi(\theta)$ will be approximately constant near $\hat{\theta}_{\text{obs}}$, so since the likelihood will fall off rapidly as θ moves away from $\hat{\theta}_{\text{obs}}$, we can assume that $\pi(\theta)$ is constant when n is large. For large n, the information in the current experiment swamps the prior information. Thus, again, $\pi(\theta \mid \mathbf{y}^{\text{obs}})$ must be close to the density of $\text{N}\left(\hat{\theta}_{\text{obs}},\, 1/I_\ell(\hat{\theta}_{\text{obs}})\right)$. This is the Bayesian view of Fact 193B.

Fa. Something to think about. Note that while the CLT was crucial to the Frequentist approach in Subsection 194D, it was *not* used in the Bayesian: in the latter, the normal distribution just appeared from Taylor expansion.

▶ **G. Bayesian reference priors.** *How should we find the appropriate prior density to represent vague prior information in general?* There is an elaborate theory of *Bayesian reference priors* (see Volume 1 of Bernardo and Smith [17]) which seeks to answer this problem. In commonly met cases, the answer provided by that theory agrees with Sir Harold Jeffreys' prescription:

$$\pi(\theta) \propto I_{f^*}(\theta)^{\frac{1}{2}} \qquad \text{('reference prior').} \tag{G1}$$

One version of the 'reference' theory is based on *entropy* and, very roughly speaking, the reference prior is the prior starting from which the experiment makes the biggest gain in information about θ. We see from Subsection 197(F2) that, for example, we might measure the gain in information about θ made by taking an observation $\mathbf{y}^{\text{obs}}$ by

$$\int \pi(\theta \mid \mathbf{y}^{\text{obs}}) \ln \frac{\pi(\theta \mid \mathbf{y}^{\text{obs}})}{\pi(\theta)} \, d\theta.$$

That's a small start on the route which you should follow up in Bernardo and Smith [17].

Entropy is 'Shannon information'. A case for the reference prior can be made on the basis of Fisher information, too. Suppose that $\alpha = u(\theta)$, where u is a strictly increasing smooth function. With the asymptotic-normality result in mind, we can try to make α behave as closely as possible to a location parameter by making the Fisher information for α a constant (say, 1) independent of α. Thus, we want (with usual misuse of information)

$$\int \left(\frac{\partial f^*}{\partial \alpha} \right)^2 f(y \mid \alpha) \, dy = 1.$$

Note that the integral is with respect to y, not α. By the chain rule for differentiation,

$$\begin{aligned} I_{f^*}(\theta) &= \int \left(\frac{\partial f^*}{\partial \theta} \right)^2 f(y \mid \theta) \, dy \\ &= u'(\theta)^2 \int \left(\frac{\partial f^*}{\partial \alpha} \right)^2 f(y \mid \alpha) \, dy = u'(\theta)^2, \end{aligned}$$

whence $u'(\theta) = I_{f^*}(\theta)^{\frac{1}{2}}$. Since the uniform prior is appropriate for the pseudo-location parameter α, and $d\alpha = u'(\theta) d\theta$, it is appropriate to use the reference prior for θ; or at least that's the way this argument goes.

▶ **Ga. Invariance of reference priors under reparametrization.** Suppose that v is a strictly increasing function of θ, and let $\theta = v(\varphi)$. Then φ is a new way of parameterizing things. Now if we regard θ and φ as Random Variables in Bayesian fashion, θ having pdf $\pi(\theta)$ and φ having pdf $\tilde{\pi}(\varphi)$, then, by the Transformation Rule at 55F, we should have

$$\tilde{\pi}(\varphi) = \pi\big(v(\theta)\big) v'(\varphi).$$

Let the pdf of an observation in terms of φ be

$$g(y \mid \varphi) := f(y \mid v(\varphi)).$$

Then the information $\tilde{I}_{g^*}(\varphi)$ about φ satisfies

$$\begin{aligned} \tilde{I}_{g^*}(\varphi) &:= \int \left(\frac{\partial g^*}{\partial \varphi} \right)^2 g(y \mid \varphi) \, dy \\ &= v'(\varphi)^2 \int \left(\frac{\partial f^*}{\partial \theta} \right)^2 f(y \mid \theta) \, dy = v'(\varphi)^2 I_{f^*}(\theta), \end{aligned}$$

and

$$\tilde{I}_{g^*}(\varphi)^{\frac{1}{2}} = v'(\varphi) I_{f^*}\big(v(\varphi)\big)^{\frac{1}{2}}.$$

In other words, the reference prior transforms correctly under change of parametrization, the strongest argument in its favour. The main dictate of Bayesian philosophy is that one must always behave consistently.

Gb. Exercise.   Prove that the reference prior density (for θ) for Bernoulli(θ) is the proper density $\{\theta(1-\theta)\}^{-\frac{1}{2}}$ illustrated on the left of Figure 209I(i) and for Poisson(θ) is the improper density $\theta^{-\frac{1}{2}}$. However persuasive the Mathematics of reference priors has been for you, are you happy that the Bernoulli reference prior attaches so much more weight to θ-regions near 0 than to regions of the same length near $\frac{1}{2}$?

▶ **Gc. Exercise.** Prove that if θ is a location parameter in that $f(y\mid\theta) = h(y-\theta)$ for some pdf h on $\mathbb{R}$, then the reference prior for θ is a constant function.

▶ **Gd. Exercise.**   For our 1-parameter situation, $\gamma > 0$ acts as a **scale parameter** if for some known constant θ_0 and some pdf g on $\mathbb{R}$, we have

$$f(y\mid\gamma) = \frac{1}{\gamma}g\left(\frac{y-\theta_0}{\gamma}\right)$$

Show that γ is a scale parameter for E(mean γ) and for N($0,\gamma^2$). Prove that

$$(\partial_\gamma f^*)(Y\mid\gamma) = -\frac{1}{\gamma}[1 + Z(g^*)'(Z)]$$

where Z has pdf g, and deduce that the Jeffreys reference prior $\pi(\gamma)$ for γ satisfies $\pi(\gamma) \propto \gamma^{-1}$. Under this prior, $\ln(\gamma)$ has uniform density on $\mathbb{R}$.

The truth is that outside of location-parameter cases and scale-parameter cases, there is controversy about the appropriateness of the reference prior at 204(G1) as a prior. The multi-parameter situation – with its Riemannian geometry and its worrying features – is discussed at 379N.

Discussion. *If one is to make any sensible inferences about θ, then*

- *either one must have reasonably good prior information, which one builds into one's prior,*

- *or, if one has only vague prior information, one must have an experiment with sample size sufficient to make one's conclusions rather robust to changes in one's prior.*

There are, however, important cases where one has neither because data are difficult or expensive to collect, or because Nature only provides examples infrequently.

▶ **H. Conjugate priors.** The idea is that in certain situations, if one chooses a prior $\pi(\cdot)$ in a family 'conjugate' to the form of the pdf $f(y \mid \theta)$ of the model, then the posterior density $\pi\left(\cdot \mid \mathbf{y}^{\mathrm{obs}}\right)$ will also be in that conjugate family. 'Conjugate priors' are mathematically very convenient; but this is not in itself a good Statistical reason for using them. However, it is usually the case that we can approximate *any* prior density arbitrarily closely (in the right sense) by mixtures of conjugate priors. See 213L for discussion of this.

Ha. Normal conjugate for $\mathrm{N}(\theta, \mathrm{prec}\ \rho_0)$ model, ρ_0 known. Recalling that **precision is the inverse of variance,**

$$\text{we write } \mathrm{N}(\theta, \mathrm{prec}\ \rho) \text{ for } \mathrm{N}\left(\theta, \tfrac{1}{\rho}\right).$$

▶ **APOLOGIES for using ρ for precision.** The symbol ρ is usually reserved for correlation. It shouldn't cause any problems that I make both uses of ρ. No correlations feature until Section 8.4. I just ran out of symbols.

Suppose that each Y_k is $N(\theta, \mathrm{prec}\ \rho_0)$ where ρ_0 is a known constant. Suppose that we take as prior for θ the density of the $\mathrm{N}(m, \mathrm{prec}\ r)$ distribution. Then

$$\ln \pi\left(\theta \mid \mathbf{y}^{\mathrm{obs}}\right)$$
$$= -\tfrac{1}{2}r(\theta - m)^2 - \tfrac{1}{2}\rho_0 n(\overline{y}_{\mathrm{obs}} - \theta)^2 + \text{function}\,(m, r, \rho_0, \overline{y}_{\mathrm{obs}})$$
$$= -\tfrac{1}{2}r_{\mathrm{new}}\left(\theta - m_{\mathrm{new}}\right)^2 + \text{function}\,(m, r, \rho_0, \overline{y}_{\mathrm{obs}})$$

where

$$r_{\mathrm{new}} = r + n\rho_0, \qquad m_{\mathrm{new}} = \frac{rm + n\rho_0 \overline{y}_{\mathrm{obs}}}{r + n\rho_0}. \tag{H1}$$

We see that

> if each Y_k has the $\mathrm{N}(\theta, \mathrm{prec}\ \rho_0)$ distribution,
> and $\pi(\cdot)$ is the density of $\mathrm{N}(m, r)$,
> then $\pi(\cdot \mid \mathbf{y}^{\mathrm{obs}})$ is the pdf of $\mathrm{N}(m_{\mathrm{new}}, \mathrm{prec}\ r_{\mathrm{new}})$.

The family of normal distributions is therefore conjugate for the $\mathrm{N}(\theta, \sigma_0^2)$ model.

Our prior estimate of θ is m with precision r. In Frequentist theory, the estimate $\overline{y}_{\mathrm{obs}}$ is an estimate of θ with precision $n\rho_0$. We see that, in going from prior to posterior,

- precisions add,

- the new mean m_{new} is the probability mixture of m and $\overline{y}_{\mathrm{obs}}$ weighted proportionately to the corresponding precisions.

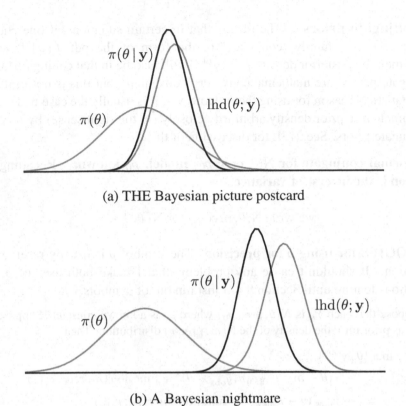

(a) THE Bayesian picture postcard

(b) A Bayesian nightmare

Figure H(i): Prior-likelihood-posterior pictures

This makes good sense; and it can help guide us in our choice of a suitable prior when we do have prior information about θ. Note that the effective prior sample size N_{pri} is r/ρ_0, the effective complete sample size is $N_{\text{comp}} = r_{\text{new}}/\rho_0$, as is checked out by

$$N_{\text{comp}} = N_{\text{pri}} + n, \qquad m_{\text{new}} = \frac{N_{\text{pri}}m + n\overline{y}}{N_{\text{pri}} + n}.$$

Formally, the uniform *vague* prior has prior precision $r = 0$ and any m. In strict Bayesian theory with proper priors, we might take as prior the density of $N(m, \text{prec } \rho_0)$ with ρ_0 very small.

Figure H(i)(a) illustrates a good case of the 'normal conjugate for normal' situation. Figure H(i)(b) is a bad case in which prior and likelihood are in serious conflict and in which therefore we have no robustness: a small change in the prior could cause a large change in the posterior density. When prior and likelihood conflict, prior × likelihood is always small, so the $\pi(\theta \,|\, \mathbf{y}^{\text{obs}})$ is obtained from something close to 0/0. You can see how lack of robustness can arise.

▶▶ **I. The Beta**(K, L) **distribution.** The important family of Beta distributions acts as a conjugate family for the Bernoulli(θ) model.

For $K > 0$, $L > 0$, a Random Variable X is said to have the Beta(K, L) distribution on $[0, 1]$ if it has pdf

$$f_X(x) \; = \; \mathrm{beta}(K, L; x) \; := \; \frac{x^{K-1}(1-x)^{L-1}}{B(K, L)} \quad (0 < x < 1) \qquad \text{(I1)}$$

where, of course, $B(K, L)$ is the *Beta function*:

$$B(K, L) \; := \; \int_0^1 x^{K-1}(1-x)^{L-1}\,\mathrm{d}x \; = \; \frac{\Gamma(K)\Gamma(L)}{\Gamma(K+L)}. \qquad \text{(I2)}$$

We saw a probabilistic proof of the Γ-function expression in the case of positive-

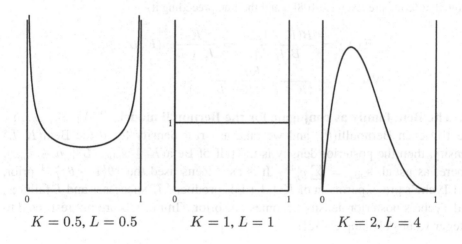

$$K = 0.5,\, L = 0.5 \qquad\qquad K = 1,\, L = 1 \qquad\qquad K = 2,\, L = 4$$

Figure I(i): Pdfs for some Beta(K, L) distributions

integer K and L at 107(L), from which a nice interpretation of the Beta(K, L) distribution follows. A probabilistic proof of the general case will be given later at 250(Cb).

Illustration of some Beta pdfs are given in Figure I(i).

Ia. Analytic proof of the Γ-function formula. This proof assumes familiarity with Jacobians, revision of which occurs in Chapter 7. We have

$$\Gamma(K)\Gamma(L) \; = \; \int_{u=0}^{\infty} e^{-u} u^{K-1}\,\mathrm{d}u \int_{v=0}^{\infty} e^{-v} v^{L-1}\,\mathrm{d}v.$$

Write $u + v = s$, $u/(u+v) = r$, so that $u = rs$, $v = (1-r)s$, and we have the Jacobian

$$\begin{vmatrix} \frac{\partial u}{\partial r} & \frac{\partial u}{\partial s} \\ \frac{\partial v}{\partial r} & \frac{\partial v}{\partial s} \end{vmatrix} = \begin{vmatrix} s & r \\ -s & 1-r \end{vmatrix} = s.$$

Hence,

$$\Gamma(K)\Gamma(L) = \int_0^\infty e^{-s} s^{K-1} s^{L-1} s \, ds \int_0^1 r^{K-1}(1-r)^{L-1} dr$$
$$= \Gamma(K+L)B(K,L).$$

Ib. Exercise. Show that the *mode* of X, the value where $f_X(x)$ is a maximum, satisfies

$$\mathrm{mode}(X) = \frac{K-1}{K+L-2} \quad \text{if } K > 1 \text{ and } L > 1.$$

Show that (compare Exercise 108La and the line preceding it)

$$\mathbb{E}(X) = \frac{B(K+1,L)}{B(K,L)} = \frac{K}{K+L} \quad (K,L > 0),$$
$$\mathrm{Var}(X) = \frac{KL}{(K+L)^2(K+L+1)}.$$

Ic. The Beta family as conjugate for the Bernoulli model. If $Y_1, Y_2, \ldots, Y_n$ are IID each Bernoulli(θ), and we take as prior density for θ the Beta(K, L) density, then the posterior density is the pdf of Beta($K + s_{\mathrm{obs}}$, $L + n - s_{\mathrm{obs}}$), where, as usual, $s_{\mathrm{obs}} = \sum y_k^{\mathrm{obs}}$. It is as if Zeus used the $\{\theta(1-\theta)\}^{-1}$ prior, and Tyche's pre-experiment of $K + L$ trials produced K successes and L failures, and Tyche's posterior density becomes our prior. (Immortals are not restricted to integer values of sample size!)

The reference prior for θ is the (proper) pdf of Beta($\frac{1}{2}, \frac{1}{2}$) illustrated on the left of Figure 209I(i), but there are, as we have seen, rival arguments for using the improper formal Beta($0, 0$) density $\pi(\theta) = \{\theta(1-\theta)\}^{-1}$ or, in some cases, the uniform prior pdf of Beta($1, 1$).

Id. Calculation of Beta distributions. This may be achieved in a way analogous to that used at 149Ea for Gamma distributions. Here's the relevant part of a program:

```
double Fbeta(double x){
  int n;
  int ans;
  double sum, frac, temp, logfrac, ratio, term;
  temp =  - loggam(K) - loggam(L) + loggam(K + L);
```

```
if (x <= 0.5){
  logfrac = K * log(x) + L * log(1-x) - log(K) + temp;
  frac = exp(logfrac);
  term = sum = frac;   n=0;
  do{
    ratio = x * (K + L + n)/(K + n+1.0);
    term = term * ratio;   sum = sum + term;
    n++;
  }while((fabs(term) > htol) || (ratio > 0.55));
  return sum;
}
else {
  logfrac =  K * log(x) + L * log(1-x) - log(L) + temp;
  frac = exp(logfrac);
  term = sum = frac;   n=0;
  do{
    ratio = (1-x) * (K + L + n)/(L + n+1.0);
    term = term * ratio;   sum = sum + term;
    n++;
  }while((fabs(term) > htol) || (ratio > 0.5));
  return (1.0 - sum);
}
}
```

▶ **J. A summary table of some conjugate families.**

$Y_1, Y_2, \ldots, Y_n$, IID each with distribution	Prior density $\pi(\theta)$ is density of distribution	Posterior density $\pi(\theta \mid \mathbf{y}^{\mathrm{obs}})$ is pdf of distribution
$N(\theta, \mathrm{prec}\ \rho_0)$	$N(m, \mathrm{prec}\ r)$	$N\left(\frac{rm+n\rho_0\bar{y}}{r+n\rho_0}, \mathrm{prec}\ r + n\rho_0\right)$
$E(\mathrm{rate}\ \theta)$	$\mathrm{Gamma}(K, \mathrm{rate}\ \alpha)$	$\mathrm{Gamma}(K + n, \mathrm{rate}\ \alpha + s_{\mathrm{obs}})$
$\mathrm{Poisson}(\theta)$	$\mathrm{Gamma}(K, \mathrm{rate}\ \alpha)$	$\mathrm{Gamma}(K + s_{\mathrm{obs}}, \mathrm{rate}\ \alpha + n)$
$\mathrm{Bernoulli}(\theta)$	$\mathrm{Beta}(K, L)$	$\mathrm{Beta}(K + s_{\mathrm{obs}},\ L + n - s_{\mathrm{obs}})$
$N(0, \mathrm{prec}\ \theta)$	$\mathrm{Gamma}(K, \mathrm{rate}\ \alpha)$	$\mathrm{Gamma}(K + \frac{1}{2}n, \mathrm{rate}\ \alpha + \frac{1}{2}\sum y_k^2)$

Table J(i): Conjugate priors

Table J(i) collects together some useful information.

▶▶ **Ja. Exercise.** Check the last four rows of the table. In each case say how the reference prior fits into the pattern.

▶▶ **K. Predicting new observations from current data.** Making predictions about future observations is a key element of Statistics.

Ka. A coin-tossing example. Suppose that our experiment consists of tossing a coin with probability θ of Heads n times, and that a Beta(K, L) prior is thereby converted to a Beta$(K + s_{\text{obs}}, L + n - s_{\text{obs}})$ posterior. If we wish to estimate the probability that the next throw will produce Heads, then, as Chapter 9 will clarify, we have to find the posterior mean of Θ, namely,

$$\mathbb{E}_{\text{post}}(\Theta) = \int \theta \pi(\theta \mid \mathbf{y}^{\text{obs}}) \, d\theta = \frac{K + s_{\text{obs}}}{K + L + n},$$

as we know from Exercise 210Ib. If we wish to estimate *instead* the probability that the next *two* throws (after the nth) will produce Heads, we must find $\mathbb{E}_{\text{post}}(\Theta^2)$, namely,

$$\frac{B(K + 2 + s_{\text{obs}}, L + n - s_{\text{obs}})}{B(K + s_{\text{obs}}, L + n - s_{\text{obs}})} = \frac{K + s_{\text{obs}}}{K + L + n} \times \frac{K + s_{\text{obs}} + 1}{K + L + n + 1}.$$

There is good recursive sense in this: Bayesian predictive probabilities display a nice coherence.

Frequentist theory is in some difficulty here. If you tell a Frequentist the results of your n tosses of a coin and ask "What is the probability that the next two results will be Heads?", he/she will say "Well, θ^2, of course, where θ is the unknown parameter. But I *can* give you an approximate CI for θ^2 (which, to make matters worse, does not take into account the fact that θ is bound to be close to $\frac{1}{2}$)." Not much recursiveness here!

Kb. A 'normal' example. Suppose that for observations $y_1^{\text{obs}}, y_2^{\text{obs}}, \ldots, y_n^{\text{obs}}$, we use the N$(\theta, \text{prec } \rho_0)$ model (ρ_0 known) with N$(m, \text{prec } r)$ prior, thereby obtaining an N$(m_{\text{new}}, \text{prec } r_{\text{new}})$ posterior. What would we say about Y_{n+1} in this situation? Intuition (made precise in the study of joint pdfs in the next chapter) would say that, conditionally on $\mathbf{Y} = \mathbf{y}^{\text{obs}}$, Y_{n+1} is $\Theta + \varepsilon_{n+1}$ where Θ is $N(m_{\text{new}}, \text{prec } r_{\text{new}})$, ε_{n+1} is N$(0, \sigma_0^2)$, where $\sigma_0^2 = 1/\rho_0$, and Θ and ε_{n+1} are independent. Thus we would say that given our current information $(Y_1, Y_2, \ldots, Y_n)$, Y_{n+1} is N$(m_{\text{new}}, r_{\text{new}}^{-1} + \sigma_0^2)$.

Kc. Exercise. Suppose that we take the 'Frequentist' uniform prior for θ in the 'normal' example just discussed. Explain that there is a certain sense in which a Frequentist might be happy about the Bayesian prediction for Y_{n+1} because of the use of a Pivotal RV.

Kd. Exercise. Suppose that $Y_1, Y_2, \ldots, Y_n, Y_{n+1}, Y_{n+2}$ are IID, exponential rate λ. Suppose that $y_1^{\text{obs}}, y_2^{\text{obs}}, \ldots, y_n^{\text{obs}}$ have been observed, and that their sum is $s = s_{\text{obs}}$. Show that a Bayesian using the 'Frequentist' prior λ^{-1} for λ would say that

$$\mathbb{P}_{\text{post}}(Y_{n+1} > ts) = \frac{1}{(1 + t)^n},$$

$$\mathbb{P}_{\text{post}}(Y_{n+1} > t_{n+1}s; \; Y_{n+2} > t_{n+2}s) = \frac{1}{(1 + t_{n+1} + t_{n+2})^n}.$$

As will be clear from Chapter 7, and clearer still after our study of the Poisson process in Chapter 9, it is true for every λ that, with $S := Y_1 + \cdots + Y_n$,

$$\mathbb{P}(Y_{n+1} > tS) = \frac{1}{(1+t)^n},$$

$$\mathbb{P}(Y_{n+1} > t_{n+1}S;\ Y_{n+2} > t_{n+2}S) = \frac{1}{(1+t_{n+1}+t_{n+2})^n}.$$

which makes Frequentist and Bayesian views tally here.

Ke. Exercise: Poisson model. Suppose that all nuclear power stations on Planet Altair 3 were installed at the same instant, exactly 1 year ago, and that accidents occur purely at random, that is in a Poisson process of rate θ per year, as studied in detail in Chapter 9. This means that the number of accidents occurring in a time interval of length t has the Poisson distribution of parameter θt independently of what happens in disjoint time intervals. The only information we have about θ is that 1 accident has occurred in the 1 year. What is our best prediction of the expected number of accidents in the next year? Frequentists would say 1, of course. Show that the Bayesian reference prior would lead one to say $1\frac{1}{2}$. This example will be discussed further in Chapter 9.

Kf. Exercise. Show that if θ has *posterior* distribution Gamma(K, rate α), then our estimate that a Poisson θ variable takes the value n is

$$\frac{(n+K-1)(n+K-2)\cdots(K+1)K}{n!} \frac{\alpha^K}{(\alpha+1)^{K+n}}$$

for $n = 0, 1, 2, \ldots$. Show that if K is a positive integer, then this is the probability that the Kth Head occurs on the $(K+n)$th toss of a coin with probability $\alpha/(\alpha+1)$ of Heads. See Exercise 53Dc. We shall discuss this further in Chapter 9.

▶▶ **L. Mixtures of Conjugates.** Suppose that our likelihood function has a conjugate family. Then we may well use a prior of the form

$$\pi(\theta) = \sum_{k=1}^{r} p(k)\pi_k(\theta), \tag{L1}$$

where each $\pi_k(\cdot)$ is in the conjugate family, and $p(k) \geq 0$, $\sum p(k) = 1$. Thus $\pi(\theta)$ *is a probabilistic ($p(k)$-weighted) mixture of conjugate pdfs*. In the 'God and the Reverend Bayes' picture, God could choose Θ directly using $\pi(\theta)$. Alternatively, we could have a hierarchical model in which God does a two-stage experiment: first, He chooses N with values in $\{1, 2, \ldots, r\}$ at random with $\mathbb{P}(N = k) = p(k)$, and then He chooses Θ with pdf π_N. He tells Bayes all the values $p(k)$ and all the functions π_k but keeps N and Θ secret.

La. Exercise. Show that

$$\pi(\theta \mid \mathbf{y}) = \sum_{k=1}^{r} p(k \mid \mathbf{y})\pi_k(\theta \mid \mathbf{y}), \tag{L2}$$

where, of course, each $\pi_k(\cdot\,|\,\mathbf{y})$ is in the conjugate family and where (*Exercise!*) the $p(k\,|\,\mathbf{y})$ satisfy

$$p(k\,|\,\mathbf{y}) \;=\; \frac{I_k}{\sum I_j} \quad \text{where} \quad I_k := \int p(k)\pi_k(\theta)\mathrm{lhd}(\theta;\mathbf{y})\,\mathrm{d}\theta. \qquad (\text{L3})$$

Of course, $p(k\,|\,\mathbf{y}) \geq 0$, $\sum p(k\,|\,\mathbf{y}) = 1$. Indeed, show that

$$p(k\,|\,\mathbf{y}) \;=\; \mathbb{P}(N = k\,|\,\mathbf{Y} = \mathbf{y}) \qquad (\text{L4})$$

in a sense which should be intuitively obvious and which is made rigorous in the next chapter. This formula is the key to **Bayesian significance testing.** We see that

$\pi(\cdot\,|\,\mathbf{y})$ is again a probabilistic mixture of conjugate pdfs.

We shall see later in Subsection 217N and Chapter 8 how this may be used.

In principle, the scope of the method is as wide as possible because ('in the cases of interest to us' is understood here)

we can approximate ANY prior DF by a mixture of conjugates.   (L5)

We know from our study of the CLT that this means: given any DF G, we can find DFs G_n, each with pdf a probabilistic mixture of conjugate pdfs, such that $G_n(\theta) \to G(\theta)$ at every point θ at which $G(\cdot)$ is continuous. The following proof of this fact is of independent interest.

Proof of (L5). The idea is as follows: first (Stage 1) we prove that G can be approximated (in an even stronger sense) by DFs of RVs each of which takes only finitely many values. But the DF of an RV taking only finitely many values is a probabilistic mixture of the DFs of deterministic (constant on Ω) RVs. Hence, to complete the argument, we need only show (Stage 2) that the DF of a deterministic RV may be approximated by a probabilistic mixture of DFs each with conjugate pdf.

Stage 1. Let G be any DF. Recall that G is right-continuous, which is why we can use min rather than inf in the following definitions. Put $\theta_0 = -\infty$, $G(\theta_0) = 0$. Let n be a positive integer. Define recursively for $k \geq 1$,

$$\theta_k := \min\{\theta : G(\theta) \geq G(\theta_{k-1}) + 1/n\}$$

provided the θ-set on the right-hand side is non-empty. Note that we can define θ_k for at most n values of k. Let a be the largest such k, and set $\theta_{a+1} = \infty$. Now define (for $\theta \in \mathbb{R}$)

$$H(\theta) := G(\theta_k) \quad \text{if } \theta_k \leq \theta < \theta_{k+1}.$$

Then

$$0 \leq G(\theta) - H(\theta) \leq 1/n \quad \text{for every } \theta.$$

This completes Stage 1.

Stage 2. It is certainly enough to show that a single DF with conjugate pdf can approximate the DF H_c of the constant variable $\Theta = c$:

$$H_c(\theta) = \begin{cases} 0 & \text{if } \theta < c, \\ 1 & \text{if } \theta \geq c. \end{cases}$$

Tchebychev's inequality shows that the DFs of

$$\mathrm{N}\left(c, \frac{1}{n}\right), \quad \mathrm{Gamma}\left(n, \text{mean } \frac{c}{n}\right), \quad \mathrm{Beta}(nc, n(1-c)),$$

which all have mean c and which have variances

$$\frac{1}{n}, \quad \frac{c}{n}, \quad \frac{c(1-c)}{n+1}$$

respectively, all converge to H_c at values of θ other than the discontinuity point c of H_c.

$\square$

Remarks. It goes without saying that if we allow ourselves complete freedom in our choice of mixtures, we have essentially the same problem as that of choosing a prior from the set of all priors. We would normally use a mixture of just a few conjugate pdfs.

▶▶ **Lb. Hierarchical models.** That the parameter Θ might itself be considered to be the result of a two-stage experiment, involving hyper-parameters $(p(k))$ and $(\pi_k(\cdot))$ in our example, is part of the theory of hierarchical modelling. Draper [67] is one of the places where this is presented with clarity and good sense. Of course, the $p(k)$ could themselves be considered random if we take another level in the hierarchy.

The idea of 'adding levels' will feature in different ways in this book: in modelling coin-tossing in Subsection 217N, in the Gibbs-sampler analysis of a Cauchy location parameter in the next chapter, in models in Chapter 8; and it does in a way form a central part of de Finetti's theorem in Subsection 220P. But the main uses of hierarchical models feature in Chapter 8.

▶▶ **M. First thoughts on choosing Your prior.** The points raised here will be illustrated in later subsections.

▶ **Ma. Subjectivity.** Bayesian Statistics is a subjective theory:

> **Everyone is entitled to his/her own prior.**
> **You have Your prior, I have Mine.**
> **But this puts a real responsibility on You to say**
> **how you arrived at Your prior.**

You may choose the very popular route of choosing one of the standard 'vague' priors or something close to it. Or you may – and whenever possible, **should** – take the business of choosing Your prior very seriously, arriving at one with characteristics approved by people working in the field of application. Those people may suggest either a prior mean and variance or a 95% prior CI, and You might choose as Your prior a conjugate prior to fit their beliefs. Draper [67] has examples you should study. See also the *use of imaginary results* in the next subsection.

▶ **Mb. Comparing 'pre' and actual sample sizes.** Suppose that there is a natural measure of effective prior sample size, the number of experiments that Tyche did before she handed over to you. (For example, for large K and L, a Beta(K, L) prior has effective sample size $K + L$, modulo irrelevant quibbles over vague priors.)

If the effective prior sample size is much larger than the experimental sample size, then there is little point in doing the experiment. At the other extreme, if the experimental sample size is much larger than the effective prior sample size, then one might as well use a vague prior. That leaves important cases where the two sample sizes are 'comparable', but it raises other points as well.

▶ **Mc. Robustness and sensitivity.** As mentioned earlier, one can only be happy about one's conclusions if they are relatively robust in that small changes in the prior result in small changes in one's conclusion. (This assumes that one is happy with the model, that is, with the likelihood function.) We lose robustness in cases where the likelihood function $\text{lhd}(\cdot; \mathbf{y}^{\text{obs}})$ is in conflict with the prior density function $\pi(\cdot)$ in the sense that whenever one is large, the other is small.

Md. Cromwell's dictum. Bayesian statisticians all quote *Cromwell's dictum*, a sentence in a letter (3 August 1650) from Oliver Cromwell to the General Assembly of the Church of Scotland begging them to consider the possibility that they might be wrong. (The actual sentence, "I beseech you, in the bowels of Christ, think it possible that you may be mistaken" reads very strangely in terms of modern language. However, it's not so much language that has changed as our knowledge of the way we function: in the time of Cromwell, the bowels were thought to play a much more important part in our lives than merely that of helping process food! The brain was very under-rated! I guess that 'bowels' in Cromwell's sentence might these days read 'innermost being' or something similar. It's interesting that we have not fully escaped the old ideas, as phrases such as 'gut reaction' demonstrate.)

Bayesians usually quote Cromwell's dictum to state that one's prior density should never be literally zero on a theoretically possible range of parameter values. But we should regard it as a warning not to have a prior density which is too localized, and which is therefore extremely small on a range of parameter values which we cannot exclude. Consider ye therefore the possibility that a prior with very large effective sample size just might be wrong! (That your model might be wrong is something you must always bear in mind.)

Using 'heavy-tailed' priors increases robustness.

▶ **N. Choosing a prior for coin tossing: use of imaginary results.** How would you decide on Your prior for the probability θ that a coin which you are just about to toss 100 times will fall Heads? What makes this case difficult is your very strong prior belief that θ is very close to $\frac{1}{2}$. But there are points in the following discussion which are relevant much more generally.

Suppose that

$$\text{you are 95\% confident that } \theta \in (0.49, 0.51). \tag{N1}$$

You might decide that you will use a Beta(N, N) distribution with large integer N. Such a distribution has mean 0.5 and is strongly concentrated about that point. It is as if in the Pre-experiment, Zeus used the density $\{\theta(1-\theta)\}^{-1}$ (the 'Frequentist prior') and that Tyche then tossed the coin $2N$ times getting N heads. But we know from our 'voting' discussion that Tyche's posterior density is approximately normal $N(\frac{1}{2}, \frac{1}{8n})$, so that, with 2 for 1.96 as usual, we must have (as nearly as makes no difference)

$$2\sqrt{\frac{1}{4 \times 2N}} = 0.01, \text{ whence } N = 5000.$$

The effective prior sample size, the size of Tyche's experiment, is therefore 10000, and if we really are going to stick to the Beta$(5000, 5000)$ prior, then there is no point in doing the '100 tosses' experiment because the effective prior sample size would dwarf the experimental sample size. If you are determined to stick to the Beta$(5000, 5000)$ prior, you need to do an experiment with a sample size of many thousands for the experiment to be worth doing.

But you may think to yourself: "But what if I *were* to toss the coin 100 times and it fell Heads only 30 times? Surely, I should begin to think that the coin is biased?!" Cromwellian doubt about your prior is setting in.

You might allow a small probability c (the 'Cromwell factor') that something in the coin's manufacture (or in its history since then) makes (say) a Beta(M, M) distribution appropriate, where M is much smaller than N. Hence, recalling that beta$(K, L; \cdot)$ denotes the density of the Beta(K, L) distribution, you might choose to use as Your prior density the function (mixture of conjugate pdfs)

$$(1 - c)\,\text{beta}(N, N; \theta) + c\,\text{beta}(M, M; \theta). \tag{N2}$$

For really small c, you will still have (N1) – very nearly.

Na. Exercise. Let H_1, T_1, H_2, T_2, h, t be integers. Show using equation 213(L2) that if you use a prior

$$(1 - c)\,\text{beta}\,(H_1, T_1; \theta) + c\,\text{beta}\,(H_2, T_2; \theta),$$

and the coin is tossed $h + t$ times, producing h Heads and t Tails, then Your posterior density for θ is given by

$$(1 - \tilde{c})\,\text{beta}\,(H_1 + h, T_1 + t; \theta) + \tilde{c}\,\text{beta}\,(H_2 + h, T_2 + t; \theta),$$

where

$$\tilde{c} = \frac{c}{A(1 - c) + c},$$

where (in a form for easy computer implementation)

$$\ln A = \ln B\,(H_1 + h, T_1 + t) - \ln B\,(H_1, T_1)$$
$$- \ln B\,(H_2 + h, T_2 + t) + \ln B\,(H_2, T_2)\,.$$

Hence show that Your posterior mean θ_c of θ (Your estimate that a further toss of the coin will produce Heads) is

$$\theta_c = \frac{VA(1-c) + Sc}{A(1-c) + c},$$

where

$$V = \frac{H_1 + h}{H_1 + h + T_1 + t}, \qquad S = \frac{H_2 + h}{H_2 + h + T_2 + t}.$$

If A is very small, then the derivative

$$\frac{d\theta_c}{dc} = \frac{A(S - V)}{\{A + c(1 - A)\}^2} \qquad \text{(N3)}$$

is huge at $c = 0$ and becomes very small very rapidly as c moves away from 0.

Use a computer to show that Your estimate of θ_c takes the values shown in Table N(i) [respectively Table N(ii)] when

$$H_1 = T_1 = N = 5000, \quad H_2 = T_2 = M, \quad h + t = 100,$$

and $M = 4$ [respectively, $M = 6$].   □

| h | A | \multicolumn{5}{c}{Value of c} |
		1	0.1000	0.0100	0.0010	0
20	0.0000001	0.2222	0.2222	0.2222	0.2222	0.4970
25	0.0000225	0.2685	0.2686	0.2690	0.2736	0.4975
30	0.0019177	0.3148	0.3179	0.3440	0.4352	0.4980
35	0.0554319	0.3611	0.4068	0.4773	0.4961	0.4985
40	0.5856032	0.4074	0.4844	0.4975	0.4989	0.4990
45	2.3684058	0.4537	0.4975	0.4993	0.4995	0.4995
50	3.7630309	0.5000	0.5000	0.5000	0.5000	0.5000

Table N(i): Posterior means θ_c of θ when $M = 4$

Note that the strange top rows of these Tables tally exactly with the discussion around (N3): for results with h far from 50, there is extreme sensitivity to c for c near 0. Table N(iii) gives the values $\tilde{c}$ for various (h, c) pairs when $M = 6$.

This subsection has illustrated the 'method of imaginary results' in which You look at posterior densities (or at summary statistics such as the posterior mean, as we have done) for hypothetical results, both to try to assess how strongly You

h	A	Value of c				
		1	0.1000	0.0100	0.0010	0
20	0.0000001	0.2321	0.2321	0.2321	0.2322	0.4970
25	0.0000298	0.2768	0.2768	0.2774	0.2832	0.4975
30	0.0021079	0.3214	0.3247	0.3519	0.4412	0.4980
35	0.0534058	0.3661	0.4091	0.4774	0.4961	0.4985
40	0.5161140	0.4107	0.4834	0.4973	0.4988	0.4990
45	1.9823827	0.4554	0.4972	0.4993	0.4995	0.4995
50	3.0968888	0.5000	0.5000	0.5000	0.5000	0.5000

Table N(ii): Posterior means θ_c of θ when $M = 6$

h	Prior value of c				
	1	0.1000	0.0100	0.0010	0
20	1.0000	1.0000	1.0000	0.9999	0.0000
25	1.0000	0.9997	0.9971	0.9711	0.0000
30	1.0000	0.9814	0.8273	0.3220	0.0000
35	1.0000	0.6754	0.1591	0.0184	0.0000
40	1.0000	0.1771	0.0192	0.0019	0.0000
45	1.0000	0.0531	0.0051	0.0005	0.0000
50	1.0000	0.0346	0.0033	0.0003	0.0000

Table N(iii): Posterior values $\tilde{c}$ of c when $M = 6$

feel about certain aspects of Your prior and to assess sensitivity. You can present people in the field of application with tables based on imaginary results and let their reactions guide you in Your choice of prior.

In the coin-tossing example, if You *do* decide to use a prior of the form 217(N2), then you could look at several possibilities for the pair (c, M), but you would need to be very cautious about interpreting Your results if h did turn out to be far from 50 because of the pathological sensitivity of θ_c to c near $c = 0$.

Our coin-tossing example is too close to the notorious *Bayesian significance testing of a sharp hypothesis* discussed in Subsection 234I below. And yet it *is* the case that we would have very strong prior belief that θ is very close to $\frac{1}{2}$.

▶ **O. Philosophy of priors.** The controversial questions:

- *Can we really measure 'degree of belief' quantitatively; and if so, is there any reason why we should be able to apply the rules of Probability to these measurements?*

have had endless debate. Many Bayesians would say that logical coherence forces us to answer yes to both questions.

Their *betting argument* for this is that a set of questions of the form 'Would You be more willing to bet on this than on that?' can determine Your prior. I am not at all persuaded by the 'betting argument', many forms of which come very close to tautology; and since I cannot present it with any conviction, I skip it.

A second argument is based on **exchangeability**. I am not convinced by this argument either; but it is an interesting argument based on a remarkable result, *de Finetti's Theorem*, the subject of the next subsection.

To become obsessed with *defining* real-world things is to miss completely the subtlety of Mathematics. We saw that it is both practically and *logically* impossible to *define* probability in terms of long-term relative frequency. In all Applied Mathematics, we build *models* for the real world, and we study the models, not the real world. We investigate how useful our models are as analogues of the real world by testing how well predictions made in the model transfer to the real world. Testing predictions is more difficult in Probability and Statistics than in many other branches of Applied Maths, but there is a lot of evidence that, used sensibly, Bayesian Statistics is very useful as a model for the (everyday, 'non-quantum') real world. (We know that all of the ideas we have seen so far in this book fail for quantum mechanics, but that there Quantum Probability works supremely well.) The fact that sensible Bayesian Statistics generally *works* is good enough for me, and cutting the discussion of philosophy *there* (except for a discussion of de Finetti's Theorem) is my contribution to world ecology. Enough paper has already been devoted to the topic.

It would, however, be dishonest of me to pretend that after decades of controversy, all conflicts between Bayesian and Frequentist schools are now resolved. Inevitably, some conflicts feature in later sections.

▶ **P. Exchangeability; de Finetti's Theorem.**   We study only the simplest case of variables $Y_1, Y_2, Y_3, \ldots$ taking values in the two-point set $\{0, 1\}$.

An *infinite* sequence $Y_1, Y_2, Y_3, \ldots$ of RVs taking values in $\{0, 1\}$ is called **exchangeable** if whenever τ is a permutation of $\mathbb{N} = \{1, 2, 3, \ldots\}$ which leaves all but finitely many elements fixed, then the sequences $Y_1, Y_2, Y_3, \ldots$ and $Y_{\tau(1)}, Y_{\tau(2)}, Y_{\tau(3)}, \ldots$ have the same probabilistic law. This means that for every $m \in \mathbb{N}$, for $y_k \in \{0, 1\}$ for $1 \leq k \leq m$, and for every permutation ρ of $\{1, 2, \ldots, m\}$,

$$\mathbb{P}(Y_1 = y_1; Y_2 = y_2; \ldots; Y_m = y_m)$$
$$= \mathbb{P}(Y_{\rho(1)} = y_1; Y_{\rho(2)} = y_2; \ldots; Y_{\rho(m)} = y_m).$$

A Bayesian might say, "If I want to model coin-tossing or indeed any situation in which a Frequentist would use an 'IID Bernoulli(θ) model', then in My model, I must surely assume that the finite-length experiment actually performed has the structure of the first part of an *infinite* exchangeable sequence $Y_1, Y_2, Y_3, \ldots$; it would be illogical to do otherwise."

▶▶ **Pa. Fact: de Finetti's Theorem.** *Let $Y_1, Y_2, Y_3, \ldots$ be an (infinite) exchangeable sequence with values in $\{0, 1\}$ in the sense described above. Then, with probability 1, the limit*

$$\Theta := \lim \frac{Y_1 + Y_2 + \cdots + Y_n}{n}$$

exists (in $[0, 1]$); and conditionally on $\Theta = \theta$, the variables $Y_1, Y_2, Y_3, \ldots$ are IID each Bernoulli(θ).

Thus, for example, for $m \in \mathbb{N}$, for $y_1, y_2, \ldots, y_m \in \{0, 1\}$ with sum r,

$$\mathbb{P}(Y_1 = y_1; \ldots; Y_r = y_r) = \int_0^1 \theta^r (1 - \theta)^{m-r} \mathbb{P}(\Theta \in d\theta).$$

Do note however, that for de Finetti's Theorem, we must have an *infinite* exchangeable sequence. See Exercise Pb below.

So, a Bayesian might say, "I must assume 'infinite' exchangeability, so the Random Variable Θ must exist, Θ plays exactly the rôle of the Frequentist's probability of success, and the distribution of Θ is what I mean by My prior distribution."

Whether or not de Finetti's Theorem justifies the *existence* of a prior distribution, it clearly does not help You decide what to take for Your prior.

We shall see later that de Finetti's Theorem and the completely general Strong Law of Large Numbers follow from the same use of the *Convergence Theorem for Reverse Martingales*. This reasoning will make it clear why an *infinite* exchangeable sequence is necessary for de Finetti's Theorem.

Pb. Exercise. Suppose that Y_1 and Y_2 are RVs with values in $\{0, 1\}$, exchangeable in that for some non-negative a, b, c with $a + 2b + c = 1$,

$$\mathbb{P}(Y_1 = Y_2 = 0) = a,$$
$$\mathbb{P}(Y_1 = 1, \ Y_2 = 0) = \mathbb{P}(Y_1 = 0, \ Y_2 = 1) = b,$$
$$\mathbb{P}(Y_1 = Y_2 = 1) = c.$$

Why need it not be the case that there exists a measure μ such that whenever $y_1, y_2 \in \{0, 1\}$ with sum r,

$$\mathbb{P}(Y_1 = y_1; Y_2 = y_2) = \int_0^1 \theta^r (1 - \theta)^{2-r} \mu(d\theta)?$$

You do not need to know about measures. What is the *simple* idea needed here?

▶ **Pc. Exercise.** Let $Y_n = 1$ if the ball drawn from Pólya's urn (76D) just before time n is Red, 0 if black. Prove that $Y_1, Y_2, Y_3, \ldots$ is an exchangeable sequence. What is the distribution of Θ when de Finetti's Theorem is applied to this case?

Q. Remark. The Bayesian theory of CIs continues at a higher level in Chapters 7 and 8. A discussion of Bayesian Hypothesis Testing features in the next section.

6.7 Hypothesis Testing – if you must

In addition to discussing Hypothesis Testing, this section also includes at Subsection 236J a first discussion of **Model Choice**.

For a long period, Frequentist Hypothesis Testing (FHT) and its associated p-values, etc, dominated Statistics; and this alone requires that we take a look at the topic. How can we not look at the F-test in Analysis of Variance?! For simple situations, FHT then went out of favour with many statisticians, for reasons I'll explain. The main point is that, for simple situations,

> **CIs usually provide a much better way of analysing data,**

and computational advances mean that CIs are available in many contexts. However, FHT perhaps retains value to all as a diagnostic aid, something which might be used to decide on whether further investigation would be of value. And it might be said to retain its value in regard to decision making, though even there it is hard to see why CIs are not better. I have not the space (and perhaps not the enthusiasm) to have a chapter on **Decision Theory**, a topic which I agree to be of importance. See Appendix D for references.

I have stressed that for *simple* situations, Hypothesis Testing is best replaced by the use of CIs. However, as I have earlier emphasized, for investigating interactions in complex ANOVA tables, for testing independence in contingency tables, for questions of goodness of fit, etc, Hypothesis Testing remains essential.

Where some Bayesians commit in their testing of sharp hypotheses the very error for which Bayesians have long criticized Frequentist theory, I shall say so. In this context I hope to lay to rest claims that Bayesian and Frequentist views are sometimes in extreme conflict for large sample sizes.

I begin by describing Frequentist Hypothesis Testing as clearly as I can. I present the full story of the **Likelihood-Ratio (LR) Test** first, and then try to bring it to life via examples. We shall see in Chapter 8 that

> **the t test, χ^2 tests, F tests, are all special cases of the LR test.**

The LR Test makes Hypothesis Testing well structured mathematically, not the shambles of *ad hoc* methods it used to be, though, of course, the brilliant intuition of the classical figures in Statistics had led them to the right answers without a general theory.

So sensible is the LR Test that a cynic might say that statisticians have struggled hard to 'prove' that it is the best test by cooking up for each situation a criterion for 'best' which ensures that the LR test *is* best. A celebrated result, the **Neyman–Pearson Lemma** (not discussed in this book), does prove that the LR Test is unquestionably best ('most powerful') in a certain extremely simple situation of little practical importance. In other situations, the LR Test might be Uniformly Most Powerful (UMP) Unbiased, or UMP Invariant, or UMP Similar or UMP whatever. I take the attitude that if one has to test a hypothesis, then the LR Test is manifestly a good way to do it, and leave it at that ... except that I think that the form of the LR Test can sometimes be improved even within the terms of Frequentist theory.

▶▶ **A. The LR Test: fundamental terminology.**   Recall that we are adopting the Frequentist viewpoint for a time. If you wish, suppose that we have very little prior information about the whereabouts of θ.

Everything in this subsection is worded so that it applies without modification to the case when θ is a multidimensional parameter or, in other language, when there is more than one parameter.

One has to think of Hypothesis Testing as follows. We have a **Null Hypothesis** H_0 which we shall only reject if there is rather strong evidence against it. We set up an **Alternative Hypothesis** H_A which indicates what kind of departure from the Null Hypothesis we are looking out for. It is important to realize straight away that in Hypothesis Testing, *the Null and Alternative Hypotheses are not placed on an equal footing*: the Null Hypothesis is 'innocent unless proven guilty' in the 'H_0 versus H_A' trial. *We behave in a way that justice is seen to be done to the Null Hypothesis: even when we reject it, we must have given it a very fair hearing.* (*Note.* The asymmetry between H_0 and H_A is inappropriate in many practical situations, but ignore this fact for now.)

For the moment, we consider the situation where the sample size n is somehow predetermined.

The Null and Alternative Hypotheses will take the form

$$H_0 : \theta \in B_0, \qquad H_A : \theta \in B_A,$$

where B_0 and B_A are disjoint subsets of the parameter space. We define

$$\text{mlhd}\left(H_0; \mathbf{y}^{\text{obs}}\right) := \sup_{\theta \in B_0} \text{lhd}\left(\theta; \mathbf{y}^{\text{obs}}\right),$$

$$\text{mlhd}\left(H_A; \mathbf{y}^{\text{obs}}\right) := \sup_{\theta \in B_A} \text{lhd}\left(\theta; \mathbf{y}^{\text{obs}}\right)$$

so that $\text{mlhd}\left(H_0; \mathbf{y}^{\text{obs}}\right)$ is the maximum likelihood of obtaining the observed data $\mathbf{y}$ consistent with Hypothesis H_0. Yes, 'maximum' should be 'supremum' ... .

We define the **likelihood-ratio**

$$\text{lr}(\mathbf{y}^{\text{obs}}) := \text{lr}(H_A, H_0; \mathbf{y}^{\text{obs}}) := \frac{\text{mlhd}\left(H_A; \mathbf{y}^{\text{obs}}\right)}{\text{mlhd}\left(H_0; \mathbf{y}^{\text{obs}}\right)}, \tag{A1}$$

which measures how much better an explanation of the observed data H_A is than H_0. **The Likelihood-Ratio Test takes the form:**

$$\text{Reject } H_0 \text{ if } \text{lr}(\mathbf{y}^{\text{obs}}) \geq \kappa \tag{A2}$$

for some suitably chosen constant κ. Thus, we reject H_0 if H_A provides at least κ times as good an explanation of the observed data. Sensible so far – yes?! (Well, actually, not always; but stick with it for now.)

Now, this strategy could lead us to reject H_0 when H_0 is in fact true. To consider this, we need to think about the Pre-Statistic $\text{LR}(\mathbf{Y})$ which crystallizes into the actual statistic $\text{lr}(\mathbf{y}^{\text{obs}})$. The **size** or **significance level of the TEST** is defined to be

$$\text{significance level of test} := \text{size} = \sup_{\theta \in B_0} \mathbb{P}\left(\text{LR}(\mathbf{Y}) \geq \kappa \,|\, \theta\right), \tag{A3}$$

the maximum probability consistent with H_0's being true that we reject H_0. To protect H_0, we insist that the size is small. Tradition has tended to set the significance level at one of the values 5% or 1%. When we have decided on the significance level of the test, we can find the appropriate constant κ, as we shall see. In discrete situations, we normally cannot fix the size to be exactly what we want for the same reason that we could not fix the exact level of a CI in discrete situations. I am going to ignore this problem, and *pretend that we can always fix the size to be a given value* α. I therefore avoid randomization tests and the like.

A related concept is the following. We define the **p-value** of the observed data, also called the **significance level of the DATA** by

$$\text{p-value}(\mathbf{y}^{\text{obs}}) := \sup_{\theta \in B_0} \mathbb{P}\left(\text{LR}(\mathbf{Y}) \geq \text{lr}(\mathbf{y}^{\text{obs}}) \,|\, \theta\right) \tag{A4}$$

the maximum probability consistent with H_0's being true that evidence against H_0 as least as strong as that provided by the data would occur by Chance. (A convoluted concept!)

We see that

> we reject H_0 if and only if the p-value$(\mathbf{y}^{\text{obs}}) \leq$ size.

Remember that it's a SMALL p-value which represents strong evidence against H_0. Traditionally, a p-value of less than 5% [respectively, 1%, 0.1%] has been said to constitute *significant* [respectively, *very significant*, *extremely significant*] evidence against H_0.

I stress that the size of the test and the p-value of the data are calculated on the assumption that the Null Hypothesis H_0 is true. Protecting the Null Hypothesis in the way we do certainly gives the theory a lopsided character, which we must abandon when we come to study model choice.

▶ **The power function.** The p-value alone does not contain enough information. We can only achieve adequate understanding of the operation of the LR Test via consideration of the power function

$$\text{power}(\theta) := \mathbb{P}(\text{reject } H_0 \,|\, \theta) = \mathbb{P}(\text{LR} \geq \kappa \,|\, \theta). \tag{A5}$$

We clearly want the power function to be 'small on B_0 and large on B_A', even though these regions will usually have a common boundary across which the power function is continuous! We note that

$$\text{size} = \sup_{\theta \in B_0} \text{power}(\theta).$$

When we plan a Hypothesis Test, *we need to choose BOTH sample size and value of κ to make the test operate reasonably.* We see this in the next subsection.

▶▶ **Non-rejection is not acceptance.** If we do not reject H_0, then all that we are saying is that there is not strong evidence in data to reject that H_0 might be a reasonable approximation to the truth. This certainly does not mean that there is strong evidence to accept H_0. I shall always use 'Reject' or 'Do not reject', **never** 'Accept'.

▶ **Test statistics.** In practice, with likelihoods arising from the exponential family, we rarely calculate the likelihood ratio. Instead, we find a SIMPLE (sufficient) *test statistic* t^{obs} such that the rejection criterion $\{\text{lr}(\mathbf{y}^{\text{obs}}) \geq \kappa\}$ takes the form $\{t^{\text{obs}} \geq c\}$ or $\{|t^{\text{obs}}| \geq c\}$ for some constant c.

However, there is a remarkable result that says that, under certain circumstances, for large samples, $2 \ln \text{LR}(\mathbf{Y})$ has approximately one of the χ^2

distributions. We shall see how to prove the one-dimensional case in Subsection 230D.

▶▶ **Aa. Deviance.**   Because of the importance of $2 \ln \mathrm{LR}(\mathbf{Y})$, we have a special name and notation for it: we refer to the Deviance

$$\mathrm{Dev}(\mathbf{Y}) := \mathrm{Dev}(H_A, H_0; \mathbf{Y}) := 2 \ln \mathrm{LR}(\mathbf{Y})$$

or deviance $\mathrm{dev}(\mathbf{y}) = 2 \ln \mathrm{lr}(\mathbf{y})$.

▶ **Ab. Type I and Type II errors.**   (You must be made aware of this terminology.) If we reject H_0 when it is true, we commit a 'Type I' error. If we fail to reject H_0 when it is false, we commit a 'Type II' error. Type I error probabilities are therefore described by the power function on B_0, and Type II error probabilities by $1-$(power function) on B_A. The size of the test is the maximum Type I error probability.

▶ **B. Hypothesis Tests and Confidence Regions.**   This linking of Hypothesis Tests and CRs will make more sense in the concrete context of the next subsection where it all works well. But here's the general story, in which we consider that the sample size n is fixed in advance. It will force us to consider modifying the LR Test (as we do in Subsection 232F) in certain circumstances.

The size α test of $\theta = \varphi$ against $\theta \neq \varphi$ has the form

$$\text{Reject } \theta = \varphi \text{ if } \frac{\mathrm{lhd}(\hat{\theta}_{\mathrm{obs}}; \mathbf{y}^{\mathrm{obs}})}{\mathrm{lhd}(\varphi; \mathbf{y}^{\mathrm{obs}})} \geq \kappa(\varphi), \tag{B1}$$

where $\hat{\theta}_{\mathrm{obs}}$ is the mle of θ and $\kappa(\varphi)$ is the appropriate constant. The *set* CR *of* φ *such that we do NOT reject* $\theta = \varphi$ *against* $\theta \neq \varphi$ *at size* α has the form

$$\mathrm{CR} = \left\{ \varphi : \mathrm{lhd}(\varphi; \mathbf{y}^{\mathrm{obs}}) > \lambda(\varphi, \mathbf{y}^{\mathrm{obs}}) \right\}, \quad \text{where} \tag{B2}$$

$$\lambda(\varphi, \mathbf{y}^{\mathrm{obs}}) := \kappa(\varphi)^{-1} \mathrm{lhd}(\hat{\theta}_{\mathrm{obs}}; \mathbf{y}^{\mathrm{obs}}).$$

For any φ,

$$\mathbb{P}(\varphi \in \mathrm{CR} \mid \varphi) = \mathbb{P}(\mathrm{lhd}(\varphi; \mathbf{Y}) > \lambda(\varphi, \mathbf{Y}) \mid \varphi)$$
$$= \mathbb{P}(\text{we do NOT reject } \theta = \varphi \mid \varphi) = 1 - \alpha,$$

so that

CR is a $100(1 - \alpha)\%$ Confidence Region for θ.

This fact is often pointed out in the literature, but numerous difficulties attach to it. I mention one now, and discuss another in Subsection 232F.

If $\kappa(\cdot)$ is a constant function, in which case $\lambda(\varphi, \mathbf{y}^{\mathrm{obs}})$ depends only on $\mathbf{y}^{\mathrm{obs}}$, not on φ, then CR at (B2) has the sensible property that if $\alpha \in \mathrm{CR}$ and $\mathrm{lhd}(\beta; \mathbf{y}^{\mathrm{obs}}) \geq \mathrm{lhd}(\alpha; \mathbf{y}^{\mathrm{obs}})$, then $\beta \in \mathrm{CR}$. If $\kappa(\cdot)$ is not constant, we have inconsistency with the intuitive idea of the LR Test.

Ba. Lemma. (a) *If θ is a location parameter so that $f(y \,|\, \theta) = h(y - \theta)$ for some pdf h on $\mathbb{R}$, then the function $\kappa(\cdot)$ is constant.*

(b) *If γ is a **scale parameter** in that $f(y \,|\, \gamma) = \gamma^{-1} f(\gamma^{-1} y \,|\, 1)$, then the function $\kappa(\cdot)$ is constant.*

Proof of (b). If $\mathbf{Y}$ has the $\mathrm{lhd}(\varphi; \cdot)$ density, then $\mathbf{Z} = \varphi^{-1}\mathbf{Y}$ has the $\mathrm{lhd}(1; \cdot)$ density. But (with the middle equality 'just logic')

$$\sup_{\gamma} \mathrm{lhd}(\gamma; \mathbf{Y}) \;=\; \sup_{\gamma} \mathrm{lhd}(\gamma; \varphi\mathbf{Z}) \;=\; \sup_{\gamma} \mathrm{lhd}(\gamma\varphi; \varphi\mathbf{Z}) \;=\; \varphi^{-n} \sup_{\gamma} \mathrm{lhd}(\gamma\varphi; \mathbf{Z})$$

while

$$\mathrm{lhd}(\varphi; \mathbf{Y}) \;=\; \mathrm{lhd}(\varphi; \varphi\mathbf{Z}) \;=\; \varphi^{-n}\mathrm{lhd}(1; \mathbf{Z}).$$

Thus,

$$\sup_{\gamma} \mathrm{lhd}(\gamma; \mathbf{Y}) \geq \kappa(1)\mathrm{lhd}(\varphi; \mathbf{Y}) \text{ if and only if}$$
$$\sup_{\gamma} \mathrm{lhd}(\gamma; \mathbf{Z}) \geq \kappa(1)\mathrm{lhd}(\varphi; \mathbf{Z}).$$

Since $\mathbf{Z}$ has the $\mathbb{P}(\cdot \,|\, 1)$ law of $\mathbf{Y}$,

$$\mathbb{P}\!\left(\sup_{\gamma} \mathrm{lhd}(\gamma; \mathbf{Y}) \geq \kappa(1)\mathrm{lhd}(\varphi; \mathbf{Y}) \,\Big|\, \varphi \right)$$
$$= \mathbb{P}\!\left(\sup_{\gamma} \mathrm{lhd}(\gamma; \mathbf{Z}) \geq \kappa(1)\mathrm{lhd}(\varphi; \mathbf{Z}) \,\Big|\, \varphi \right)$$
$$= \mathbb{P}\!\left(\sup_{\gamma} \mathrm{lhd}(\gamma; \mathbf{Y}) \geq \kappa(1)\mathrm{lhd}(\varphi; \mathbf{Y}) \,\Big|\, 1 \right) = \alpha,$$

and so $\kappa(\varphi) = \kappa(1)$. $\qquad\qquad\square$

Exercise. Prove (a), which is much easier to do.

C. Application to the $N(\theta, 1)$ case. We look at both two-sided and one-sided tests, though CIs are better!

Ca. Two-sided test. Consider testing

$$H_0 : \theta = 0 \quad \text{against} \quad H_A : \theta \neq 0$$

based on an IID sample $Y_1, Y_2, \ldots, Y_n$ from an $N(\theta, 1)$ population. Using the fact that the MLE of θ is $\overline{Y}$ and the Parallel-Axis result at 182Ca, we find that (you check!)

$$\mathrm{Dev}(\mathbf{Y}) \;=\; 2\ln LR(\mathbf{Y}) \;=\; n\overline{Y}^2. \tag{C1}$$

We shall therefore reject H_0 if $|\overline{Y}| \geq c$, where $c = \sqrt{2n^{-1}\ln\kappa}$, which is very sensible. Suppose that we want a power function such that

$$\text{size} = \text{power}(0) = 5\%, \quad \text{power}(0.4) \approx 90\%,$$

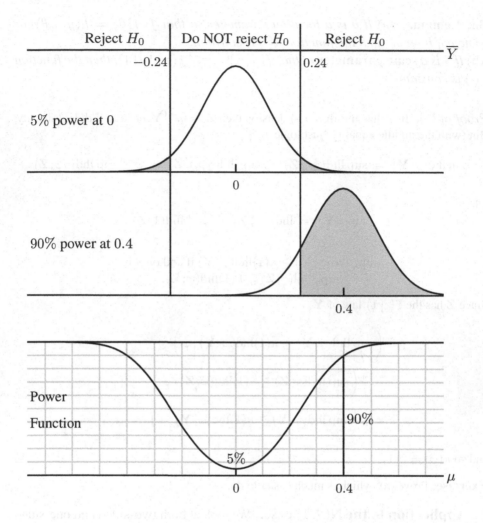

Figure C(i): Power function of a certain test

so that in particular we shall be about 90% certain of rejecting H_0 if the true value of θ is 0.4. We therefore want (look at Figure C(i))

$$c = \frac{1.96}{\sqrt{n}}, \quad 0.4 - c \approx \frac{1.28}{\sqrt{n}}, \quad 0.4 \approx \frac{3.24}{\sqrt{n}},$$

whence $n = 66$ and $c = 0.24$. We therefore take $n = 66$ and reject H_0 at the 5% level if the 95% CI

$$\left(\overline{Y} - \frac{1.96}{\sqrt{n}}, \ \overline{Y} - \frac{1.96}{\sqrt{n}} \right)$$

does NOT contain 0. This is a sensible tie-up between CIs and HTs, as we knew it would be because θ is a location parameter.

▶ **Cb. The resolution at which we examine things.** The numbers $0, 0.24$ and 0.4 in Figure 228C(i) are the 'actual real-world' numbers. But it is important to realize that the natural unit of length for the situation is $1/\sqrt{n}$ (with $n = 66$) so that the three numbers are, as we have seen, $0/\sqrt{n}$, $1.96/\sqrt{n}$ and $(1.96 + 1.28)/\sqrt{n}$.

Cc. Chi-squared formulae for $\mathrm{Dev} = 2\ln\mathrm{LR}$. Suppose again that we are working with an IID sample $Y_1, Y_2, \ldots, Y_n$, where n need not be 66. Then if H_0 is true then $n^{\frac{1}{2}}\overline{Y} \sim N(0, 1)$, so that, from 227(C1),

> if $\theta = 0$, then $\mathrm{Dev}(\mathbf{Y}) = 2\ln LR(\mathbf{Y})$ has exactly a χ_1^2 distribution.

Recalling the point about the resolution at which we examine things, we consider $\mathrm{power}(\gamma\delta)$ where $\delta = n^{-\frac{1}{2}}$. If $\theta = \gamma\delta$, then

$$\mathrm{Dev}(\mathbf{Y}) \;=\; 2\ln\mathrm{LR}(\mathbf{Y}) \;=\; (n^{\frac{1}{2}}\overline{Y})^2,$$

where $n^{\frac{1}{2}}\overline{Y} \sim N(\gamma, 1)$. Hence $\mathrm{power}(\gamma/\sqrt{n})$ is obtained exactly from the result

> if $\theta = \gamma/\sqrt{n}$, then $\mathrm{Dev}(\mathbf{Y}) \;=\; 2\ln\mathrm{LR}(\mathbf{Y}) \;\sim\;$ non-central $\chi_1^2(\gamma^2)$.

The normal case is obeying exactly what (as we shall see in the next subsection) other distributions obey approximately for large sample sizes – the usual story.

Cd. Warning. In the above discussion, $H_0 : \theta = 0$ was a *sharp* hypothesis in which θ is specified *exactly*. The problems which arise because of the absurdity of making sharp hypotheses are discussed in Subsections 232G and 234I.

Ce. One-sided test. Consider the (now-sensible) question of testing

$$H_0 : \theta \le 0 \quad \text{against} \quad H_A : \theta > 0$$

again based on an IID sample $Y_1, Y_2, \ldots, Y_n$ from an $N(\theta, 1)$ population. This time,

$$\mathrm{Dev}(\mathbf{Y}) \;=\; 2\ln LR(\mathbf{Y}) \;=\; \begin{cases} -n\overline{Y}^2 & \text{if } \overline{Y} \le 0, \\ +n\overline{Y}^2 & \text{if } \overline{Y} > 0. \end{cases}$$

Again suppose that we want a power function such that

$$\text{size} = \mathrm{power}(0) = 5\%, \quad \mathrm{power}(0.4) \approx 90\%.$$

We therefore want

$$c = \frac{1.645}{\sqrt{n}}, \quad 0.4 - c \approx \frac{1.28}{\sqrt{n}},$$

so we take $n = 54$ and $c = 0.22$.

Exercise: Sketch the diagram corresponding to Figure 228C(i) for this case.

▶▶ **D. A χ^2 approximation for the distribution of Deviance.** Suppose that n is large and that $Y_1, Y_2, \ldots, Y_n$ is a sample from a distribution with pdf $f(y \mid \theta)$ where θ is a one-dimensional parameter. Consider testing

$$H_0 : \theta = \varphi \quad \text{against} \quad H_A : \theta \neq \varphi.$$

Under H_0, we have (compare the discussion in Subsection 194D)

$$
\begin{aligned}
\mathrm{Dev}(\mathbf{Y}) = 2 \ln \mathrm{LR}(\mathbf{Y}) &= 2 \left\{ \ell(\widehat{\Theta}, \mathbf{Y}) - \ell(\varphi, \mathbf{Y}) \right\} \\
&\approx -2 \partial_\theta \ell(\widehat{\Theta}, \mathbf{Y})(\widehat{\Theta} - \varphi) + \partial_\theta^2 \ell(\widehat{\Theta}, \mathbf{Y})(\widehat{\Theta} - \varphi)^2 \\
&= \partial_\theta^2 \ell(\widehat{\Theta}, \mathbf{Y})(\widehat{\Theta} - \varphi)^2 \approx I_\ell(\varphi)(\widehat{\Theta} - \varphi)^2 \\
&\approx \chi_1^2.
\end{aligned}
$$

Here we have used the facts that

$$\partial_\theta \ell(\widehat{\Theta}, \mathbf{Y}) = 0, \quad \widehat{\Theta} - \varphi \approx \mathrm{N}\left(0, \frac{1}{I_\ell(\theta)}\right).$$

Summarizing then,

$$\boxed{2 \ln \mathrm{LR}(\mathbf{Y}) \approx \chi_1^2, \text{ under } H_0,}$$

a special case of a general result which we study later.

It is interesting to note that, from the Normal Table, p515, or the χ_1^2 Table, p517, we have

for 5% significance: $2 \ln \mathrm{LR} \geq 1.96^2 = 3.84$, so $\mathrm{LR} \geq 6.82$,
for 1% significance: $2 \ln \mathrm{LR} \geq 2.58^2 = 6.64$, so $\mathrm{LR} \geq 27.6$.

This gives some idea of the size of κ for the LR Test.

▶▶ **Da. Power function and resolution. The non-central χ^2 approximation.** In considering the power function, we must – as at 229Cb – examine the resolution at which we measure things. Here the natural scale of measurement is

$$\boxed{\delta = \{I_\ell(\varphi)\}^{-\frac{1}{2}} = \{n I_{f^*}(\varphi)\}^{-\frac{1}{2}},}$$

and we must consider power$(\varphi + \gamma \delta)$. We have

$$2 \ln \mathrm{LR}(\mathbf{Y}) \approx I_\ell(\varphi + \gamma \delta)(\widehat{\Theta} - \varphi)^2 \approx I_\ell(\varphi)(\widehat{\Theta} - \varphi)^2 = \left\{ \delta^{-1}(\widehat{\Theta} - \varphi) \right\}^2.$$

But $\widehat{\Theta} - \varphi \approx N(\gamma \delta, \delta^2)$. Hence power$(\varphi + \gamma \delta)$ is obtained approximately from the result

$$\boxed{\text{if } \theta = \varphi + \gamma \delta \text{ then } \mathrm{Dev}(\mathbf{Y}) = 2 \ln \mathrm{LR}(\mathbf{Y}) \approx \text{ non-central } \chi_1^2(\gamma^2).}$$

▶ **E. Application to coin tossing.** Let's discuss an example rather informally before we go on to the theory.

Ea. Example. Suppose that Jane tosses her coin 100 times and obtains the result

$$\text{HHTH} \dots \text{T} \tag{E1}$$

with 54 Heads in all, while John tosses his coin 100 times and obtains the result

$$\text{THHT} \dots \text{H} \tag{E2}$$

with 61 Heads in all. Do note that in each case we are given the full result of all individual tosses, not just the total number of Heads obtained. (I know that I have not typed out all the details for you, but pretend that I have.)

If we make the Null Hypothesis (a sharp one!) that Jane's coin is fair, and the Alternative that it is not, then the chance of Jane's getting the detailed result at (E1) under the Null Hypothesis is exactly $\frac{1}{2} \times \frac{1}{2} \times \dots = 2^{-100}$. The best value of p for the Alternative Hypothesis for Jane's coin, $p = 0.54$ (corresponding to the mle), does *not* provide significantly better explanation than the Null Hypothesis, and the p-value of the data is 0.484 as we shall see shortly. We would certainly *not* reject at the 5% significance level the Null Hypothesis that Jane's coin is fair.

If we make the Null Hypothesis that John's coin is fair, then the chance under the Null Hypothesis of John's getting the detailed result (E2) is again exactly 2^{-100}. Thus if both coins are fair, John's detailed result is no more unlikely than Jane's. However, as we shall see, the statement within the Alternative Hypothesis that John's coin has $p = 0.61$ *does* provide a significantly better explanation of John's result, and the p-value of the data is 0.035. Note that, with lots of entropy floating around,

$$\text{lr}(\mathbf{y}^{\text{obs}}) = \frac{(0.61)^{61}(0.39)^{39}}{2^{-100}}, \quad 2\ln\text{lr}(\mathbf{y}^{\text{obs}}) = 4.88 > 1.96^2.$$

We *would* therefore reject at the 5% significance level the hypothesis that John's coin is fair.

Saying whether or not we would reject the Null Hypothesis that the coin is fair is *not* the best way to analyse these data. Plotting the posterior pdf $\pi(\cdot \mid \mathbf{y}^{\text{obs}})$ for the Frequentist prior $\theta^{-1}(1-\theta)^{-1}$ (or the reference prior $\theta^{-\frac{1}{2}}(1-\theta)^{-\frac{1}{2}}$) and giving $C\%$ CIs for a range of C values would be MUCH better.

Eb. The theory for the Bernoulli case. So, suppose that $Y_1, Y_2, \dots, Y_n$ are IID each Bernoulli(θ). Then for testing $\theta = \varphi$ against $\theta \neq \varphi$, we have

$$\text{Dev}(\mathbf{Y}) = 2\ln\text{LR}(\mathbf{Y}) = nh(\overline{Y}),$$

where

$$h(y) = y\ln y + (1-y)\ln(1-y) - y\ln\varphi - (1-y)\ln(1-\varphi).$$

Now,

$$h'(y) = \ln \frac{y}{1-y} \frac{1-\varphi}{\varphi} = \begin{cases} +\text{ve} & \text{if } y > \varphi, \\ 0 & \text{if } y = \varphi, \\ -\text{ve} & \text{if } y < \varphi. \end{cases}$$

So,

$$\text{LR}(\mathbf{Y}) \geq \kappa \text{ is equivalent to } \overline{Y} \notin [a, b]$$

for some interval $[a, b]$ around φ. One needs a computer to carry out the test accurately.

For large, or even moderately large, n, we effectively reject H_0 at size α if the $100(1 - \alpha)\%$ CI at 178(E1) does not contain φ.

Exercise. Use this to check the p-values for Jane and John. The ones listed above were calculated to 3 places on the computer. The CLT with integer correction is here spectacularly good because of the symmetry of the binomial pdf when $\varphi = \frac{1}{2}$.

F. Frequentist priors and the LR Test.
We have seen that Frequentist CIs are often based on Frequentist priors which are not constant. For example, the Frequentist prior for $E(\text{mean } \theta)$ or for any scale parameter is θ^{-1}.

Surely, in such a case, the correct form of the LR Test for Frequentists is

$$\text{Reject } H_0 \text{ if } \frac{\sup_{\theta \in B_A} \pi(\theta \,|\, \mathbf{y}^{\text{obs}})}{\sup_{\theta \in B_0} \pi(\theta \,|\, \mathbf{y}^{\text{obs}})} \geq \kappa,$$

where we use the Frequentist prior. Only in this way will we get the correct tie-up between the LR Test and Frequentist CIs, and then only in cases where the $\kappa(\cdot)$ function in Subsection 226B is constant. But at least, both location and scale parameters would be covered.

For Bernoulli(θ), where the Frequentist prior is $\theta^{-1}(1 - \theta)^{-1}$, we are in trouble if there are either no successes or no failures. The reference prior $\theta^{-\frac{1}{2}}(1 - \theta)^{-\frac{1}{2}}$ does not share this difficulty.

It really is impossible to obtain a tidy all-embracing Frequentist theory, except for asymptotic large-sample results.

▶ G. Discussion of Frequentist Hypothesis Testing.
This discussion will be continued at later stages.

Ga. Having to make decisions.
One of the areas in which FHT is often still considered of value is that in which Yes/No decisions have to made.

Suppose that a drug company (or a company making agricultural fertilizers) has to decide on the basis of an experiment on a proposed new treatment whether to switch from an established product to the new one, the switch entailing substantial costs. It *knows* how effective the established treatment is. It may well formulate this as 'testing $\theta_{\text{new}} \leq \theta_{\text{old}}$ against $\theta_{\text{new}} > \theta_{\text{old}}$' where θ_{new} represents the effectiveness of the new treatment. The company will however, if it is sensible, pay close attention to the power function by asking 'how much better' makes the switch worthwhile. So, to all intents and purposes we are back with the more informative CIs.

I really should again mention Decision Theory, utility, loss functions, etc, at this point. I am not fond of that material, and I leave its presentation to others. See Appendix D.

Gb. Sharp hypotheses. A sharp Null Hypothesis for the one-dimensional situation which we are considering states that θ takes some definite value θ_0: $\theta = \theta_0$. **Now the point about a sharp Null Hypothesis is that (in every case of interest to us) it is false.** (Newton's theory is immensely more precisely correct than any Null Hypothesis in Statistics; but it is wrong.) If we set a fixed size α for our Test of H_0 (against the Alternative that H_0 is false), then if the sample size is large enough, we shall definitely reject the false hypothesis H_0. So, why bother to take a small sample? This objection has continually been raised by Bayesians.

However, *sensible Frequentists know that they are only testing whether there is strong evidence that H_0 is not an acceptably good approximation to the truth.* (After all, as remarked earlier, Newton's theory is good enough for manned flights to the Moon and back.) *The power function specifies what one means by 'acceptably good'.* Again, *everything hinges on the resolution at which we look at things.* The discussion in Subsection 227C where we chose the sample size to achieve desired power at a certain level of deviation from the Null Hypothesis shows the ideal way to do things.

We have to be particularly careful not to reject an acceptable Null Hypothesis these days when vast data sets are available on computers. One also needs to be careful that one might reject an exactly true conjecture in Probability because one takes a vast simulation (affected by the inevitable approximations computer methods entail and perhaps defects in even good random-number generators) and very likely uses tests which are much too sensitive. Computer simulation once persuaded me that a result is false, but I later proved that it is true.

We take up the problem of sharp hypotheses again in Subsection I and later in multidimensional situations where it is more difficult to deal with.

Note. In my first draft, I deliberately omitted mentioning Absence of Extra-Sensory Perception, which seems to some of us a sharp Null Hypothesis which may be true. Then David Cox mentioned it in his comments. We do not, however, regard this as disproof of the Null Hypothesis.

Gc. When Hypothesis Testing is definitely inappropriate. In the example of Subsection 178F, there is no reason whatever to protect a Null Hypothesis that $p_A = p_B$. That is why the χ^2 test is totally inappropriate.

Gd. When Hypothesis Testing is helpful. I have kept mentioning cases – for complex ANOVA models, for contingency tables, etc – where Hypothesis Testing serves a very useful purpose. See especially Chapter 8.

H. Bayesian Hypothesis Testing (BHT). Bayesian Hypothesis Testing is obvious. In Bayesian theory, we can speak of the (posterior) probability that a Null Hypothesis is true: it is just

$$\mathbb{P}(H_0 \text{ is true} \mid \mathbf{y}^{\mathrm{obs}}) = \int_{\theta \in B_0} \pi(\theta \mid \mathbf{y}^{\mathrm{obs}}) d\theta.$$

That's the end of the story! The difficulty for the theory is, as usual, caused by the beginning of the story: how does one choose one's prior? There are, of course, often serious difficulties in implementing the formula in practice, though to a large extent these are now overcome by MCMC methods.

▶▶ **I. Bayesian significance testing of a sharp hypothesis: a plea for sanity.**

As stated at 233Gb, it was Bayesians who correctly emphasized that if the sample size is large, Frequentists will inevitably reject a sharp hypothesis at the 1% level because a sharp hypothesis will not be exactly true. It astonishes me therefore that some Bayesians now assign non-zero prior probability that a sharp hypothesis is exactly true to obtain results which seem to support strongly Null Hypotheses which Frequentists would very definitely reject. (Of course, it is blindingly obvious that such results must follow.)

Suppose that $Y_1, Y_2, \ldots, Y_n$ is an IID sample from an $N(\theta, 1)$ population. We are again interested in testing

$$H_0 : \theta = 0 \quad \text{against} \quad H_A : \theta \neq 0.$$

However, now, we think like this. Let us assume that God tossed a coin with probability p_n of Heads, chose $\Theta = 0$ (so H_0 is true) if it fell Heads, and chose Θ according to the $N(m, \mathrm{prec}\ r)$ distribution (in which case H_A is true (with probability 1)) if the coin fell Tails. Then using 213(L2), we find that

$$\mathbb{P}(H_0 \mid \mathbf{y}) = \frac{I_0}{I_0 + I_A},$$

where

$$I_0 = p_n \left(\frac{1}{2\pi}\right)^{\frac{1}{2}n} \exp\left(-\tfrac{1}{2}\sum y_k^2\right),$$

$$I_A = (1 - p_n) \int_\theta \left(\frac{r}{2\pi}\right)^{\frac{1}{2}} \exp(-\tfrac{1}{2}r\theta^2) \left(\frac{1}{2\pi}\right)^{\frac{1}{2}n} \exp\left\{-\tfrac{1}{2}\sum(y_k - \theta)^2\right\} d\theta.$$

Now use the Parallel-Axis result

$$\sum(y_k - \beta)^2 = \sum(y_k - \overline{y})^2 + n(\beta - \overline{y})^2$$

with $\beta = 0$ and $\beta = \theta$ to obtain

$$
\frac{I_A}{I_0} = \frac{1 - p_n}{p_n \exp\left(-\frac{1}{2}n\bar{y}^2\right)} \int_\theta \left(\frac{r}{2\pi}\right)^{\frac{1}{2}} \exp(-\tfrac{1}{2}r\theta^2) \exp\left\{-\tfrac{1}{2}n(t - \bar{y})^2\right\} d\theta
$$

$$
= \frac{1 - p_n}{p_n} \exp\left\{\frac{n\bar{y}^2}{2(1 + r/n)}\right\} \sqrt{\frac{r}{r + n}}. \tag{I1}
$$

You can work out the integral now if you wish, but we shall see why the calculation is correct at 251(Da) below.

If we allow n to tend to infinity and keep p_n fixed at p, and if $y = c/\sqrt{n}$, then for fixed c,

$$
\frac{I_A}{I_0} = \frac{1 - p}{p} \exp\left\{\frac{c^2}{2(1 + r/n)}\right\} \sqrt{\frac{r}{r + n}} \sim \frac{\text{constant}}{\sqrt{n}}.
$$

So, in investigating what happens as $n \to \infty$, if we make what I regard as the serious mistake of keeping p_n fixed, then for $\bar{y} = c/\sqrt{n}$ (which Frequentists would regard as significance evidence against H_0 if $c \geq 1.96$), we would have $\mathbb{P}(H_0 \mid \mathbf{y}) \to 1$.

Let me explain why I think that keeping p_n fixed as $n \to \infty$ is a serious mistake. Do remember that we are considering the situation where $n \to \infty$. We just do NOT really believe that θ is EXACTLY 0. If we keep p_n fixed at p as $n \to \infty$, then we are saying that we have a prior probability p that θ is EXACTLY 0; and this is plain silly. Think of the resolution at which we are looking at things, remembering the discussions at 229Cb and 230Da. Doing an experiment of sample size n is looking at things measured on the scale where the natural unit of length is $1/\sqrt{n}$. What we should say therefore is that H_0 represents the hypothesis that θ belongs to the little atom of natural length $1/\sqrt{n}$ centered at 0. Thus it is natural to take $p_n = a/\sqrt{n}$ for some a, and then for $y = c/\sqrt{n}$,

$$
\mathbb{P}(H_0 \mid \mathbf{y}) \to \frac{1}{1 + (\sqrt{r}/a) \exp(c^2/2)},
$$

and sanity returns, with close agreement between Bayesian and Frequentist theories. Note that a has dimension $(\text{length})^{-1}$ and that r as the inverse of variance has dimension $(\text{length})^{-2}$ whence $\sqrt{r}/a$ is dimensionless as it should be.

What I am saying is that *we CANNOT ignore the degree of resolution of the experiment when choosing our prior.* Bayesians normally claim that one's prior must not depend on the nature of one's experiment, but many of them are willing to compromise this when it comes to reference priors. This point will be explained in Chapter 9. I hope that I have persuaded some that they must adapt their prior to resolution too.

I regret having felt it necessary to take up so much of your time with a situation where there is only one sensible thing to do: **Give a Confidence Interval**.

I am not claiming any originality here (or anywhere else in regard to Statistics). See, for example, the discussion in Cox and Hinkley [49] for similar remarks on sharp hypotheses. Bartlett may have been the first to counter rather silly claims made by some Bayesian statisticians who on all other matters have shown great wisdom.

But what about ESP, you say! I knew you would.

▶▶ **J. Frequentist Model Choice: parsimony, AIC and all that – in the simplest case.**   The Principle of Parsimony encourages us to choose an acceptably good model with as few parameters as possible. There are two reasons for this. Firstly, *a model with more parameters leads to predictions with less bias but with higher variance, and we have to balance one against the other.* Secondly, *a model with more parameters tends to be more sensitive to small changes in the data.* The first point will be illustrated in this subsection, and the second mentioned in Chapter 8.

Because a good model will yield a high maximum likelihood for the data, it is becoming ever more common to choose a model which maximizes the value of

$$2\ln(\text{maximum likelihood under model})$$
$$- \text{function}(\text{sample size, number of parameters in model}).$$

We have the usual fascinating situation for Statistics. There are no completely satisfactory answers: each of the various proposals for the 'function' has advantages and disadvantages.

A popular practice is to choose a model which maximizes the **Akaike Information Criterion (AIC)**

$$\text{AIC} := 2\ln(\text{maximum likelihood under model})$$
$$-2(\text{number of parameters in model}). \tag{J1}$$

I am a great believer in Dynkin's advice: 'Always discuss the simplest case'. In this subsection, I discuss an almost ridiculously simple case (truly the simplest possible). I use it to illustrate the point about balancing lower bias against increased variance, and to present two arguments which help motivate the AIC criterion, one based on Decision Theory, the other (Akaike's own) on the Kullback–Leibler measure of how badly one pdf approximates another.

Our simple situation.   Suppose that a Random Variable Y has the $\text{N}(\mu, 1)$ distribution for some μ. Let M_0 be the model that $\mu = 0$ and M_A the 'full' model that $\mu \in \mathbb{R}$. We take the view that how good a model is depends on how well it makes predictions about a Random Variable X, independent of Y, and with the same distribution as Y.

Suppose that y^{obs} is the observed value of Y. On the basis that M_0 is true, we predict that $X \sim \text{N}(0, 1)$, a biased prediction in that X has mean 0 rather than μ. On the basis of M_A, we predict that $X \sim \text{N}(y^{\text{obs}}, 1)$, which is 'unbiased' in the sense that on average y^{obs} gives the correct value μ.

▶ **A Decision-Theory argument.** Concentrate first on the observed value y^{obs} rather than Y. The mean squared error in predicting X incurred by using the prediction $\mu = 0$ associated with M_0 is

$$\mathbb{E}\left\{(X - 0)^2\right\} = \mu^2 + \text{Var}(X) = \mu^2 + 1.$$

The mean square error in predicting X incurred by using the prediction $\mu = y^{\text{obs}}$ associated with M_A is

$$\mathbb{E}\left\{(X - y^{\text{obs}})^2\right\} = (\mu - y^{\text{obs}})^2 + 1 = \mu^2 + 1 + (y^{\text{obs}})^2 - 2\mu y^{\text{obs}}.$$

We now average these over possible values of y^{obs} leading to results

$$\mu^2 + 1 \text{ for } M_0,$$
$$\mu^2 + 1 + \mathbb{E}(Y^2) - 2\mu\mathbb{E}(Y) = (\mu^2 + 1) + (\mu^2 + 1 - 2\mu^2) \text{ for } M_A.$$

Yes, we *could* simply have used independence to say that

$$\mathbb{E}\left\{(X - Y)^2\right\} = \text{Var}(X - Y) = \text{Var}(X) + \text{Var}(Y) = 2 \ (!).$$

On average, therefore, using M_A rather than M_0 will reduce the mean square error by $\mu^2 - 1$. This value (unknown to us) is the 'average amount by which M_A is better than M_0' in terms of Decision Theory with mean square error as loss function.

Now concentrate only on the Y experiment. Twice the maximum-likelihood of obtaining a result y under M_0 is

$$2 \ln \text{mlhd}(M_0; y) = -\ln(2\pi) - y^2,$$

while

$$2 \ln \text{mlhd}(M_A; y) = -\ln(2\pi) - 0.$$

Hence,

$$\mathbb{E}\,\text{Dev}(M_A, M_0; Y) := \mathbb{E}\,2 \ln \text{mlhd}(M_A; Y) - \mathbb{E}\,2 \ln \text{mlhd}(M_0; Y)$$
$$= \mathbb{E}(Y^2) = \mu^2 + 1,$$

2 more than $\mu^2 - 1$. On average, therefore, the Deviance is favouring M_A over M_0 by 2 more than the Decision Theory suggests is appropriate. On average, to

get consistency with Decision Theory, we need to take 2, which is here twice the difference in number of parameters, from the deviance; and this is in agreement with AIC.

We note that AIC suggests that if $(y^{\text{obs}})^2 < 2$, we choose Model M_0; otherwise, we choose Model M_A. (We ignore the case when $(y^{\text{obs}})^2 = 2$.)

▶ **Akaike's idea.** Akaike arrived at what he called An Information Criterion by the following argument which again involves making predictions about X, but now the prediction of the whole pdf of X. See [2] for a more recent paper by the man himself.

Let φ_μ be the pdf of $N(\mu, 1)$. Then, as we know from Exercise 198Fb,

$$\text{App}\big(\varphi_\mu(\cdot) \leftarrow \varphi_0(\cdot)\big) = \tfrac{1}{2}\mu^2,$$
$$\text{App}\big(\varphi_\mu(\cdot) \leftarrow \varphi_{y^{\text{obs}}}(\cdot)\big) = \tfrac{1}{2}(\mu - y^{\text{obs}})^2.$$

The difference between these two expressions measures how much better $\varphi_{y^{\text{obs}}}$ (the estimate from M_A) describes the true φ_μ than does φ_0 (the estimate from M_0). But in this simple case, this difference is exactly the same as the difference we had between the mean square errors; and the rest of the argument is now identical. We have, however, replaced the quadratic loss function, which is specially related to normal distributions, by a Kullback–Leibler loss function which applies in much more general cases.

▶ **Ja. Exercise.** Do the n-variable case where $Y_1, Y_2, \ldots, Y_n$ are IID each $N(\mu, 1)$, $X_1, X_2, \ldots, X_n$ are also IID each $N(\mu, 1)$, and independent of $Y_1, Y_2, \ldots, Y_n$, and where the loss function for the Decision-Theory approach is $\sum (X_k - \hat\mu)^2$, where $\hat\mu$ is an estimate of μ. Show that AIC would suggest choosing M_0 as our model if $n\bar{y}_{\text{obs}}^2 < 2$. Show that if M_0 is indeed correct, then the probability of accepting it as our model does not change with n, whereas we would have wished it to tend to 1 as $n \to \infty$.

Jb. Discussion. Other criteria, for example that of Schwarz [209], do not show the 'lack of consistency' described in the last sentence of the above exercise.

But all the familiar problems associated with sharp hypotheses surface here; and remember that, whatever the context, a lower-dimensional hypothesis will be sharp.

In our simple case, M_0 will not be exactly true, so (check!) if we take a large sample, Akaike's criterion will lead us to choose M_A. Why then bother to take a small sample? Here we go again. To make sense of things, we have to think of alternative hypotheses of the form $\mu = \delta n^{-\frac{1}{2}}$, working in the natural scale for the sample size. You will not want me to go through all this again in the new context.

I should not need to add that the only sensible thing to do in the situation described by our simplest case, is to give a Confidence Interval for μ.

K. It's a complex world. The point I wish to make here is that:

real Statistics is not primarily about the Mathematics which underlies it: common sense and scientific judgement are more important. (But, as stated earlier, this is no excuse for not using the right Mathematics when it is available.)

Suppose that it has been reported that a certain town has more sufferers from a certain disease than would be expected in a place of that population size, and it is thought that a nearby industrial site of a certain kind is the cause. One might set up a Null Hypothesis that the number of cases is Poisson(θ_0) where θ_0 is a 'national average' for a population of the given size, and test this against an Alternative that the number of cases is Poisson(θ) where $\theta > \theta_0$.

But an enormous number of questions arise. Shouldn't we look at the total population living near this *and other similar* industrial sites? Should we then compare these results with 'control' regions similar in many ways, but away from such industrial sites? Were there perhaps always more cases of the disease in the particular town? Did a journalist discover this case by looking through the Atlas of Cancer Mortality, finding a 'cluster', realizing that there was an industrial site near, and sensing a story? And so on.

CONDITIONAL pdfs AND
MULTI-PARAMETER BAYESIAN STATISTICS

I have deferred the topic of joint and conditional pdfs for as long as possible. Since we are now equipped to appreciate at least some of its usefulness, its study is now much more interesting than it would have been earlier. The fundamental 'F' and 't' distributions of Statistics are introduced in this chapter. Bayesian Statistics is extended to cover multi-parameter situations, and we see how it may be made effective by **Gibbs sampling**.

One of the reasons for my deferring the topic of joint pdfs is that it can look a bit complicated at first sight. It isn't really. Amongst the *theory* as it applies to Probability and Statistics, only the 'Jacobian' Theorem 249C has any substance; and you can, if you like, take that for granted. (Gibbs sampling certainly has substance too!) What is a little offputting is the notation. Wherever possible, I simplify the appearance of things by using vector notation.

Note. We defer the study of the most important joint pdf, that of the *multivariate normal distribution*, until Chapter 8. □

We start with the simple situation for joint pmfs, which proves a valuable guide to that for joint pdfs.

7.1 Joint and conditional pmfs

▶ **A. Definition of joint pmf.** Suppose that each of $X_1, X_2, \ldots, X_n$ takes values in $\mathbb{Z}$. The 'joint' pmf of $\mathbf{X} = (X_1, X_2, \ldots, X_n)$ is the function $p_{\mathbf{X}} = p_{X_1, X_2, \ldots, X_n}$ on $\mathbb{Z}^n$, where, for $\mathbf{x} = (x_1, x_2, \ldots, x_n)$ in $\mathbb{Z}^n$,

$$p_{\mathbf{X}}(\mathbf{x}) \; := \; p_{X_1, X_2, \ldots, X_n}(x_1, x_2, \ldots, x_n)$$

$$:= \; \mathbb{P}(\mathbf{X} = \mathbf{x}) = \mathbb{P}(X_1 = x_1; \ldots; X_n = x_n).$$

Really, $p_{\mathbf{X}}(\cdot)$ is the pdf of $\mathbf{X}$ and $p_{X_1, X_2, \ldots, X_n}(\cdots)$ is the joint pdf of $X_1, X_2, \ldots, X_n$. Of course the pmf $p_{X_k}(\cdot)$ of an individual X_k, now often called the **marginal pmf** of X_k, is obtained as

$$p_{X_k}(x) = \sum p_{X_1, X_2, \ldots, X_n}(i_1, \ldots, i_{k-1}, x, i_{k+1}, \ldots, i_n),$$

the sum being over all possible $(n-1)$-tuples $(i_1, \ldots, i_{k-1}, i_{k+1}, \ldots, i_n)$.

Aa. Exercise. This concerns the same situation as at Exercise 52Da. A fair coin is tossed twice, the number N of Heads noted, and then the coin is tossed N more times. Let X be the *total* number of Heads obtained, and Y the total number of Tails obtained. We have Table Aa for the joint pmf $p_{X,Y}$ of X and Y in units of $\frac{1}{16}$. Thus, for example, $p_{X,Y}(3,1) = 2/16$.

			y			
		0	1	2		
	0	0	0	4	4	
	1	0	0	4	4	
x	2	0	4	1	5	$p_X(x)$
	3	0	2	0	2	
	4	1	0	0	1	
		1	6	9		
			$p_Y(y)$			

Table A(i): Joint and marginal pmfs in units of $\frac{1}{16}$

You should check the table. Note the appropriateness of the term 'marginal'.

▶ **B. Independence.** Suppose that $\mathbf{X} = (X_1, X_2, \ldots, X_n)$ has joint pmf $p_{\mathbf{X}} = p_{X_1, X_2, \ldots, X_n}$. Then, $X_1, X_2, \ldots, X_n$ are independent if and only if

$$p_{X_1, X_2, \ldots, X_n}(x_1, x_2, \ldots, x_n) = p_{X_1}(x_1)p_{X_2}(x_2)\cdots p_{X_n}(x_n).$$

Ba. Exercise. Suppose that, for some functions $g_1, g_2, \ldots, g_n$ each on $\mathbb{Z}$, we have

$$p_{X_1, X_2, \ldots, X_n}(x_1, x_2, \ldots, x_n) = g_1(x_1)g_2(x_2)\cdots g_n(x_n).$$

Prove that $X_1, X_2, \ldots, X_n$ are independent, X_k having pmf $g_k(\cdot)/G_k$, where $G_k := \sum_x g_k(x)$.

Bb. Example. Consider an experiment which consists of two parts. First an integer N is chosen at random in $\mathbb{Z}^+$ with pmf p_N. Then a coin with probability θ of Heads is thrown N times, producing X Heads and Y Tails. Then, for non-negative integers x and y,

$$p_{X,Y}(x,y) = p_N(x+y)\binom{x+y}{x}\theta^x(1-\theta)^y.$$

[Note that if $p_N(\cdot)$ is known, then if we wish to make inferences about θ, then N is an Ancillary Statistic, so we must condition on the value of N in the Frequentist theory. The same is true if $p_N(\cdot)$ is not known, but we can be sure that 'the choice of N cannot influence θ'. In other words, if we obtain x Heads and y Tails, we must behave exactly as if the number of tosses had been fixed at $x + y$ in advance.]

We know from the 'Boys and Girls' Problem 100I that if N has the Poisson(λ) distribution, then X and Y are independent, X having the Poisson($\lambda\theta$) distribution and Y the Poisson($\lambda(1 - \theta)$) distribution.

▶▶ **C. Conditional pmfs.** Suppose that $\mathbf{X} = (X_1, X_2, \ldots, X_r)$ takes values in $\mathbb{Z}^r$ and $\mathbf{Y} = (Y_1, Y_2, \ldots, Y_s)$ takes values in $\mathbb{Z}^s$. Then the 'joint' vector $(\mathbf{X}, \mathbf{Y}) = (X_1, X_2, \ldots, X_r, Y_1, Y_2, \ldots, Y_s)$ takes values in $\mathbb{Z}^{r+s}$; let $p_{\mathbf{X},\mathbf{Y}}$ be its pmf.

Then the conditional pmf of $\mathbf{X}$ given $\mathbf{Y}$ is the function $p_{\mathbf{X}|\mathbf{Y}}(\cdot \,|\, \cdot)$ defined via

$$p_{\mathbf{X}|\mathbf{Y}}(\mathbf{x}\,|\,\mathbf{y}) \;:=\; \mathbb{P}(\mathbf{X} = \mathbf{x}\,|\,\mathbf{Y} = \mathbf{y})$$

$$= \frac{\mathbb{P}(\mathbf{X}=\mathbf{x}\,;\mathbf{Y}=\mathbf{y})}{\mathbb{P}(\mathbf{Y}=\mathbf{y})} \;=\; \frac{p_{\mathbf{X},\mathbf{Y}}(\mathbf{x},\mathbf{y})}{p_{\mathbf{Y}}(\mathbf{y})}. \tag{C1}$$

How this is defined when $p_{\mathbf{Y}}(\mathbf{y}) = 0$ in which case $p_{\mathbf{X},\mathbf{Y}}(\mathbf{x}, \mathbf{y})$ is also 0, does not matter here.

In Exercise 241Aa, $p_{X|Y}(x\,|\,2)$ takes the values $\frac{4}{9}, \frac{4}{9}, \frac{1}{9}, 0, 0$ for $x = 0, 1, 2, 3, 4$ respectively. In Example 241Bb, we used intuitively

$$\mathbb{P}(X = x; Y = y) = \mathbb{P}(X = x; Y = y; N = x + y)$$
$$= \mathbb{P}(N = x + y)\mathbb{P}(X = x; Y = y\,|\,N = x + y).$$

Of course,

$$\mathbb{P}(\mathbf{X} \in A) = \sum_y \mathbb{P}(\mathbf{X} \in A; \mathbf{Y} = y)$$

$$= \sum_{\mathbf{x}\in A} \sum_y p_{\mathbf{Y}}(\mathbf{y})p_{\mathbf{X}|\mathbf{Y}}(\mathbf{x}\,|\,\mathbf{y}), \tag{C2}$$

the y-sums being over all possible values of $\mathbf{y}$.

It is all very similar to what we have seen before in connection with Bayes' Theorem in Probability.

D. Conditional independence for discrete Variables.　　　　Let $X_1, X_2, \ldots, X_n, Y$ be discrete RVs.　　Then $X_1, X_2, \ldots, X_n$ are called conditionally independent given Y if

$$p_{X_1, X_2, \ldots, X_n | Y} = p_{X_1 | Y} p_{X_2 | Y} \cdots p_{X_n | Y}.$$

Da. Exercise. (a) Give an example of three RVs X_1, X_2, Y where X_1 and X_2 are independent but are not conditionally independent given Y.

(b) Give an example based on the experiment of tossing a fair coin once(!) of three RVs X_1, X_2, Y where X_1 and X_2 are conditionally independent given Y but not independent.

7.2 Jacobians

Here we prepare for Theorem 249C on the behaviour of pdfs under transformations. If you know about Jacobians, or are willing to take that theorem completely on trust, you can jump on to the next section. Meanwhile, I am going to explain the heuristic idea which underlies the rôle of Jacobians.

A. Oriented parallelograms. Let $\mathbf{a} = (a_1, a_2) = a_1\mathbf{i} + a_2\mathbf{j}$ and $\mathbf{b}$ be vectors in the plane $\mathbb{R}^2$, $\mathbf{i}$ and $\mathbf{j}$ being unit vectors along the x-axis and y-axis as usual. Let $\mathrm{OrPar}(\mathbf{a}, \mathbf{b})$ be the oriented parallelogram which is described by a particle which moves (in a straight line) from $\mathbf{0}$ to $\mathbf{a}$, then moves from $\mathbf{a}$ to $\mathbf{a} + \mathbf{b}$, then from $\mathbf{a} + \mathbf{b}$ to $\mathbf{b}$, then from $\mathbf{b}$ to $\mathbf{0}$. See Figure A(i). The parallelogram is oriented in that 'it has an arrow going around it' showing the 'sense' in which the particle moves. The unoriented parallelogram

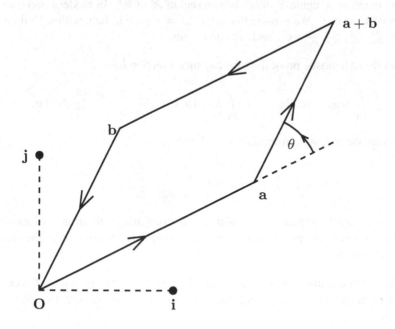

Figure A(i): The oriented parallelogram $\mathrm{OrPar}(\mathbf{a}, \mathbf{b})$

which is the track of the particle is denoted by $\mathrm{Par}(\mathbf{a}, \mathbf{b})$.

The **signed area** of OrPar($\mathbf{a}, \mathbf{b}$) is $+$Area (Par($\mathbf{a}, \mathbf{b}$)) if OrPar($\mathbf{a}, \mathbf{b}$) is traversed anticlockwise (as it is in Figure A(i)), $-$Area (Par($\mathbf{a}, \mathbf{b}$)) if OrPar($\mathbf{a}, \mathbf{b}$) is traversed clockwise. An anticlockwise 'arrow' thus corresponds to positive orientation, a clockwise arrow to negative orientation.

Let θ be the angle needed to rotate the vector $\mathbf{a}$ anticlockwise until it lies along $\mathbf{b}$, as shown in Figure 243A(i). Introduce the vector product $\mathbf{k} = \mathbf{i} \times \mathbf{j}$, so that $\mathbf{k}$ is a unit vector which projects out from the page towards you. You know that the vector product $\mathbf{a} \times \mathbf{b}$ has the expressions

$$\mathbf{a} \times \mathbf{b} = (\|\mathbf{a}\| \|\mathbf{b}\| \sin \theta) \, \mathbf{k} = \det \begin{pmatrix} \mathbf{i} & \mathbf{j} & \mathbf{k} \\ a_1 & a_2 & 0 \\ b_1 & b_2 & 0 \end{pmatrix} = \det \begin{pmatrix} a_1 & a_2 \\ b_1 & b_2 \end{pmatrix} \mathbf{k},$$

so that

$$\text{SignedArea (OrPar}(\mathbf{a}, \mathbf{b})) = \det \begin{pmatrix} a_1 & a_2 \\ b_1 & b_2 \end{pmatrix}. \tag{A1}$$

Note. Vector products work only in $\mathbb{R}^3$. However, there is a theory of **exterior products** which gives generalizations to all dimensions of such results as equation (A1).

B. The Jacobian formula in two dimensions. Suppose that we have a *one-one* map ψ mapping a region $\mathcal{V}$ of $\mathbb{R}^2$ *onto* a region $\mathcal{X}$ of $\mathbb{R}^2$. In modern parlance, ψ is a *bijection* of $\mathcal{V}$ onto $\mathcal{X}$. We assume that if $(x, y) = \psi(v, w)$, then, within $\mathcal{V}$, the partial derivatives $\frac{\partial x}{\partial v}, \frac{\partial y}{\partial v}, \frac{\partial x}{\partial w}, \frac{\partial y}{\partial w}$ exist and are continuous.

We develop a heuristic proof that, for any nice function h,

$$\int\int_{\mathcal{X}} h(x, y) \, \mathrm{d}x \, \mathrm{d}y = \int\int_{\mathcal{V}} h(\psi(u, v)) \left| J \begin{pmatrix} x & y \\ v & w \end{pmatrix} \right| \mathrm{d}v \, \mathrm{d}w, \tag{B1}$$

where we have the following formula for the Jacobian J:

$$J \begin{pmatrix} x & y \\ v & w \end{pmatrix} = \det \begin{pmatrix} \frac{\partial x}{\partial v} & \frac{\partial y}{\partial v} \\ \frac{\partial x}{\partial w} & \frac{\partial y}{\partial w} \end{pmatrix},$$

and $|J(\cdot)|$ is the absolute value of $J(\cdot)$ as usual. In comparing definitions of Jacobians in different books, remember that the determinant of a matrix is invariant under the operation of taking transposes.

Imagine splitting the region $\mathcal{V}$ into lots of disjoint positively oriented rectangles, the typical rectangle (see the left-hand side of Figure B(i)) having corners $1, 2, 3, 4$ with coordinates

$$1 : (v, w); \quad 2 : (v + \delta v, w); \quad 3 : (v + \delta v, w + \delta w); \quad 4 : (v, w + \delta w).$$

This is the oriented parallelogram OrPar $((\delta v)\mathbf{i}, (\delta w)\mathbf{j})$ based at (v, w) rather than at $(0, 0)$.

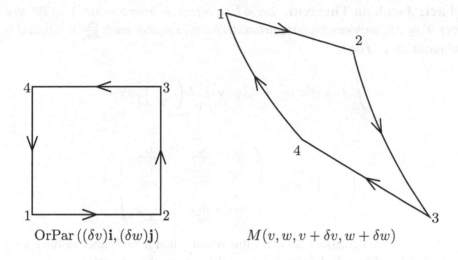

OrPar $((\delta v)\mathbf{i}, (\delta w)\mathbf{j})$ $\qquad\qquad$ $M(v, w, v + \delta v, w + \delta w)$

Figure B(i): Effect of map on small rectangle

Under ψ, the oriented parallelogram just described will map into a miniregion $M(v, w, v + \delta v, w + \delta w)$ closely approximating OrPar$(\mathbf{a}, \mathbf{b})$ based at $(x, y) = \psi(v, w)$ (see the right-hand side of Figure B(i)) where

$$\mathbf{a} = \psi(v + \delta v, w) - \psi(v, w) \approx (\delta v) \left(\frac{\partial x}{\partial v}, \frac{\partial y}{\partial v} \right),$$

$$\mathbf{b} = \psi(v, w + \delta w) - \psi(v, w) \approx (\delta w) \left(\frac{\partial x}{\partial w}, \frac{\partial y}{\partial w} \right).$$

Thus,

$$\text{SignedArea} \left(M(v, w, v + \delta v, w + \delta w) \right) \approx (\delta v)(\delta w) J \left(\begin{array}{cc} x & y \\ v & w \end{array} \right).$$

Obviously, Figure B(i) refers to a situation where J is negative.

The integral

$$\int \int h(x, y) \, dx \, dy \approx \sum \sum h(x, y) \text{Area} \left(M(v, w, v + \delta v, w + \delta w) \right),$$

summed over all the little miniregions in $\mathcal{X}$,

$$\approx \sum \sum h \left(\psi(v, w) \right) \left| J \left(\begin{array}{cc} x & y \\ v & w \end{array} \right) \right| \delta v \, \delta w,$$

summed over all the little rectangles in $\mathcal{V}$. The good sense of result 244(B1) is now clear.

▶▶ **C. Fact: Jacobian Theorem.** *Let ψ be a bijection from a subset $\mathcal{V}$ of $\mathbb{R}^n$ to a subset $\mathcal{X}$ of $\mathbb{R}^n$, with continuous partial derivatives in that each $\frac{\partial x_i}{\partial v_j}$ exists and is continuous on $\mathcal{V}$. Then*

$$\int_{\mathcal{X}} h(\mathbf{x})\mathrm{d}\mathbf{x} \;=\; \int_{\mathcal{V}} h(\psi(\mathbf{v}))\left| J\left(\frac{\mathbf{x}}{\mathbf{v}}\right)\right|\mathrm{d}\mathbf{v},$$

where

$$J\left(\frac{\mathbf{x}}{\mathbf{v}}\right) \;=\; \det\begin{pmatrix} \frac{\partial x_1}{\partial v_1} & \frac{\partial x_2}{\partial v_1} & \cdot & \frac{\partial x_n}{\partial v_1} \\ \frac{\partial x_1}{\partial v_2} & \frac{\partial x_2}{\partial v_2} & \cdot & \frac{\partial x_n}{\partial v_2} \\ \cdot & \cdot & \cdot & \cdot \\ \frac{\partial x_1}{\partial v_n} & \frac{\partial x_2}{\partial v_n} & \cdot & \frac{\partial x_n}{\partial v_n} \end{pmatrix}.$$

We have seen the heuristic reason for this result when $n = 2$, and we have long known it as the 'Chain Rule' when $n = 1$. We assume the general case.

7.3   Joint pdfs; transformations

▶▶ **A. Definition of a joint pdf.**   Let $\mathbf{X} = (X_1, X_2, \ldots, X_n)$ be an RV with values in $\mathbb{R}^n$.   Then $\mathbf{X}$ is called 'continuous' with a pdf $f_{\mathbf{X}}$ (a joint pdf of $X_1, X_2, \ldots, X_n$) if, for every nice (Borel) subset A of $\mathbb{R}^n$,

$$\mathbb{P}(\mathbf{X} \in A) \;=\; \int_A f_{\mathbf{X}}(\mathbf{x})\,\mathrm{d}\mathbf{x}$$

$$= \int\int\cdots\int f_{X_1, X_2, \ldots, X_n}(x_1, x_2, \ldots, x_n)\,\mathrm{d}x_1 \mathrm{d}x_2 \ldots \mathrm{d}x_n;$$

or, in shorthand,

$$\mathbb{P}(\mathbf{X} \in \mathrm{d}\mathbf{x}) \;=\; f_{\mathbf{X}}(\mathbf{x})\,\mathrm{d}\mathbf{x}.$$

Then, for example,

$$f_1(x) := \int_{x_2}\int_{x_3}\cdots\int_{x_n} f_{X_1, X_2, \ldots, X_n}(x, x_2, x_3, \ldots, x_n)\,\mathrm{d}x_2 \mathrm{d}x_3 \ldots \mathrm{d}x_n$$

is a pdf for X_1. Moreover $X_1, X_2, \ldots, X_n$ are independent if and only if

$$f_{X_1}(x_1)f_{X_2}(x_2)\ldots f_{X_n}(x_n)$$

is a joint pdf for $X_1, X_2, \ldots, X_n$.

Aa. Discussion. "What's all this '*a* pdf'? Why not '*the* pdf'?'", you say. Well, it is just to emphasize that we have to remember that a pdf is not uniquely defined: it is only defined 'modulo sets of measure zero'.

If, for example, a point with coordinates (X, Y) is chosen uniformly in the unit disc of radius 1 centred at the origin in the plane, then we would set

$$f_{X,Y}(x, y) = \begin{cases} \pi^{-1} & \text{if } x^2 + y^2 < 1, \\ 0 & \text{if } x^2 + y^2 > 1, \end{cases}$$

but the value of $f_{X,Y}$ on the circle $x^2 + y^2 = 1$ is arbitrary. (Yes, purists, we could change $f_{X,Y}$ on other sets of measure zero, too.)

If (X, Y) is instead chosen uniformly in the unit square $[0, 1] \times [0, 1]$, then $I_{[0,1]}$ acts as a pdf for each of X and Y, but $I_{(0,1) \times (0,1)}$ is a perfectly good pdf for (X, Y) and

$$I_{(0,1) \times (0,1)}(x, y) \neq I_{[0,1]}(x) I_{[0,1]}(y)$$

on the boundary of the square. But X and Y are most certainly independent, and

$$(x, y) \mapsto I_{[0,1]}(x) I_{[0,1]}(y) = I_{[0,1] \times [0,1]}(x, y)$$

is a pdf for (X, Y).

Note, while we are discussing such points that if (X, Y) is chosen uniformly at random on the circle $\{(x, y) : x^2 + y^2 = 1\}$, then (X, Y) is not 'continuous': it has no pdf on $\mathbb{R}^2$. Of course, each of X and Y has a pdf, namely $\pi^{-1}(1 - x^2)^{-\frac{1}{2}}$ (*Exercise!*). [You can see why the 'continuous' terminology is silly. In this example. we would take $\Omega = \{(x, y) : x^2 + y^2 = 1\}$, $X(x, y) = x$, $Y(x, y) = y$. Then (X, Y) is a continuous function on Ω, so a continuous RV, but not a 'continuous' RV!]

Ab. Example. *Problem.* Two people A and B plan to meet outside Bath Abbey. Each is prepared to wait for 10 minutes for the other. The arrival times of A and B are independent RVs, each with the uniform distribution on $[0, 1]$, 0 signifying 12.00 noon and 1 signifying 1 pm. What is the chance that A and B meet?

Solution. Let X be the arrival time for A, Y that for B. Then (X, Y) is uniformly distributed in the unit square, so the answer is the area of that portion of the unit square corresponding to 'A meets B', namely, the portion

$$\{(x, y) : 0 \leq x \leq 1, \ 0 \leq y \leq 1; \ |x - y| \leq \tfrac{1}{6}\}.$$

This is the shaded area in Figure A(i). The two unshaded triangles would form a square of side $\frac{5}{6}$, so the answer is

$$1 - \left(\tfrac{5}{6}\right)^2 = \tfrac{11}{36}.$$

Ac. Exercise. Two points are chosen independently on $[0, 1]$, each according to the uniform distribution. Imagine that the interval $[0, 1]$ is cut at these two points making three pieces. Show that the probability that these three pieces can be made to form the sides of a triangle is $1/4$. *Hint.* Shade in the relevant region in the unit square.

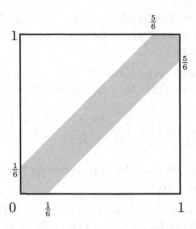

Figure A(i): Shaded region corresponds to 'A meets B'

Ad. Exercise.   (Compare Exercise 241Ba.)   Suppose that, for some functions $g_1, g_2, \ldots, g_n$ each on $\mathbb{R}$,

$$g_1(x_1)g_2(x_2)\cdots g_n(x_n)$$

acts as a pdf for $\mathbf{X} = (X_1, X_2, \ldots, X_n)$. Prove that $X_1, X_2, \ldots, X_n$ are independent, X_k having pdf $g_k(\cdot)/G_k$, where $G_k := \int_{x \in \mathbb{R}} g_k(x)\mathrm{d}x$.

B. Example.   Suppose that we have the situation where

$$X \text{ and } Y \text{ are independent RVs, } X \sim \mathrm{E}(\text{rate } \lambda) \text{ and } Y \sim \mathrm{E}(\text{rate } \mu).$$

Let

$$T := \min(X, Y), \quad Z = \max(X, Y), \quad A = Z - T.$$

Let us find $\mathbb{P}(A \geq a)$.

Solution. We have $A \geq a$ if and only if (X, Y) belongs to one of the shaded triangular regions in Figure B(i). Looking at the top region, we see that

$$\mathbb{P}(A \geq a;\, Y \geq X) = \int_{x=0}^{\infty} \int_{y=a+x}^{\infty} \mathbb{P}(X \in \mathrm{d}x, Y \in \mathrm{d}y)$$

$$= \int_{x=0}^{\infty} \int_{y=a+x}^{\infty} \lambda e^{-\lambda x} \mu e^{-\mu y}\, \mathrm{d}x \mathrm{d}y = \int_{x=0}^{\infty} \mathrm{d}x\, \lambda e^{-\lambda x} \int_{y=a+x}^{\infty} \mu e^{-\mu y}\, \mathrm{d}y$$

$$= \int_{x=0}^{\infty} \lambda e^{-\lambda x} e^{-\mu(a+x)} \mathrm{d}x = \lambda e^{-\mu a} \int_{x=0}^{\infty} e^{-(\lambda+\mu)x} \mathrm{d}x = \frac{\lambda e^{-\mu a}}{\lambda + \mu}.$$

Combining this with the corresponding formula for the lower triangle, we have

$$\mathbb{P}(A \geq a) = \frac{\lambda e^{-\mu a} + \mu e^{-\lambda a}}{\lambda + \mu}.$$

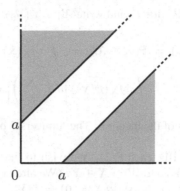

Figure B(i): The region $\{A \geq a\}$

Ba. Exercise. A man and woman go into two next-door shops at the same time. The times they spend in the shops are independent and are exponentially distributed with MEANS 5 minutes for the man, 10 minutes for the more discerning woman. **(a)** How many minutes on average until the first emerges from a shop? **(b)** What is the probability that the man emerges from his shop first? **(c)** Given that the man emerges from his shop first, how long on average does he have to wait for the woman? **(d)** How long on average does the first have to wait for the second? **(e)** Is the time which the first has to wait for the second independent of the time when the first emerges from a shop?

▶▶▶ **C. Theorem: effect of transformation on pdfs.** Suppose that **X** is a 'continuous' $\mathbb{R}^n$-valued Random Variable with values in some (Borel) subregion $\mathcal{X}$ of $\mathbb{R}^n$ and with a pdf f_X (zero off $\mathcal{X}$). (Of course, $\mathcal{X}$ may well be the whole of $\mathbb{R}^n$.) Suppose that ψ is a bijection from a (Borel) subset $\mathcal{V}$ of $\mathbb{R}^n$ onto $\mathcal{X}$ with continuous partial derivatives within $\mathcal{V}$. If we write

$$\mathbf{X} = \psi(\mathbf{V}), \text{ so } \mathbf{V} = \psi^{-1}(\mathbf{X}),$$

then

$$f_{\mathbf{V}}(\mathbf{v}) = f_X(\psi(\mathbf{v})) \left| J\left(\begin{matrix} \mathbf{x} \\ \mathbf{v} \end{matrix} \right) \right|$$

is a pdf for **V** on $\mathcal{V}$.

'Conversely', if $\mathcal{V}$ has pdf $f_{\mathbf{V}}$ on $\mathcal{V}$ and we put $\mathbf{X} = \psi(\mathbf{V})$, then f_X is a pdf for **X** on $\mathcal{X}$.

Proof. Suppose that $\mathcal{V}_0 \subseteq \mathcal{V}$ ($\mathcal{V}_0$ Borel), and write $\mathcal{X}_0 = \psi(\mathcal{V}_0)$. Then, using Fact 246C,

$$\mathbb{P}(\mathbf{V} \in \mathcal{V}_0) = \mathbb{P}(\mathbf{X} \in \mathcal{X}_0) = \int_{\mathcal{X}_0} f_{\mathbf{X}}(\mathbf{x})\,d\mathbf{x}$$

$$= \int_{\mathcal{V}_0} f_X(\psi(\mathbf{v})) \left| J\left(\begin{matrix} \mathbf{x} \\ \mathbf{v} \end{matrix}\right) \right| d\mathbf{v};$$

and this clinches the direct part of the theorem. The 'converse' part follows similarly. □

Ca. Example. Let X, Y be IID, each $E(1)$. We wish to find the joint distribution of (V, W), where $V = X/(X + Y)$ and $W = X + Y$. (We already know that the marginal distribution of W is Gamma$(2, 1)$.) First, $\mathbb{P}(X = 0) = \mathbb{P}(Y = 0) = 0$, so we can take $\mathcal{X} = (0, \infty)^2$. We take $\mathcal{V} = (0, 1) \times (0, \infty)$ and (since we want $v = x/(x + y)$ and $w = x + y$) we take

$$\psi(v, w) = (x, y), \text{ where } x = vw, \ y = w(1 - v).$$

That ψ is a bijection from $\mathcal{V}$ to $\mathcal{X}$ is now obvious. We have

$$\det\left(\begin{matrix} \frac{\partial x}{\partial v} & \frac{\partial y}{\partial v} \\ \frac{\partial x}{\partial w} & \frac{\partial y}{\partial w} \end{matrix}\right) = \left(\begin{matrix} w & -w \\ v & 1 - v \end{matrix}\right) = w > 0,$$

so $|J(\cdot)| = w$. Thus, since $f_{X,Y}(x, y) = e^{-(x+y)}$ on $\mathcal{X}$ and $x + y = w$, we have as a pdf for (V, W):

$$f_{V,W}(v, w) = we^{-w} I_{(0,\infty)}(w) I_{(0,1)}(v).$$

Thus, V and W are independent, $W \sim$ Gamma$(2, 1)$, $V \sim \text{U}(0, 1)$.

▶▶ **Cb. Exercise: Relation between Gamma and Beta variables.** Generalize the above Example to show that if

$$X \sim \text{Gamma}(K, 1), \quad Y \sim \text{Gamma}(L, 1), \quad \text{and } X \text{ and } Y \text{ are independent,}$$

and if we put

$$V = \frac{X}{X + Y}, \quad W = X + Y,$$

then

$$V \sim \text{Beta}(K, L), \quad W \sim \text{Gamma}(K + L, 1),$$
$$\text{and } V \text{ and } W \text{ are independent.}$$

This gives a probabilistic interpretation to the proof of the relation between Gamma and Beta *functions* given at 209Ia. It also gives **a method for simulating Beta RVs.**

▶ **Cc. Exercise: 'Polar' method of simulating N$(0, 1)$ variables.** Suppose that

$$X \text{ and } Y \text{ are IID RVs, each with the N}(0, 1) \text{ distribution.}$$

The probability that $(X, Y) = (0, 0)$ is zero, and we ignore this possibility. So, take $\mathcal{X} = \mathbb{R}^2 \setminus \{(0,0)\}$, the plane with the origin removed. Take $\mathcal{V} = (0, \infty) \times [0, 2\pi)$, and write

$$x = \sqrt{v} \cos w, \quad y = \sqrt{v} \sin w, \quad \text{so } v = x^2 + y^2, \quad w = \arctan(y/x).$$

Prove that

$$J\left(\begin{array}{cc} x & y \\ v & w \end{array}\right) = \frac{1}{2}.$$

Show that if

$$X = \sqrt{V} \cos W, \quad Y = \sqrt{V} \sin W,$$

then

V and W are independent, $\quad V \sim \text{E(mean 2)}, \quad W \sim \text{U}[0, 2\pi)$.

Hence, if U_1 and U_2 are IID each U[0, 1], and we set

$$\xi := \sqrt{2\ln(1/U_1)} \cos(2\pi U_2), \quad \eta := \sqrt{2\ln(1/U_1)} \sin(2\pi U_2),$$

then ξ and η are IID each N(0, 1). This is a widely used method of simulating normal variables. That it has led to difficulties in the past is explained in Ripley's book.

▶▶ **D. Pdf of a sum.** Let X and Y be RVs with joint pdf $f_{X,Y}$. Let $V = X$ and $W = X + Y$. Then $x = v$, $y = w - v$, and

$$J\left(\begin{array}{cc} x & y \\ v & w \end{array}\right) = \det\left(\begin{array}{cc} 1 & -1 \\ 0 & 1 \end{array}\right) = 1,$$

so that

$$f_{X,X+Y}(v, w) = f_{X,Y}(v, w - v), \qquad f_{X+Y}(w) = \int_v f_{X,Y}(v, w - v)\, dv.$$

In particular, **if X and Y are independent**, then we have the convolution formula:

$$f_{X+Y}(w) = (f_X * f_Y)(w) := \int_v f_X(v) f_Y(w - v) dv, \qquad \text{(D1)}$$

$f_X * f_Y$ being the so-called 'convolution' of f_X and f_Y.

Da. Exercise. We know from our study of MGFs that if X and Y are independent and $X \sim \text{N}(\mu_1, \sigma_1^2)$, $Y \sim \text{N}(\mu_2, \sigma_2^2)$ then $X + Y \sim \text{N}(\mu_1 + \mu_2, \sigma_1^2 + \sigma_2^2)$. Combine this with equation (D1) to evaluate the integral at 235(I1).

E. Pdf of a ratio. Let X and Y be RVs with joint pdf $f_{X,Y}$. Let $V = X$ and $W = X/Y$. Then $x = v$, $y = v/w$, and (Check!) $|J| = |v/w^2|$. Hence,

$$f_{X,X/Y}(v, w) = f_{X,Y}\left(v, \frac{v}{w}\right)\left|\frac{v}{w^2}\right|, \qquad f_W(w) = \int f_{X,Y}\left(v, \frac{v}{w}\right)\left|\frac{v}{w^2}\right| dv.$$

Ea. Exercise. Check that if X and Y are IID each E(1), then X/Y has pdf $(1+w)^{-2}$ $(0 < w < \infty)$. Note that

$$\mathbb{E}(X) = \mathbb{E}(Y) = 1, \quad \mathbb{E}(X/Y) = \infty.$$

Do remember that if X and Y are independent, then

$$\mathbb{E}\left(\frac{X}{Y}\right) = \mathbb{E}(X)\mathbb{E}\left(\frac{1}{Y}\right), \quad \textbf{NOT } \mathbb{E}\left(\frac{X}{Y}\right) = \frac{\mathbb{E}(X)}{\mathbb{E}(Y)}.$$

▶▶ **F. The Fisher $F_{r,s}$ distribution.** The F distributions of this subsection and the t distributions of the next are amongst the most fundamental distributions for Statistics.

Suppose that

$$X \sim \chi_r^2, \quad Y \sim \chi_s^2, \quad X \text{ and } Y \text{ are independent.}$$

Then the $F_{r,s}$ distribution (with r degrees of freedom in the numerator, and s degrees of freedom in the denominator) is the distribution of

$$W := \frac{X/r}{Y/s} = \frac{sX}{rY}.$$

We use the mnemonic (aid to memory)

$$F_{r,s} \sim \frac{\chi_r^2/r}{\chi_s^2/s}, \quad \text{Num, Den independent.}$$

Now, $\tfrac{1}{2}X \sim \text{Gamma}(\tfrac{1}{2}r, 1)$, $\tfrac{1}{2}Y \sim \text{Gamma}(\tfrac{1}{2}s, 1)$, so that

$$\frac{1}{\frac{r}{s}W + 1} = \frac{Y}{X+Y} \sim \text{Beta}(\tfrac{1}{2}s, \tfrac{1}{2}r).$$

Thus the computer can calculate the $F_{r,s}$ distribution function from that of a Beta distribution; and that's the way the F-table, p518, was done.

We have (with beta as the pdf of the Beta distribution and B as the Beta *function*)

$$
\begin{aligned}
f_W(w) &= \frac{r}{s\left(1 + \frac{r}{s}w\right)^2}\,\text{beta}\left(\tfrac{1}{2}s, \tfrac{1}{2}r; \frac{1}{1 + \frac{r}{s}w}\right) \\
&= \frac{(r/s)^{\frac{1}{2}r} w^{\frac{1}{2}r - 1}}{B(\tfrac{1}{2}s, \tfrac{1}{2}r)\left(1 + \frac{r}{s}w\right)^{\frac{1}{2}(r+s)}}.
\end{aligned}
\tag{F1}
$$

Since $\chi_2^2 \sim \text{E(mean 2)}$, it follows from Exercise Ea that the $F_{2,2}$ distribution has pdf $(1+w)^{-2}$. This checks out with the above formula. Note that if $W \sim F_{2,2}$, then $\mathbb{P}(W > 19) = 1/20$ *exactly*. This tallies with the table on p518.

▶▶ **G. Student's t_ν distribution.** ('Student' was the pseudonym of W S Gosset, who, as an employee of the Guinness brewery, had to publish under a pseudonym.) Suppose that $X \sim N(0,1)$, $Y \sim \chi_\nu^2$ and X and Y are independent. Then Student's t_ν distribution with ν degrees of freedom is the distribution of

$$T := \frac{X}{\sqrt{Y/\nu}}, \qquad t_\nu \sim \frac{N(0,1)}{\sqrt{\chi_\nu^2/\nu}} \quad \text{Num, Den independent.} \tag{G1}$$

Now, T is clearly symmetric about 0, so $f_T(-t) = f_T(t)$. Also, since $X^2 \sim \chi_1^2$ and is independent of Y, we have $S := T^2 \sim F(1,\nu)$. This allows us to calculate the t-table on page 516. With the first $\frac{1}{2}$ in (G2) taking into account the fact that T is symmetric and $(-t)^2 = t^2$, we have for $t > 0$,

$$f_T(t) = \tfrac{1}{2}.2t f_S(t^2) = \frac{1}{\nu^{\frac{1}{2}} B(\frac{1}{2}, \frac{1}{2}\nu)} \left(1 + \frac{t^2}{\nu}\right)^{-\frac{1}{2}(\nu+1)}. \tag{G2}$$

This also holds, of course, for $t < 0$. In particular, the t_1 distribution is the standard Cauchy distribution with density $\pi^{-1}(1 + t^2)^{-1}$.

As $\nu \to \infty$, the distribution of Y/ν concentrates heavily about 1 (by the Strong or Weak Law), so the t_∞ distribution, the limit of t_ν as $\nu \to \infty$, is the $N(0,1)$ distribution. Hence,

$$t_1 = \text{Cauchy}, \qquad t_\infty = N(0,1).$$

Ga. Exercise. Show that if $\nu \geq 3$, $T \sim t_\nu$, and $F \sim F_{r,\nu}$ ($r \in \mathbb{N}$), then

$$\mathbb{E}(T) = 0, \quad \text{Var}(T) = \frac{\nu}{\nu - 2}, \quad \mathbb{E}(F) = \frac{\nu}{\nu - 2}.$$

Hint. Your calculations should include

$$\mathbb{E}(T^2) = \mathbb{E}(X^2)\mathbb{E}\left(\frac{\nu}{Y}\right),$$

where $Y \sim \text{Gamma}(\frac{1}{2}\nu, \text{rate } \frac{1}{2})$, and you should employ the 'Change pdf' trick, considering the $\text{Gamma}(\frac{1}{2}\nu - 1, \text{rate } \frac{1}{2})$ density.

▶ **Gb. StandT$_\nu$ ($\nu \geq 3$), Standardized t_ν.** We define the standardized t_ν distribution StandT$_\nu$ ($\nu \geq 3$) to be the distribution of $\{(\nu - 2)/\nu\}^{\frac{1}{2}} T_\nu$ where $T_\nu \sim t_\nu$. Then the StandT$_\nu$ distribution has mean 0 and variance 1. Figure G(i) plots the pdf of StandT$_4$ in black with the pdf of $N(0,1)$ in grey.

Gc. Importance in modelling. The t_ν densities for finite ν tail off much more slowly than the Normal, and are important in modelling the many situations where the very fast Normal tail off is unrealistic.

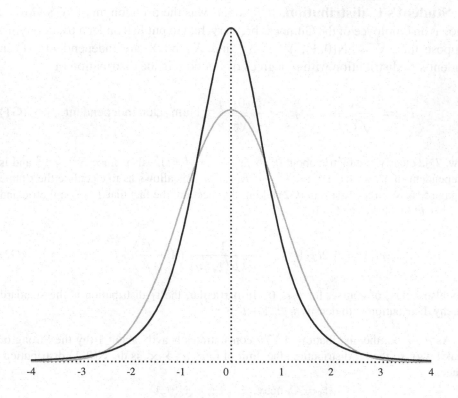

Figure G(i): Pdfs of StandT$_4$ (black) and N$(0, 1)$ (grey)

▶ **Gd. Simulating from t distributions.** It is important to be able to simulate from t distributions for the modelling reasons described above.

The **Kinderman–Monahan method** is as follows. Let $U \sim U[0, 1]$. Define

$$X = \begin{cases} 1/(4U - 1) & \text{if } U < \frac{1}{2}, \\ 4U - 3 & \text{if } U \geq \frac{1}{2}. \end{cases}$$

Then (check!) X has pdf $\frac{1}{4}\min(1, x^{-2})$. By the rejection-sampling idea, we therefore accept X with probability

$$\left(1 + \frac{X^2}{\nu}\right)^{-\frac{\nu+1}{2}} \max(1, X^2).$$

But, of course, we have to check that this expression does not exceed 1. I skip this.

The above idea is incorporated in the `SimGamT` programs in the next subsection but one. A 'speeding-up' trick (due to K&M) utilizes the idea that

$$1 - \tfrac12|X| \le \left(1 + \frac{X^2}{\nu}\right)^{-\frac{\nu+1}{2}},$$

which allows quick acceptance in a good proportion of cases.

```
Test of KMStudT() at various quantiles of t(4) distribution
for run of length 100000:

exact        0.50000  0.70000  0.90000  0.95000  0.99000  0.99500
empirical  0.49985  0.70141  0.90090  0.95052  0.99026  0.99493
```

Thus if $P(t_4 \le x) = 0.7$, then 70141 of the simulated values were less than or equal to x.

▶ **Ge. Exercise.** Show from 253(G1) that we could also simulate a Y-value from the t_ν distribution as follows. Choose

$$R \sim \mathrm{Gamma}(\tfrac12\nu, \mathrm{rate}\ \tfrac12\nu), \quad \text{and then} \quad Y \sim \mathrm{N}(0, \mathrm{prec}\ R).$$

Note that we can simulate a Y-value from the StandT$_\nu$ distribution via

$$R \sim \mathrm{Gamma}(\tfrac12\nu, \mathrm{rate}\ (\tfrac12\nu - 1)), \quad \text{and then} \quad Y \sim \mathrm{N}(0, \mathrm{prec}\ R).$$

▶ **H. Ratio method of simulation (Kinderman–Monahan).** A very useful method of simulation is based on the following lemma.

Ha. Lemma. *Suppose that h is a non-negative function with finite integral H over* $\mathbb{R}$. *Let*

$$A := \left\{ (x, y) : 0 \le x \le \sqrt{h(y/x)} \right\}.$$

Suppose that (X, Y) *is chosen uniformly over A, so that*

$$f_{X,Y}(x, y) = \frac{I_A(x, y)}{\mathrm{Area}(A)}$$

acts as a pdf for (X, Y). *Then*

$$V := Y/X \text{ has pdf } H^{-1}h(v).$$

Use. Suppose that $A \subseteq B$ where B is a region from which it is easy to pick a point uniformly at random. We obtain points uniformly within A by selecting those chosen independently and uniformly within B which lie in A.

Proof. Let $u = x$, $v = y/x$, so that $x = u$, $y = uv$. Then, by the Jacobian result,

$$f_{U,V}(u, v) = \frac{I_A(u, uv)}{\mathrm{Area}(A)}\, u,$$

and

$$f_V(v) = \text{Area}(A)^{-1} \int_{u=0}^{\sqrt{h(v)}} u\,du = \tfrac{1}{2}\text{Area}(A)^{-1}h(v).$$

But

$$1 = \int f_V(v)dv = \tfrac{1}{2}\text{Area}(A)^{-1}H,$$

so the desired result follows.   □

▶▶ **I. Simulation of Gamma and t variables.** Here now are the programs I use to simulate t and Gamma variables. The KMSTudT part follows the K–M method of 254Gd. The AD parts relate to the Ahrens–Dieter algorithm from Subsection 150F. After stating the programs, I explain the relation of the CF parts (the **Cheng–Feast algorithm**) to the above Lemma.

Here's a header file in which the 'set' and 'prepare' bits assign values to static variables:

```
/* SimGamT.h                          DW and William Browne
   for simulation of Gamma(K, rate alpha) and t(nuT) variables
*/
#if defined SimGamT_h
#else
#define SimGamT_h

void setKalpha(double KK, double aa);
void ADprepare(); void CFprepare();
double ADgam(); double CFgam();
double rGrate();
   /* for doing many simulations with the same (K,alpha); */
double rnewGrate(double KK, double aa);
   /* for dealing with a different (K,alpha) each time */
void prepKM_T(int nu1); double KMStudT();
#endif
```

And here's the program file, partly based on a program of Bill Browne's:

```
/* SimGamT.c
   for simulation of Gamma(K, rate alpha) and t(nuT) variables
   cc -c SimGamT.c RNG.o -o SimGamT.o -lm          to compile
*/
#include <stdio.h>
#include <math.h>
#include "RNG.h"
#include "SimGamT.h"
static int nuT; static double K,alpha,a,b,c,d,sqrtK;
void setKalpha(double KK, double aa){ K = KK; alpha = aa;}
```

```
void ADprepare(){a = E/(K+E); b = 1/a;}
void CFprepare(){a = K-1; b = (K - 1/(6*K))/a; c = 2/a;
                 d = c+2; sqrtK = sqrt(K); }
double ADgam(){ int accept = 0;    double X,U;
  do{U = Unif();
    if (U>a){
      X = - log(b*(1-U)/K); if (Unif() < pow(X, K-1)) accept = 1;
    }
    else{X = pow(b*U,1/K); if (Unif() < exp(-X)) accept = 1;}
  }while (accept == 0);
  return X;
}
double CFgam(){ int done;  double U,U1,U2,W;
  do{ if (K>2.5){
        do{U1 = Unif(); U2 = Unif();
           U = U1 + (1 - 1.86*U2)/sqrtK;
        }while ((U >= 1)||(U<=0));
      }
      else{U1 = Unif(); U = Unif();}
    W = b*U1/U;   done = 1;
    if (c*U + W + 1/W > d){
      if (c*log(U) - log(W) + W > 1) done = 0; }
  }while (done == 0);
  return a*W;
}
double rGrate(){
  if (K==1) return -log(Unif())/alpha;
  if (K>1) return CFgam()/alpha; else return ADgam()/alpha;
}
double rnewGrate(double KK, double aa){
  setKalpha(KK,aa);
  if (K==1) return -log(Unif())/alpha;
  if (K>1) {CFprepare(); return CFgam()/alpha;}
  else     {ADprepare(); return ADgam()/alpha;}
}
void prepKM_T(int nu1){  nuT = nu1; }
double KMStudT(){ int done = 0; double U, U1, V, X;
  do{
    U = Unif(); U1 = Unif();
    if (U < 0.5) {X = 1.0/(4.0*U - 1.0); V = U1/(X*X);}
    else {X = 4.0*U - 3.0; V = U1;}
    if (V < 1.0 - 0.5* fabs(X)) done = 1;
    else if (V < pow((1.0 + X*X/nuT),
        - 0.5*(nuT + 1.0))) done = 1;
  }while (done == 0);
  return X; }
```

The 'if K>2.5 ...' part of the Cheng–Feast algorithm cleverly trims down the 'B' region. See Ripley [196] for pictures.

If $V = aW$, then

$$c \ln U - \ln W + W - 1 < 0$$

is equivalent to the Kinderman–Monahan condition

$$U \le \sqrt{V^{K-1} e^{-V}} \, \text{function}(K).$$

For any $R > 0$, $\ln r \le r - 1$, so

$$c \ln U + \ln(1/W) + W - 1 \le cU - d + W + 1/W.$$

Hence if the quickly-checked first condition within the last 'if' in CFgam fails to hold, $V = aW$ is returned.

7.4　Conditional pdfs

▶▶ **A. Definition.**　Let $\mathbf{X} = (X_1, X_2, \dots, X_r)$ be an $\mathbb{R}^r$-valued RV and $\mathbf{Y} = (Y_1, Y_2, \dots, Y_s)$ an $\mathbb{R}^s$-valued RV such that $(\mathbf{X}, \mathbf{Y})$ is a 'continuous' $\mathbb{R}^{r+s}$-valued RV with pdf $f_{\mathbf{X}, \mathbf{Y}}$. Intuitively,

$$f_{\mathbf{X}}(\mathbf{x}) d\mathbf{x} = \mathbb{P}(\mathbf{X} \in d\mathbf{x}),$$

the probability that $\mathbf{X}$ will lie in a little volume $d\mathbf{x}$ around $\mathbf{x}$. In analogy with the discrete case 242(C1), we want

$$f_{\mathbf{X}|\mathbf{Y}}(\mathbf{x}|\mathbf{y}) d\mathbf{x} = \mathbb{P}(\mathbf{X} \in d\mathbf{x} \mid \mathbf{Y} \in d\mathbf{y})$$

$$= \frac{\mathbb{P}(\mathbf{X} \in d\mathbf{x}; \mathbf{Y} \in d\mathbf{y})}{\mathbb{P}(\mathbf{Y} \in d\mathbf{y})} = \frac{f_{\mathbf{X}, \mathbf{Y}}(\mathbf{x}, \mathbf{y}) \, d\mathbf{x} d\mathbf{y}}{f_{\mathbf{Y}}(\mathbf{y}) d\mathbf{y}},$$

so that we define *the conditional pdf* $f_{\mathbf{X}|\mathbf{Y}}(\cdot|\cdot)$ *of* $\mathbf{X}$ *given* $\mathbf{Y}$ as

$$f_{\mathbf{X}|\mathbf{Y}}(\mathbf{x}|\mathbf{y}) := \frac{f_{\mathbf{X}, \mathbf{Y}}(\mathbf{x}, \mathbf{y})}{f_{\mathbf{Y}}(\mathbf{y})}. \tag{A1}$$

(We need only bother about situations where $f_{\mathbf{Y}}(\mathbf{y}) > 0$.)

Then the heuristic idea that for a nice (Borel) subset A of $\mathbb{R}^r$,

$$\mathbb{P}(\mathbf{X} \in A \mid \mathbf{Y} \in d\mathbf{y}) = \int_{\mathbf{x} \in A} f_{\mathbf{X}|\mathbf{Y}}(\mathbf{x}|\mathbf{y}) \, d\mathbf{x},$$

tallies in the following calculation (where B is a nice subset of $\mathbb{R}^s$):

$$
\begin{aligned}
\mathbb{P}(\mathbf{X} \in A; \mathbf{Y} \in B) &= \int_{\mathbf{y} \in B} \mathbb{P}(\mathbf{X} \in A; \mathbf{Y} \in d\mathbf{y}) \\
&= \int_{\mathbf{y} \in B} \mathbb{P}(\mathbf{X} \in A \mid \mathbf{Y} \in d\mathbf{y}) \mathbb{P}(\mathbf{Y} \in d\mathbf{y}) \\
&= \int_{\mathbf{y} \in B} \int_{\mathbf{x} \in A} f_{\mathbf{X}|\mathbf{Y}}(\mathbf{x} \mid \mathbf{y}) \, d\mathbf{x} \, f_{\mathbf{Y}}(\mathbf{y}) \, d\mathbf{y} \\
&= \int_{\mathbf{y} \in B} \int_{\mathbf{x} \in A} f_{\mathbf{X},\mathbf{Y}}(\mathbf{x}, \mathbf{y}) \, d\mathbf{x} d\mathbf{y} \quad \text{(correct!).}
\end{aligned}
$$

For real-valued RVs X and Y for which (X, Y) is 'continuous',

$$
\mathbb{P}(X \in A \mid Y \in dy) = \lim_{\delta \downarrow 0} \mathbb{P}(X \in A \mid Y \in (y - \delta, y + \delta)), \tag{A2}
$$

except perhaps for boundary effects as discussed at 247Aa.

Note *Bayes' formula*: for fixed $\mathbf{y}$, as functions of $\mathbf{x}$,

$$
f_{\mathbf{X}|\mathbf{Y}}(\mathbf{x} \mid \mathbf{y}) = \frac{1}{f_{\mathbf{Y}}(\mathbf{y})} f_{\mathbf{X}}(\mathbf{x}) f_{\mathbf{Y}|\mathbf{X}}(\mathbf{y} \mid \mathbf{x}) \propto f_{\mathbf{X}}(\mathbf{x}) f_{\mathbf{Y}|\mathbf{X}}(\mathbf{y} \mid \mathbf{x}), \tag{A3}
$$

always with a proviso that $f_{\mathbf{X}|\mathbf{Y}}(\mathbf{x} \mid \mathbf{y})$ and $f_{\mathbf{X},\mathbf{Y}}(\mathbf{x}, \mathbf{y})$ are only defined modulo (x, y) subsets of measure zero.

Aa. Convention. Unless there is some point to be made, I am not going to continue mentioning this 'modulo sets of measure zero' business: henceforth, it is understood. We shall therefore blur the distinction between '*a* pdf' and '*the* pdf'.

Very often, there is an obvious smooth **canonical** choice of $f_{\mathbf{X}|\mathbf{Y}}(\mathbf{x} \mid \mathbf{y})$ away from boundaries and given by (A2).

Ab. Exercise. Suppose that X and Y are independent, $X \sim \mathrm{E}(\text{rate } \lambda)$, $Y \sim \mathrm{E}(\text{rate } \mu)$, where $\lambda \neq \mu$. Let $V = X$ and $W = X + Y$. Prove that

$$
f_{V|W}(v \mid w) = \frac{(\mu - \lambda) e^{(\mu - \lambda)v}}{e^{(\mu - \lambda)w} - 1} I_{[0,w]}(v).
$$

Note (and *discuss*) the contrast between the cases $\mu > \lambda$ and $\mu < \lambda$. Continuity correctly suggests that if $\lambda = \mu$, then $f_{V|W}(v \mid w) = w^{-1} I_{[0,w]}(v)$.

▶▶ **Ac. Exercise.** Suppose that M and ε are independent RVs, $M \sim \mathrm{N}(0, \sigma^2)$, $\varepsilon \sim N(0, 1)$. Let $V = M$, $W = M + \varepsilon$. Prove that $f_{V|W}(\cdot | w)$ is the pdf of the $\mathrm{N}(\beta w, \beta)$ distribution, where $\beta^2 = \sigma^2/(\sigma^2 + 1)$. We write

$$
\text{Shorthand notation: } (V \mid W) \sim \mathrm{N}(\beta W, \beta) \tag{A4}
$$

to summarize this. Of course, $(W \mid V) \sim N(V, 1)$. The full multivariate-normal generalization of these results will feature later.

Explain why this exercise could have been deduced from the table 211 J(i) of conjugate distributions.

Ad. Exercise – Warning! Consider the following modification of the problem in Example 247Ab. Two people A and B plan to meet outside Bath Abbey. Each is prepared to wait for 10 minutes for the other. The arrival times X and Y of A and B are independent RVs, each with the uniform distribution on $[0, 1]$, 0 signifying 12.00 noon and 1 signifying 1 pm. What is the chance that A and B meet given that A arrives before B? You must correct the following wrong argument. *Wrong argument.* If A arrives before B, then the pdf of X given Y is $y^{-1} I_{[0,y]}$. So, if we know that $Y \in dy$, then the chance that A and B meet is 1 if $y \le 1/6$, $(6y)^{-1}$ if $y \ge 1/6$. Hence, since Y is uniform on $[0, 1]$, the answer is

$$\int_0^{1/6} 1 \, dy + \int_{1/6}^1 (6y)^{-1} dy \;=\; \tfrac{1}{6} + \tfrac{1}{6} \ln 6. \quad \text{Wrong!}$$

Note. The answer *has* to be wrong because the areas of triangles with rational coordinates cannot possibly involve the transcendental number $\ln 6$.

You *know* that the correct answer is $11/36$. Your task now is to make the above idea into a correct argument.

▶ **B. (Further) Need for care with conditioning.** Suppose that (X, Y) is a 'continuous' RV in $\mathbb{R}^2$. It is usual to write

$$\mathbb{P}(X \in dx \mid Y = y) \;=\; f_{X|Y}(x|y) dx, \tag{B1}$$

so that $f_{X,Y}(\cdot \mid y)$ is read as *the conditional pdf of X given that $Y = y$*. But though this notation is sound (modulo sets of measure zero), one needs to handle it with a degree of care. The better notation

$$f_{X|Y}(x \mid y) dx \;=\; \mathbb{P}(X \in dx \mid Y \in dy)$$

correctly suggests the 'limit' interpretation at 259(A2).

As an event, '$Y = y$', namely, $\{\omega : Y(\omega) = y\}$, has probability zero; and we cannot condition on an event of probability zero. If E and F are events, then $\mathbb{P}(E \mid F) = \mathbb{P}(F \cap E)/\mathbb{P}(F)$ makes little sense if $\mathbb{P}(F) = 0$, in which case $\mathbb{P}(F \cap E) = 0$ also. The potential for creating 'paradoxes' by working out $0/0$ in various ways clearly exists.

Suppose that (X, Y) is chosen uniformly at random within the quarter-disc

$$Q := \{(x, y) : x > 0, \; y > 0, \; x^2 + y^2 < 1\},$$

so that (X, Y) has pdf $4/\pi$ on Q.

Let $V = X$, $W = Y - X$. Then $J = 1$ and V, W has pdf $4/\pi$ on the region

$$\left\{ (v, w) : 0 < v < 1, \; -v < w < -v + \sqrt{1 - v^2} \right\}.$$

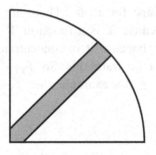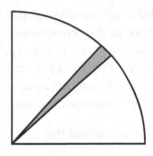

Figure B(i): $Q \cap \{-\delta < Y - X < \delta\}$ and $Q \cap \{1 - \delta < (Y/X) < 1 + \delta\}$

The canonical conditional pdf of V given W when $W = 0$ is $\sqrt{2}I_{(0,1/\sqrt{2})}$; and this is the pdf of X given that $Y - X = 0$.

However, if we put $S = X$ and $T = Y/X$, then (check via pdfs!) the canonical choice of pdf of S given T when $T = 1$ is $4s$ on $(0, 1/\sqrt{2})$; and $4x$ is the pdf of X given that $Y/X = 1$.

Thus the pdf of X given that $Y - X = 0$ is different from the pdf of X given that $Y/X = 1$. Figure B(i) shows the regions

$$Q \cap \{-\delta < Y - X < \delta\} \text{ and } Q \cap \{1 - \delta < (Y/X) < 1 + \delta\}.$$

We *see* that as $\delta \downarrow 0$, the pdf of X given that $-\delta < Y - X < \delta$ will tend to a constant, and that the pdf of X given that $1 - \delta < (Y/X) < 1 + \delta$ will tend to a constant multiple of x (and since we know that in each case the range of X is $(0, 1/\sqrt{2})$, the constants take care of themselves).

What emerges then is that **it is meaningless to speak of**

$$\mathbb{P}(X \in dx \mid X = Y)$$

because we *rightly* get different answers according to whether we interpret it as $\mathbb{P}(X \in dx \mid Y - X = 0)$ or $\mathbb{P}(X \in dx \mid Y/X = 1)$ or ... .

Ba. Repeated Warning. Bayesian theory with improper priors uses 'conditional probabilities' where the underlying measure is not finite. This can cause paradoxes, and has led me into serious error on at least one occasion.

▶ **C. Sufficient Statistics, 2.** Please re-read Subsections 183D and 185F. As usual, we assume that $Y_1, Y_2, \ldots, Y_n$ are IID, each with pmf/pdf $f(y \mid \theta)$, where θ is a vector-valued parameter. The usual definition of a (vector-valued) Sufficient Statistic $\mathbf{T}$ for θ in Frequentist theory is that

$$f_{\mathbf{Y}|\mathbf{T}}(\mathbf{y} \mid \mathbf{t}, \boldsymbol{\theta}) \text{ does not involve } \boldsymbol{\theta};$$

in other words, this conditional pdf is the same for all $\boldsymbol{\theta}$. There are technical difficulties here in the 'continuous' case, because $\mathbf{T}$ is a function $\mathbf{T} = \tau(\mathbf{Y})$ of $\mathbf{Y}$, and $f_{\mathbf{Y}|\mathbf{T}}(\mathbf{y}\,|\,\mathbf{t},\boldsymbol{\theta})$ is not a pdf on $\mathbb{R}^n$ because it is concentrated on the submanifold $\{\mathbf{y} : \tau(\mathbf{Y}) = \mathbf{t}\}$. Example 185Fa illustrates this. So, $f_{\mathbf{Y}|\mathbf{T}}(\cdot\,|\,\mathbf{t},\boldsymbol{\theta})$ is what is called a *Schwartz distribution*. If $\mathbf{T} = \mathbf{Y}$, for example, then $f_{\mathbf{Y}|\mathbf{T}}(\cdot\,|\,\mathbf{t},\boldsymbol{\theta})$ is the Dirac delta function concentrated at $\mathbf{t}$.

Still, the intuitive idea that

$$f_{\mathbf{Y}}(\mathbf{y}\,|\,\boldsymbol{\theta}) \;=\; f_{\mathbf{T}}(\mathbf{t}\,|\,\boldsymbol{\theta})f_{\mathbf{Y}|\mathbf{T}}(\mathbf{y}\,|\,\mathbf{t},\boldsymbol{\theta}) \;=\; f_{\mathbf{T}}(\mathbf{t}\,|\,\boldsymbol{\theta})f_{\mathbf{Y}|\mathbf{T}}(\mathbf{y}\,|\,\mathbf{t})$$

can be interpreted sensibly, and it obviously corresponds closely to the factorization criterion (see 183Da) that

$$f_{\mathbf{Y}}(\mathbf{y}\,|\,\boldsymbol{\theta}) \;=\; g(\mathbf{t},\boldsymbol{\theta})h(\mathbf{y}).$$

Formal proof of the analogue of Fact 184Db has to be skipped.

Ca. Exercise. Suppose that μ, v are unknown parameters, and that

$$Y_1, Y_2, \ldots, Y_n \text{ are IID each } \mathrm{N}(\mu, v).$$

Show that the pair $\mathbf{T} = (\overline{Y}, R^2)$, where $R^2 = \sum(Y_k - \overline{Y})^2 = \sum Y_k^2 - n\overline{Y}^2$ is sufficient for the pair (μ, v). *Important Note.* After Zeus chooses the value $\mathbf{t}^{\mathrm{obs}}$ of $\mathbf{T}$ in a way which we shall understand fully in Chapter 8, Tyche chooses $\mathbf{y}^{\mathrm{obs}}$ uniformly on the $(n-2)$-dimensional sphere in which the hyperplane through $\overline{y}_{\mathrm{obs}}\mathbf{1}$ perpendicular to $\mathbf{1}$ cuts the $(n-1)$-dimensional sphere $\{\mathbf{y} \in \mathbb{R}^n : \sum(y_k - \overline{y}_{\mathrm{obs}})^2 = r_{\mathrm{obs}}^2\}$. Wait for 308Bd where this is discussed again.

▶ **D. Minimal sufficiency.** Continue with the general situation of the preceding subsection. Recall that the crucial point about a Sufficient Statistic $\mathbf{T}$ is that it contains all information in the sample $\mathbf{Y}$ relevant to making inferences about $\boldsymbol{\theta}$. We clearly want $\mathbf{T}$ to be of as low a dimension as possible. Taking $\mathbf{T} = \mathbf{Y}$ does not achieve much, though it is the best we can do in the generic case (that is for a case without some special structure).

A Sufficient Statistic $\mathbf{T}$ is called *Minimal Sufficient* if it is a function of every other Sufficient Statistic. When I say that $\mathbf{T}$ is a function ψ of $\mathbf{U}$ (say), I mean of course that $\mathbf{T}(\omega) = \psi(\mathbf{U}(\omega))$ for every ω. Minimal Sufficient Statistics $\mathbf{S}$ and $\mathbf{T}$ are therefore functions of each other: in this sense, a Minimal Sufficient Statistic is essentially unique.

The 1-dimensional Sufficient Statistics at 184E are clearly Minimal Sufficient, and it is 'obvious' that the pair $(\overline{Y}, R)$ in Exercise Ca is Minimal Sufficient.

▶ **E. Ancillary Statistics.** Suppose that $\mathbf{Y}$ is a vector of observations and that $(\mathbf{T}, \mathbf{A})$ is a Minimal Sufficient Statistic for the parameter θ. Suppose further that the distribution of $\mathbf{A}$ does not involve θ, and that $\mathbf{A}$ is maximal in this regard in that any function of $(\mathbf{T}, \mathbf{A})$ the distribution of which does not involve θ, is a function of $\mathbf{A}$. *Please note that such a pair* $(\mathbf{T}, \mathbf{A})$ *may not exist.* If the pair *does* exist, we call $\mathbf{A}$ an Ancillary Statistic for θ for $\mathbf{Y}$.

Assume that a pair $(\mathbf{T}, \mathbf{A})$ with the above properties exists. Then,

$$
\begin{aligned}
f_{\mathbf{Y}}(\mathbf{y} \mid \theta) &= f_{\mathbf{T},\mathbf{A}}(\mathbf{t}, \mathbf{a} \mid \theta) f_{\mathbf{Y}}(\mathbf{y} \mid \mathbf{t}, \mathbf{a}, \theta) \\
&= f_{\mathbf{A}}(\mathbf{a} \mid \theta) f_{\mathbf{T}\mid\mathbf{A}}(\mathbf{t} \mid \mathbf{a}, \theta) f_{\mathbf{Y}}(\mathbf{y} \mid \mathbf{t}, \mathbf{a}) \\
&= f_{\mathbf{A}}(\mathbf{a}) f_{\mathbf{T}\mid\mathbf{A}}(\mathbf{t} \mid \mathbf{a}, \theta) f_{\mathbf{Y}}(\mathbf{y} \mid \mathbf{t}, \mathbf{a}).
\end{aligned}
$$

We are using the last density heuristically: it will generally be restricted to a submanifold of $\mathbb{R}^n$.

So, we can envisage the following situation in which Zeus knows the value of θ, but Tyche does not.

Stage 1: Tyche chooses the actual value $\mathbf{a}^{\mathrm{obs}}$ of $\mathbf{A}$ with pdf $f_{\mathbf{A}}(\mathbf{a})$ without knowing θ.

Stage 2: Zeus then chooses the actual value $\mathbf{t}^{\mathrm{obs}}$ of $\mathbf{T}$ according to the pdf $f_{\mathbf{T}\mid\mathbf{A}}(\mathbf{t} \mid \mathbf{a}^{\mathrm{obs}}, \theta)$, and tells Tyche the chosen value.

Stage 3: Finally, Tyche chooses $\mathbf{y}^{\mathrm{obs}}$ according to the density $f_{\mathbf{Y}}(\mathbf{y} \mid \mathbf{t}^{\mathrm{obs}}, \mathbf{a}^{\mathrm{obs}})$ which does not involve θ.

The only stage at which θ was involved was in Zeus's choice of $\mathbf{t}^{\mathrm{obs}}$ given the value $\mathbf{a}^{\mathrm{obs}}$ of $\mathbf{A}$. Hence we should base our inference on that stage, that is on the function

$$
\mathbf{t} \mapsto f_{\mathbf{T}\mid\mathbf{A}}(\mathbf{t} \mid \mathbf{a}^{\mathrm{obs}}, \theta).
$$

▶ **Ea. Exercise*.** This was discussed at 187Hb as an illustration of the general rule for location parameters (for derivation of which, see Eb below). We now give a more careful study of this case and its special features, but arrive at the same answer.

Suppose that $Y_1, Y_2, \ldots, Y_n$ are IID, each with the $\mathrm{U}[\theta - \tfrac{1}{2}, \theta + \tfrac{1}{2}]$ distribution. Let $T = \min Y_k$ and $A = \max Y_k - \min Y_k$. Show that (T, A) satisfies the conditions described above. Show that

$$
f_A(a) = n(n-1)a^{n-2}(1-a)I_{[0,1]}(a).
$$

Explain exactly how Zeus and Tyche perform Stages 2 and 3 for this example. For calculating Confidence Intervals, we may regard θ as having the uniform $\mathrm{U}[\max Y_k - \tfrac{1}{2}, \min Y_k + \tfrac{1}{2}]$ distribution.

▶ **Eb. Ancillary Statistics and location parameters.**   Suppose that θ is a location parameter, so that $f(y \mid \theta) = h(y - \theta)$. Assume that f is 'generic', so that $\mathbf{Y}$ is Minimal Sufficient for θ for $\mathbf{Y}$. Recall Order-Statistic notation (Subsection 107L) and let

$$\mathbf{A} = (U_2, U_3, \ldots, U_n) \quad \text{where} \quad U_k := Y_{(k)} - Y_{(k-1)}, \tag{E1}$$

and $T := U_1 := Y_{(1)}$. Clearly,

$$f_{Y_{(1)}, \ldots, Y_{(n)}}\left(y_{(1)}, \ldots, y_{(n)}\right) = n! f\left(y_{(1)} \mid \theta\right) \ldots f\left(y_{(n)} \mid \theta\right) I_{y_{(1)} < \cdots < y_{(n)}}.$$

So, since $y_{(k)} = u_1 + \cdots + u_k$ and $J = 1$,

$$\begin{aligned} &f_{U_1, U_2, \ldots, U_n}\left(u_1, u_2, \ldots, u_n\right) \\ &= n! f\left(u_1 \mid \theta\right) f\left(u_1 + u_2 \mid \theta\right) \ldots f\left(u_1 + u_2 + \cdots + u_n \mid \theta\right) I_{\{u_k \geq 0, \ k \geq 2\}}. \end{aligned}$$

Now, the integral of the right-hand side relative to u_1 is

$$\begin{aligned} &n! \int_{u_1 \in \mathbb{R}} h(u_1 - \theta) h(u_1 + u_2 - \theta) \ldots h(u_1 + \cdots + u_n - \theta) \mathrm{d}u_1 \\ &= n! \int_{x \in \mathbb{R}} h(x) h(x + u_2) \ldots h(x + u_2 + \cdots + u_n) \mathrm{d}x \\ &=: I(u_2, \ldots, u_n), \end{aligned}$$

which is independent of θ – the whole point about Ancillary Statistics! If we condition on $U_2, \ldots, U_n$, then we make our inferences about θ depend only on

$$f_{U_1 \mid U_2, \ldots, U_n}\left(u_1 \mid u_2, \ldots, u_n\right) = n! I(u_2, \ldots, u_n)^{-1} \mathrm{lhd}(\theta; \mathbf{y}),$$

as claimed in Subsection 186Ha.

Tyche produces a sequence with the same probabilistic law as that at (E1) without knowing the value of θ (but, of course, *knowing* the function h). Given Tyche's results, Zeus can choose the value $y_{(1)}^{\mathrm{obs}}$ of $Y_{(1)}$ to be observed (or else, for example, the value $\overline{y}_{\mathrm{obs}}$ of $\overline{Y}$ to be observed) according to the conditional pdf of $Y_{(1)}$ (or of $\overline{Y}$) given the actual sequence

$$\mathbf{a}^{\mathrm{obs}} : y_{(2)}^{\mathrm{obs}} - y_{(1)}^{\mathrm{obs}}, \ y_{(3)}^{\mathrm{obs}} - y_{(2)}^{\mathrm{obs}}, \ \ldots, \ y_{(n)}^{\mathrm{obs}} - y_{(n-1)}^{\mathrm{obs}}, \tag{E2}$$

produced by Tyche; and then the entire actual sequence $y_{(1)}^{\mathrm{obs}}, y_{(2)}^{\mathrm{obs}}, \ldots y_{(n)}^{\mathrm{obs}}$ of order statistics is determined. Utilizing a random permutation of these, Tyche finally produces the actual set $y_1^{\mathrm{obs}}, y_2^{\mathrm{obs}}, \ldots, y_n^{\mathrm{obs}}$ of observations. We therefore should condition on the sequence at (E2) in producing a CI for θ. The random permutation employed by Tyche can never be known to us.

▶ **Ec. Important comment.**   Of course, when the pdf $f(y \mid \theta)$ is of a special structure which supports lower-dimensional Sufficient Statistics, we can approximate this pdf arbitrarily closely by a 'generic' one. So, *if any kind of continuity applies, it cannot matter what the Minimal Sufficient Statistic is for a* *location-parameter model: we must get the right answer from the formula for the generic case. Note that in all contexts, not just location-parameter ones, this point gives Bayesian Statistics a great advantage over Frequentist in that Bayesian theory handles every distribution the same way, not being concerned with fragile properties such as possessing a sufficient statistic, properties lost under more-or-less any small perturbation of the model.*

People would wish me to remind you again at this point of the 'problem of priors', always the difficulty with the Bayesian approach.

▶ **F. Conditional independence for 'continuous' Variables.** Suppose that $(X_1, X_2, \ldots, X_r, Y_1, Y_2, \ldots, Y_s)$ is a 'continuous' $\mathbb{R}^{r+s}$-valued RV. We say that $X_1, X_2, \ldots, X_r$ are conditionally independent given **Y** if

$$f_{X_1, X_2, \ldots, X_r | \mathbf{Y}}(x_1, x_2, \ldots, x_r \,|\, \mathbf{y})$$
$$= f_{X_1|\mathbf{Y}}(x_1 \,|\, \mathbf{y}) f_{X_2|\mathbf{Y}}(x_2 \,|\, \mathbf{y}) \ldots f_{X_r|\mathbf{Y}}(x_r \,|\, \mathbf{y}).$$

Fa. Exercise. We take Y to be 1-dimensional for simplicity. Show that we can easily recognize conditionally independence because it arises if and only if we have a factorization of type

$$f_{X_1, X_2, \ldots, X_r, Y}(x_1, x_2, \ldots, x_r \,|\, y) = g_1(x_1, y) g_2(x_2, y) \ldots g_r(x_r, y) h(y)$$

for some functions $g_1, g_2, \ldots, g_r, h$. (I do know that the h can be absorbed.) Show that then

$$f_{X_k|Y}(x_k \,|\, y) = \frac{g_k(x_k, y)}{G_k(y)}, \quad \text{where } G_k(y) = \int g_k(x_k, y) \mathrm{d}x_k,$$

and

$$f_Y(y) = G_1(y) G_2(y) \ldots G_r(y) h(y).$$

7.5   Multi-parameter Bayesian Statistics

From one point of view, theoretical Bayesian Statistics still consists of just the one formula, even in the multi-parameter context. I shall explain **Gibbs sampling** as one important technique amongst a number of brilliant Markov Chain Monte Carlo (**MCMC**) techniques which may be used to put THE formula into practice, allowing us to estimate the posterior distribution of a multi-dimensional parameter θ. I shall also describe the impressive BUGS program which gives easy access to the use of such techniques. This program allows one to study models of almost limitless complexity, something I find as disturbing as it is exciting. But there are new factors to be borne in mind when dealing with multi-parameter situations; and these alert one to dangers of misuse of programs.

▶▶ **A. THE formula, again.**   The situation is this. We have Pre-Statistics $Y_1, Y_2, \ldots, Y_n$ which in the Frequentist approach are IID RVs each with pdf $f(\cdot \mid \boldsymbol{\theta})$, where $\boldsymbol{\theta} = (\theta_1, \theta_2, \ldots, \theta_c)$ is a multi-dimensional parameter. (For example, $Y_1, Y_2, \ldots, Y_n$ might be IID each $\mathrm{N}(\mu, \mathrm{prec}\ \rho)$ and $\boldsymbol{\theta} = (\mu, \rho)$.) In the Bayesian approach, we consider the parameter as being an RV $\boldsymbol{\Theta}$ with prior pdf $\pi(\boldsymbol{\theta})$, and use the model that, conditionally on the value $\boldsymbol{\theta}$ of $\boldsymbol{\Theta}$, $Y_1, Y_2, \ldots, Y_n$ are IID each with pdf $f(y \mid \boldsymbol{\theta})$, so that

$$f_{\mathbf{Y}|\boldsymbol{\Theta}}(\mathbf{y}|\boldsymbol{\theta}) = \mathrm{lhd}(\boldsymbol{\theta}; \mathbf{y}) := \prod_k f(y_k \mid \boldsymbol{\theta}).$$

Then THE formula, a special case of 259(A3) asserts that

$$\pi(\boldsymbol{\theta} \mid \mathbf{y}) \propto \pi(\boldsymbol{\theta})\,\mathrm{lhd}(\boldsymbol{\theta}; \mathbf{y}). \qquad \textbf{(THE formula!)} \qquad (\mathrm{A1})$$

But now, we understand that

$$\pi(\boldsymbol{\theta} \mid \mathbf{y}) = f_{\boldsymbol{\Theta}|\mathbf{Y}}(\boldsymbol{\theta} \mid \mathbf{y}).$$

All this was discussed in Subsection 200A.

The difficulty in implementing this idea is that the 'constant of proportionality' (actually, a function of $\mathbf{y}$) involved in (A1) is $1/K(\mathbf{y})$, where

$$K(\mathbf{y}) = \int_{\boldsymbol{\theta}} \pi(\boldsymbol{\theta})\,\mathrm{lhd}(\boldsymbol{\theta}; \mathbf{y})\,\mathrm{d}\boldsymbol{\theta},$$

(what physicists call the *partition function*); and this integral might be very hard to evaluate – even numerically – in complex cases. Gibbs sampling and other similar techniques allow us to estimate $\int_\Gamma \pi(\boldsymbol{\theta} \mid \mathbf{y})\mathrm{d}\boldsymbol{\theta}$ for any nice set Γ *without* evaluating the integral $K(\mathbf{y})$. Of course, $K(\mathbf{y})$ is the 'prior predictive pdf' for $\mathbf{Y}$, the absolute pdf for $\mathbf{Y}$ if $\boldsymbol{\Theta}$ were indeed chosen by God according to the pdf $\pi(\cdot)$.

We look at a number of examples – in this chapter *and the next*.

Aa. Sufficient Statistics in Bayesian theory.   Suppose that $\boldsymbol{\theta}$ is a multi-dimensional parameter, that $\boldsymbol{\tau} : \mathbb{R}^n \to \mathbb{R}^k$ and that for $\mathbf{y} \in \mathbb{R}^n$ (compare 183D1)

$$\mathrm{lhd}(\boldsymbol{\theta}; \mathbf{y}) = g(\boldsymbol{\theta}, \mathbf{t})h(\mathbf{y}) = g(\boldsymbol{\theta}, \boldsymbol{\tau}(\mathbf{y}))h(\mathbf{y}),$$

where $\mathbf{t} = \boldsymbol{\tau}(\mathbf{y})$ and where $h(\mathbf{y})$ does not involve $\boldsymbol{\theta}$. Then, with functions of $\mathbf{y}$ being absorbed into the $\propto$,

$$f_{\boldsymbol{\Theta}|\mathbf{Y}}(\boldsymbol{\theta} \mid \mathbf{y}) \propto \pi(\boldsymbol{\theta})\,\mathrm{lhd}(\boldsymbol{\theta}; \mathbf{y}) \propto \pi(\boldsymbol{\theta})g(\boldsymbol{\theta}, \mathbf{t}),$$

so that *the conditional pdf of $\boldsymbol{\Theta}$ given $\mathbf{Y}$ depends on $\mathbf{y}$ only via $\mathbf{t}$.* Our wish to formulate that

$$f_{\boldsymbol{\Theta}|\mathbf{Y}}(\boldsymbol{\theta} \mid \mathbf{y}) = f_{\boldsymbol{\Theta}|\mathbf{T}}(\boldsymbol{\theta} \mid \mathbf{t})$$

is frustrated by technical obstacles related to those mentioned at the start of Subsection 261C. These can be overcome – but not here!

▶ **B. The $N(\mu, \text{prec } \rho)$ case when μ and ρ are both unknown.** (The Frequentist study of this case needs extra technology and has to wait until the next chapter, though our use of the 'Frequentist prior' will tell us what results to expect.)

We take the

Model: $Y_1, Y_2, \ldots, Y_n$ are IID, each $N(\mu, \text{prec } \rho)$. Then,

$$\text{lhd}\left(\mu, \rho; y_1^{\text{obs}}, y_2^{\text{obs}}, \ldots, y_n^{\text{obs}}\right) = (2\pi)^{-\frac{1}{2}n}\rho^{\frac{1}{2}n}\exp\left\{-\tfrac{1}{2}\rho\sum\left(y_k^{\text{obs}} - \mu\right)^2\right\}$$

$$= (2\pi)^{-\frac{1}{2}n}\rho^{\frac{1}{2}n}\exp\left\{-\tfrac{1}{2}\rho\, n\left(\overline{y}_{\text{obs}} - \mu\right)^2\right\}\exp\left\{-\tfrac{1}{2}\rho\, q(\mathbf{y}^{\text{obs}}, \mathbf{y}^{\text{obs}})\right\},$$

where

$$q(\mathbf{y}^{\text{obs}}, \mathbf{y}^{\text{obs}}) := \sum\left(y_k^{\text{obs}} - \overline{y}_{\text{obs}}\right)^2. \tag{B1}$$

This is part of a systematic notation we shall develop for 'quadratics'.

Let M and R be μ and ρ, viewed as RVs in Bayesian fashion. Let us take as

Prior: $M \sim N(m, \text{prec } r)$, $R \sim \text{Gamma}(K, \text{rate } \alpha)$, M, R independent.

Then, the posterior pdf $\pi(\mu, \rho \mid y^{\text{obs}})$ is proportional as a function of (μ, ρ) to

$$\exp\left\{-\tfrac{1}{2}r(\mu - m)^2\right\}\rho^{K-1}e^{-\alpha\rho} \times \rho^{\frac{1}{2}n}\exp\left\{-\tfrac{1}{2}\rho\sum\left(y_k^{\text{obs}} - \mu\right)^2\right\}. \tag{B2}$$

Ba. Taking the Frequentist prior. If we take the Frequentist prior with $m = r = K = \alpha = 0$, then

$$\pi(\mu, \rho \mid \mathbf{y}^{\text{obs}}) \propto \rho^{\frac{1}{2}n-1}\exp\left\{-\tfrac{1}{2}n\rho\left(\overline{y}_{\text{obs}} - \mu\right)^2 - \tfrac{1}{2}\rho q(\mathbf{y}^{\text{obs}}, \mathbf{y}^{\text{obs}})\right\}. \tag{B3}$$

Integrate out μ (keeping the $N(\overline{y}_{\text{obs}}, \text{prec } n\rho)$ pdf in mind) to see that

$$\pi(\rho \mid \mathbf{y}^{\text{obs}}) = \int_\mu \pi(\mu, \rho \mid \mathbf{y}^{\text{obs}})\,d\mu \propto \rho^{\frac{1}{2}(n-3)}\exp\left\{-\tfrac{1}{2}\rho q(\mathbf{y}^{\text{obs}}, \mathbf{y}^{\text{obs}})\right\}.$$

This gives the following result (corresponding to a central result in the Frequentist theory of the next chapter)

$$q(\mathbf{y}^{\text{obs}}, \mathbf{y}^{\text{obs}})R \sim \chi^2_{n-1} = \text{Gamma}(\tfrac{1}{2}(n-1), \text{rate } \tfrac{1}{2}). \tag{B4}$$

If we integrate out ρ in (B3) (with the pdf of the $\text{Gamma}(\frac{1}{2}n, \text{rate } \frac{1}{2}\sum(y_k^{\text{obs}} - \mu)^2)$ distribution in mind), we obtain

$$\pi(\mu \mid \mathbf{y}^{\text{obs}}) \propto \left\{n\left(\overline{y}_{\text{obs}} - \mu\right)^2 + q(\mathbf{y}^{\text{obs}}, \mathbf{y}^{\text{obs}})\right\}^{-\frac{1}{2}n},$$

which, as looking at 253(G2) makes us realize, states that

$$\frac{(M - \overline{y}_{\text{obs}})\sqrt{n}}{\hat{\sigma}_{n-1}^{\text{obs}}} \sim t_{n-1}, \tag{B5}$$

where $\hat{\sigma}_{n-1}^{\text{obs}}$ is the estimate of standard deviation σ (of which more later):

$$\hat{\sigma}_{n-1}^{\text{obs}} := \left(\frac{q(\mathbf{y}^{\text{obs}}, \mathbf{y}^{\text{obs}})}{n-1} \right)^{\frac{1}{2}}. \tag{B6}$$

Results (B5) and (B6) give us the CIs associated with the standard 't' method, of which more in the next chapter.

▶▶ C. The Gibbs sampler for this case.

The function at 267(B2) is a little complicated, and we would need to integrate it over (μ, ρ) to get the normalizing function of $\mathbf{y}^{\text{obs}}$.

However, on looking at 267(B2), we see that, with obvious notation,

Conditionally on $R = \rho$ (and the known $\mathbf{y}^{\text{obs}}$),

$$M \sim \mathrm{N}\left(\frac{rm + n\rho \overline{y}_{\text{obs}}}{r + n\rho}, \text{ prec } (r + n\rho) \right). \tag{C1}$$

Compare the top row of Table 211 J(i). Also,

Conditionally on $M = \mu$,

$$R \sim \mathrm{Gamma}\left(K + \tfrac{1}{2}n, \text{ rate } \left[\alpha + \tfrac{1}{2} \sum (y_k^{\text{obs}} - \mu)^2 \right] \right). \tag{C2}$$

Compare the last row of Table 211 J(i).

Thus, the conditional probabilities are rather simple.

The idea of the Gibbs sampler, as for all MCMCs, is that **being given a sample with Empirical Distribution close to a desired distribution is nearly as good as being given the distribution itself.**

For our present case, the Gibbs sampler works as follows.

Choose an initial value $\rho[0]$ arbitrarily, say $\rho[0] = K/\alpha$.
Set $c = 0$.
REPEAT
Choose $\mu[c + 1]$ according to the $\pi(\cdot \mid \rho[c])$ density described at (C1),
Choose $\rho[c + 1]$ according to the $\pi(\cdot \mid \mu[c + 1])$ density described at (C2),
UNTIL $c = C$.

Here, we have always 'updated' μ before ρ. We could randomize the order of updating.

Since we have 'C' programs for simulating both Normal and Gamma variables, it is easy to write a bare-hands 'C' program to implement this Gibbs

sampler. Such bare-hands programs are given for other examples later. You can
return and do this one as exercise.

▶▶▶ **Ca. Fact.** For a 'burn-in' b and a 'gap' g, the Empirical Distribution of

$$\{(\mu[c], \rho[c]) : c = b + tg, \ t = 1, 2, \ldots, N\}$$

is close to the distribution with pdf $\pi(\mu, \rho \,|\, \mathbf{y}^{\text{obs}})$. More precisely, the following
ergodic property holds: for a nice function h, we have with probability 1 as
$N \to \infty$,

$$\frac{1}{N} \sum_{t=1}^{N} h\big(\mu[b + tg], \rho[b + tg]\big) \ \to \ \int h(\mu, \rho)\pi(\mu, \rho \,|\, \mathbf{y}^{\text{obs}})\mathrm{d}\mu\mathrm{d}\rho.$$

See the introduction to Gibbs sampling in Norris [176] and the articles by
Roberts and by Tierney in Gilks, Richardson and Spiegelhalter [94]. Ideally, of
course, we want a result which gives with probability 1 the required convergence
simultaneously for a wide class of functions h; and the appropriate 'ergodic'
theorems exist.

The burn-in time is meant to arrange that the sampler is close to reaching
the desired 'steady state'. A nice feature of the Gibbs sampler is that if our
primary concern is with μ (say), in which case ρ is a **nuisance parameter**, then
the Empirical Distribution Function of

$$\{\mu[c] : c = b + tg, \ t = 1, 2, \ldots, N\}$$

will be close to the DF with $\pi(\mu \,|\, \mathbf{y}^{\text{obs}})$ as pdf.

The idea of the gap is to stop a high degree of correlation between successive values
of the pair $(\mu[c], \rho[c])$, but it usually does not matter if we take $g = 1$. Diagnostics
from the theory of **time series** are often used to assess the degree of correlation between
$(\mu[c_1], \rho[c_1])$ and $(\mu[c_2], \rho[c_2])$ when c_1 and c_2 are moderately spread apart. The real
point however is that *we want to be sure that the values $(\mu[c], \rho[c])$ wander over the full
range of reasonably likely values*. We shall see later that we can fail to get the desired
'wandering' even for simple situations.

MCMC theory, originally developed by physicists, has blossomed into a huge field
within Bayesian Statistics. The seminal papers are those by Metropolis *et al* [164] and by
Hastings [110]. We take only the briefest look to get the flavour, and we only consider
the (perhaps most intuitive) Gibbs-sampler algorithm, a special case of the Metropolis–
Hastings algorithm, use of which in Statistics owes much to Ripley [195] (utilising work
of Kelly and Ripley, and of Preston) and to Geman and Geman [91]. I have neither the
space nor the expertise to give you a full treatment of Gibbs sampling here. We shall look
at a number of examples, and learn via them.

▶ **D. `WinBUGS` applied to our $N(\mu, \text{prec } \rho)$ model.**   The `BUGS` program (Bayesian analysis Using the Gibbs Sampler) was developed by a team (led by David Spiegelhalter) at the Medical Research Council Biostatistics Unit at Cambridge University, and is now being further developed by them in collaboration with the Department of Epidemiology and Public Health, Imperial College School of Medicine, London.   Numerous people (DS, Wally Gilks, Nicky Best, Andrew Thomas, David Clayton, ...) have contributed significantly to the program. As mentioned in Appendix D, the program is structured along the lines of graphical models and Markov random fields.

Since (I make no comment!) Windows is currently the most widely used operating system (though not the one I use), I describe `WinBUGS`, the version for Windows which may be downloaded – FREE when I got mine, but possibly not by the time you read this – via [238].

`WinBUGS` **likes to simulate from a density** (of one parameter given the data and all other parameters) **which is logconcave,** so that it can utilize the Gilks–Wild Adaptive Rejection Sampling technique mentioned at 139Ib. But the program uses a variety of other sampling techniques, including Metropolis sampling. Brilliant as `WinBUGS` is, it is a good strategy to try out your programs first on data **simulated** using the model – where you *know* what the answers are. This will indicate if something is going wrong.

▶ **Two other good things about** `WinBUGS`. An excellent set of examples comes with the package. The package also enables you to join a User Group. If you do so, you can see problems posed by others and their solutions, and, if necessary, you can pose your own problems. As always you should do this last only after studying the online manual and examples carefully.

Let's get the computer to simulate a random sample of size 10 from a $N(0,1)$ distribution:

$$1.21, \ 0.51. \ 0.14, \ 1.62, \ -0.80, \ 0.72, \ -1.71, \ 0.84, \ 0.02, \ -0.12, \qquad \text{(D1)}$$

which we shall use as data. I am taking a proper prior 'close' to the improper Frequentist prior for which $m = r = K = \alpha = 0$, choosing my values of these parameters as suggested by the 'online' `WinBUGS` manual.

What you type in is in `typewriter` font; so are the results. Comments made by the machine are in *italic* font. Things you do with mouse are <u>underlined</u>. I have put some headings in **bold** font to structure things. It is always possible that slight modifications of my instructions will be needed for more recent versions of `WinBUGS`.

Model:
Do <u>File – New</u> to create a new window. In it, type
```
model{
    for(i in 1:n){
        y[i] ~ dnorm(mu, rho)                        Yᵢ ~ N(μ, prec ρ)
```
$Y_i \sim N(\mu, \text{prec } \rho)$

```
    }
    mu ~ dnorm(m, r)                        M ~ N(m, r)
    rho ~ dgamma(K, alpha)                  R ~ Gamma(K, rate α)
               (Independence of M and R under prior assumed by machine)
    }
```

Model – Check model *(model is syntactically correct)*

Data:
Make another new window in which you type
```
    list(n=10, y=c(1.21, 0.51, ... , 0.02, -0.12),
        m=0.0, r = 1.0E-6, K=1.0E-3, alpha = 1.0E-3)
```
$$\text{So, } n = 10, \text{ y is as at 270(D1)},$$
$$m = 0, r = 10^{-6}, K = \alpha = 10^{-3}.$$

Do Model – Data *(data loaded)*

Compilation:
Model – Compile *(model compiled)*

Every 'stochastic element' described in a '`name   d...`' statement and not already initialized in the 'Data' statement, must now be initialized.

Initial values of parameters:
Make a new window, and type
```
    list(mu = -1, rho = 2)       (arbitrarily chosen)
```
Model – Inits *(initial values loaded: model initialized)*

(Note: The machine can be asked to carry out the 'non-data' initialization step automatically via Gen inits.)

Do a burn-in:
Model – Update
(updates =) 1000 (refresh=) 100   (**refresh** explained below)
update
Note: Update Tool window remains in view

Say what to monitor:
Statistics – Samples (node =) mu (in dialog box)
set
Notes: You can ask WinBUGS also to monitor other nodes simultaneously. Sample Monitor Tool window remains in view.

Begin the monitored sampling used in the analysis:
Model Update
(updates =) 10000 (refresh=) 100 update
stats   (not all of which I quote)

```
node    mean      sd     2.5%  97.5%  start   sample
  mu   0.239   0.356  -0.4607 0.9469   1001    10000
```

The `mean` is the mean of the sample of the $\mu[c]$, and the 2.5% and 97.5% percentiles are obtained from the empirical distribution of the $\mu[c]$ values. Thus BUGS is suggesting the 95% Confidence Interval $(-0.4607, 0.9469)$ for μ. The Frequentist CI obtained by the famous 't' method explained in the next chapter (already obvious from 267(B5)) is $(-0.456, 0.942)$ which may be obtained from `Minitab`. The `Minitab` commands are

```
MTB > set c1                          Put in column 1 the following data.
DATA> 1.21 0.51 0.14 ... 0.02 -0.12
DATA> end
MTB > tint c1                         Obtain a 95% CI for μ using 't' method.
      N    MEAN  STDEV  SE MEAN   95.0 PERCENT CI
     10   0.243  0.976   0.309   (-0.456, 0.942)
```

`Minitab`'s `STDEV` is the value $\hat{\sigma}_{n-1}^{\text{obs}}$ on the left of 268(B6), and `SE MEAN` denotes $\hat{\sigma}_{n-1}^{\text{obs}}/\sqrt{n}$.

You could also use R as follows:
```
x <- c(1.21, 0.51, ..., -0.12)
ttest(x)
```
to get the same results.

WinBUGS will show you the evolution of the sampled values of $\mu[c]$, $\rho[c]$ in windows of 'width' `refresh`. Try different values of this parameter. WinBUGS will plot the estimated posterior pdf for μ or for ρ, etc, etc. For people who wish to study correlations, etc, in detail, BUGS may be used in conjunction with the CODA diagnostics package which may also be downloaded via the Internet (same address). This package is designed to work with R and/or S-PLUS.

Da. The MLwiN package. The very useful package MLwiN mentioned in the Preface is available from [166] – at a cost.

E. WinBUGS for our Cauchy example. The '$n = 3$' case in Figure 196E(i) shows a tri-modal (three-peaked) posterior pdf for the Frequentist uniform prior $\pi(\theta) = 1$ for the Cauchy density

$$f(y \,|\, \theta) = \left\{ \pi \left(1 + (y - \theta)^2 \right) \right\}^{-1}. \tag{E1}$$

It is an interesting case on which to test out WinBUGS , to see how soon the program recognizes that there are three peaks. Now, my current version of WinBUGS doesn't seem to accept the Cauchy (or t_1) distribution.

However, we can utilize Exercise 255Ge, with $\nu = 1$. We can produce a variable Y with density as at (E1) via

$$R \sim \text{Gamma}(\tfrac{1}{2}, \text{rate } \tfrac{1}{2}), \qquad (Y|R, \Theta) \sim \text{N}(\theta, \text{prec } R).$$

So, we set up a WinBUGS model

```
model{
  theta ~ dnorm(0.0, 1.0E-6)                              (Vague prior)
  for (i in 1:N){
    R[i] ~ dgamma(0.5,0.5)                          $R_i \sim \text{Gamma}(\frac{1}{2}, \frac{1}{2})$
    Y[i] ~ dnorm(theta, R[i])              $(Y_i \mid R_i, \Theta) \sim \text{N}(\Theta, \text{prec } R_i)$
  }
}
```

with data

```
list(N = 3, Y=c(-5.01, 0.4, 8.75))
```

and initial values

```
list(theta = 0,  R = c(1, 1, 1))
```

We use WinBUGS exactly as in Subsection 270D. The tri-modal nature of the posterior pdf very soon emerges.

Ea. 'Bare-hands' Gibbs sampling in 'C'. Suppose that we take the improper prior $\pi(\theta) = 1$ for θ. Then (you check!)

$$f_{\Theta,\mathbf{R},\mathbf{Y}}(\theta, \mathbf{r}, \mathbf{y}) \;=\; \prod \left\{ (2\pi)^{-1} \exp\left(-\tfrac{1}{2}r_i[1 + (y_i - \theta)^2]\right) \right\},$$

so that *given* Θ *and* $\mathbf{Y} = \mathbf{y}$, *the variables* R_1, R_2, R_3 *are independent,*

$$(R_i \mid \Theta, \mathbf{Y}) \;\sim\; \text{E}\left(\text{rate } \tfrac{1}{2}[1 + (Y_i - \theta)^2]\right),$$

and

$$(\Theta \mid \mathbf{R}, \mathbf{Y}) \;\sim\; \text{N}\left(\frac{\sum r_i Y_i}{\sum r_i}, \text{prec } \sum r_i \right).$$

You should check that the following will therefore work. (Recall from Subsection 136H that Expl(lambda) simulates an E(rate λ) variable, and, from Subsection 160Ca that PRECnormal(mu, rho) simulates an N(μ, prec ρ) variable.)

```
theta = 0.0;
for(c=1; c<=Nobs + 100; c++){
  rysum = rsum = 0.0;
  for(n=0; n<3; n++)
    r[n] = Expl(0.5 * (1.0 + (y[n] - theta)*(y[n] - theta)));
  for(n=0; n<3; n++){
    rysum = rysum + r[n]*y[n]; rsum = rsum + r[n];
  }
  mu = rysum/rsum,
  theta = PRECnormal(mu, rsum);
}
```

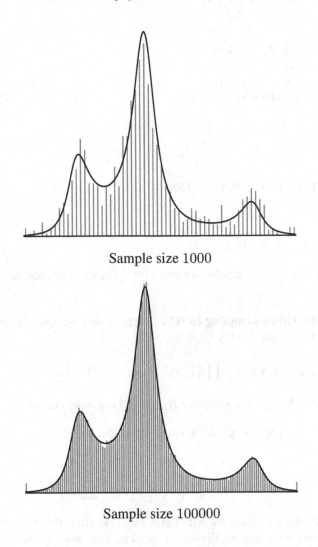

Sample size 1000

Sample size 100000

Figure E(i): 'Bare-hands in C' Gibbs sampler for a Cauchy model

the various theta values for $c > 100$ giving the desired sample. The results for a sample of size 1000 with a 64-bin histogram and for a sample of size 100000 with a 128-bin histogram are shown in Figure E(i). The curve shows the exact posterior pdf for the uniform prior. (The pictures are from my C-to-Postscript converter.) Even the '100000' case took at most 3 seconds on a 200Mhz PC.

As we saw at Figure 138I(i), the end bins of the histogram carry estimated tail probabilities. If concerned with just a few parameters, I prefer to watch histograms evolving than to look at the time series which show the successive simulated values of parameters.

▶ **Eb. Important Warning.** Before you become too excited about the success of Figure 274E(i), consider the following. We have added as many nodes R_i to the sampler as there observations Y_i, an extremely 'non-parsimonious' strategy in general. We should not therefore be surprised if there are problems even for moderately large sample sizes. See 353Hf below.

F. Exercise. Suppose that

$$X_1, X_2, \ldots, X_n \sim \text{Poisson}(\lambda),$$
$$Y_1, Y_2, \ldots, Y_r \sim \text{Poisson}(\mu),$$

$X_1, X_2, \ldots, X_n, Y_1, Y_2, \ldots, Y_r$ being independent. Assuming the reference priors for λ and μ (you can only do this to a high degree of approximation), find an approximate 95% CI for λ/μ on WinBUGS .

Looking at Exercise 100Ia for a clue, determine the distribution of $S/(S+T)$, where $S = \sum X_i$ and $T = \sum Y_j$. Does that give you a Frequentist method of dealing with this problem?

▶▶ **G. Advantages of, and problems with, MCMC methods.** *One of the great advantages of MCMC methods is that they can deal with much more realistic models than will the classical methods.* However there is still value in the old idea of using *parsimonious* models: models with no more parameters than is really necessary: there is nothing clever in introducing loads of new parameters just for the sake of it. (And you have already been warned about the difficulties which can arise if you introduce lots of new nodes to implement the Gibbs sampler.) On the other hand, in image restoration for example, there just *have* to be loads of unknown parameters (the colours of all the pixels. etc).

An obvious problem with MCMC methods is that they *are* simulations, not exact answers. In some situations, convergence to the posterior distribution can be very slow. Worse, in others, it is quite possible for the system to appear to have converged to some distribution, but the wrong one! This is when the simulated values of θ get stuck within some region of the parameter space. One should always do a number of runs – with different initial values and different seeds. One should also always investigate how robust things are to small changes in priors. See Gilks, Richardson & Spiegelhalter [94], especially the article by Gelman, and Besag [18] (the text of a well-attended lecture to the Royal Statistical Society).

In fact, there is a remarkable new technique, called *perfect simulation*, discovered by Propp and Wilson at M.I.T., which can produce one value of θ definitely chosen from exactly the correct posterior distribution. When it works acceptably quickly (which it does not always do), it certainly resolves questions regarding burn-in time in MCMC methods. But it is usually hardly feasible to run the whole Propp-Wilson algorithm independently a huge number of times to get the desired perfect sample. Nor can I see how it avoids the 'sticking' phenomenon: what matters is *ergodicity* rather than correctness of distribution of an individual element of a sample. See Propp & Wilson [189], and the informative site

> http : //dimacs.rutgers.edu/ dbwilson/exact.html/

As mentioned previously, Gilks, Richardson and Spiegelhalter [94] and the series [16] edited by Bernardo, Berger, Dawid, and Smith make for excellent reading of theory and practice.

▶ **H. Warning: priors for precisions.**   In hierarchical (or random-effect) models, the usual parameters are themselves regarded as 'random' with distributions determined by hyperparameters which themselves have prior distributions. These hyperparameters are therefore two steps away from the Observations; and we have to be careful about assigning vague priors to hyperparameters in the way we have done for parameters.

It is fairly standard to use $\pi(\rho) = \rho^{-1}$ for the vague prior for a parameter representing a precision, so that $\ln(\rho)$ is uniform on $\mathbb{R}$. However, assigning prior $\pi(\tau) = \tau^{-1}$ to a hyperparameter representing a precision can lead to improper posterior densities, something we certainly wish to avoid.

Of course, we should never use improper priors. However, it is as well to be aware of situations where posterior densities can have similar tail-off to prior ones, and which could lead to undesirable behaviour of the Gibbs sampler in certain (probably rather extreme) cases. Even when our chief concern is with parameters, we should monitor what is happening to hyperparameters.

In the following discussion, I shall treat certain parameters as Random Variables, using Θ rather than θ. This has the effect of promoting hyperparameters to parameters.

Let's first consider a ridiculous, but instructive, situation. Suppose that an Observation Y is given by

$$Y = \Theta + \varepsilon, \quad \Theta \sim \mathrm{N}(\mu, \operatorname{prec}\tau), \quad \varepsilon \sim \mathrm{N}(0, 1),$$

Θ and ε being independent. Suppose that the prior $\pi(\mu, \tau)$ for (μ, τ) has the form $\pi(\tau)$ so that μ has uniform prior. Since we cannot then hope to pick up any information about precision from one observation, we would expect

$$\pi(\tau \,|\, y) = \pi(\tau).$$

You calculate $\pi(\mu, \tau \,|\, y)$ and confirm our conjecture. In particular, if $\pi(\tau)$ is improper, so will be $\pi(\tau \,|\, y)$. Here, the true precision of the Observation Y is $\tau/(\tau + 1)$, so it is not surprising that the reference prior for τ is $\{\tau(\tau + 1)\}^{-1}$, corresponding to the fact that $\ln(\tau/(\tau + 1))$ has uniform density on $(-\infty, 0)$; and this reference prior leads to a proper posterior density.

▶ **Ha. Exercise.** Consider the very important model (where $1 \leq j \leq J$ and $1 \leq k \leq K$)

$$Y_{jk} = \Theta_j + \varepsilon_{jk}, \qquad \Theta_j \sim \mathrm{N}(\mu, \text{ prec } \tau), \qquad \varepsilon_{jk} \sim \mathrm{N}(0, \text{ prec } \rho) \qquad \text{(H1)}$$

where all variables Θ_j $(j \leq J)$, ε_{jk} $(j \leq J, k \leq K)$ are independent.

Prove that for the situation described at (H1), the maximum-likelihood estimate $(\hat{\mu}, \hat{\tau}, \hat{\theta}_j, \hat{\rho})$ corresponding to possible observations $(y_{jk} : 1 \leq j \leq J, 1 \leq k \leq K)$ satisfy the unsurprising relations:

$$\hat{\mu} = \overline{y}_{**}, \quad \tau^{-1} = \frac{1}{J} \sum (\hat{\theta}_j - \hat{\mu})^2, \quad \hat{\rho}^{-1} = \frac{1}{JK} \sum\sum (y_{jk} - \hat{\theta}_j)^2,$$

$$\hat{\theta}_j = \frac{\hat{\tau}\hat{\mu} + K\hat{\rho}\overline{y}_{j*}}{\hat{\tau} + K\hat{\rho}};$$

but these have to be solved numerically.

Prove that if we take a prior $\pi(\mu, \tau, \rho) = \pi(\tau, \rho)$, then

$$(\theta_j \mid \mu, \tau, \rho, \mathbf{y}) \sim \mathrm{N}\left(\frac{\tau\mu + K\rho\overline{y}_{j*}}{\tau + k\rho}, \text{ prec } (\tau + K\rho)\right),$$

$$(\mu \mid \tau, \rho, \mathbf{y}) \sim \mathrm{N}\left(\overline{y}_{**}, \text{ prec } \frac{\tau K\rho}{\tau + K\rho}\right),$$

$$\pi(\tau, \rho \mid \mathbf{y}) \propto \pi(\tau, \rho) \frac{\tau^{\frac{1}{2}(J-1)} \rho^{\frac{1}{2}(JK-1)}}{(\tau + K\rho)^{\frac{1}{2}(J-1)}} \exp\left\{-\frac{1}{2}\frac{\tau K\rho}{\tau + K\rho} B(\mathbf{y}) - \frac{1}{2}\rho W(\mathbf{y})\right\}$$

where

$$W(\mathbf{y}) := \sum\sum (y_{jk} - \overline{y}_{j*})^2, \qquad B(\mathbf{y}) := \sum (\overline{y}_{j*} - \overline{y}_{**})^2.$$

Note that we will not obtain a proper posterior density if we take $\pi(\tau, \rho)$ of the form $\tau^{-1}\pi(\rho)$. This is because as $\tau \to \infty$,

$$\frac{\pi(\tau, \rho \mid \mathbf{y})}{\pi(\tau, \rho)} \to \text{function}(\rho, \mathbf{y}),$$

so that $\int \pi(\tau, \rho \mid \mathbf{y}) \mathrm{d}\tau$ will be infinite.

Remark. In analogous situations to this, the standard (Jeffreys) reference priors can cause substantially worse difficulties even for parameters (rather than hyperparameters), as we shall see later.

▶ **I. Simultaneous versus individual Confidence Intervals.** Many packages, including BUGS, will produce 95% CIs for each of many parameters separately: we refer to 'individual CIs'. This 'separately' must be remembered. If for $i \in \{1, 2, \ldots, n\}$, we are 95% confident that parameter θ_i is in interval I_i, then we may have nothing like 95% confidence in the 'simultaneous' statement that '$\theta_i \in I_i$ for all i'. (If the statements $\theta_i \in I_i$ may be regarded as 'independent', our confidence in the 'simultaneous' statement is only $(0.95)^n$; but, in theory, without independence, it could be as low as 0 for $n \geq 20$ – Yes?!)

On those occasions when we require a Confidence Region for a vector parameter $\boldsymbol{\theta} = (\theta_1, \theta_2, \ldots, \theta_n)$ there are two serious difficulties. Firstly, it can be rather expensive in computer time and memory to calculate such a CR. Secondly, it is usually difficult to convey the results in a comprehensible way. The best CR will have the form

$$\{\boldsymbol{\theta} : \pi(\boldsymbol{\theta} \,|\, \mathbf{y}^{\text{obs}}) \geq c\} \tag{I1}$$

and this will not have the comprehensible product form

$$\{\boldsymbol{\theta} : \theta_i \in I_i, \ i = 1, 2, \ldots, n\}. \tag{I2}$$

Since product forms, simultaneous CIs, are the only easy-to-understand ones, we often use non-optimal CRs of product form.

As mentioned earlier, *if we only have two parameters* $\boldsymbol{\theta} = (\theta_1, \theta_2)$, *then if we find 97.5% CIs* I_1 *and* I_2 *for* θ_1 *and* θ_2 *separately, then* $I_1 \times I_2$ *is an at-least-95% CR for* $\boldsymbol{\theta}$. So, we can get *conservative CRs* this way.

Simultaneous CIs will be mentioned in the next subsection and discussed in the next chapter.

▶ **J. Extreme data viewed after an experiment.** Suppose that $M_1, M_2, \ldots, M_n$ are 'performance indicators' which represent on some scale how good hospitals in cities $1, 2, \ldots, n$ are. (Well, ... , assume that such a concept makes some sense. Of course, a performance indicator must take into account the severity of cases brought to the hospital in question, etc.) Quantity M_i is subject to a measurement error ε_i, so that the observed value (measurement) for city i is $Y_i = M_i + \varepsilon_i$. The ε_i might represent the fact that people may recover or die for reasons which have little to do with their treatment. A journalist might focus on the hospital with the lowest *measurement* (which is not necessarily the one with the lowest M-value).

Ja. One aspect. Just suppose that God knows that $M_i = 0$ for $1 \leq i \leq n$, and that, as we assume, the ε_i are IID each $N(0, 1)$. Then,

$$\mathbb{P}(Y_i \leq -1.645) = 0.05 \text{ for every } i,$$

but

$$\mathbb{P}(\min_i Y_i \leq -1.645) = 1 - (0.95)^n,$$

which will be close to 1 for large n. In this case we would clearly be unfair if we choose the hospital with the lowest measurement and treat it as if it were the only hospital. (*We do not know that all M_i are zero.*)

Jb. An inappropriate use of 'absolute probabilities'. Suppose instead that experience suggests that we use a model where

$$M_1, M_2, \ldots, M_n, \varepsilon_1, \varepsilon_2, \ldots, \varepsilon_n$$

are independent, $M_i \sim N(0, \sigma^2)$, $\varepsilon_i \sim N(0, 1)$, σ^2 being known. Let $Y_i = M_i + \varepsilon_i$, and let W be the value of i which produces the minimum *measurement*:

$$Y_W = \min Y_i.$$

We want a 95% CI for $m_w = M_W(\omega^{\text{act}})$ of the form $(-\infty, c)$.

Here is one way to do this. We can evaluate a constant c such that

$$\mathbb{P}(M_W \le Y_W + c) = 95\%.$$

Then there is no argument about the fact that $(-\infty, y_w^{\text{obs}} + c)$ is a 95% CI for m_w. *But is it the one we should use?* At first sight, it seems to get round the 'unfairness' mentioned in regard to example 278Ja by the use of a kind of simultaneous CI. But it is *not* the appropriate thing to do.

▶ **Jc. 'Conditional' study of the last example.** Bayesian theory tells us that we should condition on the fact that $\mathbf{Y} = \mathbf{y}^{\text{obs}}$. If we know $\mathbf{y}^{\text{obs}}$, then, of course, we know the corresponding w and the lowest observation y_w^{obs}. From the Frequentist point of view, the fact that $Y_1, Y_2, \ldots, Y_n$ are IID RVs each with the $N(0, 1 + \sigma^2)$ distribution which does not depend on the values of $M_1, M_2, \ldots, M_n$ might incline us to regard $Y_1, Y_2, \ldots, Y_n$ as *Ancillary Statistics* and (as in the Bayesian theory) condition on their observed values. See Subsection 263E. Now the pairs $(M_1, Y_1), \ldots, (M_n, Y_n)$ are independent, and for each k,

$$(M_k \mid Y_k) \sim N(\beta Y_k, \beta), \quad \beta := \sigma^2 / (\sigma^2 + 1); \tag{J1}$$

see Exercise 259Ac. It follows that

$$\mathbb{P}\left(W = k; \ M_k \le \beta y_k^{\text{obs}} + 1.645\sqrt{\beta} \ \middle| \ \mathbf{Y} = \mathbf{y}^{\text{obs}} \right) = 0.95 I_{\{y_k^{\text{obs}} < y_j^{\text{obs}}, \ j \ne k\}}.$$

Summing over k yields

$$\mathbb{P}\left(M_W \le \beta y_w^{\text{obs}} + 1.645\sqrt{\beta} \ \middle| \ \mathbf{Y} = \mathbf{y}^{\text{obs}} \right) = 95\%. \tag{J2}$$

Thus, the conditional approach tells us to use

$$\left(-\infty, \ \beta y_w^{\text{obs}} + 1.645\sqrt{\beta} \right]$$

as a CI for m_w, exactly as if the lowest observation had been the only one!! Do note however that the amount $\beta y_w^{\text{obs}} - y_w^{\text{obs}}$ by which y_w^{obs} is increased to estimate the mean of M_W is maximized at the lowest observation.

The discussion at 278Ja may make us somewhat worried about the conclusion just obtained. But remember that in the example there, we essentially had $\sigma = 0$, and the conditional method is giving us $(-\infty, 0]$ as a CI for m_w and $[0, 0]$ for a two-sided CI.

As we shall better appreciate in Chapter 9, result (J2) tells us that we have the *absolute*-probability result:

$$\mathbb{P}\left(M_W \leq \beta Y_W + 1.645\sqrt{\beta} \right) = 95\%,$$

and this is the 'absolute' result we should use, agreeing with the 'conditional' approach to getting CIs.

In case you are worried, you can do a cross-check when $n = 2$. The independence structure yields the fact that

$$f_{M_1, Y_1, Y_2}(m_1, y_1, y_2) = f_{Y_2}(y_2) f_{Y_1}(y_1) f_{M_1|Y_1}(m_1|y_1),$$

and result 279(J1) shows that

$$\int_{m_1 < \beta y_1 + 1.645\sqrt{\beta}} f_{M_1|Y_1}(m_1|y_1) \, dm_1 = 95\%.$$

Thus, we have

$$\mathbb{P}\left(M_W \leq \beta Y_W + 1.645\sqrt{\beta} \right) = 2\mathbb{P}\left(M_1 \leq Y_1 + 1.645\sqrt{\beta}; \; Y_1 < Y_2 \right)$$

$$= 2 \int_{y_2} \int_{y_1 < y_2} \int_{m_1 < \beta y_1 + 1.645\sqrt{\beta}} f_{Y_2}(y_2) f_{Y_1}(y_1) f_{M_1|Y_1}(m_1|y_1) \, dm_1 dy_1 dy_2$$

$$= 2 \int_{y_2} dy_2 \, f_{Y_2}(y_2) \int_{y_1 < y_2} dy_1 \, f_{Y_1}(y_1) \int_{m_1 < \beta y_1 + 1.645\sqrt{\beta}} dm_1 f_{M_1|Y_1}(m_1|y_1)$$

$$= 2 \int_{y_2} dy_2 \, f_{Y_2}(y_2) \int_{y_1 < y_2} dy_1 \, f_{Y_1}(y_1) \times 95\%$$

$$= 2\mathbb{P}(Y_1 < Y_2) \times 95\% = 95\%.$$

Part of the correct intuition is that if Y_W is significantly less than 0, then it is likely that M_W is roughly βY_W and ε_W is roughly $(1 - \beta)Y_W$.

▶ **Jd. Important Warning: a 'Bayesian–Frequentist conflict'.** We can think of the situation just described as corresponding to that in which $Y_i \sim \mathrm{N}(\mu_i, 1)$ and we do a Bayesian analysis supposing that each μ_i is a Random Variable M_i, the prior-distribution statement being that the M_i's are IID, each $\mathrm{N}(0, \sigma^2)$, where σ^2 is taken as known from experience.

Now, we know that, for any fixed i, the Bayesian CI for μ_i will, as $\sigma^2 \to \infty$, converge to the Frequentist CI for μ_i: as $\sigma^2 \to \infty$, we have $\beta \to 1$. However, the Bayesian CI for $\mu_w = m_w$ will converge to

$$(-\infty, y_w^{\mathrm{obs}} + 1.645]$$

which is most certainly **not** a Frequentist CI. The 'Frequentist' probability

$$\mathbb{P}(\mu_W \leq Y_W + 1.645 \,|\, \mu_1, \mu_2, \ldots, \mu_n)$$

is a horrible function of $(\mu_1, \mu_2, \ldots, \mu_n)$ equal (as we saw earlier) to 0.95^n when all μ_i's are zero.

So, what is going wrong? Well, letting $\sigma^2 \to \infty$ for the Bayesian prior is tending to pull the μ_i's far apart *from one another*, something which does not correspond to what the Frequentist or sensible Bayesian wants. Though this does something to resolve the 'Bayesian–Frequentist conflict', the problem still leaves me with a feeling of some unease about the 'Bayesian' result for fixed σ^2, not so much for this example as for its implications for other situations.

Of course, *we should estimate σ^2 from the data and what reliable prior information we have.* This remark does not fully resolve things by any means; but to take this discussion further and to extend to more complex situations is outside the scope of this book. You are alerted to the need for care.

▶ **K. A perplexing Exercise: more 'Bayesian–Frequentist conflict'.**
Suppose that $n \geq 3$ and that $Y_1, Y_2, \ldots, Y_n$ are IID each E(rate θ). Let

$$S := Y_1 + Y_2 + \cdots + Y_n.$$

Show that
(a) the MLE for θ is n/S,
(b) the estimator $(n-1)/S$ is unbiased for θ,
(c) the value α which minimizes the mean-square error

$$\mathbb{E}\left\{(\theta - \alpha/S)^2 \,|\, \theta\right\}$$

is, for every θ, given by $\alpha = n - 2$.

According to what you have so far been led to believe, if we use the vague prior θ^{-1}, the Frequentist prior(!), for this example, then Bayesians and Frequentists agree. This is not entirely true.

Show that if we view θ as a 'random variable' Θ in the usual Bayesian sense, then, conditionally on **Y**, Θ has the Gamma(n, rate S) distribution of mean n/S and variance n/S^2. Hence (see Lemma 69Ja) a Bayesian must regard n/S as the best estimator of Θ in the least-squares (lowest mean-square error) sense.

Let us take $n = 3$ for definiteness. Then a Frequentist will say:

$$\mathbb{E}\left\{\left(\theta - \frac{1}{S}\right)^2 \,\middle|\, \theta\right\} = \frac{1}{5}\mathbb{E}\left\{\left(\theta - \frac{3}{S}\right)^2 \,\middle|\, \theta\right\} \qquad \text{for every } \theta,$$

so that $1/S$ is 5 times better an Estimator of θ than $3/S$ in the least-squares sense. A Bayesian using the Frequentist prior will say that

$$\mathbb{E}\left\{\left(\theta - \frac{3}{S}\right)^2 \,\middle|\, S\right\} = \frac{3}{7}\mathbb{E}\left\{\left(\theta - \frac{1}{S}\right)^2 \,\middle|\, S\right\} \qquad \text{for every } S,$$

so that $3/S$ is $2\frac{1}{3}$ times better an Estimator of θ than $1/S$ in the least-squares sense. Check out these calculations.

But surely, you say, something must be wrong here: things just do not tally. We discuss this in Chapter 9. Use of improper priors is always prone to this kind of 'paradox'. However, there is more involved than improper priors as the Two-Envelopes Problem will illustrate spectacularly.

L. Final Remark. I just wish to reiterate that study of multiparameter Bayesian Statistics, `WinBUGS`, simultaneous CIs, etc, continues within the next chapter.

8

LINEAR MODELS, ANOVA, etc

8.1   Overview and motivation

The main theme of this chapter concerns a completely different aspect of the perfection of the normal distribution from that provided by the CLT. Classical Linear-Model theory hinges on the fact that

- the normal distribution,

- independence, and

- Pythagoras's Theorem

are linked in a fascinating way. Let's meet a simple case first.

▶ **A. The Orthonormality Principle for two dimensions.**   (This story is probably due to Gauss and James Clerk Maxwell.) Let us suppose that X and Y are IID Random Variables each 'continuous' with pdf f which we assume continuously differentiable and strictly positive on $\mathbb{R}$. Suppose that the law of (X, Y) is invariant under rotations about the origin in that for any such rotation τ, $\tau((X, Y))$ has the same pdf on $\mathbb{R}^2$ as (X, Y). It is intuitively clear that the joint pdf $f_{X,Y}(x, y) = f(x)f(y)$ must be a function of the distance from (x, y) to the origin, whence

$$f(x)f(y) \;=\; g(x^2 + y^2).$$

But then, $f(x)f(0) = g(x^2)$, $f(0)^2 = g(0)$, and

$$h(x^2)h(y^2) = h(x^2 + y^2), \quad \text{where } h(x) = g(x)/g(0).$$

However, since $h(r)h(s) = h(r + s)$ for $r, s \geq 0$, then, differentiating with respect to r keeping s fixed, we have $h'(r)h(s) = h'(r + s)$, whence

$$\frac{h'(r + s)}{h(r + s)} = \frac{h'(r)}{h(r)} = \text{constant} = c \text{ (say).}$$

(*Technical Note*. I have assumed that h is differentiable, but by the theory of one-parameter semigroups, this follows already from the measurability of h.) We have shown that $h(r) = h(0)e^{cr}$, and $f(x) = f(0)e^{cx^2}$, whence f must be the pdf of the $N(0, \sigma^2)$ distribution for some $\sigma > 0$. Conversely, *if* (X, Y) *are IID each* $N(0, \sigma^2)$*, then the law of* (X, Y) *is invariant under rotations about the origin.*

We require various n-dimensional generalizations of this **Orthonormality Principle**. All relevant algebra and geometry are revised/explained as they are needed.

▶▶ **B. What we study in this chapter.** The theory of Linear Models is the central part of traditional Statistics culture.

The Orthonormality Principle allows us really to understand (Captain Kirk: 'to really understand') the 't' result derived via Bayes' formula in Subsection 267Ba. It also allows us to study all the **regression/ANOVA models** mentioned in Subsection 28C. As stated there, the Orthonormality Principle marks the perfect tie-up which often exists between likelihood geometry and least-squares methods. We **shall** study ANOVA tables and the associated tests, but we shall be more concerned with Confidence Intervals, 'individual' and 'simultaneous'. Of course, we look at the Bayesian theory too; and, of course, the geometry remains relevant. We look at how `Minitab` and `BUGS` do the calculations, and at

how MCMC methods allow more realistic models to be studied,

though I have to add that things do not always work out as we might hope.

We give some thought to **assessing whether our models are reasonable**, and, because of this, look at **goodness-of-fit** methods: 'chi-squared', quantile-quantile plots, and Kolmogorov(–Smirnov) tests; I prefer the chi-squared amongst these.

We study the **Multivariate Normal (MVN) Distribution** and **correlation as a measure of association**.

Where the tie-up between likelihood geometry and least-squares methods breaks down, and unfortunately it *does* do so, our intuition takes a savage jolt. The most celebrated example, the **shrinkage** phenomenon discovered by Charles Stein, receives illuminating discussion in Stigler [218].

The Mathematics of ANOVA and of the MVN is by far the best way to understand finite-dimensional inner-product spaces. It is sad that courses on

Linear Algebra so often ignore tensor products, partitioned matrices, etc. *I want to emphasize that in this chapter we use Linear Algebra only as a language. No deep result from Linear Algebra is needed here – not that there are that many in that subject!* The 'Extension to Basis' Principle, assumed as a fact, is the deepest result used until we need the diagonalization property of symmetric matrices for the 'Multivariate Normal' section.

The algebra which we learn here will equip us to take a quick look in Chapter 10 at the really mind-blowing situation related to *Bell's inequality* and the *Aspect experiment*. It perplexes *me* that Probability at the level of elementary particles requires very different methods of calculation from those which we have so far studied; and it should worry *you* too.

Conventions. Henceforth, **all vectors are column vectors,** ^{T} stands for 'transposed', **1** stands for the vector

$$\mathbf{1} := (1, 1, \ldots, 1)^T$$

and $\|\mathbf{v}\|$ stands for the length of a vector $\mathbf{v}$: $\|v\|^2 = v_1^2 + v_2^2 + \cdots + v_n^2$ for $\mathbf{v} \in \mathbb{R}^n$.

▶ **C. Providing more motivation.** Many people have suggested that before launching into the (easy but 'abstract') Linear Algebra, I should provide more motivation: that some numbers and pictures would help. David Cox made the good suggestion: "Why don't you consider, for example, measurements on the three angles of a triangle?" So, here's my set of 'trailers' for what's coming. The best bits are in the 'main features', though the trailers contain some nice geometry not studied later.

In Subsection 28C, we considered the Model:

$$\text{Observation} = (\text{true mean}) + (\text{True Error})$$

where the true means are structured in some way, and where the Errors corresponding to different Observations are IID, each $N(0, \sigma^2)$ for some unknown parameter σ^2. As explained in that subsection, after the experiment, we mirror this with

$$\text{observation} = (\text{fitted value}) + \text{residual}$$

where the fitted values are chosen on the basis of likelihood considerations. The assumption of IID normal Errors means that the likelihood geometry and least-squares methods tie up perfectly and that the Orthonormality Principle holds. Here are some key points:

- the fitted values reflect the algebraic structure of the true means (in a sense which will become clear later); and subject to that, are chosen to minimize the residual sum of squares, rss, the sum of the squares of the residuals;

- we estimate σ^2 by dividing rss by the associated 'number of degrees of freedom' (which notorious concept will be nailed down thoroughly by our treatment);

- by combining these ideas with others, we obtain Confidence Intervals or Confidence Regions for the various parameters.

Please re-read Subsection 28C now.

Do not miss out the trailers, to which we shall be referring back.

▶▶ **D. A trailer for the Normal-Sampling Theorem.**   The simplest situation considered in this chapter is where we obtain a Confidence Interval for the mean μ of a sample assumed taken from a normal distribution $N(\mu, \sigma^2)$, where both μ and σ^2 are unknown. Here every true mean is μ, and the fitted values must all be equal. In Subsection 270D, we looked at the data

$$1.21, \ 0.51. \ 0.14, \ 1.62, \ -0.80, \ 0.72, \ -1.71, \ 0.84, \ 0.02, \ -0.12.$$

I said that the Frequentist CI obtained by the famous 't' method is $(-0.456, 0.942)$ which may be obtained from Minitab. The full theory of how this is done will be explained in Subsection 305B, but some pictures and remarks now might help indicate the way in which we shall be thinking.

Diagram (a) in Figure D(i) shows the observations (•) and the fitted values (○), the latter forming a horizontal line $y = \overline{y}_{obs}$ since the value m^{obs} which minimizes the residual sum of squares $\sum(y_k^{obs} - m^{obs})^2$ is $\overline{y}_{obs}$ as we have often seen. Thus $\overline{y}_{obs}$ is the 'best estimate' of μ. The vertical lines represent the 'residuals', the discrepancies between the fits and the observations. We have

$$\text{rss}^{obs} := q(\mathbf{y}^{obs}, \mathbf{y}^{obs}) := \sum(y_k^{obs} - \overline{y})^2 = \left(\sum(y_k^{obs})^2\right) - n\overline{y}_{obs}^2.$$

Diagram (b) in Figure D(i) shows the way in which we must rethink Diagram (a). This diagram shows what is happening in $\mathbb{R}^n$. There, $\mathbf{y}^{obs}$ is the vector of observations, $\overline{y}_{obs}\mathbf{1} = (\overline{y}_{obs}, \overline{y}_{obs}, \dots, \overline{y}_{obs})^T$ is the vector of fitted values, and $\mathbf{y}^{obs} - \overline{y}_{obs}\mathbf{1}$ is the vector of residuals. The value $\overline{y}_{obs}\mathbf{1}$ is the value $m^{obs}\mathbf{1}$ closest to $\mathbf{y}^{obs}$, that is, the **perpendicular projection** of $\mathbf{y}^{obs}$ onto the line joining $\mathbf{0}$ to $\mathbf{1}$. In particular, the vector of residuals, $\mathbf{y}^{obs} - \overline{y}_{obs}\mathbf{1}$ is in the $(n-1)$-dimensional space perpendicular to $\mathbf{1}$, and this helps explain intuitively why the appropriate number of degrees of freedom for the Residual Sum of Squares, RSS, is here $n-1$. We therefore use

$$\hat{\sigma}_{n-1}^2(\mathbf{Y}) := \frac{\text{RSS}}{n-1} = \frac{Q(\mathbf{Y}, \mathbf{Y})}{n-1} = \frac{1}{n-1}\sum(Y_k - \overline{Y})^2$$

as our Estimator of σ^2. We know that

$$\frac{n^{\frac{1}{2}}(\overline{Y} - \mu)}{\sigma} \sim N(0, 1),$$

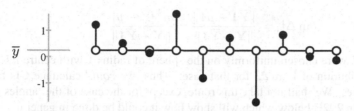

(a) The observations (•), fits (○) and residuals (verticals)

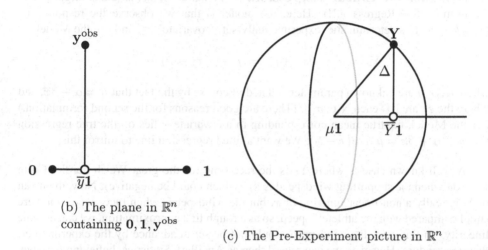

(b) The plane in $\mathbb{R}^n$ containing $\mathbf{0}, \mathbf{1}, \mathbf{y}^{\mathrm{obs}}$

(c) The Pre-Experiment picture in $\mathbb{R}^n$

Figure D(i): Illustrations for the Normal-Sampling Theorem

a fact that we could have used to obtain a CI for μ had σ been known. The appropriate modification for unknown σ is, as we shall prove at 305B,

$$\frac{n^{\frac{1}{2}}(\overline{Y} - \mu)}{\hat{\sigma}_{n-1}(\mathbf{Y})} \sim t_{n-1}.$$

Da. Exercise. Use this to confirm that `Minitab` did the calculation of the 95% CI for μ correctly. There are t-tables on page 516.

Diagram (c) in Figure D(i) is meant to depict the Pre-Experiment situation, where $\mathbf{Y}$ is the vector of observations as PreStatistics. By the Orthonormality Principle, the law of $\mathbf{Y}$ is invariant under rotations about the 'true mean' vector $\mu \mathbf{1}$. On combining this with the obvious 'scaling' property, we see (do think about this carefully) that the distribution of the ratio

$$|\tan \Delta| = \frac{\|\overline{Y}\mathbf{1} - \mu\mathbf{1}\|}{\|\mathbf{Y} - \overline{Y}\mathbf{1}\|} = \frac{n^{\frac{1}{2}}|\overline{Y} - \mu|}{\|\mathbf{Y} - \overline{Y}\mathbf{1}\|}$$

is the same as if $\mathbf{Y}$ were chosen uniformly on the sphere of radius 1 with centre $\mu\mathbf{1}$: we just need the distribution of $|\tan \Delta|$ for that case. Thus, we *could* calculate CIs from properties of spheres. We shall not take this route, except for the case of the 'angles of a triangle' problem in 292G below which will show how it could be done in general.

▶▶ **E. A trailer for Linear Regression.**   (Pedantic mathematicians might prefer the term 'Affine Regression'!)   Here, the model is that we observe the response Y_k ($1 \le k \le n$) to a deterministic, experimentally-set, 'covariate' x_k, and use the Model

$$Y_k = \alpha + \beta x_k + \varepsilon_k = \mu + \beta(x_k - \overline{x}) + \varepsilon_k,$$

where α, β, μ are unknown parameters related of course by the fact that $\mu = \alpha + \beta\overline{x}$, and where the ε_k are IID each $N(0, \sigma^2)$. (There are good reasons for the second formulation.) For the Model, the true mean corresponding to a covariate x lies on the true regression line $y = \alpha + \beta x = \mu + \beta(x - \overline{x})$. We want a fitted regression line to mirror this.

A well-known case is where Y_k is the race time for the great Welsh hurdler Colin Jackson when the supporting windspeed is x_k (which could be negative). Now, the mean of Y is really a non-linear function of x, but the windspeeds when races were run were small compared with the athlete's speed, so as a rough first approximation, we can assume linearity. In this case, the windspeeds could not be set in advance by the experimenter: they are random. However, we can regard then as Ancillary Statistics, believing that they do not (much) affect the ε_k values, and condition on them: hence, their randomness does not matter. Of course, the α, β, μ refer to this particular athlete, and depend amongst other things on his mean running speed; and of course, his fitness at the time of a race and other factors are relevant, but if we only want a rough approximation, we can use the linear-regression model.

What do we want to do for a general linear-regression model? The answer is provided by Figure E(i), In that figure, the top picture shows some data considered later and also the fitted regression line

$$y = \overline{y}_{\text{obs}} + b^{\text{obs}}(x - \overline{x}),$$

and marks the residuals as vertical lines. The values $\overline{y}_{\text{obs}}$ for the estimate of μ and b^{obs} for the estimated slope are obtained from the fact that the residual sum of squares has to be minimized. The bottom picture repeats the observations and fitted regression line in grey, and in black shows 'individual confidence bands for fit and for prediction'. With 95% confidence, the true mean $\mu + \beta(x - \overline{x})$ corresponding to a single x-value x will lie between the inner curves. With 95% confidence, a new measurement Y corresponding to a single new value x will lie between the outer curves.

All the theory is in Subsection 312D, and Figure 314D(ii) there will provide the 'likelihood geometry' picture corresponding to Figure E(i).

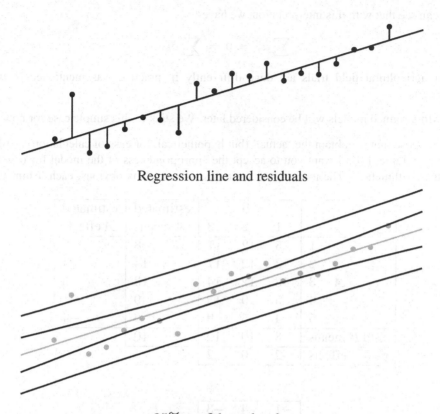

Regression line and residuals

95% confidence bands:
inner for mean $\bar{y} + \beta(x - \bar{x})$ for a single x;
outer for new observation for a single x-value x.

Figure E(i): Linear regression

▶▶ **F. A trailer for ANOVA.** The model we shall now consider is that in which we have a pair of 'factors' $\mathcal{A}$ and $\mathcal{B}$ (type of crop and type of soil, type of car and type of fuel, which workman/woman and which machine. There are J 'levels' of $\mathcal{A}$ and K of $\mathcal{B}$. For $1 \le j \le J$ and $1 \le k \le K$, we measure a response Y_{jk} (crop weight, output of pollution, number of goods produced, etc) when we use level j of $\mathcal{A}$ and level k of $\mathcal{B}$. We assume that

$$Y_{jk} = \mu + \alpha_j + \beta_k + \varepsilon_{jk},$$

where the ε_{jk} are IID each $N(0, \sigma^2)$. Thus μ is an overall mean, α_j is the amount above the mean by which use of level j of $\mathcal{A}$ raises the mean, and β_k is described analogously.

You can see that with this interpretation, we have

$$\sum \alpha_j = 0 = \sum \beta_k.$$

(That agricultural field trials are done differently in practice was mentioned in the Preface.)

More refined models will be considered later. We stick to this simple case for now.

Suppose that we obtain the 'actual' (but hypothetical, for ease of calculation) results shown in Table F(i). I want you to accept the appropriateness of the model for now to see the Arithmetic. The meaning of 'estimated $\mathcal{A}$ means' is obvious, each 'estimated

		$\mathcal{B}$			estimated	estimated
		1	2	3	$\mathcal{A}$ means	$\mathcal{A}$ effects
	1	5	12	7	8	-2
	2	13	12	17	14	4
$\mathcal{A}$	3	13	12	14	13	3
	4	5	9	13	9	-1
	5	4	5	9	6	-4
estd $\mathcal{B}$ means		8	10	12	10	
estd $\mathcal{B}$ effects		-2	0	2		

Residuals:

-1	4	-3
1	-2	1
2	-1	-1
-2	0	2
0	-1	1

Table F(i): Results for a two-factor ANOVA model

$\mathcal{A}$ effect' a_j is the corresponding mean minus the estimated overall mean m, and the estimated residual e_{jk} is given by

$$e_{jk} = y_{jk} - a_j - b_k - m,$$

so that the model is mirrored by

$$y_{jk} = m + a_j + b_j + e_{jk},$$

as explained in Subsection 28C.

All row sums and all column sums of the e_{jk} are zero. However, if all row sums are zero and all column sums except the last are zero, then the last column sum must be zero.

Thus the vector of estimated residuals satisfies $J + K - 1$ independent linear constraints and therefore lies in a subspace of $\mathbb{R}^{JK}$ of dimension

$$\nu := JK - (J + K - 1) = (J - 1)(K - 1);$$

and this is the number of degrees of freedom for the residual sum of squares rss. In our example it is 8. We find that rss = 48, whence the estimate of σ^2 based on rss is $\sigma_\nu = 48/8 = 6$.

We are going to be interested here only in the $\mathcal{A}$ factor. (The $\mathcal{B}$ factor may be handled similarly.) We find that the sum of squares of the estimated $\mathcal{A}$ effects is 46, with degrees of freedom 4. The estimate of the variance of the mean of an $\mathcal{A}$ row will be $46/4 = 11.5$. But the variance of the mean of an $\mathcal{A}$ row will be $\sigma^2/3$, so that our estimate of σ^2 based on $\mathcal{A}$-means is $3 \times 11.5 = 34.5$ if the Null Hypothesis that all α_i are equal (to 0) is true. (*Note.* The calculation would normally be done as $(3 \times 46)/4 = 138/4 = 34.5$, as you will see in the Minitab output below.) Thus, it is natural to consider the

$$\frac{\text{estimate of } \sigma^2 \text{ based on estimated } \mathcal{A} \text{ means}}{\text{estimate of } \sigma^2 \text{ based on estimated residuals}} = \frac{34.5}{6} = 5.75,$$

and to reject the Null Hypothesis that all α_i are equal (to 0) if this ratio is too big. How we decide on 'too big' is explained later. More importantly, we discuss CIs, CRs and simultaneous CIs for the various parameters.

Fa. Minitab **study of our example.** The following is an executable file myANOVA.MTB for studying our example. The command additive signifies 'no interaction'.

```
name C1 'y' C2 'A' C3 'B'
set 'y'
 5 12  7
13 12 17
13 12 14
 5  9 13
 4  5  9
end
set 'A'
(1:5)3
end
set 'B'
5(1:3)
end

print 'y' 'A' 'B';
format(3I3).

twoway 'y' 'A' 'B';
additive;
means 'A' 'B'.
```

When run via `exec 'myANOVA'` it produces the output:

```
Executing from file: myANOVA.MTB
  5  1  1
 12  1  2
  7  1  3
 13  2  1
etc

ANALYSIS OF VARIANCE  y

SOURCE          DF        SS        MS
A                4    138.00     34.50
B                2     40.00     20.00
ERROR            8     48.00      6.00
TOTAL           14    226.00
```

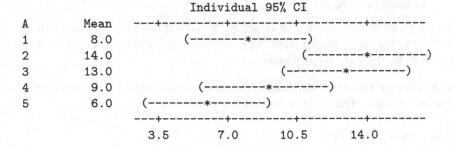

```
                          Individual 95% CI
A        Mean    ---+---------+---------+---------+---------
1         8.0        (--------*--------)
2        14.0                   (---------*---------)
3        13.0               (---------*---------)
4         9.0          (----------*--------)
5         6.0      (--------*---------)
                ---+---------+---------+---------+---------
                  3.5       7.0      10.5      14.0


                          Individual 95% CI
B        Mean    ---------+---------+---------+---------+--
1         8.0     (---------*---------)
2        10.0        (---------*---------)
3        12.0           (---------*---------)
                ---------+---------+---------+---------+--
                        7.5      10.0     12.5      15.0
```

Of course, you can get much fancier pictures in `Minitab` for Windows. `Minitab` ignores the 'Sum of Squares associated to the overall mean', as it is perfectly entitled to do. We shall study the table in Subsection 326L and see there why the half-length of an individual 95% CI is 3.27 for $\mathcal{A}$ means, 2.52 for $\mathcal{B}$ means, as is indicated in the pictures.

▶ **G. Measuring the angles of a triangle.** Suppose that $y_1^{\mathrm{obs}}, y_2^{\mathrm{obs}}, y_3^{\mathrm{obs}}$ are measurements of the true angles $\theta_1, \theta_2, \theta_3$ of a triangle. We use the Model:

$$Y_k = \theta_k + \varepsilon_k, \quad \text{where } \varepsilon_1, \varepsilon_2, \varepsilon_3 \text{ are IID each N}(0, \sigma^2).$$

Of course, we need to discuss with the experimenter whether this Model is reasonable. Amongst other things, σ must be small compared with the smallest angle measured. We have the constraint $\theta_1 + \theta_2 + \theta_3 = \pi$ (in radians). Now,

$$\mathrm{lhd}(\boldsymbol{\theta}; \mathbf{y}) = (2\pi\sigma^2)^{-3/2} \exp\{-\tfrac{1}{2}\|\boldsymbol{\theta} - \mathbf{y}\|^2/\sigma^2\},$$

and the mle $\mathbf{t}^{\mathrm{obs}}$ of $\boldsymbol{\theta}$ must minimize $\|\mathbf{t}^{\mathrm{obs}} - \mathbf{y}^{\mathrm{obs}}\|^2$ subject to the constraint $\sum t_k^{\mathrm{obs}} = \pi$. Thus, $\mathbf{t}^{\mathrm{obs}}$ must be the foot of the perpendicular from $\mathbf{y}^{\mathrm{obs}}$ to the plane $\{\mathbf{t} : t_1 + t_2 + t_3 = \pi\}$. This plane is perpendicular to $\mathbf{1}$, so that $\mathbf{t} = \mathbf{y}^{\mathrm{obs}} + c\mathbf{1}$ for some c^{obs}, and the sum constraint implies that

$$c^{\mathrm{obs}} = \tfrac{1}{3}\{\pi - (y_1^{\mathrm{obs}} + y_2^{\mathrm{obs}} + y_3^{\mathrm{obs}})\},$$

exactly as you could have guessed without knowing any Statistics. With this c^{obs}, we use $\mathbf{t}^{\mathrm{obs}} = \mathbf{y}^{\mathrm{obs}} + c^{\mathrm{obs}}\mathbf{1}$ as our best estimator for $\boldsymbol{\theta}$. However, we want a Confidence Region for $\boldsymbol{\theta}$, with $|c^{\mathrm{obs}}|$ as our natural guide to accuracy.

So, let $\mathbf{Y} - \boldsymbol{\theta}$ have IID components, each $N(0, \sigma^2)$. Let $\mathbf{T}$ be the foot of the perpendicular from $\mathbf{Y}$ to the plane $\{\mathbf{t} : t_1 + t_2 + t_3 = \pi\}$. Let Γ be the angle between $\boldsymbol{\theta} - \mathbf{Y}$ and $\mathbf{T} - \mathbf{Y}$.

Now we use the argument from Subsection 286D. Because the law of $\mathbf{Y}$ is invariant under rotations about $\mathbf{t}$, and by obvious scaling properties, we see that the distribution of the ratio

$$|\tan \Gamma| = \frac{\|\boldsymbol{\theta} - \mathbf{T}\|}{\|\mathbf{Y} - \mathbf{T}\|}$$

is the same as the distribution of the corresponding ratio if $\mathbf{Y}$ were chosen on the sphere of radius 1.

But, by Archimedes' Theorem (the one on his tombstone), $U := |\cos \Gamma|$ has the $U[0, 1]$ distribution, so, if $A = \tan^2 \Gamma$, then

$$\frac{1}{\sqrt{1 + A}} = |\cos \Gamma| = U \sim U[0, 1].$$

Hence, we can find a 95% Confidence Region for $\boldsymbol{\theta}$, namely,

$$\frac{\|\boldsymbol{\theta} - \mathbf{t}^{\mathrm{obs}}\|^2}{\|\mathbf{y}^{\mathrm{obs}} - \mathbf{t}^{\mathrm{obs}}\|^2} < K,$$

where $(1 + K)^{-\frac{1}{2}} = 5\%$, so that $K = 399$. This agrees exactly with the F-distribution approach, as you will be asked to check later. The value of K is exactly twice the $\nu_1 = 2$, $\mu_2 = 1$ point in the F-table, p518.

We have found a 95% Confidence Region

$$\|\boldsymbol{\theta} - \mathbf{t}^{\mathrm{obs}}\| < \sqrt{133}\,|y_1^{\mathrm{obs}} + y_2^{\mathrm{obs}} + y_3^{\mathrm{obs}} - \pi|, \qquad \sum \theta_i = \pi, \qquad \text{(G1)}$$

for $\boldsymbol{\theta}$, stating that $\boldsymbol{\theta}$ lies in a 2-dimensional disc centre $\mathbf{t}^{\mathrm{obs}}$ with the described radius. We could calculate on the computer a number b^{obs} such that we can be 95% confident that

simultaneously for $k = 1, 2, 3$, we have $|\theta_k - t_k^{\mathrm{obs}}| < b^{\mathrm{obs}}$,

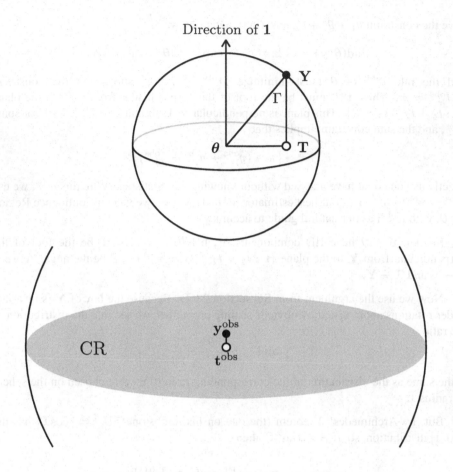

Figure G(i): Illustration for triangle problem

thus obtaining **simultaneous CIs**, something easier to understand (though in many ways not as good) as our CR.

Because the CR at (G1) is surprisingly large, the real picture being suggested by the bottom picture in Figure G(i), I checked my calculation by simulation as follows:

```
/*triangle.c
   To compile:    cc triangle.c RNG.o -o triangle -lm
*/
#include "RNG.h"
int main(){
long int i, kount=0, N=100000;
double y1,y2,y3,t1,t2,t3,c,fract;
  setseeds();
```

```
for(i=1;i<=N;i++){
  y1 = aGauss(); y2 = aGauss();
  y3 = aGauss(); c = (y1 + y2 + y3)/3.0;
  t1 = y1 - c; t2 = y2 - c, t3 = y3 - c;
  if (t1*t1 + t2*t2 + t3*t3 < 1197*c*c) kount++;
}
fract = kount/((float) N);
printf("\nfraction was %8.6f\n\n", fract);
return 0;
}
```

Ga. Individual CIs. The RV $(3/\sqrt{2})(T_1 - \theta_1)/(Y_1 + Y_2 + Y_3 - \pi)$ has the standard Cauchy distribution. Prove this later, when you have more technology.

8.2   The Orthonormality Principle and the F-test

Much of the linear algebra in this section will be familiar to you, so I only give a brief resumé of it. Where things may well be new (for example, perpendicular projections, and in a later section, tensor products), I move carefully through the theory.

A. $\mathbb{R}^n$ and its subspaces; span, spanning set.   We work with the standard representation of points in $\mathbb{R}^n$, a point of $\mathbb{R}^n$ being represented by a **column** vector

$$\mathbf{x} = (x_1, x_2, \ldots, x_n)^T = \begin{pmatrix} x_1 \\ x_2 \\ \cdot \\ x_n \end{pmatrix},$$

the superscript T standing for 'transposed'. Previously, we have written row vectors for vectors of data for typographical reasons. Henceforth, data vectors are always column vectors so $\mathbf{Y} = (Y_1, Y_2, \ldots, Y_n)^T$.

$\mathbb{R}^n$ is, of course, a vector space over $\mathbb{R}$. We can add vectors, multiply a vector $\mathbf{x}$ on the left by a scalar c to form $c\mathbf{x}$, and we have rules

$$\mathbf{x} + \mathbf{y} = \mathbf{y} + \mathbf{x}, \quad \mathbf{x} + (\mathbf{y} + \mathbf{z}) = (\mathbf{x} + \mathbf{y}) + \mathbf{z},$$
$$(c_1 + c_2)\mathbf{x} = c_1\mathbf{x} + c_2\mathbf{x}, \quad c(\mathbf{x} + \mathbf{y}) = c\mathbf{x} + c\mathbf{y},$$

etc.

A subspace U of $\mathbb{R}^n$ is a subset which is also a vector space over $\mathbb{R}$ in its own right. It is enough that if $\mathbf{x}$ and $\mathbf{y}$ are in U and $c \in \mathbb{R}$, then $c\mathbf{x}$ and $\mathbf{x} + \mathbf{y}$ are in U. As a picture, a line through the origin is a '1-dimensional' subspace of $\mathbb{R}^3$, a plane through the origin is a 2-dimensional subspace. The single point $\mathbf{0}$ is a 0-dimensional subspace. What 'dimension' is, will now be recalled.

The *span* of a set of vectors is the smallest subspace of $\mathbb{R}^n$ containing each vector in the given set. If $\mathbf{z}^{(1)}, \mathbf{z}^{(2)}, \ldots, \mathbf{z}^{(k)}$ are vectors in $\mathbb{R}^n$, we write $[\mathbf{z}^{(1)}, \mathbf{z}^{(2)}, \ldots, \mathbf{z}^{(k)}]$ for the space spanned by $\{\mathbf{z}^{(1)}, \mathbf{z}^{(2)}, \ldots, \mathbf{z}^{(k)}\}$, that is the space of all vectors of the form

$$c_1 \mathbf{z}^{(1)} + c_2 \mathbf{z}^{(2)} + \cdots + c_k \mathbf{z}^{(k)}.$$

We say that $\{\mathbf{z}^{(1)}, \mathbf{z}^{(2)}, \ldots, \mathbf{z}^{(k)}\}$ is a spanning set for a subspace U if $[\mathbf{z}^{(1)}, \mathbf{z}^{(2)}, \ldots, \mathbf{z}^{(k)}] = U$.

The space $[U, V]$ spanned by two subspaces U and V, the smallest subspace with $U \cup V$ as a subset, is $U + V$, the set of all vectors of the form $\mathbf{u} + \mathbf{v}$ where $\mathbf{u} \in U$ and $\mathbf{v} \in V$.

If U and V are subspaces of $\mathbb{R}^n$, then $U \cap V$ is a subspace.

B. Linear Dependence; basis; dimension; coordinates.   Vectors
$\mathbf{x}^{(1)}, \mathbf{x}^{(2)}, \ldots, \mathbf{x}^{(k)}$ are called *Linearly Dependent (LD)* if there exist constants (elements of $\mathbb{R}$) $c_1, c_2, \ldots, c_k$, not all zero, such that

$$c_1 \mathbf{x}^{(1)} + \cdots + c_k \mathbf{x}^{(k)} = \mathbf{0};$$

otherwise, $\mathbf{x}^{(1)}, \mathbf{x}^{(2)}, \ldots, \mathbf{x}^{(k)}$ are called *Linearly Independent (LI)*. A *basis* for a subspace U is an *LI spanning set* for U: it is important however that we regard a basis as an *ordered* set. Every subspace U of $\mathbb{R}^n$ has a basis; and (non-trivial Fact) any two bases for U have the same number of elements, this number being the *dimension* of U. If $\{\mathbf{u}^{(1)}, \mathbf{u}^{(2)}, \ldots, \mathbf{u}^{(m)}\}$ is a basis for U (so that $\dim(U) = m$), and $\mathbf{u} \in U$, then the ordered set of numbers $c_1, c_2, \ldots, c_m$ such that

$$\mathbf{u} = c_1 \mathbf{u}^{(1)} + c_2 \mathbf{u}^{(2)} + \cdots + c_m \mathbf{u}^{(m)} \tag{B1}$$

is uniquely specified: we say that $c_1, c_2, \ldots, c_m$ are the *coordinates* of $\mathbf{u}$ relative to the basis $\{\mathbf{u}^{(1)}, \mathbf{u}^{(2)}, \ldots, \mathbf{u}^{(m)}\}$ of U.

Let

$$\mathbf{e}^{(j)} := (0, \cdots, 0, 1, 0, \cdots, 0)^T,$$

the 1 being in the jth place. The vectors $\{\mathbf{e}^{(1)}, \mathbf{e}^{(2)}, \ldots, \mathbf{e}^{(n)}\}$ form the *standard basis* for $\mathbb{R}^n$.

C. Fact: the 'Extension to Basis' Principle *Suppose that U and W are subspaces of $\mathbb{R}^n$ with $U \subseteq W$. Let $m := \dim(U)$ and $s := \dim(W)$. Then given any basis $\{\mathbf{u}^{(1)}, \mathbf{u}^{(2)}, \ldots, \mathbf{u}^{(m)}\}$ for U, we can find vectors $\mathbf{u}^{(m+1)}, \ldots, \mathbf{u}^{(n)}$ such that $\{\mathbf{u}^{(1)}, \mathbf{u}^{(2)}, \ldots, \mathbf{u}^{(s)}\}$ is a basis for W and $\{\mathbf{u}^{(1)}, \mathbf{u}^{(2)}, \ldots, \mathbf{u}^{(n)}\}$ is a basis for $\mathbb{R}^n$.*

D. Inner (or scalar) products; orthonormal bases; etc.   The space $\mathbb{R}^n$ is
equipped with the standard inner (or scalar) product $\langle \mathbf{x}, \mathbf{y} \rangle$ defined by

$$\langle \mathbf{x}, \mathbf{y} \rangle := \mathbf{x}^T \mathbf{y} = \sum_{i=1}^{n} x_i y_i. \tag{D1}$$

Note that $\langle \mathbf{y}, \mathbf{x} \rangle = \langle \mathbf{x}, \mathbf{y} \rangle$ and that $\langle \mathbf{x}, \mathbf{y} + \mathbf{z} \rangle = \langle \mathbf{x}, \mathbf{y} \rangle + \langle \mathbf{x}, \mathbf{z} \rangle$. We define the length or norm $\|\mathbf{x}\|$ of $\mathbf{x}$ via (the fact that it is non-negative and)

$$\|\mathbf{x}\|^2 := \langle \mathbf{x}, \mathbf{x} \rangle = \mathbf{x}^T \mathbf{x} = \sum_{i=1}^{n} x_i^2. \tag{D2}$$

We say that $\mathbf{x}$ and $\mathbf{y}$ are perpendicular (or *orthogonal*) and write $\mathbf{x} \perp \mathbf{y}$ if $\langle \mathbf{x}, \mathbf{y} \rangle = 0$. *Pythagoras's Theorem* states that (as you can easily prove)

$$\text{if } \mathbf{x} \perp \mathbf{y} \text{ then } \|\mathbf{x} \pm \mathbf{y}\|^2 = \|\mathbf{x}\|^2 + \|\mathbf{y}\|^2.$$

A basis $\{\mathbf{u}^{(1)}, \mathbf{u}^{(2)}, \ldots, \mathbf{u}^{(m)}\}$ for a subspace U of $\mathbb{R}^n$ is called an *orthonormal basis* for U if

$$\|\mathbf{u}^{(j)}\| = 1, \quad \langle \mathbf{u}^{(i)}, \mathbf{u}^{(j)} \rangle = 0 \quad (i \neq j).$$

The vectors in an orthonormal basis are therefore normalized in that each has unit length, and they are perpendicular (orthogonal) to one another. Of course, $\{\mathbf{e}^{(1)}, \mathbf{e}^{(2)}, \ldots, \mathbf{e}^{(n)}\}$ is an orthonormal basis for $\mathbb{R}^n$.

Fact: *Every subspace U of $\mathbb{R}^n$ possesses an orthonormal basis; and the obvious 'orthonormal' analogue of the Extension-to-Basis Principle holds: replace 'basis' in Fact 296C by 'orthonormal basis'.*

The techniques for constructing an orthonormal basis for U and extending it to an orthonormal basis for $\mathbb{R}^n$ ('Gram-Schmidt orthogonalization') will develop naturally when needed. Just accept the Fact for now.

If $\{\mathbf{u}^{(1)}, \mathbf{u}^{(2)}, \ldots, \mathbf{u}^{(m)}\}$ is an orthonormal basis for a subspace U of $\mathbb{R}^n$, then, in the coordinate expression at 296(B1), we have

$$\mathbf{u} = \sum_{j=1}^{m} c_j \mathbf{u}^{(j)}, \qquad c_j = \langle \mathbf{u}^{(j)}, \mathbf{u} \rangle, \qquad \|\mathbf{u}\|^2 = \sum_{j=1}^{m} c_j^2. \tag{D3}$$

Exercise: Prove this.

E. Orthonormality Principle; $\mathrm{SN}(U)$.

We can move quickly to a form of the Orthonormality Principle if we assume the natural extension of what we did about characteristic functions in Section 5.5.

Recall that if X is an $\mathbb{R}$-valued RV, then the distribution of X is uniquely determined by the characteristic function $\mathbb{R} \ni \alpha \mapsto \mathbb{E} \, e^{i\alpha X}$ and that in particular, X has the standard normal $N(0, 1)$ distribution if (and only if) $\mathbb{E} \, e^{i\alpha X} = e^{-\frac{1}{2}\alpha^2}$ for $\alpha \in \mathbb{R}$.

It is likewise true that if $\mathbf{C}$ is an $\mathbb{R}^m$-valued RV, then the distribution of $\mathbf{C}$ is determined by the 'multivariate' characteristic function

$$\mathbb{R}^m \ni \boldsymbol{\alpha} \mapsto \mathbb{E} \, e^{i\langle \boldsymbol{\alpha}, \mathbf{C} \rangle},$$

$\langle \cdot, \cdot \rangle$ here denoting the inner product in $\mathbb{R}^m$ of course. In particular, $C_1, C_2, \ldots, C_m$ are IID each $N(0, 1)$ if and only if

$$\mathbb{E}\, e^{i\langle \boldsymbol{\alpha}, \mathbf{C} \rangle} = e^{-\frac{1}{2}\|\boldsymbol{\alpha}\|^2}, \qquad \boldsymbol{\alpha} \in \mathbb{R}^m.$$

The 'only if' result follows from the calculation

$$
\begin{aligned}
\mathbb{E}\, e^{i\langle \boldsymbol{\alpha}, \mathbf{C} \rangle} &= \mathbb{E}\, e^{i(\alpha_1 C_1 + \cdots + \alpha_m C_m)} = \mathbb{E}\left(e^{i\alpha_1 C_1} \ldots e^{i\alpha_m C_m}\right) \\
&= \left(\mathbb{E}\, e^{i\alpha_1 C_1}\right) \cdots \left(\mathbb{E}\, e^{i\alpha_m C_m}\right) \quad \text{(by independence)} \\
&= e^{-\frac{1}{2}\alpha_1^2} \ldots e^{-\frac{1}{2}\alpha_m^2} = \mathbb{E}\, e^{-\frac{1}{2}\|\boldsymbol{\alpha}\|^2};
\end{aligned}
$$

and the 'if' part now follows from the uniqueness result described above.

We say that an RV $\mathbf{X}$ has the **standard normal distribution on a subspace** U of $\mathbb{R}^n$ and write

$$\mathbf{X} \sim \mathrm{SN}(U)$$

if $\mathbf{X}$ takes values in U and for every $\mathbf{u} \in U$,

$$\mathbb{E}\, e^{i\langle \mathbf{u}, \mathbf{X} \rangle} = \mathbb{E}\, e^{-\frac{1}{2}\|\mathbf{u}\|^2}. \tag{E1}$$

We need to prove that there is such a distribution and that it is unique. Let $\{\mathbf{u}^{(1)}, \mathbf{u}^{(2)}, \ldots, \mathbf{u}^{(m)}\}$ be any orthonormal basis for U. Let $\alpha_1, \alpha_2, \ldots, \alpha_m$ be the coordinates of a vector $\mathbf{u}$ in U relative to this basis, and let $C_1, C_2, \ldots, C_m$ be the coordinates of $\mathbf{X}$ relative to this basis. Then (*Exercise.* Use the orthonormality of the basis to check this)

$$\langle \mathbf{u}, \mathbf{X} \rangle_U := \langle \mathbf{u}, \mathbf{X} \rangle_{\mathbb{R}^n} = \langle \boldsymbol{\alpha}, \mathbf{C} \rangle_{\mathbb{R}^m}, \qquad \|\mathbf{u}\|^2 = \|\boldsymbol{\alpha}\|^2,$$

where subscripts indicate the spaces in which inner products are taken. Hence, property (E1) amounts to

$$\mathbb{E}\, e^{i\langle \boldsymbol{\alpha}, \mathbf{C} \rangle} = e^{-\frac{1}{2}\|\boldsymbol{\alpha}\|^2},$$

in other words, to the statement that $C_1, C_2, \ldots, C_m$ are IID each $N(0, 1)$. The existence and uniqueness of the $\mathrm{SN}(U)$ distribution follows.

▶▶▶ **Ea. Orthonormality Principle, 1:**   I emphasize that a U-valued RV $\mathbf{X}$ has the $\mathrm{SN}(U)$ distribution if and only if the coordinates of $\mathbf{X}$ relative to *SOME* orthonormal basis for U are IID each $N(0, 1)$, and then the coordinates of $\mathbf{X}$ relative to *ANY* orthonormal basis for U are IID each $N(0, 1)$.

Note that if $\mathbf{X} \sim \mathrm{SN}(U)$, then $\|\mathbf{X}\|^2 \sim \chi_m^2$.

Eb. Note on the non-central χ^2 distribution. Recall the non-central χ^2 distribution from 153Ja. Suppose that $\mathbf{X} \sim \mathrm{SN}(\mathbb{R}^n)$ and that $\mathbf{a} \in \mathbb{R}$, $\mathbf{a} \neq \mathbf{0}$. Then we can choose an orthonormal basis $\mathcal{U} = \{\mathbf{u}^{(1)}, \mathbf{u}^{(2)}, \ldots, \mathbf{u}^{(n)}\}$ of $\mathbb{R}^n$ with $\mathbf{u}^{(1)} = \|\mathbf{a}\|^{-1}\mathbf{a}$. Let $\mathbf{Y}$ be the vector of coordinates of $\mathbf{X}$ relative to $\mathcal{U}$. Then the coordinates of $\mathbf{X} + \mathbf{a}$ relative to $\mathcal{U}$ form the vector $\mathbf{Z}$, where $Z_1 = Y_1 + \|\mathbf{a}\|$ and $Z_k = Y_k$ for $k \geq 2$. Since $Y_1, Y_2, \ldots, Y_n$ are IID each $\mathrm{N}(0, 1)$, the distribution of $\|\mathbf{X} + \mathbf{a}\|^2 = \|\mathbf{Z}\|^2$ depends on $\mathbf{a}$ only via $\|\mathbf{a}\|$.

F. Perpendicular projections. I use the term 'perpendicular' projection rather than 'orthogonal' projection because a non-trivial 'orthogonal' projection is not an 'orthogonal transformation'.

Let U be a subspace of $\mathbb{R}^n$. For $\mathbf{x}$ in $\mathbb{R}^n$, we write $P_U\mathbf{x}$ for the foot of the perpendicular from $\mathbf{x}$ to U, so that $P_U\mathbf{x} \in U$ and $\mathbf{x} - P_U\mathbf{x} \perp \mathbf{u}$ for every $\mathbf{u} \in U$. How do we know that there is such a vector, and that it is unique?

Fa. Exercise. Check that if $\{\mathbf{u}^{(1)}, \mathbf{u}^{(2)}, \ldots, \mathbf{u}^{(m)}\}$ is any orthonormal basis for U, extended to an orthonormal basis $\{\mathbf{u}^{(1)}, \mathbf{u}^{(2)}, \ldots, \mathbf{u}^{(n)}\}$ of $\mathbb{R}^n$, then

$$P_U\mathbf{x} := \sum_{j=1}^{m} \langle \mathbf{u}^{(k)}, \mathbf{x}\rangle \mathbf{u}^{(k)}$$

does the trick. Check also that $P_U\mathbf{x}$ *is the closest point of U to* $\mathbf{x}$. Figure F(i) illustrates Exercise Fa where $n = 3$ and $U = [\mathbf{u}^{(1)}, \mathbf{u}^{(2)}]$, and P_k is projection onto $[\mathbf{u}^{(k)}]$. $\square$

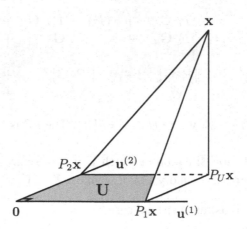

Figure F(i): To illustrate perpendicular projections

The space $U^\perp$ spanned by $\mathbf{u}^{(m+1)}, \ldots, \mathbf{u}^{(n)}$ in the notation of the exercise, is the space of all vectors perpendicular to U. We have

$$P_U^\perp := P_{U^\perp} = I - P_U,$$

where I is the identity transformation on $\mathbb{R}^n$.

▶▶▶ **G. Theorem: Orthonormality Principle for Projections.** Suppose that U and W are subspaces of $\mathbb{R}^n$ with $U \subseteq W$. Set

$$m := \dim(U), \quad s := \dim(W).$$

We have

$$P_W - P_U = P_Z, \quad \text{where } Z = U^\perp \cap W.$$

If $\mathbf{G}$ has the $\text{SN}(\mathbb{R}^n)$ distribution, then

$$P_U\mathbf{G}, \ (P_W - P_U)\mathbf{G} \text{ and } (I - P_W)\mathbf{G}$$

are independent variables, and

$$P_U\mathbf{G} \sim \text{SN}(U), \quad (P_W - P_U)\mathbf{G} = P_Z\mathbf{G} \sim \text{SN}(Z),$$
$$(I - P_W)\mathbf{G} = P_W^\perp\mathbf{G} \sim \text{SN}(W^\perp).$$

We have

$$
\begin{aligned}
\mathbf{G} &= P_U\mathbf{G} &+& \ (P_W - P_U)\mathbf{G} &+& \ (I - P_W)\mathbf{G} \\
&= P_U\mathbf{G} &+& \ P_Z\mathbf{G} &+& \ P_W^\perp\mathbf{G},
\end{aligned}
$$

$$
\begin{aligned}
\chi_n^2 \sim \|\mathbf{G}\|^2 &= \|P_U\mathbf{G}\|^2 &+& \ \|P_Z\mathbf{G}\|^2 &+& \ \|P_W^\perp\mathbf{G}\|^2 \\
&\sim \quad \chi_m^2 &+& \quad \chi_{s-m}^2 &+& \quad \chi_{n-s}^2.
\end{aligned}
$$

Suppose that $W = U + V$, where $U \cap V = \{\mathbf{0}\}$. Then Z is 'V corrected for U'.

In the 'orthogonal factor' situation in which V is a subspace orthogonal to U (in that $\mathbf{u} \perp \mathbf{v}$ whenever $\mathbf{u} \in U$ and $\mathbf{v} \in V$), we have $Z = V$.

Figure 314D(ii) will illustrate a special case.

Ga. Important Exercise. Prove the above theorem by letting $\{\mathbf{u}^{(1)}, \mathbf{u}^{(2)}, \ldots, \mathbf{u}^{(n)}\}$ be an orthonormal basis for $\mathbb{R}^n$ such that $\{\mathbf{u}^{(1)}, \mathbf{u}^{(2)}, \ldots, \mathbf{u}^{(m)}\}$ is a basis for U and $\{\mathbf{u}^{(1)}, \mathbf{u}^{(2)}, \ldots, \mathbf{u}^{(s)}\}$ is a basis for W. Since the variables $\langle \mathbf{u}^{(i)}, \mathbf{G} \rangle$ $(1 \le i \le n)$ are IID each $\text{N}(0, 1)$ ....

Gb. Exercise. Prove that if U and V are subspaces of $\mathbb{R}^n$, then the statements

$$\text{(a) } P_{U+V} = P_U + P_V, \quad \text{(b) } U \perp V, \quad \text{(c) } P_U P_V = 0 = P_V P_U,$$

are equivalent. Show that if $P_U P_V = 0$, then $P_V P_U = 0$.

H. The Classical F-test. Here's the result which dominated Statistics for a long period. We shall see later how it adapts to yield CIs. Note that another way of saying that

$$Y_1, Y_2, \ldots, Y_n \text{ are IID, with } Y_i \sim N(\mu_i, \sigma^2),$$

is to say that $\mathbf{Y} = \boldsymbol{\mu} + \sigma\mathbf{G}$, where $\boldsymbol{\mu} = (\mu_1, \mu_2, \ldots, \mu_n)^T$ and $\mathbf{G} \sim SN(\mathbb{R}^n)$.

In the following theorem, H_0 actually implies H_A. But this does not really matter. You have to remember that we know that H_0 is not exactly true: we are only testing whether or not there is strong evidence that H_0 is not an acceptable approximation to the truth. Of course, CRs will provide the correct language for simple situations.

▶▶▶ **Ha. Theorem: the F-test for Linear Models (Fisher, ...).** Suppose that

$$\mathbf{Y} = \boldsymbol{\mu} + \sigma\mathbf{G}, \text{where } \boldsymbol{\mu} \in \mathbb{R}^n \text{ and } \mathbf{G} \sim SN(\mathbb{R}^n).$$

Let U and W be subspaces of $\mathbb{R}^n$ with $U \subset W$, $\dim(U) = m$, $\dim(W) = s$. Let $Z := U^\perp \cap W$, as before. Then the size η Likelihood-Ratio Test of

$$\boldsymbol{\mu} \in U \text{ against } \boldsymbol{\mu} \in W,$$

that is, of

$$H_0 : \boldsymbol{\mu} \in U, \ \sigma > 0 \text{ against } H_A : \boldsymbol{\mu} \in W, \ \sigma > 0,$$

takes the form

$$\text{Reject } H_0 \text{ if } R := \frac{\|P_Z\mathbf{Y}\|^2/(s-m)}{\|P_W^\perp\mathbf{Y}\|^2/(n-s)} > c,$$

where $P(F > c) = \eta$ if $F \sim F_{s-m, n-s}$. For every $\boldsymbol{\mu}_0 \in U$, the probability of rejecting H_0 if the true value of $\boldsymbol{\mu}$ is $\boldsymbol{\mu}_0$ is exactly η.

If n is large, then, if H_0 is true, the deviance $\text{Dev}(\mathbf{Y}) = 2\ln \text{LR}(\mathbf{Y})$ has approximately the χ^2_{s-m} distribution.

Proof. With $\ell := \ln \text{lhd}$ as usual, we have

$$\ell(\boldsymbol{\mu}, \sigma; \mathbf{y}) = -\frac{\|\mathbf{y} - \boldsymbol{\mu}\|^2}{2\sigma^2} - n\ln\sigma - \tfrac{1}{2}n\ln(2\pi).$$

The maximum likelihood when $\boldsymbol{\mu} \in U$ is achieved when $\boldsymbol{\mu} = P_U\mathbf{y}$ and

$$\frac{\partial}{\partial\sigma}\left(-\frac{\|\mathbf{y} - P_U\mathbf{y}\|^2}{2\sigma^2} - n\ln\sigma\right) = 0, \text{ so, } \sigma^2 = n^{-1}\|\mathbf{y} - P_U\mathbf{y}\|^2.$$

Thus,
$$\ln \mathrm{mlhd}(H_0; \mathbf{y}) = -\tfrac{1}{2}n - \tfrac{1}{2}n \ln \left(n^{-1} \| \mathbf{y} - P_U \mathbf{y} \|^2 \right) - \tfrac{1}{2}n \ln(2\pi).$$

Similarly,

$$\ln \mathrm{mlhd}(H_A; \mathbf{y}) = -\tfrac{1}{2}n - \tfrac{1}{2}n \ln \left(n^{-1} \| \mathbf{y} - P_W \mathbf{y} \|^2 \right) - \tfrac{1}{2}n \ln(2\pi).$$

Thus,

$$\mathrm{dev}(\mathbf{y}) = 2 \ln \mathrm{lr}(\mathbf{y}) = n \ln \frac{\| \mathbf{y} - P_U \mathbf{y} \|^2}{\| \mathbf{y} - P_W \mathbf{y} \|^2} = n \ln \left(1 + \frac{\| P_Z \mathbf{y} \|^2}{\| P_W^\perp \mathbf{y} \|^2} \right),$$

whence the form of the LR Test is obvious.

Before reading the next sentences, revise the definition of the F-distribution from Subsection 252F.

If H_0 is true, then $\boldsymbol{\mu} \in U$ and $P_Z \boldsymbol{\mu} = P_W^\perp \boldsymbol{\mu} = \mathbf{0}$, whence

$$P_Z \mathbf{Y} = \sigma P_Z \mathbf{G}, \quad P_W^\perp \mathbf{Y} = \sigma P_W^\perp \mathbf{G}, \quad R = \frac{\| P_Z \mathbf{G} \|^2 / (s - m)}{\| P_W^\perp \mathbf{G} \|^2 / (n - s)}.$$

The 'F' result now follows since $\| P_Z \mathbf{G} \|^2$ and $\| P_W^\perp \mathbf{G} \|^2$ are independent and distributed as χ_{s-m}^2 and χ_{n-s}^2 respectively.

If n is large, as we now assume that it is, and H_A is true, then $n\sigma^2 / \| P_W^\perp \mathbf{Y} \|^2 \sim n / \chi_{n-s}^2$ is reasonably close to 1 with high probability, so, since $\ln(1 + \Delta) \approx \Delta$ for Δ small,
$$\text{if } H_A \text{ is true, then } \mathrm{Dev}(\mathbf{Y}) = 2 \ln \mathrm{LR}(\mathbf{Y}) \approx \sigma^{-2} \| P_Z \mathbf{Y} \|^2. \tag{H1}$$

Now, if H_0 is true, then $P_Z \boldsymbol{\mu} = \mathbf{0}$, so $P_Z \mathbf{Y} = \sigma P_Z \mathbf{G}$ and hence

$$\text{if } H_0 \text{ is true, then } \mathrm{Dev}(\mathbf{Y}) \approx \| P_Z \mathbf{G} \|^2 \sim \chi_{s-m}^2; \tag{H2}$$

and this allows us to assign approximate p-values if we so wish.

It will be explained later that this χ^2 result is part of a very general principle. Note that $s - m$ is the dimension of H_A minus the dimension of H_0 in that H_A corresponds to an $(s + 1)$-dimensional region of parameter space and H_0 to an $(m + 1)$-dimensional one, the '1' arising from the σ. $\qquad\square$

I. The 'sharp-hypothesis' nature of the F-test.

It is important to realize that the Null Hypothesis H_0 in the F-test is a sharp hypothesis stating that $\boldsymbol{\mu}$ lies in a lower-dimensional subspace. All the usual problems with sharp hypotheses reappear: H_0 is false, and a large sample would inevitably lead to its rejection at the 1% level; we must consider the natural scale at which we examine things, etc.

As usual, Confidence Intervals (and simultaneous CIs when appropriate) provide a much better language for analyzing things. However, the geometry of the F-test is very relevant to the obtaining of CIs.

J. The F-test and choice of model. As explained in Subsection 236J, in Statistics, we seek acceptably good models which are **parsimonious** in that they involve few unknown parameters. We might wish to protect the Null Hypothesis, rejecting it only if there is rather strong evidence against it as measured by the deviance $\mathrm{dev}(\mathbf{y}^{\mathrm{obs}})$. This is the point of view of Hypothesis Testing, where, when n is large, as we assume that it is, we use (H2) to assign a p-value and to determine the size of the test, both being calculated on the assumption that H_0 is true.

Ja. AIC for this situation. Even though you will have gathered from the discussion in Subsection 236J that I am somewhat uneasy about AIC because of its 'sharp-hypothesis' problems (amongst other things), the fairly wide use of AIC in practice persuades me to give the relevant geometry in this context.

We saw before 302(H1) that, since n is large, we essentially know σ^2, and so we shall assume that σ^2 is a known constant, not an unknown parameter. Suppose that $\mathbf{X}$ is independent of $\mathbf{Y}$ with the same law as $\mathbf{Y}$. Exactly as in Subsection 236J, the Decision-Theory and Kullback–Leibler approaches lead us to say that M_A (corresponding to H_A) is better than M_0 (corresponding to H_0) on average by an amount

$$\sigma^{-2}\mathbb{E}\left\{\|\mathbf{X}-P_U\mathbf{Y}\|^2-\|\mathbf{X}-P_W\mathbf{Y}\|^2\right\},$$

the σ^{-2} appearing automatically in the Kullback–Leibler approach and being a natural scaling in the Decision-Theory approach.

But our geometry tells us that

$$\|\mathbf{X}-P_U\mathbf{Y}\|^2-\|\mathbf{X}-P_W\mathbf{Y}\|^2 \;=\; 2\langle\mathbf{X},P_Z\mathbf{Y}\rangle-\|P_Z\mathbf{Y}\|^2,$$

and, since $\mathbf{X}$ is independent of $\mathbf{Y}$, we have

$$\mathbb{E}\langle\mathbf{X},P_Z\mathbf{Y}\rangle \;=\; \langle\mathbb{E}\mathbf{X},\mathbb{E}P_Z\mathbf{Y}\rangle \;=\; \langle\boldsymbol{\mu},P_Z\boldsymbol{\mu}\rangle \;=\; \|P_Z\boldsymbol{\mu}\|^2.$$

Also,

$$\mathbb{E}\|P_Z\mathbf{Y}\|^2 \;=\; \mathbb{E}\|P_Z\boldsymbol{\mu}+P_Z\mathbf{G}\|^2 \;=\; \|P_Z\boldsymbol{\mu}\|^2+\sigma^2(s-m),$$

whence

$$\sigma^{-2}\mathbb{E}\left\{\|\mathbf{X}-P_U\mathbf{Y}\|^2-\|\mathbf{X}-P_W\mathbf{Y}\|^2\right\} \;=\; \sigma^{-2}\|P_Z\boldsymbol{\mu}\|^2-(s-m).$$

However, from 302(H1),

$$\mathbb{E}\,\mathrm{Dev}(\mathbf{Y}) \;=\; \sigma^{-2}\|P_Z\boldsymbol{\mu}\|^2+(s-m).$$

Thus, to get 'on average' agreement with the Decision–Theory and Kullback–Leibler ideas, we need to subtract $2(s-m)$ from the deviance when comparing models M_A and M_0, exactly in agreement with AIC.

If we wish to compare a model $M_1 : \boldsymbol{\mu} \in U_1$ with $M_2 : \boldsymbol{\mu} \in U_2$, where U_1 and U_2 are not nested, we can compare each with $M : \boldsymbol{\mu} \in W$ where $W = [U_1, U_2]$ is the space spanned by U_1, U_2. In this way, provided we believe the AIC philosophy, we can use AIC to compare M_1 and M_2 directly.

8.3   Five basic models: the Mathematics

The models at which we look are 'Sampling from a normal distribution', Linear Regression, two forms of ANOVA, and (at the end of the section) the General Linear Model. First, we look only at the Mathematics of the Frequentist theory, the most elegant Maths in Stats, which deserves not to be interrupted. Then we look at the Bayesian approach and at how WinBUGS allows us to study more realistic models. In the important next section, we give some consideration to *goodness-of-fit* and *robustness* questions, and how to extend to more realistic models. But first, the Mathematics.

Advice: Please reread Subsection 28C before continuing.

▶▶ **A. Important notation.**
Quadratic forms. For vectors $\mathbf{x}, \mathbf{y} \in \mathbb{R}^n$ and $\mathbb{R}^n$-valued RVs $\mathbf{X}$ and $\mathbf{Y}$, we again write

$$q(\mathbf{x}, \mathbf{y}) := \sum_{i=1}^{n} (x_i - \overline{x})(y_i - \overline{y}), \quad Q(\mathbf{X}, \mathbf{Y}) := \sum_{i=1}^{n} (X_i - \overline{X})(Y_i - \overline{Y}),$$

the 'upper-case' notation being used for Pre-Statistics as usual. Check that

$$q(\mathbf{x}, \mathbf{y}) = \left(\sum x_i y_i \right) - n \overline{x} \, \overline{y}.$$

A special vector. Recall that we define

$$\mathbf{1} := (1, 1, \ldots, 1)^T.$$

Re distributions. If we write $R \sim \sigma^2 \chi_\nu^2$, we mean that $\sigma^{-2} R \sim \chi_\nu^2$; and we use obvious extensions of this idea. We define

$$\mathbb{P}\left(\chi_\nu^2 \in [c_-, c_+] \right) := \mathbb{P}(Z \in [c_-, c_+]) \text{ where } Z \sim \chi_\nu^2,$$

and use obvious extensions of *this* idea too.

Upper percentage points. Let $F_{r,s}^*(\eta)$ and $|t|_\nu^*(\eta)$ be the numbers such that

$$\mathbb{P}\left\{ F_{r,s} > F_{r,s}^*(\eta) \right\} = \eta, \qquad \mathbb{P}\left\{ |t_\nu| > |t|_\nu^*(\eta) \right\} = \eta,$$

that is, such that

$$\mathbb{P}\{W > F_{r,s}^*(\eta)\} = \eta, \qquad \mathbb{P}\{|T| > |t|_\nu^*(\eta)\} = \eta,$$

where $W \sim F_{r,s}$ and $T \sim t_\nu$. Because (see Subsection 253G)

$$F_{1,\nu} \sim t_\nu^2, \tag{A1}$$

we have $F_{1,\nu}^*(\eta) = |t|_\nu^*(\eta)^2$.

Intervals. For intervals, we use shorthands:

$$c + [a, b] := [c + a, c + b], \qquad c \pm \delta := [c - \delta, c + \delta],$$
$$c[a, b] := [ca, cb] \quad (c \geq 0).$$

▶▶▶ **B. The 'Normal Sampling' Theorem.** Let

$$Y_1, Y_2, \ldots, Y_n \text{ be IID each } N(\mu, \sigma^2),$$

where μ and σ^2 are unknown parameters. Then

$$\overline{Y} \sim N\left(\mu, \frac{\sigma^2}{n}\right),$$

$$\text{RSS} := Q(\mathbf{Y}, \mathbf{Y}) := \sum (Y_k - \overline{Y})^2 \sim \sigma^2 \chi_{n-1}^2,$$

and

$$\overline{Y} \text{ and } Q(\mathbf{Y}, \mathbf{Y}) \text{ are independent.}$$

The residuals $Y_k - \overline{Y}$ are the components of a vector with the $\text{SN}(W^\perp)$ distribution, $W^\perp$ being an $(n-1)$-dimensional subspace of $\mathbb{R}^n$ described below.

Note. The independence of $\overline{Y}$ and $Q(\mathbf{Y}, \mathbf{Y})$ actually characterizes normal distributions.

Note. RSS always stands for the Residual Sum of Squares appropriate to the model currently being used. □

We saw motivation and pictures relevant to the theorem in Subsection 286D. The theorem is one of the mainstays of classical Statistics. Before proving it, we look at some famous consequences. Define

$$\hat{\sigma}_{n-1}^2(\mathbf{Y}) := \frac{\text{RSS}}{n - 1},$$

the Estimator of σ^2 based on the Residuals $Y_k - \overline{Y}$, with, of course, $\hat{\sigma}_{n-1}(\mathbf{Y})$ the non-negative square root of $\hat{\sigma}_{n-1}^2(\mathbf{Y})$. This Estimator takes the standard form

$$\text{ANOVA Estimator of } \sigma^2 \;=\; \frac{\text{Residual Sum of Squares}}{\text{Residual degrees of freedom}}.$$

Then (see the definition of the t distributions at Subsection 253G)

$$\frac{n^{\frac{1}{2}}\left(\overline{Y} - \mu\right)}{\hat{\sigma}_{n-1}(\mathbf{Y})} \;\sim\; t_{n-1}, \tag{B1}$$

so that, if $\mathbb{P}(t_{n-1} \in [b_-, b_+]) = C\%$, then

$$\overline{y}_{\text{obs}} + \frac{\hat{\sigma}_{n-1}(\mathbf{y}^{\text{obs}})}{\sqrt{n}}[-b_+, -b_-]$$

is a $C\%$ CI for μ, as (because of the symmetry of t) is

$$\overline{y}_{\text{obs}} + \frac{\hat{\sigma}_{n-1}(\mathbf{y}^{\text{obs}})}{\sqrt{n}}[b_-, b_+].$$

If $\mathbb{P}\left(\chi_{n-1}^2 \in [c_-, c_+]\right) = C\%$, then $q(\mathbf{y}^{\text{obs}}, \mathbf{y}^{\text{obs}})[c_+^{-1}, c_-^{-1}]$ is a $C\%$ CI for σ^2.

The 'simultaneous CI' problem is here best dealt with by saying that if $[\mu_-, \mu_+]$ is a 97.5% CI for μ and $[\sigma_-, \sigma_+]$ is a 97.5% CI for σ, then we can be at least 95% confident that

$$(\mu, \sigma) \in [\mu_-, \mu_+] \times [\sigma_-, \sigma_+].$$

Do note that the exact degree of confidence is *not* 0.975^2 because the left-hand side of (B1) is not independent of $Q(\mathbf{Y}, \mathbf{Y})$.

Proof of Theorem 305B. We have, in the notation of Theorems 301Ha (for testing $\mu = 0$ against $\mu \neq 0$) and 300G,

$$\mathbf{Y} = \mu\mathbf{1} + \sigma\mathbf{G}, \quad U = \{\mathbf{0}\}, \quad W = [\mathbf{1}], \quad Z = [\mathbf{1}].$$

The likelihood function is

$$-\frac{n}{2}\ln(2\pi) - \frac{\|\mathbf{y} - \mu\mathbf{1}\|^2}{2\sigma^2} - n\ln\sigma,$$

and for a given $\mathbf{y}$, this is maximized when $\mu\mathbf{1} = P_{[\mathbf{1}]}\mathbf{y}$, $\sigma^2 = n^{-1}\|\mathbf{y} - P_{[\mathbf{1}]}\mathbf{y}\|^2$. (Don't worry about n rather than $n - 1$.) Now, an orthonormal basis for Z is provided by the single vector $\mathbf{z} = n^{-\frac{1}{2}}\mathbf{1}$, and

$$P_Z\mathbf{Y} = \langle\mathbf{z}, \mathbf{Y}\rangle\mathbf{z} = \overline{Y}\mathbf{1}.$$

But, since $\mathbf{Y} = \mu\mathbf{1} + \sigma\mathbf{G}$ and $P_Z\mathbf{1} = \mathbf{1}$,

$$P_Z\mathbf{Y} = \mu\mathbf{1} + \sigma P_Z\mathbf{G}, \quad \text{so } n^{\frac{1}{2}}\sigma^{-1}\left(\overline{Y} - \mu\right)\mathbf{z} = P_Z\mathbf{G},$$

whence, as we already know, $n^{\frac{1}{2}}\sigma^{-1}\left(\overline{Y} - \mu\right) \sim N(0, 1)$. The point here is that $P_Z\mathbf{G}$ has $N(0, 1)$ component relative to the unit vector $\mathbf{z}$.

We have

$$\sigma P_W^\perp\mathbf{G} = P_W^\perp\mathbf{Y} = P_Z^\perp\mathbf{Y} = (I - P_Z)\mathbf{Y} = \mathbf{Y} - \overline{Y}\mathbf{1},$$

so that, since $P_W^\perp\mathbf{G} \sim \text{SN}(W^\perp)$,

$$\chi_{n-1}^2 \sim \|P_W^\perp\mathbf{G}\|^2 = \sigma^{-2}\|P_W^\perp\mathbf{Y}\|^2 = \sigma^{-2}\|\mathbf{Y} - \overline{Y}\mathbf{1}\|^2 = \sigma^{-2}Q(\mathbf{Y}, \mathbf{Y}).$$

The desired independence property follows because $P_Z\mathbf{G}$ and $P_W^\perp\mathbf{G}$ are independent.

Do note that $P_W^\perp\mathbf{Y}$, which has the $\text{SN}(W^\perp)$ distribution, has ith component $Y_i - \overline{Y}$, the ith Residual. □

The Parallel-Axis consequence of $\|\mathbf{Y}\|^2 = \|P_W\mathbf{Y}\|^2 + \|P_W^\perp\mathbf{Y}\|^2$, namely

$$Q(\mathbf{Y}, \mathbf{Y}) = \left(\sum Y_i^2\right) - n\overline{Y}^2$$

is reflected in the ANOVA table, Table B(i), for testing $\mu = 0$ against $\mu \neq 0$ in this case

Source of variation	Sum of Squares	Degrees of freedom	Variance Estimator
Mean	$\|P_{[1]}\mathbf{Y}\|^2 = n\overline{Y}^2$	1	$n\overline{Y}^2/1$
Residual	$\|P_{[1]}^\perp\mathbf{Y}\|^2 = Q(\mathbf{Y}, \mathbf{Y})$	$n - 1$	$Q(\mathbf{Y}, \mathbf{Y})/(n - 1)$
Total	$\|\mathbf{Y}\|^2$	n	

Table B(i): ANOVA for 'normal sampling'

Ba. Exercise. Check that

$$q(\mathbf{x}, \mathbf{y}) = \langle \mathbf{x} - \overline{x}\mathbf{1}, \mathbf{y} - \overline{y}\mathbf{1}\rangle = \langle \mathbf{x} - \overline{x}\mathbf{1}, \mathbf{y}\rangle = \langle \mathbf{x}, \mathbf{y}\rangle - n\overline{x}\,\overline{y}.$$

Bb. Exercise. Check that `Minitab` found the correct symmetric 95% CI $(-0.456, 0.942)$ for μ from the data at 270(D1). You will find t-tables on page 516.

▶ **Bc. Exercise.** After values $y_1^{\text{obs}}, y_2^{\text{obs}}, \ldots, y_n^{\text{obs}}$ of IID RVs $Y_1, Y_2, \ldots, Y_n$ have been observed, a new observation Y_{new} is to be made. Describe the natural two-sided 95% CI for Y_{new}.

In Hypothesis Testing of $H_0 : \mu = 0$ against $H_A : \mu \neq 0$, we would reject H_0 at size η if the variance estimate arising from the mean is at least $F^*_{1,n-1}(\eta)$ times the variance estimate based on the residuals, that is, if

$$\frac{n\bar{y}^2_{obs}}{1} > F^*_{1,n-1}(\eta)\frac{q(\mathbf{y}^{obs}, \mathbf{y}^{obs})}{(n-1)}.$$

Because of 305(A1), this tallies with the fact that we reject H_0 if the symmetric $100(1 - \eta)\%$ CI for μ does not contain 0.

Bd. A Zeus–Tyche double act.   The pair $(\overline{Y}, Q(\mathbf{Y}, \mathbf{Y}))$ is sufficient for the pair (μ, σ^2). Zeus chooses $\overline{Y}$ and $Q(\mathbf{Y}, \mathbf{Y})$ independently, with $\overline{Y} \sim N(\mu, \sigma^2/n)$ and $Q(\mathbf{Y}, \mathbf{Y}) \sim \sigma^2\chi^2_{n-1}$. He tells Tyche the values of $\overline{Y}$ and $Q(\mathbf{Y}, \mathbf{Y})$. Tyche then knows that $\mathbf{Y}$ must lie on the $(n-1)$-dimensional sphere S_{n-1} centre $\overline{Y}\mathbf{1}$ and with radius squared equal to $Q(\mathbf{Y}, \mathbf{Y})$. She knows that $\mathbf{Y}$ also lies on the hyperplane specified by the fact that $\mathbf{Y} - \overline{Y}\mathbf{1}$ is perpendicular to $\mathbf{1}$. Thus, she knows that $\mathbf{Y}$ lies on the $(n - 2)$-dimensional sphere S_{n-2} where the hyperplane cuts S_{n-1}, and she chooses $\mathbf{Y}$ uniformly on S_{n-2}. We saw parts of this earlier.

▶ **Be. Exercise.**   In a *real* experiment to compare the amount of wear occurring in shoes with two different types of sole, each of 10 boys was given a pair of shoes, one with Material A, the other with Material B, which of 'left' and 'right' being of which material having been chosen randomly.

The results were as in Table B(ii):

boy	1	2	3	4	5	6	7	8	9	10
A	13.2	8.2	10.9	14.3	10.7	6.6	9.5	10.8	8.8	13.3
B	14.0	8.8	11.2	14.2	11.8	6.4	9.8	11.3	9.3	13.6

Table B(ii): Amount of wear for shoes trial

What is your model? (*Note:* Logarithms would feature in mine.)

Is there significant evidence that one type of material is better than the other? And if so, use a CI to quantify the difference.

▶ **C. Behrens–Fisher problem: a key Bayesian-Frequentist conflict.**

The Behrens–Fisher problem, which has received much discussion in the literature, presents a simple and *important* context which emphasizes differences between Bayesian and Frequentist philosophies. Suppose that

$X_1, X_2, \ldots, X_m$ are IID, each $N(\mu_x, \sigma^2_x)$,

$Y_1, Y_2, \ldots, Y_r$ are IID, each $N(\mu_y, \sigma^2_y)$,

the Variables $X_1, X_2, \ldots, X_m, Y_1, Y_2, \ldots, Y_r$ being independent.   The two samples could be thought of as obtained via different, independent, experiments. We are interested in Confidence Intervals for $\mu_x - \mu_y$.

Write $B_{m-1}(\mathbf{X}) := \hat{\sigma}_{m-1}(\mathbf{X})/\sqrt{m}$ and $B_{r-1}(\mathbf{Y}) := \hat{\sigma}_{r-1}(\mathbf{Y})/\sqrt{r}$, so that $B_{m-1}(\mathbf{X})$ is the usual Estimator of the SD of $\overline{X}$. We know that in Frequentist theory,

$$T_{m-1}(\mathbf{X}) := \frac{\overline{X} - \mu_x}{B_{m-1}(\mathbf{X})} \sim t_{m-1}, \qquad T_{r-1}(\mathbf{Y}) := \frac{\overline{Y} - \mu_y}{B_{r-1}(\mathbf{Y})} \sim t_{r-1}.$$

For Bayesian theory with the Frequentist prior $\pi(\mu_x, \sigma_x, \mu_y, \sigma_y) \propto \sigma_x^{-1}\sigma_y^{-1}$ which is the reference prior (see Subsection 379N), we have corresponding results

$$(\mu_x \,|\, \mathbf{x}^{\mathrm{obs}}) \sim \overline{x}_{\mathrm{obs}} + b_{m-1}(\mathbf{x}^{\mathrm{obs}})t_{m-1}, \qquad (\mu_y \,|\, \mathbf{y}^{\mathrm{obs}}) \sim \overline{y}_{\mathrm{obs}} + b_{r-1}(\mathbf{y}^{\mathrm{obs}})t_{r-1}.$$
(C1)

Recall that in Bayesian theory, $\mathbf{x}^{\mathrm{obs}}$ is the m-vector of numbers obtained in the one and only experiment actually performed in the real world, and $b_{m-1}(\mathbf{x}^{\mathrm{obs}})$ is a number calculated from $\mathbf{x}^{\mathrm{obs}}$. Recall too that our Bayesian will regard μ_x and μ_y after the two independent experiments as independent Variables with distributions as at (C1). It therefore makes sense to a Bayesian to calculate the function $a_{m-1,r-1}(\cdot,\cdot)$ on $(0,\infty)^2$ such that if $T_{m-1}^{(1)} \sim t_{m-1}$ and $T_{r-1}^{(2)} \sim t_{r-1}$ are independent RVs, then

$$\mathbb{P}\big(|uT_{m-1}^{(1)} - vT_{r-1}^{(2)}| \le a_{m-1,r-1}(u,v)\big) = 95\% \tag{C2}$$

and to assert that

$$\mathbb{P}_{\mathrm{post}}\left\{\big|(\mu_x - \overline{x}_{\mathrm{obs}}) - (\mu_y - \overline{y}_{\mathrm{obs}})\big| \le a_{m-1,r-1}(b_{m-1}(\mathbf{x}^{\mathrm{obs}}), b_{r-1}(\mathbf{y}^{\mathrm{obs}}))\right\} = 95\%.$$

Hence,

$$(\overline{x}_{\mathrm{obs}} - \overline{y}_{\mathrm{obs}}) \pm a_{m-1,r-1}\big(b_{m-1}(\mathbf{x}), b_{r-1}(\mathbf{y}^{\mathrm{obs}})\big)$$

gives a natural 95% *Bayesian* Confidence Interval for $\mu_x - \mu_y$.

However, this will be a 95% *Frequentist* Confidence Interval if and only if we have, for all $\mu_x, \sigma_x^2, \mu_y, \sigma_y^2$,

$$\mathbb{P}\big\{\big|(\overline{X} - \mu_x) - (\overline{Y} - \mu_y)\big| \le a_{m-1,r-1}(B_{m-1}(\mathbf{X}), B_{r-1}(\mathbf{Y}))\big\} = 95\%, \quad (??) \tag{C3}$$

a statement which, if true as *Mathematics*, a Frequentist would interpret in LTRF terms.

The problem is that (C3) is false, and the fundamental reason is that $B_{m-1}(\mathbf{X})$ is not independent of $T_{m-1}(\mathbf{X})$. The left-hand side of (C3) is

$$\mathbb{P}\big\{\big|B_{m-1}(\mathbf{X})T_{m-1}(\mathbf{X}) - B_{r-1}(\mathbf{Y})T_{r-1}(\mathbf{Y})\big| \le a_{m-1,r-1}(B_{m-1}(\mathbf{X}), B_{r-1}(\mathbf{Y}))\big\}.$$

If the Variables $B_{m-1}(\mathbf{X}), T_{m-1}(\mathbf{X}), B_{r-1}(\mathbf{Y}), T_{r-1}(\mathbf{Y})$ were independent, then we could deduce (C3) from (C2) by conditioning on the values u, v of $B_{m-1}(\mathbf{X}), B_{r-1}(\mathbf{Y})$. But the listed variables are *not* independent. This in itself does not actually disprove (C3), but it makes that result very doubtful. And it would be absolutely amazing if (C3) were to hold for every value of the ratio σ_x/σ_y.

Let us restrict attention to the case where, as we now assume,

$$m = 2, \quad r = 2,$$

so that we can easily work out everything in detail. The t_1 distribution is the standard Cauchy distribution (Subsection 253G), so, by 167B and the 'Independence means Multiply' property for CFs, we have, for $u, v \geq 0$,

$$\mathbb{E} \exp\{i\alpha(uT_1^{(1)} - vT_1^{(2)})\} = e^{-(u+v)|\alpha|},$$

so that $uT_1^{(1)} - vT_1^{(2)} \sim (u + v)t_1$. Hence, for $u, v \geq 0$,

$$a_{1,1}(u, v) = (u + v)K,$$

where $\mathbb{P}(|t_1| \leq K) = 95\%$, so that $K = (2/\pi)\tan^{-1}(0.95)$. Now, $\overline{X}, \overline{Y}, B_1(\mathbf{X}), B_1(\mathbf{Y})$ are independent. We have

$$(\overline{X} - \mu_x) - (\overline{Y} - \mu_y) \sim \mathrm{N}(0, \tfrac{1}{2}(\sigma_x^2 + \sigma_y^2)).$$

Next, $B_1(\mathbf{X}) \sim |\mathrm{N}(0, \tfrac{1}{2}\sigma_x^2)|$, $B_1(\mathbf{Y}) \sim |\mathrm{N}(0, \tfrac{1}{2}\sigma_y^2)|$, so that the left-hand side of 309(C3) has the form

$$\mathbb{P}(|N| \leq (|L| + |M|)K),$$

where L, M, N are independent with

$$N \sim \mathrm{N}(0, \tfrac{1}{2}(\sigma_x^2 + \sigma_y^2)), \quad L \sim \mathrm{N}(0, \tfrac{1}{2}\sigma_x^2), \quad M \sim \mathrm{N}(0, \tfrac{1}{2}\sigma_y^2).$$

But (check!) $N/(L + M) \sim t_1$, so that

$$\mathbb{P}(|N| \leq |L + M|K) = 95\%.$$

Thus, since $|L + M| \leq |L| + |M|$ and L and M can have different signs, we see that for this '$m = r = 2$' case, the probability on the left-hand side of 309(C3) is always strictly greater than 95% whatever the values of the unknown parameters.

That in this '$m = r = 2$' case the Frequentist *always* has higher degree of confidence in the Bayesian Confidence Interval than the Bayesian has degree of belief may not seem to 'balance out'. However, the whole point is that *this example is emphasizing that Bayesian degree of belief and Frequentist (or Mathematical) probability are not the same thing,* even for situations where there is a universally agreed 'vague prior density'. Well, the prior we have discussed would be universally agreed if we assumed that the two variances 'had nothing to do with each other', reasonable in some models, not at all reasonable in others. Let us assume for now that the prior is reasonable.

Bayesian: "I have based my Credible Interval on the one and only actual performance of the experiment. I do not wish to consider long-term relative frequencies over hypothetical experiments which will never be performed."

Frequentist: "But you cannot escape the fact that if those experiments were to be performed and you always used that same strategy for getting your Credible Interval, then the long-term proportion of times when your interval would contain the true value of $\mu_x - \mu_y$ would not be 95%."

To be sure, there was a long period when furious arguments between Bayesians and Frequentists did Statistics no favours. These days, the politically correct thing is to ignore the controversy. But there have inevitably been some situations in this book (of which this is the most serious) when the philosophies disagree and even the numbers disagree. The Behrens–Fisher example will suggest to you many other important analogous situations.

I have generally left you to form your own opinions. In writing most sections of this book (especially the Frequentist ones!), I have felt strongly Bayesian. However, in other sections, especially those where there is conflict, my sympathies have tended to lie with the Frequentist school. A 'schizophrenic' attitude is quite common.

Ca. The Fiducial school. I should add that there is a third school of Statistics, the Fiducial school, based on ideas introduced by Fisher and studied deeply over many years by Barnard. I am regarding Fiducial as 'Bayesian with Frequentist prior', thereby undoubtedly again offending everyone.

Cb. The Welch Statistic. Welch discovered a method of getting approximate CIs for the Behrens–Fisher problem which works well for most situations met in practice. This is used by `Minitab` and other packages.

▶ **Cc. Exercise: the pooled-variance model.** The theory is easy and uncontroversial if we assume a model
$$X_1, X_2, \ldots, X_m \text{ are IID, each } N(\mu_x, \sigma^2),$$
$$Y_1, Y_2, \ldots, Y_r \text{ are IID, each } N(\mu_y, \sigma^2),$$
the Variables $X_1, X_2, \ldots, X_m, Y_1, Y_2, \ldots, Y_r$ being independent. The point now is that the two samples are assumed to have the same variance. Check that a Confidence Interval for $\mu_1 - \mu_2$ will have the form

$$\overline{x}_{\text{obs}} - \overline{y}_{\text{obs}} + |t_\nu| * (0.05)\hat{\sigma}_\nu \sqrt{\frac{1}{m} + \frac{1}{r}},$$

where $\nu = (m-1) + (r-1)$ and

$$\hat{\sigma}_\nu := \frac{q(\mathbf{x}^{\text{obs}}, \mathbf{x}^{\text{obs}}) + q(\mathbf{y}^{\text{obs}}, \mathbf{y}^{\text{obs}})}{\nu}.$$

Crosscheck with the F-test.

Explain *why* use of this model would be *so* disastrously wrong if applied to the 'shoes' data at Exercise 308Be.

D. Linear Regression. Motivation and pictures were provided in Subsection 288E.

As explained there, the model is that we observe the response Y_k $(1 \leq k \leq n)$ to a deterministic, experimentally-set, 'covariate' x_k, and use the model

$$Y_k = \alpha + \beta x_k + \varepsilon_k = \mu + \beta(x_k - \overline{x}) + \varepsilon_k,$$

where α, β, μ are unknown parameters related of course by the fact that $\mu = \alpha + \beta \overline{x}$, and where the ε_k are IID each $N(0, \sigma^2)$.

We return to the general case. Some computer programs use the

$$Y_k = \alpha + \beta x_k + \varepsilon_k$$

formulation. This is often rather silly for commonsense reasons. The value α is the mean value of Y when $x = 0$. But the x-value 0 may be far away from the values featuring in the experiment. Linearity may well have completely broken down. Moreover, there will obviously be greater uncertainty about the value of α than of the mean μ of Y when $x = \overline{x}$. Hence, for commonsense reasons, we use the model

$$Y_k = \mu + \beta(x_k - \overline{x}) + \varepsilon_k. \tag{D1}$$

We shall see that this is the right formulation in the mathematics too. There *are* situations where the value of α matters, and we study how to deal with the pair (α, β) for such cases.

You are reminded that in the next section, we shall consider whether the model tallies reasonably well with the data.

▶▶ **Da. Theorem.** *Take the model (D1), where the ε_k are IID each $N(0, \sigma^2)$. Then*

$$\overline{Y} \sim N\left(\mu, \frac{\sigma^2}{n}\right),$$

$$B := \frac{Q(\mathbf{x}, \mathbf{Y})}{q(\mathbf{x}, \mathbf{x})} \sim N\left(\beta, \frac{\sigma^2}{q(\mathbf{x}, \mathbf{x})}\right),$$

$$\mathrm{RSS} := \sum_{k=1}^{n} \left(Y_k - \overline{Y} - B(x_k - \overline{x})\right)^2 \sim \sigma^2 \chi_{n-2}^2.$$

Moreover, $\overline{Y}$, B and RSS are independent. The value $Y_k - \overline{Y} - B(x_k - \overline{x})$ is now the kth Residual.

We have an ANOVA table, Table D(i).

Source of variation	Sum of Squares	Degrees of freedom	Variance Estimator
Mean μ	$n\overline{Y}^2$	1	$n\overline{Y}^2$
Slope β	$q(\mathbf{x},\mathbf{x})B^2$	1	$q(\mathbf{x},\mathbf{x})B^2$
Residual	RSS	$n-2$	$\text{RSS}/(n-2)$
Total	$\|\mathbf{Y}\|^2$	n	

Table D(i): ANOVA for Linear Regression

▶ **Db. Exercise.** Assuming the theorem, check that if

$$\hat{\sigma}_{n-2}^2(\mathbf{x},\mathbf{Y}) := \frac{\text{RSS}}{n-2},$$

then

$$\frac{n^{\frac{1}{2}}\left(\overline{Y}-\mu\right)}{\hat{\sigma}_{n-2}(\mathbf{x},\mathbf{Y})} \sim t_{n-2}, \qquad \frac{q(\mathbf{x},\mathbf{x})^{\frac{1}{2}}B}{\hat{\sigma}_{n-2}(\mathbf{x},\mathbf{Y})} \sim t_{n-2}.$$

(These variables are not independent of course.) Check that the symmetric $100(1-\eta)\%$ CI for β fails to contain 0 if and only if we would reject the hypothesis $\beta = 0$ against $\beta \neq 0$ at size η in the sense that

$$q(\mathbf{x},\mathbf{x})B^2 > F_{1,n-2}^*(\eta)\frac{\text{RSS}}{n-2}.$$

Proof of Theorem 312Da. Figure D(ii) shows the 3-dimensional subspace of $\mathbb{R}^n$ spanned by $\mathbf{Y}, \mathbf{1}, \mathbf{x}$, or equally, and it is much better to think this way, by $\mathbf{Y}, \mathbf{1}, \mathbf{z}$, where $\mathbf{z} = P_{[1]}^{\perp}\mathbf{x} = \mathbf{x} - \overline{x}\mathbf{1}$. Then we have a situation:

$$U = [\mathbf{1}], \quad Z = [\mathbf{z}], \quad U \perp Z, \quad W := U + Z = [\mathbf{1},\mathbf{z}] = [\mathbf{1},\mathbf{x}].$$

In vector notation, we have

$$\mathbf{Y} = \mu\mathbf{1} + \beta(\mathbf{x} - \overline{x}\mathbf{1}) + \sigma\mathbf{G},$$

where $\mathbf{G} \sim \text{SN}(\mathbb{R}^n)$. An orthonormal basis for Z is $\{\mathbf{e}\}$, where $\mathbf{e} = q(\mathbf{x},\mathbf{x})^{-\frac{1}{2}}\mathbf{z}$. As before, an orthonormal basis for U is $\{\mathbf{u}\}$, where $\mathbf{u} = n^{-\frac{1}{2}}\mathbf{1}$. Using $P_U\mathbf{Y} = \langle \mathbf{u},\mathbf{Y}\rangle\mathbf{u}$, $P_U\mathbf{1} = \mathbf{1}$, $P_U\mathbf{z} = \mathbf{0}$, etc, we have

$$n^{-\frac{1}{2}}P_U\mathbf{Y} = \overline{Y}\mathbf{u} = \mu\mathbf{u} + n^{-\frac{1}{2}}\sigma P_U\mathbf{G},$$

$$q(\mathbf{x},\mathbf{x})^{-\frac{1}{2}}P_Z\mathbf{Y} = B\mathbf{e} = \beta\mathbf{e} + q(\mathbf{x},\mathbf{x})^{-\frac{1}{2}}\sigma\mathbf{G},$$

$$P_{\text{res}}\mathbf{Y} = P_W^{\perp}\mathbf{Y} = \sigma P_W^{\perp}\mathbf{G}, \quad \|P_W^{\perp}\mathbf{Y}\|^2 = \text{RSS}.$$

The theorem now follows immediately from the Orthonormality Principle 300G for Projections. □

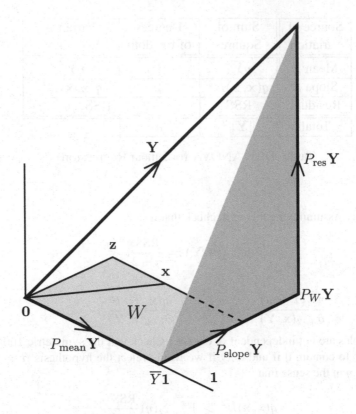

Figure D(ii): $\mathbb{R}^n$ geometry of linear regression

Note that

$$P_{\text{res}}\mathbf{Y} = P_W^\perp\mathbf{Y} = (I - P_U - P_Z)\mathbf{Y} = \mathbf{Y} - \overline{Y}\mathbf{1} - B(\mathbf{x} - \overline{x}\mathbf{1}),$$

so that

$$(P_{\text{res}}\mathbf{Y})_k = Y_k - \overline{Y} - B(x_k - \overline{x}),$$

the kth Residual. Recall that

$$\sigma^{-1}P_{\text{res}}\mathbf{Y} = P_{\text{res}}\mathbf{G} = \sigma^{-1}P_W^\perp\mathbf{Y} \sim \text{SN}(W^\perp)$$

and that $W^\perp$ is an $(n - 2)$-dimensional subspace of $\mathbb{R}^n$.

▶ **Dc. Exercise: Predicting new observations.**   Suppose that we have made our n observations $Y_1, Y_2, \ldots, Y_n$ corresponding to x-values $x_1, x_2, \ldots, x_n$. As a Pre-Statistic, our best Estimator of the true *mean* $\mu + \beta(x - \overline{x})$ of an observation corresponding to x is $\overline{Y} + B(x - \overline{x})$. Show that

$$\frac{\overline{Y} + B(x - \overline{x}) - \{\mu + \beta(x - \overline{x})\}}{\hat{\sigma}_{n-2}(\mathbf{x}, \mathbf{Y})\left\{\frac{1}{n} + \frac{(x-\overline{x})^2}{q(\mathbf{x},\mathbf{x})}\right\}^{\frac{1}{2}}} \sim t_{n-2}.$$

Suppose now that, having made our n observations, we are going to make a new observation Y_{new} corresponding to a new value x_{new}. Show that

$$\frac{Y_{new} - \{\overline{Y} + B(x_{new} - \overline{x})\}}{\hat{\sigma}_{n-2}(\mathbf{x}, \mathbf{Y}) \left\{ 1 + \frac{1}{n} + \frac{(x_{new} - \overline{x})^2}{q(\mathbf{x}, \mathbf{x})} \right\}^{\frac{1}{2}}} \sim t_{n-2}.$$

This gives confidence bounds for our prediction of Y_{new}.

Dd. Example. Minitab would do an example like this. In this case, the one depicted in Figure 289E(i), $n = 20$.

```
MTB > set c1                              Column 1 to contain x values
DATA> (0:19)                              xₖ = k − 1 for k = 1, 2, ..., 20
DATA> end
MTB > set c2                              Column 2 to contain y values
DATA> 1.30  2.65  0.89  1.09  1.60  1.96  1.89  1.50  2.99  2.77
DATA> 3.30  3.22  3.61  3.09  3.32  3.28  3.61  4.04  4.10  4.74
DATA> end
MTB > regr c2 1 c1                         Perform the regression

The regression equation is
C2 = 1.17 + 0.166 C1                       Oops! Not a good formulation
```

Predictor	Coef	Stdev	t-ratio	p
Constant	1.1696	0.2145	5.45	0.000
C1	0.16610	0.01930	8.61	0.000

The t-ratio and p are not very meaningful.

```
s = 0.4977                                 estimate σ̂ₙ₋₂(x, yᵒᵇˢ) of σ
```

$s = 0.4977$ — estimate $\hat{\sigma}_{n-2}(\mathbf{x}, \mathbf{y}^{obs})$ of σ

```
Unusual Observations
Obs.  C1     C2     Fit  Stdev.Fit  Residual  St.Resid
  2   1.0   2.650  1.336    0.198      1.314     2.88R

R denotes an obs. with a large st. resid.
```

My own regression program in 'C' gives:

```
   ybar   its SD      b  its SD
 2.7475  0.1113  0.1661  0.0193

    Sig      b-      b+   ybar-    ybar+
 0.4977  0.1256  0.2066  2.5137   2.9813


Possible outlier(s):
i = 2, x = 1.0000,  y = 2.6500,  crude SDs = 2.6408
```

For my program, ybar $= \overline{y}$ and 'its SD' is $\hat{\sigma}_{n-2}(\mathbf{x}, \mathbf{y}^{\mathrm{obs}})/\sqrt{n}$, b $= B(\omega^{\mathrm{act}})$ and 'its SD' is $\hat{\sigma}_{n-2}(\mathbf{x}, \mathbf{y}^{\mathrm{obs}})/\sqrt{q(\mathbf{x}, \mathbf{x})}$, Sig $= \hat{\sigma}_{n-2}(\mathbf{x}, \mathbf{y}^{\mathrm{obs}})$, (b−, b+) is a 95% CI for β, (ybar−, ybar+) is a 95% CI for μ. Both programs draw attention to the unusual 2nd observation. More on this later. The bottom part of Figure 289E(i) is just a case of programming Exercise 314Dc. The outer curves in this case are nearly linear along the range shown, but would become clearly curved if the range of x-values were extended. All of the curves are hyperbolas, as is obvious from Exercise 314Dc.

The second observation is an outlier, a point discussed in the next section.

Think about how Figures 314D(ii) and 289E(i) are connected. The fact that *the best regression line*

$$y = m + b(x - \overline{x}) \ \text{ with } \ m = \overline{y}_{\mathrm{obs}} \ \text{ and } \ b = b_{\mathrm{obs}} = B(\omega^{\mathrm{act}})$$

minimizes the sum of squares of the residuals tallies with the fact that $P_W \mathbf{Y}$ is the nearest point of W to $\mathbf{Y}$.

I am playing down least-squares ideas because it is *likelihood* ideas which are fundamental in our approach. Here they tally with least-squares ideas because of the link between the normal distribution and Pythagoras's theorem.

▶ **De. Leverage.** The leverage of the observation (X_k, y_k) measures how much a change in y_k would affect the fit

$$m + b(x_k - \overline{x})$$

at x_k. We see that a change Δy_k in y_k would cause a change

$$\left(\frac{1}{n} + \frac{(x_k - \overline{x})^2}{q(\mathbf{x}, \mathbf{x})} \right) \Delta y_k$$

in the fit. Here, the sum of all the leverage coefficients is the number, 2, of parameters. Note how leverage has in effect already featured in Exercise 314Dc.

▶ **Df. Exercise on non-orthogonality.** Consider a situation in which our model is

$$Y_k = \alpha + \beta x_k + \varepsilon, \quad \mathbf{Y} = \alpha \mathbf{1} + \beta \mathbf{x} + \sigma \mathbf{G},$$

where our concern is with the intercept α of the regression line on the y-axis. Of course, in our previous notation $\alpha = \mu - \beta \overline{x}$, so that it is natural to use $A := \overline{Y} - \overline{x} B$ as an Estimator for α. Prove that

$$\mathbb{E} A = \alpha, \qquad \mathrm{Var}(A) = \sigma^2 \left(\frac{1}{n} + \frac{\overline{x}^2}{q(\mathbf{x}, \mathbf{x})} \right),$$

$$\left(\frac{1}{n} + \frac{\overline{x}^2}{q(\mathbf{x}, \mathbf{x})} \right)^{-\frac{1}{2}} \frac{A - \alpha}{\hat{\sigma}_{n-2}(\mathbf{x}, \mathbf{Y})} \sim t_{n-2}.$$

This leads us to think that in a test of

$$H_0 : \alpha = 0, \ \beta \in \mathbb{R} \quad \text{against} \quad H_A : \alpha \in \mathbb{R}, \ \beta \in \mathbb{R},$$

we would reject H_0 at size η if

$$\left(\frac{1}{n} + \frac{\overline{x}^2}{q(\mathbf{x}, \mathbf{x})} \right)^{-\frac{1}{2}} \frac{|A|}{\hat{\sigma}_{n-2}(\mathbf{x}, \mathbf{Y})} > |t|^*_{n-2}(\eta).$$

Check this out from the F-test in Theorem 301Ha by reversing the rôles of $\mathbf{1}$ and $\mathbf{x}$ in our previous work. Show that if $H = [\mathbf{1}, \mathbf{x}] \cap [\mathbf{x}]^{\perp}$, then

$$P_H \mathbf{Y} = \frac{\langle \mathbf{h}, \mathbf{Y} \rangle \mathbf{h}}{\|\mathbf{h}\|^2}, \quad \text{where } \mathbf{h} = \mathbf{1} - \frac{\langle \mathbf{x}, \mathbf{1} \rangle \mathbf{x}}{\langle \mathbf{x}, \mathbf{x} \rangle},$$

and

$$\|\mathbf{h}\|^2 = \frac{nq(\mathbf{x}, \mathbf{x})}{n\overline{x}^2 + q(\mathbf{x}, \mathbf{x})}, \quad P_H \mathbf{Y} = A\mathbf{h}, \quad \|P_H \mathbf{Y}\| = |A| \|\mathbf{h}\|.$$

Of course, since $[\mathbf{x}] \perp H$ and $[\mathbf{1}] \perp Z$,

$$\|P_W \mathbf{Y}\|^2 = \|P_{[\mathbf{x}]} \mathbf{Y}\|^2 + \|P_H \mathbf{Y}\|^2 = \|P_{[\mathbf{1}]} \mathbf{Y}\|^2 + \|P_Z \mathbf{Y}\|^2.$$

We say that $\|P_H \mathbf{Y}\|^2$ is the Sum of Squares for intercept corrected for (or adjusted for) slope, that $\|P_{[\mathbf{x}]} \mathbf{Y}\|^2$ is the unadjusted SS for slope, that $\|P_{[\mathbf{1}]} \mathbf{Y}\|^2$ is the unadjusted SS for intercept. As we know, $\|P_{[\mathbf{1}]} \mathbf{Y}\|^2$ is the Sum of Squares appropriate for testing $\alpha = 0$ in the model $Y_k = \alpha + \varepsilon_k$.

▶ **E. Non-parametric Statistics; distribution-free tests.** We consider just one example.

▶ **Ea. Exercise: Spearman's 'Rank-Correlation Coefficient'.** Let the components of $\mathbf{Z} = (Z_1, Z_2, \ldots, Z_n)^T$ be a random permutation of $\{1, 2, \ldots, n\}$, all permutations being equally likely. Let $x_k = k$ $(1 \leq k \leq n)$. Show that if we do a least-squares fit with a line $Z = \mu + \beta(x - \overline{x})$, then the Estimator B of β will satisfy

$$B = \frac{Q(\mathbf{x}, \mathbf{Z})}{q(\mathbf{x}, \mathbf{x})} = R_S := \frac{\left(\frac{1}{n} \sum k Z_k \right) - \frac{1}{4}(n+1)^2}{\frac{1}{12}(n^2 - 1)}.$$

The Statistic R_S is called Spearman's 'Rank-Correlation Coefficient'. Show that

$$\mathbb{E}(R_S) = 0; \quad \text{Var}(R_S) = \frac{1}{n-1}. \qquad \square$$

Suppose that one wishes to test whether there is a drift up or down in a sequence of Observations $W_1, W_2, \ldots, W_n$ when one has no real knowledge of the types of distribution involved. One could make a Null Hypothesis H_0 that the variables $W_1, W_2, \ldots, W_n$ are exchangeable, a hypothesis which makes no other assertion about the distribution of $\mathbf{W} = (W_1, W_2, \ldots, W_n)$. For our convenience here only, we assume that with probability 1, no two W_k-values are equal. The Alternative Hypothesis is merely a vague one that there is some sort of 'drift'. Let Z_k be the *rank* of W_k in that $W_{(Z_j)} = W_j$, where the $W_{(\cdot)}$ are the usual Order Statistics. Intuitively, a positive value of R_S suggests a drift up, a negative one a drift down.

The distribution of R_S under the Null Hypothesis is the same as that in Exercise 317Ea, and so a 'p-value for H_0' based on the observed value r_s^{obs} of R_S, namely $P(|R_S| \geq r_s^{\mathrm{obs}})$, may be calculated knowing only n and without any further assumption on the distribution of $\mathbf{W}$. (You can find p-values for R_S for given n in most sets of statistical tables, all computer packages, and on the Web.) You can see why we are in a 'non-parametric' or 'distribution-free' situation.

Assigning a p-value in this way amounts to a 'pure test of significance'. It harks back to Fisher's idea that if the p-value is very small, then one can say, 'Either the Null Hypothesis is false or the Null Hypothesis is true and an event of very small probability has occurred' and count this as evidence against the Null Hypothesis. We know that this reasoning is incomplete: one usually wants to know, 'Does the Alternative Hypothesis provide a significantly better explanation?'; but here, one has no well-defined Alternative Hypothesis (and no sufficiently precisely defined Null Hypothesis, come to that) for using Likelihood-Ratio ideas.

Still, the use of such 'non-parametric tests' as that based on Spearman's coefficient can be at least a useful first step (especially if the p-value is not small!). They are widely used in psychology, Spearman's own field.

Another use of Spearman's coefficient is when the Null Hypothesis states that $(X_1, Y_1), (X_2, Y_2), \ldots, (X_n, Y_n)$ are n exchangeable RVs in $\mathbb{R}^2$, and when Z_k is the rank of the Y-value corresponding to the X-value of rank k. You can see the point of this when doing a kind of 'regression' without any of the assumptions of the linear-regression model.

Note that Bayesian theory is not equipped to do such non-parametric things.

▶▶ **F. Polynomial Regression.**   As well as the Mathematics of Polynomial Regression (which is much easier than it looks), this subsection contains several fundamental Statistical points.

Sometimes the data points $(x_1, y_1^{\mathrm{obs}}), (x_2, y_2^{\mathrm{obs}}), \ldots, (x_n, y_n^{\mathrm{obs}})$ associated with regression clearly lie close to a curve rather than a straight line. We might then choose to use a model

$$Y_k = g(x_k) + \varepsilon_k, \quad \varepsilon_k\text{'s IID each N}(0, \sigma^2),$$

where g is a polynomial of degree r:

$$g(t) = \alpha_0 + \alpha_1 t + \cdots + \alpha_r t^r,$$

where r is less than the number of distinct x_k values. For simplicity, we shall assume that all the x_k-values are distinct. Note that since this model is *linear in the parameters* $\alpha_0, \alpha_1, \ldots, \alpha_r$, it is not considered a 'non-linear' model in Statistics.

Note that if we shift the origin of the x-values, the parameters $\alpha_0, \alpha_1, \ldots, \alpha_{r-1}$ will change in a rather complicated way. It is therefore difficult to assign a clear meaning to these parameters individually.

Fa. Nelder's dictum. This asks you to *bear in mind* an important related point. A model such as

$$Y_k = \alpha + \gamma x_k^2 + \varepsilon_k$$

rarely makes sense (though it sometimes does) because an 'affine' change of covariate $x \mapsto \tilde{x} = c_0 + c_1 x$ would lead to a model

$$Y_k = \tilde{\alpha} + \tilde{\beta}\tilde{x}_k + \tilde{\gamma}\tilde{x}_k^2 + \varepsilon_k,$$

with non-zero $\tilde{\beta}$ coefficient. If, to use a Nelder example, each x_k is a temperature, then we do not wish to change the nature of our model according to whether we measure in Centigrade or Fahrenheit.

Fb. Some Mathematics. For $0 \leq q \leq r$, let $\mathbf{x}^{(q)}$ be the vector in $\mathbb{R}^n$ with kth component x_k^q, the qth power of x_k. Thus our model reads

$$\mathbf{Y} = \alpha_0 \mathbf{x}^{(0)} + \alpha_1 \mathbf{x}^{(1)} + \cdots + \alpha_r \mathbf{x}^{(r)} + \sigma \mathbf{G}. \tag{F1}$$

Except in freak situations, the vectors $\mathbf{x}^{(0)}, \mathbf{x}^{(1)}, \ldots, \mathbf{x}^{(r)}$ will not be orthogonal in $\mathbb{R}^n$. We therefore seek orthogonal vectors $\mathbf{z}^{(0)}, \mathbf{z}^{(1)}, \ldots, \mathbf{z}^{(r)}$ such that for every q,

$$[\mathbf{z}^{(0)}, \mathbf{z}^{(1)}, \ldots, \mathbf{z}^{(q)}] = [\mathbf{x}^{(0)}, \mathbf{x}^{(1)}, \ldots, \mathbf{x}^{(q)}], \tag{F2}$$

and then we can write our model as

$$\mathbf{Y} = \gamma_0 \mathbf{z}^{(0)} + \gamma_1 \mathbf{z}^{(1)} + \cdots + \gamma_r \mathbf{z}^{(r)} + \sigma \mathbf{G}.$$

This idea will also guarantee that there exist polynomials $g^{(0)}, g^{(1)}, \ldots, g^{(r)}$, with $g^{(q)}$ of degree q such that $z_k^{(q)} = g^{(q)}(x_k)$ and

$$\gamma_0 g^{(0)}(t) + \gamma_1 g^{(1)}(t) + \cdots + \gamma_r g^{(r)}(t) = \alpha_0 + \alpha_1 t + \cdots + \alpha_r t^r = g(t).$$

To achieve this we use the inductive Gram–Schmidt procedure, the first steps of which we have used previously. Use the notation

$$P_q := P_{[\mathbf{z}^{(q)}]}, \quad K_{q,s} := \frac{\langle \mathbf{z}^{(s)}, \mathbf{x}^{(q)} \rangle}{\langle \mathbf{z}^{(s)}, \mathbf{z}^{(s)} \rangle} \quad (s < q).$$

We take

$$
\mathbf{z}^{(0)} := \mathbf{x}^{(0)} = \mathbf{1},
$$
$$
\mathbf{z}^{(1)} := \mathbf{x}^{(1)} - P_0 \mathbf{x}^{(1)} = \mathbf{x}^{(1)} - K_{1,0}\, \mathbf{z}^{(0)},
$$
$$
\mathbf{z}^{(2)} := \mathbf{x}^{(2)} - P_0 \mathbf{x}^{(2)} - P_1 \mathbf{x}^{(2)} = \mathbf{x}^{(2)} - K_{2,0}\, \mathbf{z}^{(0)} - K_{2,1}\, \mathbf{z}^{(1)},
$$

etc, and, correspondingly,

$$
g^{(0)}(t) := 1,
$$
$$
g^{(1)}(t) := t - K_{1,0}\, g^{(0)}(t),
$$
$$
g^{(2)}(t) := t - K_{2,0}\, g^{(0)}(t) - K_{2,1}\, g^{(1)}(t),
$$

etc. For every q, $\mathbf{x}^{(q)}$ clearly belongs to $[\mathbf{z}^{(0)}, \mathbf{z}^{(1)}, \ldots, \mathbf{z}^{(q)}]$ and $\mathbf{z}^{(q)}$ clearly belongs to $[\mathbf{x}^{(0)}, \mathbf{x}^{(1)}, \ldots, \mathbf{x}^{(q)}]$. Equation 319(F2) follows.

Now, everything is just as in the case of Linear Regression. We use

$$
C_q := \frac{\langle \mathbf{z}^{(q)}, \mathbf{Y} \rangle}{\langle \mathbf{z}^{(q)}, \mathbf{z}^{(q)} \rangle} \sim \mathrm{N}\left(\gamma_q, \frac{\sigma^2}{\|\mathbf{z}^{(q)}\|^2}\right)
$$

as our (Unbiased) Estimator of γ_q. We set

$$
\mathrm{RSS} := \|P_{\mathrm{res}} \mathbf{Y}\|^2, \quad P_{\mathrm{res}} := I - P_0 - P_1 - \cdots - P_r
$$

in forming the Estimator

$$
\frac{\mathrm{RSS}}{n - r - 1}
$$

of σ^2. Our Estimator of $g(x_{\mathrm{new}})$ will be normally distributed with the correct mean and with variance

$$
\sigma^2 \sum_{q=1}^{r} \left(\frac{g^{(q)}(x_{\mathrm{new}})}{\|\mathbf{z}^{(q)}\|} \right)^2. \tag{F3}
$$

Fc. Model choice. So far, we have assumed that we have been given the correct model. We now consider a more relevant question for Statistics:

Given some data, how should one decide on the degree r of the polynomial to be used in modelling?

Obviously, one has to respect any underlying Physics, etc. (Perhaps, as in the next subsection, we should be using (say) exponential models rather than polynomial ones; but ignore that for now.) Generally, we try to comply with the Principle of Parsimony in that *we seek to use the simplest model consistent with any scientific laws which apply to the case being considered.*

In the case of polynomial modelling, *we try to use a model with r as small as possible.* One reason, related to the 'more parameters means greater variance for predictions' phenomenon we have studied before, is that *the variance at (F3) grows with r, particularly when* x_{new} *is close to the edge of the range of data x-values (or even more so when* x_{new} *is outside that range).*

Notice the way that the use of orthogonality means that the models are *nested* in that the $C_0, C_1, C_2, \ldots$ do not depend on the degree of the model. The reduction in the deviance resulting from adding another degree is easily determined. People sometimes look at the first value of s where an LR Test would not reject $\gamma_s = 0$ at the 5% level, and then take $r = s - 1$. But there are no 'hard and fast' rules. You clearly could have a case where there is no particular advantage in replacing a linear by a quadratic model, but where the data plot has the characteristic 'S' shape of a cubic. (At this point, do remember Nelder's dictum.) You simply have to use common sense.

Of course, the question of deciding on the degree of the polynomials comes under the heading of Model Choice; and the comments in 236J and, more especially, 303J, are relevant here. However, criteria have been designed specifically with Regression in mind. See, for example, the Mallows criterion in Mallows [155], Christensen [42], Draper and Smith [68] Gilmour [96], Montgomery and Peck [167], Myers [170]. See also Schwarz [209].

▶ **Fd. More on the Principle of Parsimony: sensitivity to data.** There is always a danger that computer packages will entice people into using over-complex models. You must bear in mind that (in addition to the 'larger variance for predictions' effect, *conclusions made using models with lots of parameters may well be very sensitive to small changes in data.* Here's an extreme example to illustrate the point.

Suppose that we are given $2n + 1$ data pairs

$$(x_{-n}, y_{-n}), \ (x_{-n+1}, y_{-n+1}), \ \ldots \ , \ (x_0, y_0), \ \ldots \ , (x_{n-1}, y_{n-1}), \ (x_n, y_n).$$

Let us also suppose that $x_k = k$ for $-n \leq k \leq n$.

If we use linear regression to make a single best predictor of the y-value corresponding to x-value $n + 1$, then changing y_0 by Δy_0 will cause a change $\Delta y_0 / (2n + 1)$ in our prediction.

If we are so stupid as to find the lowest-degree polynomial f such that $f(x_k) = y_k$ for $-n \leq k \leq n$, then (Lagrange's interpolation formula)

$$f(t) = \sum_i y_i \left(\prod_{j \neq i} \frac{t - x_j}{x_i - x_j} \right).$$

From one point of view, this is a perfect fit to the data. However, a change Δy_0 in y_0 will now cause a change

$$(-1)^n \frac{(2n+1)!}{(n!)^2} \Delta y_0 \approx (-1)^n 2^{2n+1} \left(\sqrt{n/\pi} \right) \Delta y_0$$

in our prediction $f(n+1)$. We used Stirling's formula via Exercise 11Na.

Fe. Models necessitating the use of many parameters.   Some models, by their very nature, demand a lot of parameters. (See, for example, those at 358Ka and 280Jd.) In regard to them, one important thing is not to get lulled by beautiful printouts into confusing individual CIs with simultaneous ones. The latter give the fuller picture, especially if the experiment is on new phenomena; and in models with wide-tailed distributions, simultaneous CIs could be markedly different from individual CIs.

G. Remark on Multilinear Regression An important area of regression is where we may have several different regressors or covariates $x, s, t, u, \ldots$ so that the model is

$$Y_k = \alpha + \beta x_k + \gamma s_k + \delta t_k + \lambda u_k + \cdots + \varepsilon_k.$$

In Polynomial Regression, we would have relations $r_k = x_k^2$, $s_k = x_k^3$, etc. But we are now considering the case where x, s, t, u are covariates (or *regressors*) with no mathematical relations connecting them. In our hurdler example, x might be windspeed, s the time during the season, t the height above sea-level at which the race was run, u the temperature, ... Obviously the present model may be written in the form at 319(F1). but there is now no natural order in which to place the regressors, and hence no canonical way to play a Gram–Schmidt game. You can, if you wish, use AIC or the Mallows statistic to decide on how many regressors to keep in your model. See the books and papers listed at the end of the discussion in 320Fc.

H. Non-linear Regression; exponential-decay models.   If Y_k is my measurement of the temperature at time x_k of my coffee, which I always forget to drink, then Newton tells me to use a model (non-linear in the parameters, and therefore outside the scope of what we have done)

$$Y_k = \alpha + \beta e^{-\gamma x_k} + \varepsilon_k, \tag{H1}$$

where α, β, γ are unknown parameters. The fact that γ is unknown puts this outside the scope of what we have done. This type of 'exponential decay' model is important in a number of contexts. WinBUGS doesn't handle it willingly, but can be persuaded to, as the 'dugongs' example which comes with the package indicates.

Ha. Exercise. Contrast the (H1) model with

$$\ln(Y_k - \alpha) = \ln\beta - \gamma x_k + \varepsilon_k. \tag{H2}$$

For which of (H1) and (H2) is the assumption of IID normal errors more reasonable?

I. More motivation. I (hope and) trust that these ideas are giving a much clearer idea of the underlying geometry than the '$\mathbf{Y} = X\beta + \varepsilon$' treatment usually presented, where, in the case of ANOVA, the matrix X is too hideous to contemplate. (Yes, I know why it is done that way for complete generality. And see Subsection 336Q!) I wish to show how Mathematics can handle the geometry of some important cases of ANOVA in a way which properly reflects the structure of those cases. We need to extend our Linear Algebra to cover this topic.

J. Matrices and linear transformations; adjoints. An $n \times n$ matrix

$$A = \begin{pmatrix} A_{11} & A_{12} & \cdots & A_{1n} \\ A_{21} & A_{22} & \cdots & A_{2n} \\ \cdot & \cdot & \cdots & \cdot \\ A_{n1} & A_{n2} & \cdots & A_{nn} \end{pmatrix}$$

induces a linear transformation $\varphi_A : \mathbb{R}^n \to \mathbb{R}^n$ defined by $\varphi_A(\mathbf{x}) = A\mathbf{x}$. By a linear transformation of $\mathbb{R}^n$, we mean a map $\varphi : \mathbb{R}^n \to \mathbb{R}^n$ such that for $\lambda, \mu \in \mathbb{R}$ and $\mathbf{x}, \mathbf{y} \in \mathbb{R}^n$,

$$\varphi(\lambda\mathbf{x} + \mu\mathbf{y}) = \lambda\varphi(\mathbf{x}) + \mu\varphi(\mathbf{y}).$$

If φ is a linear transformation of $\mathbb{R}^n$, then $\varphi = \varphi_A$ for some matrix A, namely the matrix with $A_{ij} = \varphi\left(\mathbf{e}^{(j)}\right)_i$, the ith component of $\varphi(\mathbf{e}^{(j)})$. We say that A represents φ with respect to the standard basis $\{\mathbf{e}^{(1)}, \mathbf{e}^{(2)}, \ldots, \mathbf{e}^{(n)}\}$ of $\mathbb{R}^n$.

Here's the general story – which we shall continue in our study of the MVN distributions. If $\mathcal{U} = \{\mathbf{u}^{(1)}, \mathbf{u}^{(2)}, \ldots, \mathbf{u}^{(n)}\}$ is ANY basis of $\mathbb{R}^n$, and φ is a linear transformation of $\mathbb{R}^n$, then for some matrix B,

$$\varphi(c_1\mathbf{u}_1 + \cdots + c_n\mathbf{u}_n) = d_1\mathbf{u}_1 + \cdots + d_n\mathbf{u}_n,$$

where, if $\mathbf{c} = (c_1, c_2, \ldots, c_n)$ and $\mathbf{d} = (d_1, d_2, \ldots, d_n)$, we have

$$\mathbf{d} = B\mathbf{c}.$$

We say that B represents φ relative to the basis $\mathcal{U}$.

The **range** $\mathcal{R}(\varphi)$ of a linear transformation φ is the set of all vectors of the form $\varphi\mathbf{w}$ where $\mathbf{w} \in \mathbb{R}^n$. Check that $\mathcal{R}(\varphi)$ is a subspace of $\mathbb{R}^n$. The dimension of $\mathcal{R}(\varphi)$ is called the **rank** of φ. By the range or rank of a matrix A, we mean the corresponding entity for φ_A.

Now, $\mathbb{R}^n$ carries the inner product $\langle \cdot, \cdot \rangle$. The **adjoint** φ^* of a linear transformation φ is defined via

$$\langle \mathbf{x}, \varphi\mathbf{y} \rangle = \langle \varphi^*\mathbf{x}, \mathbf{y} \rangle. \tag{J1}$$

Ja. Exercise.   Show that if φ is represented by B relative to an *orthonormal* basis $\mathcal{U}$, then φ^* is represented by B^T relative to $\mathcal{U}$. Of course, B^T denotes the transpose of B:

$$\left(B^T\right)_{ij} := (B)_{ji}.$$

$\square$

A transformation φ is called **self-adjoint** if $\varphi = \varphi^*$.

Jb. Lemma.   *Let U be a subspace of $\mathbb{R}^n$. Then the perpendicular projection $P = P_U$ is a linear transformation of $\mathbb{R}^n$ such that*

$$P^2 = P = P^*. \tag{J2}$$

Conversely, if P is a linear transformation of $\mathbb{R}^n$ such that equation (J2) holds then P is the perpendicular projection onto $\mathcal{R}(P)$. Proof. I'll prove the first part; you prove the 'converse'.

Let U be a subspace of $\mathbb{R}^n$. If $\lambda, \mu \in \mathbb{R}$ and $\mathbf{x}, \mathbf{y} \in \mathbb{R}^n$, then

$$\lambda P_U \mathbf{x} + \mu P_U \mathbf{y} \in U,$$

and, for $\mathbf{u} \in U$,

$$\langle \lambda \mathbf{x} + \mu \mathbf{y} - \lambda P_U \mathbf{x} - \mu P_U \mathbf{y}, \mathbf{u} \rangle = \lambda \langle \mathbf{x} - P_U \mathbf{x}, \mathbf{u} \rangle + \mu \langle \mathbf{y} - P_U \mathbf{y}, \mathbf{u} \rangle = 0.$$

Hence,

$$P_U(\lambda \mathbf{x} + \mu \mathbf{y}) = \lambda P_U(\mathbf{x}) + \mu P_U(\mathbf{y}).$$

Next, $\langle \mathbf{x} - P_U \mathbf{x}, P_U \mathbf{y} \rangle = 0$, so

$$\langle \mathbf{x}, P_U \mathbf{y} \rangle = \langle P_U \mathbf{x}, P_U \mathbf{y} \rangle = \langle P_U \mathbf{x}, \mathbf{y} \rangle$$

by symmetry. Hence $P^* = P$. Manifestly, $P_U^2 \mathbf{x} := P_U(P_U \mathbf{x}) = P_U \mathbf{x}$, since $P_U \mathbf{x} \in U$.

$\square$

▶ **K. Tensor products.**   Here is the natural Linear Algebra to describe the Mathematics of the ANOVA model of Subsection 289F.

The good thing is that *'nothing goes wrong' for tensor products*: every result you could wish for is true.

The tensor product $\mathbb{R}^J \otimes \mathbb{R}^K$ is the JK-dimensional space of vectors Θ of the form

$$\Theta = \{\Theta_{jk} : 1 \leq j \leq J,\ 1 \leq k \leq K\}.$$

In other words, Θ looks just like a matrix, but we must think of it as a *vector* in which the components are laid out in rows and columns instead of all being put in a single column. Of course, for such vectors Θ and Φ,

$$(\Theta + \Phi)_{jk} := \Theta_{jk} + \Phi_{jk}, \quad (\lambda\Theta)_{jk} := \lambda\Theta_{jk}.$$

For $\mathbf{x} \in \mathbb{R}^J$ and $\mathbf{y} \in \mathbb{R}^K$, we define $\mathbf{x} \otimes \mathbf{y}$ to be the vector in $\mathbb{R}^J \otimes \mathbb{R}^K$ with

$$(\mathbf{x} \otimes \mathbf{y})_{jk} := x_j y_k.$$

We have, for $\mathbf{x} \in \mathbb{R}^J$ and $\mathbf{y}$ in $\mathbb{R}^K$,

$$\mathbf{x} \otimes (\lambda \mathbf{y} + \mu \mathbf{z}) = \lambda (\mathbf{x} \otimes \mathbf{y}) + \mu (\mathbf{x} \otimes \mathbf{z}), \quad \text{etc.}$$

Not every vector in $\mathbb{R}^J \otimes \mathbb{R}^K$ is of the form $\mathbf{x} \otimes \mathbf{y}$ (can you see why?), a fact which is the source of the entanglement phenomenon in Quantum Mechanics, as we shall see in Chapter 10. However, vectors $\mathbf{x} \otimes \mathbf{y}$ span $\mathbb{R}^J \otimes \mathbb{R}^K$: if $\mathcal{U} = \{\mathbf{u}^{(1)}, \mathbf{u}^{(2)}, \dots, \mathbf{u}^{(J)}\}$ is a basis for $\mathbb{R}^J$ and $\mathcal{V} = \{\mathbf{v}^{(1)}, \mathbf{v}^{(2)}, \dots, \mathbf{v}^{(K)}\}$ is a basis for $\mathbb{R}^K$, then the set $\mathcal{U} \otimes \mathcal{V}$ of vectors of the form $\mathbf{u}^{(j)} \otimes \mathbf{v}^{(k)}$ is a basis for $\mathbb{R}^J \otimes \mathbb{R}^K$. (Moreover, if $\mathbf{x}$ has coordinates $c_1, c_2, \dots, c_J$ relative to $\mathcal{U}$ and $\mathbf{y}$ has coordinates $d_1, d_2, \dots, d_K$ relative to $\mathcal{V}$, then $\mathbf{x} \otimes \mathbf{y}$ has coordinates $c_j d_k$ relative to $\mathcal{U} \otimes \mathcal{V}$. This makes the tensor product intrinsic, basis-independent, so it's good Mathematics.)

For a $J \times J$ matrix A and a $K \times K$ matrix B, we define the tensor product $JK \times JK$ matrix, the $((j_1, k_1), (j_2, k_2))$th component of which is

$$(A \otimes B)_{(j_1, k_1),\, (j_2, k_2)} := A_{j_1 j_2} B_{k_1 k_2}.$$

Then (intrinsic characterization)

$$(A \otimes B)(\mathbf{x} \otimes \mathbf{y}) = (A\mathbf{x}) \otimes (B\mathbf{y}), \tag{K1}$$

since the (j_1, k_1)th component of the left-hand side is

$$\sum_{j_2} \sum_{k_2} (A \otimes B)_{(j_1, k_1),\, (j_2, k_2)} (\mathbf{x} \otimes \mathbf{y})_{j_2 k_2}$$

$$= \sum_{j_2} \sum_{k_2} A_{j_1 j_2} B_{k_1 k_2} x_{j_2} y_{k_2}$$

$$= \left(\sum_{j_2} A_{j_1 j_2} x_{j_2} \right) \left(\sum_{k_2} B_{k_1 k_2} y_{k_2} \right) = (A\mathbf{x})_{j_1} (B\mathbf{y})_{k_1}.$$

Because vectors of the form $(\mathbf{x} \otimes \mathbf{y})$ span $\mathbb{R}^J \otimes \mathbb{R}^K$, property (K1) characterizes $A \otimes B$.

The standard inner product on $\mathbb{R}^J \otimes \mathbb{R}^K$ is, of course,

$$\langle \Theta, \Phi \rangle := \sum_j \sum_k \Theta_{jk} \Phi_{jk}.$$

We have (intrinsic characterization)

$$\langle \mathbf{x} \otimes \mathbf{y}, \mathbf{w} \otimes \mathbf{z} \rangle = \langle \mathbf{x}, \mathbf{w} \rangle \langle \mathbf{y}, \mathbf{z} \rangle,$$

because

$$\sum_j \sum_k x_j y_k w_j z_k = \left(\sum_j x_j w_j \right) \left(\sum_k y_k z_k \right).$$

Finally suppose that U is a subspace of $\mathbb{R}^J$ and V is a subspace of $\mathbb{R}^K$. Then $U \otimes V$ is defined to be the subspace of $\mathbb{R}^J \otimes \mathbb{R}^K$ spanned by the vectors $\mathbf{u} \otimes \mathbf{v}$, where $\mathbf{u} \in U$ and $\mathbf{v} \in V$. We have

$$P_{U \otimes V} = P_U \otimes P_V, \tag{K2}$$

where, of course, $P_{U \otimes V}$ acts on $\mathbb{R}^J \otimes \mathbb{R}^K$, P_U on $\mathbb{R}^J$ and P_V on $\mathbb{R}^K$.

Proof of equation (K2). Let $\mathbf{x} \in \mathbb{R}^J, \mathbf{y} \in \mathbb{R}^K, \mathbf{u} \in U, \mathbf{v} \in V$. Firstly, we have

$$(P_U \mathbf{x}) \otimes (P_V \mathbf{y}) \in U \otimes V,$$

so that, since $P_U \otimes P_V$ is linear, we have

$$(P_U \otimes P_V) : \mathbb{R}^J \otimes \mathbb{R}^K \to U \otimes V.$$

Secondly,

$$\begin{aligned}\langle \mathbf{x} \otimes \mathbf{y} - (P_U \otimes P_V)(\mathbf{x} \otimes \mathbf{y}), \ \mathbf{u} \otimes \mathbf{v} \rangle \\ = \ \langle \mathbf{x}, \mathbf{u} \rangle \langle \mathbf{y}, \mathbf{v} \rangle - \langle P_U \mathbf{x}, \mathbf{u} \rangle \langle P_V \mathbf{y}, \mathbf{v} \rangle \ = \ 0,\end{aligned}$$

because $\langle \mathbf{x} - P_U \mathbf{x}, \mathbf{u} \rangle = 0 = \langle \mathbf{y} - P_V \mathbf{y}, \mathbf{v} \rangle$. Hence, $\mathbf{x} \otimes \mathbf{y} - (P_U \otimes P_V)(\mathbf{x} \otimes \mathbf{y})$ is orthogonal to $U \otimes V$; and the proof is complete. $\square$

Ka. Exercise. Prove that $\dim(U \otimes V) = (\dim U)(\dim V)$.

▶▶ **L. Two-factor ANOVA without interaction: theory.** Recall the situation in Subsection 289F. We have J levels for $\mathcal{A}$, K for $\mathcal{B}$ and the model

$$Y_{jk} = \mu + \alpha_j + \beta_k + \varepsilon_{jk},$$

where μ is the true overall mean, α_j the true jth $\mathcal{A}$-effect, β_k the true kth $\mathcal{B}$-effect, and

$$\sum \alpha_j = 0 = \sum \beta_k, \qquad \varepsilon_{jk} \text{ are IID each } N(0, \sigma^2).$$

Define

$$Q := P_{[1]} \text{ on } \mathbb{R}^J, \qquad R := P_{[1]} \text{ on } \mathbb{R}^K,$$

and, on $\mathbb{R}^J \otimes \mathbb{R}^K$, define projections as in the 'Defn' part of the following table:

Projection	Defn	SS	num	dof
P_{mean}	$Q \otimes R$	JKM^2	1500	1
$P_{\mathcal{A} \text{ eff}}$	$Q^{\perp} \otimes R$	$K \sum A_j^2$	138	$J - 1$
$P_{\mathcal{B} \text{ eff}}$	$Q \otimes R^{\perp}$	$J \sum B_k^2$	40	$K - 1$
P_{res}	$Q^{\perp} \otimes R^{\perp}$	$\sum \sum E_{jk}^2$	48	$(J-1)(K-1)$
P_{total}	$I \otimes I$	$\sum \sum Y_{jk}^2$	1726	JK

The 'num' column shows the 'observed values' for our numerical example in Subsection 289F. Other parts of the table are explained below.

Let

$$\overline{Y}_{**} := \frac{1}{JK} \sum_j \sum_k Y_{jk},$$

$$\overline{Y}_{j*} := \frac{1}{K} \sum_k Y_{jk},$$

$$\overline{Y}_{*k} := \frac{1}{J} \sum_j Y_{jk}.$$

Then,

$$
\begin{aligned}
(P_{\text{mean}} \mathbf{Y})_{jk} &= M := \overline{Y}_{**} && \text{(Estd Overall Mean)}, \\
(P_{\mathcal{A} \text{ eff}} \mathbf{Y})_{jk} &= A_j := \overline{Y}_{j*} - \overline{Y}_{**} && \text{(Estd } j\text{th } \mathcal{A} \text{ Effect)}, \\
(P_{\mathcal{B} \text{ eff}} \mathbf{Y})_{jk} &= B_k := \overline{Y}_{*k} - \overline{Y}_{**} && \text{(Estd } k\text{th } \mathcal{B} \text{ Effect)}, \\
(P_{\text{res}} \mathbf{Y})_{jk} &= E_{jk} := Y_{jk} - \overline{Y}_{j*} - \overline{Y}_{*k} + \overline{Y}_{**} && \text{(Estd } jk\text{th Residual)}.
\end{aligned}
$$

But

$$P_{\text{mean}}, \quad P_{\mathcal{A} \text{ eff}}, \quad P_{\mathcal{B} \text{ eff}}, \quad P_{\text{res}}$$

are four projections onto the four mutually orthogonal subspaces

$$[\mathbf{1}] \otimes [\mathbf{1}], \quad [\mathbf{1}]^\perp \otimes [\mathbf{1}], \quad [\mathbf{1}] \otimes [\mathbf{1}]^\perp, \quad [\mathbf{1}]^\perp \otimes [\mathbf{1}]^\perp,$$

of dimensions

$$1 \times 1, \quad (J-1) \times 1, \quad 1 \times (K-1), \quad (J-1) \times (K-1),$$

which together span $\mathbb{R}^J \otimes \mathbb{R}^K$. We set $\nu = (J-1)(K-1)$.

▶ **La. Exercise.** Identifying clearly the spaces U, Z, W, etc, for the application of Theorem 301Ha, prove that the F-test for H_0 : 'all $\mathcal{A}$ effects zero' rejects H_0 at size η if

$$\frac{K \left(\sum A_j^2 \right) / (J-1)}{\text{RSS}/\nu} > F^*_{J-1, \nu}(\eta). \qquad \square$$

For the example in Subsection 289F,

$$\frac{\text{estimate of } \sigma^2 \text{ based on estimated } \mathcal{A} \text{ means}}{\text{estimate of } \sigma^2 \text{ based on estimated residuals}} = \frac{34.5}{6} = 5.75.$$

Because $5.75 > F^*_{4,8}(0.05) = 4.46$, we reject the Null Hypothesis that all α_j are equal to 0 at the 5% significance level.

It is not enough to be told that we would reject the Null Hypothesis: we want to quantify the differences between the α_j's, the true $\mathcal{A}$ effects.

Confidence Intervals. We continue to set

$$\nu = (J-1)(K-1).$$

Let us now concentrate on obtaining CIs for true $\mathcal{A}$ means. We write

$$\theta_j := \mu + \alpha_j$$

for the true mean corresponding to the jth level of $\mathcal{A}$. Then, with

$$P_{\mathcal{A} \text{ mean}} := P_{\text{mean}} + P_{\mathcal{A} \text{ eff}},$$

we have three mutually orthogonal projections (the product of any two is zero)

$$P_{\mathcal{A} \text{ mean}}, \quad P_{\mathcal{B} \text{ eff}}, \quad P_{\text{res}}$$

summing to I on $\mathbb{R}^J \otimes \mathbb{R}^K$.

▶ **Lb. Exercise: Individual Confidence Intervals.** Prove that, for an individual j,

$$\overline{y}_{j*}^{\text{obs}} \pm |t|_\nu^*(\eta) \left(\frac{\text{rss}^{\text{obs}}}{K\nu} \right)^{\frac{1}{2}}$$

is a $100(1-\eta)\%$ CI for θ_j, and check `Minitab` out for the example in Subsection 289F. Show that, for a fixed pair (j_1, j_2),

$$\overline{y}_{j_1*}^{\text{obs}} - \overline{y}_{j_2*}^{\text{obs}} \pm |t|_\nu^*(\eta) \left(\frac{2\,\text{rss}^{\text{obs}}}{K\nu} \right)^{\frac{1}{2}}$$

is a $100(1-\eta)\%$ CI for $\theta_{j_1} - \theta_{j_2} = \alpha_{j_1} - \alpha_{j_2}$.

▶ **Lc. 'Individual versus Simultaneous' for Confidence Intervals.**   Which of individual and simultaneous CIs we use depends on the context. Of course, **one can always give both.**

If the experiment relates to a well-understood situation, we may identify before the experiment a particular 'contrast' $\alpha_{j_1} - \alpha_{j_2}$ of interest, where (j_1, j_2) is a fixed pair of distinct $\mathcal{A}$ levels. The individual CI for $\alpha_{j_1} - \alpha_{j_2}$ is then appropriate for study. If, however, the experiment is one on a new situation and we wish to describe the overall picture, then there is a good case for using simultaneous CIs.

There are complex issues here: for example picking out after the experiment a pair (j_1, j_2) corresponding to extreme features of the data should bring back memories of the later parts of Subsection 278J. As stated in that subsection, a study of such problems is outside the scope of this book. Giving both individual and simultaneous CIs (or a CR) is of course a safe strategy.

Please note that my reference to STRANGE and MAD below is in the spirit of physicists' terminology, and certainly implies no disrespect for the methods.

Note too that the technical definition of a **contrast** involving the α_j's is an expression $\sum c_j \alpha_j$ where $\sum c_j = 0$.

▶ **Ld. An 'F' Confidence Region.** The usual arguments now show that if we write

$$\tau_{jk} := \theta_j = \mu + \alpha_j,$$

then

$$\sigma^{-1}\left(P_{\mathcal{A} \text{ mean}}\mathbf{Y} - \boldsymbol{\tau}\right) \quad \sim \quad \text{SN}\left(\mathbb{R}^J \otimes [\mathbf{1}]\right), \qquad (\dim J),$$
$$\sigma^{-1}P_{\text{res}}\mathbf{Y} \quad \sim \quad \text{SN}\left([\mathbf{1}]^\perp \otimes [\mathbf{1}]^\perp\right), \qquad (\dim (J-1)(K-1)),$$

these variables being independent. We shall have

$$\text{RSS} = \|P_{\text{res}}\mathbf{Y}\|^2 \sim \sigma^2 \chi_\nu^2, \quad \nu := (J-1)(K-1).$$

If $\{\mathbf{e}^{(1)}, \mathbf{e}^{(2)}, \ldots, \mathbf{e}^{(J)}\}$ is the standard basis for $\mathbb{R}^J$, and $\{\mathbf{v}\}$, where $\mathbf{v} = K^{-\frac{1}{2}}\mathbf{1}$, is an orthonormal basis for $[\mathbf{1}]$ in $\mathbb{R}^K$, then the $\mathbf{e}^{(j)} \otimes \mathbf{v}$ form an orthonormal basis for $\mathbb{R}^J \otimes [\mathbf{1}]$. But

$$\sigma^{-1}\langle P_{\mathcal{A} \text{ mean}}\mathbf{Y} - \boldsymbol{\tau}, \mathbf{e}^{(j)} \otimes \mathbf{v}\rangle = \sigma^{-1}\{\overline{Y}_{j*} - \theta_j\}K^{\frac{1}{2}}$$

whence, as it is easy to prove directly, the variables

$$\sigma^{-1}\{\overline{Y}_{j*} - \theta_j\}K^{\frac{1}{2}} \quad (1 \leq j \leq J) \tag{L1}$$

are IID each $N(0,1)$. The Orthonormality Principle implies that these variables at (L1) are independent of RSS. Hence,

$$\frac{K\sum_j \left(\overline{Y}_{j*} - \theta_j\right)^2 / J}{\text{RSS}/\nu} \sim F_{J,\nu};$$

and this gives us as $100(1 - \eta)\%$ Confidence Region for the vector $\boldsymbol{\theta}$ a ball in $\mathbb{R}^J$,

$$\text{centre } \{\overline{y}_{j*}^{\text{obs}} : 1 \leq j \leq J\}, \qquad \text{radius } \left\{\frac{J F_{J,\nu}^*(\eta)\text{rss}^{\text{obs}}}{K\nu}\right\}^{\frac{1}{2}}.$$

Because it is not that easy to think of this CR, we tend to prefer simultaneous CIs.

▶ **Le. Studentized-Range (STRANGE) $\text{SR}_{J,\nu}$.** The *Studentized-Range* distribution $\text{SR}_{J,\nu}$ is defined to be the distribution of

$$\frac{\max\{X_j : 1 \leq j \leq J\} - \min\{X_j : 1 \leq j \leq J\}}{\sqrt{R/\nu}}$$

where $X_1, X_2, \ldots, X_J, R$ are independent RVs, each X_j being $N(0,1)$ and R having the χ_ν^2 distribution. The STRANGE distributions are not easy to calculate. Many computer packages these days have them available, but be careful to check the conventions used.

If $\mathrm{SR}^*_{J,\nu}(\eta)$ is such that $\mathbb{P}\left(\mathrm{SR}_{J,\nu} > \mathrm{SR}^*_{J,\nu}(\eta)\right) = \eta$, then there is $100(1-\eta)\%$ probability that

$$X_{j_1} - X_{j_2} \le \mathrm{SR}^*_{J,\nu}(\eta)\sqrt{\frac{R}{\nu}} \qquad \text{simultaneously for all } (j_1, j_2).$$

In our ANOVA situation, there is $100(1-\eta)\%$ probability that

$$\overline{Y}_{j_1*} - \overline{Y}_{j_2*} - (\theta_{j_1} - \theta_{j_2}) \le \mathrm{SR}^*_{J,\nu}(\eta)\sqrt{\frac{\mathrm{RSS}}{K\nu}}$$

simultaneously for all pairs (j_1, j_2). Hence, we can be $100(1-\eta)\%$ confident that, **simultaneously for all pairs** (j_1, j_2),

$$\theta_{j_1} - \theta_{j_2} \in (\overline{y}^{\mathrm{obs}}_{j_1*} - \overline{y}^{\mathrm{obs}}_{j_2*}) \pm \mathrm{SR}^*_{J,\nu}(\eta)\sqrt{\frac{\mathrm{rss}^{\mathrm{obs}}}{K\nu}}.$$

This is a good way to compare differences in $\mathcal{A}$ effects for small J.

Note that in the example in Subsection 289F,

$$\frac{\mathrm{SR}^*_{5,8}(0.05)}{2^{\frac{1}{2}}|t|^*_8(0.05)} = \frac{4.89}{2^{\frac{1}{2}} \times 2.31} = 1.50,$$

so that the 95% simultaneous CIs are here 1.50 times wider than the 95% individual CIs.

Lf. Exercise. Explain why $\mathrm{SR}^*_{2,\nu}(\eta) = 2^{\frac{1}{2}}|t|^*_\nu(\eta)$. ◻

There *can* be 'conflicts'. One could find that one rejects at the 5% significance level the Null Hypothesis that all $\mathcal{A}$ effects are zero, while finding that all intervals in the 95% simultaneous CIs for differences $\theta_{j_1} - \theta_{j_2}$ obtained via STRANGE contain 0. And it can work 'the other way round'. The intuitive reason is surely obvious: a ball is not a cube!

▶ **Lg. Maximum Absolute Deviation (MAD).** If J is large (6 or more), there are $\frac{1}{2}J(J-1)$ differences $\theta_{j_1} - \theta_{j_2}$, too many to contemplate. So the *Maximum Absolute Deviation* method focuses on the distribution of

$$\frac{\max_j |X_j - \overline{X}|}{\sqrt{R/\nu}}$$

under the same conditions as those in the definition of STRANGE. You can translate this into a method of getting simultaneous CIs for the $\theta_j - \mu = \alpha_j$ in ANOVA.

Again the MAD distributions are not easy to calculate. ◻

Note. Tukey's paper [230] suggests many lines of investigation which we do not have space to pursue here.

▶▶ **M. Two-factor ANOVA with interaction and replication.** As explained in Section 28C, we use a model

$$Y_{jk,l} = \mu + \alpha_j + \beta_k + (\alpha\beta)_{jk} + \varepsilon_{jk,l}$$

to investigate the interaction between $\mathcal{A}$ and $\mathcal{B}$ factors. The notation $(\alpha\beta)_{jk}$ just signifies a constant: there is no multiplication involved in this standard notation. We have

$$\sum \alpha_j = 0 = \sum \beta_k,$$

$$\sum_k (\alpha\beta)_{jk} = 0 \text{ for each } j, \qquad \sum_j (\alpha\beta)_{jk} = 0 \text{ for each } k,$$

and the $\varepsilon_{jk,l}$ are IID each $N(0, \sigma^2)$.

Tensor products handle things nicely. We have

Projection	Defn	SS	dof
P_{mean}	$Q \otimes R \otimes S$	$JKLM^2$	1
$P_{\mathcal{A}\text{ eff}}$	$Q^\perp \otimes R \otimes S$	$KL \sum A_j^2$	$J - 1$
$P_{\mathcal{B}\text{ eff}}$	$Q \otimes R^\perp \otimes S$	$JL \sum B_k^2$	$K - 1$
$P_{(\mathcal{AB})\text{ eff}}$	$Q^\perp \otimes R^\perp \otimes S$	$L \sum\sum (AB)_{jk}^2$	$(J-1)(K-1)$
P_{res}	$I \otimes I \otimes S^\perp$	$\sum\sum\sum E_{jk,l}^2$	$JK(L-1)$
P_{total}	$I \otimes I \otimes I$	$\sum\sum Y_{jkl}^2$	JKL

Here,

$$M = \overline{Y}_{***}, \quad A_j = \overline{Y}_{j**} - M,$$

etc, etc.

Really, you can derive for yourself any information you wish from this table, and you can see the relevance of the F-test for complex models.

Ma. Exercise. Show that if, with $\nu_1 = (J-1)(K-1)$ and $\nu_2 = JK(L-1)$, we have

$$\frac{\left\| P_{(\mathcal{AB})\text{ eff}} \mathbf{Y} \right\|^2}{\nu_1} > F^*_{\nu_1, \nu_2}(0.05) \frac{\left\| P_{\text{res}} \mathbf{Y} \right\|^2}{\nu_2},$$

then we reject at the 5% level the hypothesis that all $\mathcal{AB}$ effects are zero. We might therefore choose to investigate these interaction effects further.

If there is not evidence at the 5% level against the hypothesis that all $\mathcal{AB}$ effects are zero, we might well stick with the additive (no-interaction) model.

Mb. Exercise. Why are we not interested in the projection $Q \otimes R \otimes S^\perp$? What 'effect' does it relate to?

▶ **N. The Bayesian view of linear regression.**   Return to the situation of Subsection 312D, so that

$$\mathbf{Y} = \mu\mathbf{1} + \beta\mathbf{z} + \sigma\mathbf{G}, \quad \mathbf{z} = \mathbf{x} - \bar{x}\mathbf{1}, \quad \mathbf{G} \sim \mathrm{SN}(\mathbb{R}^n).$$

With $\rho = 1/\sigma^2$ as usual, the log-likelihood satisfies

$$\ell(\mu, \beta, \rho; \mathbf{y}) = -\tfrac{1}{2}n\ln(2\pi) + \tfrac{1}{2}n\ln\rho - \tfrac{1}{2}\rho\|\mathbf{y} - \mu\mathbf{1} - \beta\mathbf{z}\|^2.$$

However, if

$$\mathbf{v} = \mathbf{y} - \mu\mathbf{1} - \beta\mathbf{z},$$

then, we can decompose $\mathbf{v}$ 'orthogonally' as

$$\mathbf{v} = P_{[\mathbf{1}]}\mathbf{v} + P_{[\mathbf{z}]}\mathbf{v} + P_{[\mathbf{1},\mathbf{z}]}^{\perp}\mathbf{v},$$

whence (you check!)

$$\tfrac{1}{2}\rho\|\mathbf{y} - \mu\mathbf{1} - \beta\mathbf{z}\|^2 = -\tfrac{1}{2}\rho n\,(\mu - \bar{y})^2 - \tfrac{1}{2}\rho q(\mathbf{x},\mathbf{x})(\beta - b)^2 - \tfrac{1}{2}\rho(\mathrm{rss}),$$

where $b = q(\mathbf{x},\mathbf{y})/q(\mathbf{x},\mathbf{x})$ and rss is as usual.

If we assume a prior for (μ, β, ρ) where, with obvious misuse of notation,

$$\mu \sim \mathrm{N}\,(m_\mu, \,\mathrm{prec}\; r_\mu), \quad \beta \sim \mathrm{N}\,(m_\beta, \,\mathrm{prec}\; r_\beta), \quad \rho \sim \mathrm{Gamma}(K, \,\mathrm{rate}\; \gamma),$$

these being 'independent', then we would have

$$\ln\pi(\mu, \beta, \rho)$$
$$= \mathrm{constant} - \tfrac{1}{2}r_\mu(\mu - m_\mu)^2 - \tfrac{1}{2}r_\beta(\beta - m_\beta)^2 + (K - 1)\ln\rho - \gamma\rho.$$

Na. Exercise. Show that for the posterior density $\pi(\mu, \beta, \rho \,|\, \mathbf{y})$, we have

$$(\mu \,|\, \beta, \rho) \sim \mathrm{N}(?,?), \quad (\beta \,|\, \mu, \rho) \sim \mathrm{N}(?,?),$$
$$(\rho \,|\, \mu, \beta) \sim \mathrm{Gamma}\left(K + \tfrac{1}{2}n, \,\mathrm{rate}\; \gamma + \tfrac{1}{2}\rho\|\mathbf{y}^{\mathrm{obs}} - \mu\mathbf{1} - \beta\mathbf{z}\|^2\right),$$

filling in the gaps where the '?' signs appear. Table 211 J(i) of 'conjugates' should help greatly.   □

The above exercise tells one how to apply the Gibbs sampler to this situation. Here's the Example 315Dd done in WinBUGS .

```
model
const N = 20;
var mu[N], Y[N], alpha, beta, rho;
{
```

```
   x.bar <- mean(x[])
   for(i in 1:N){
     mu[i] <- alpha + beta * (x[i] - x.bar)
        Y[i] ~ dnorm(mu[i], rho)
   }
   alpha ~ dnorm(0, 1.0E-6)
   beta ~ dnorm(0, 1.0E-6)
   rho ~ dgamma(1.0E-3, 1.0E-3)
}
#data
list( N = 20,
   x = c(   0,    1,    2,    3,    4,    5,    6,    7,    8,    9,
           10,   11,   12,   13,   14,   15,   16,   17,   18,   19),
   Y = c(1.30,2.65,0.89,1.09,1.60,1.96,1.89,1.50,2.99,2.77,
           3.30,3.22,3.61,3.09,3.32,3.28,3.61,4.04,4.10,4.74)
)
#inits
list(alpha = 0, beta = 0, rho = 1)

95% CI for beta after BUGS run of 100000
     WinBUGS                 Frequentist
   (0.1249, 0.2065)   (0.1256, 0.2066)
```

There are so few parameters that we can obtain conservative simultaneous CIs in the usual way.

▶ **O. The Bayesian view of ANOVA.** You can see that the Bayesian theory is somewhat less elegant than the Frequentist for classical linear models. (However, the Bayesian approach has the advantage of allowing much more general models – as well as that of building in prior information). I am not going to bore you with all the details of the Bayesian version of ANOVA. I'm sure that you can see in principle how it would all go.

There is a difficulty however. Consider the model

$$Y_{jk} = \theta_j + \beta_k + \varepsilon_{jk},$$

where the ε_{jk} are IID each $N(0, \sigma^2)$. We have to be rather careful about possible instabilities caused by the '**aliasing**' problem that we can add an arbitrary constant C to every θ_j provided that we subtract the same constant C from every β_k. If we restrict the β_k by imposing the condition that $\sum \beta_k = 0$, then we are restricting the vector β to lie on a submanifold (which in this case is a hyperplane) of $\mathbb{R}^K$, so β will not have pdf on $\mathbb{R}^K$.

Still, WinBUGS seems to handle things well. The numerical example from Subsection 289F is now studied via WinBUGS via two variants of a program. Here's the first.

```
model ANfreq;
var
  Y[5,3], Ymean[5,3],
  tau[5], gamma[3], theta[5],
  mu, rho, gbar;

{
  mu ~ dnorm(0, 1.0E-6);
  for(j in 1:5){                    # Setting up taus
    tau[j] ~ dnorm(mu, 1.0E-6);     # Vague tau[]
  }
  for(k in 1:3){                    # Setting up gammas
    gamma[k] ~ dnorm(0, 1.0E-6);    # Vague gamma[]
  }
  rho ~ dgamma(1.0E-3, 1.0E-3);     # Vague rho
  for(j in 1:5){                    # Setting up Y's
    for(k in 1:3){
      Ymean[j,k] <- tau[j] + gamma[k];
                                    # Aliasing problem here
      Y[j,k] ~ dnorm(Ymean[j,k], rho);
    }
  }
  gbar <- mean(gamma[]);            # For
  for(j in 1:5){                    # finding thetas
    theta[j] <- tau[j] + gbar;
  }
}
#data
list(
  Y = structure(
    .Data = c(5,12,7,13,12,17,13,12,14,5,9,13,4,5,9),
    .Dim = c(5,3)
  )
)
#inits
list(mu = 0, tau = c(0,0,0,0,0), rho = 1, gamma = c(0,0,0))
```

Individual 95% CIs after WinBUGS run of length 10000

```
      node      WinBUGS          Frequentist
  theta[1]  ( 4.74, 11.29)    ( 4.73, 11.27)
  theta[2]  (10.69, 17.19)    (10.73, 17.27)
  theta[3]  ( 9.80, 16.29)    ( 9.73, 16.27)
  theta[4]  ( 5.71, 12.19)    ( 5.73, 12.27)
  theta[5]  ( 2.60,  9.39)    ( 2.73,  9.27)
```

However, gbar and mu oscillate very wildly because of

```
the aliasing problem.
```

And here's the second.

```
model ANalt;
var
  Y[5,3], Ymean[5,3],
  gamma[3], theta[5],
  mu, rho, gbar;
{
  mu ~ dnorm(0, 1.0E-6);
  for(j in 1:5){                    # Setting up thetas
    theta[j] ~ dnorm(mu, 1.0E-6);
  }
  for(k in 1:3){                    # Setting up gammas
    gamma[k] ~ dnorm(0, 1.0E-6);
  }
  gbar <- mean(gamma[]);
  for(k in 1:3){                    # Setting up betas
    beta[k] <- gamma[k] - gbar # to have zero sum
  }
  rho ~ dgamma(1.0E-3, 1.0E-3);
  for(j in 1:5){                    # Setting up Y's
    for(k in 1:3){
      Ymean[j,k] <- theta[j] + beta[k];
      Y[j,k] ~ dnorm(Ymean[j,k], rho);
    }
  }
}
#data
list(
  Y = structure(
    .Data = c(5,12,7,13,12,17,13,12,14,5,9,13,4,5,9),
    .Dim = c(5,3)
)
#inits
list(mu = 0, theta = c(0,0,0,0,0), gamma = c(0,0,0), rho = 1)
```

95% individual CIs for theta[j] values as good as before.

95% CI for mu found as (-814.0, 822.4) -- not surprising.

So, WinBUGS can obtain the Frequentist answers. However, WinBUGS can do more, and we have to consider for each situation whether the Frequentist answers are appropriate. We discuss this in Subsection 358L.

For important discussion of these and related matters, see Nobile and Green [175] and Besag, Green, Higdon and Mengerson [19].

▶ **P. An ANCOVA example.** Suppose that

$$Y_{km} \qquad (1 \leq k \leq \ell, 1 \leq m \leq 12)$$

are RVs representing the number of unemployed people in some country in month m of year k, or else the amount of gas consumed in that country in that month.

Consider two models, each of which builds in seasonal variation:

Model 1:

$$Y_{km} = \mu + \beta(k - \overline{k}) + \gamma_m + \varepsilon_{km}, \quad \sum \gamma_m = 0;$$

Model 2:

$$Y_{km} = \tilde{\mu} + \tilde{\beta}\left(k - \overline{k} + \tfrac{1}{12}(m - \overline{m})\right) + \tilde{\gamma}_m + \varepsilon_{km}, \quad \sum \tilde{\gamma}_m = \sum m\tilde{\gamma}_m = 0.$$

In each case, the ε_{km} are (for now) assumed to be IID each $N(0, \sigma^2)$.

▶ **Pa. Exercise.** Show that each model is easily analyzed via our geometry. Comment on the difference between the models. *Criticize the independence assumptions on the ε_{km}.*

Note. ANCOVA stands for ANalysis of COVAriance. The examples just considered were ones in which the m was regarded as a factor (and any one-one function of m would have done as well) but there was linear regression on k, so the actual value of k mattered. This can hint at what more general ANCOVA models look like.

▶▶ **Q. The General Linear Model.** Suppose that

$$\mathbf{Y} = X\boldsymbol{\beta} + \sigma\mathbf{G}, \qquad (\mathbf{Y} : n \times 1; \ X : n \times s; \ \boldsymbol{\beta} : s \times 1; \ \mathbf{G} : n \times 1),$$

where X is a deterministic 'design' matrix, $\boldsymbol{\beta}$ is a vector of parameters, $\mathbf{G} \sim SN(\mathbb{R}^n)$, and $\sigma > 0$.

Make the 'full-rank' assumption that $X^T X$ is invertible,

and note that

$$\mathbf{v}^T X^T X \mathbf{v} = (X\mathbf{v})^T X\mathbf{v} = \|X\mathbf{v}\|^2,$$

so that $C := X^T X$ (known to be invertible, and obviously symmetric) is positive-definite. The MLE Estimator $(\mathbf{B}, \hat{\sigma})$ of $(\boldsymbol{\beta}, \sigma)$ will minimize

$$\|\mathbf{Y} - X\mathbf{B}\|^2 = \|\mathbf{Y}\|^2 - 2\mathbf{Y}^T X\mathbf{B} + \mathbf{B}^T X^T X\mathbf{B}.$$

Let $\mathbf{W} = X^T \mathbf{Y}$. We want to minimize

$$\mathbf{B}^T C\mathbf{B} - 2\mathbf{W}^T \mathbf{B}.$$

Differentiating with respect to each B_i suggests that $\mathbf{B} = C^{-1}\mathbf{W}$. Indeed, if $\mathbf{B} = C^{-1}\mathbf{W} + \mathbf{D}$, then, using the symmetry of C,

$$\mathbf{B}^T C \mathbf{B} - 2\mathbf{W}^T \mathbf{B}$$
$$= \mathbf{W}^T C^{-1} C C^{-1} \mathbf{W} + \mathbf{D}^T \mathbf{W} + \mathbf{W}^T \mathbf{D} + \mathbf{D}^T C \mathbf{D} - 2\mathbf{W}^T C^{-1}\mathbf{W} - 2\mathbf{W}^T \mathbf{D}$$
$$= -\mathbf{W}^T C^{-1}\mathbf{W} + \mathbf{D}^T C \mathbf{D}.$$

Hence, by this algebra rather than by calculus, we see that

$$\|\mathbf{Y} - X\mathbf{B}\|^2$$

is minimized when $\mathbf{B} = C^{-1}\mathbf{W} = (X^T X)^{-1} X^T \mathbf{Y}$, with minimum value

$$\text{RSS} = \|\mathbf{Y}\|^2 - \mathbf{W}^T C^{-1}\mathbf{W} = \|\mathbf{Y}\|^2 - \mathbf{Y}^T X (X^T X)^{-1} X^T \mathbf{Y}.$$

Of course, there's meaningful geometry behind all this (rather meaningless) algebra; and we examine it below. First, though, let's look at an example.

Qa. The case of Linear Regression. For the familiar Linear Regression model from Subsection 312D, we have

$$X = \begin{pmatrix} 1 & x_1 - \overline{x} \\ 1 & x_2 - \overline{x} \\ \cdot & \cdot \\ 1 & x_n - \overline{x} \end{pmatrix}, \qquad \beta = \begin{pmatrix} \mu \\ \beta \end{pmatrix},$$

$$X^T X = \begin{pmatrix} n & 0 \\ 0 & q(\mathbf{x}, \mathbf{x}) \end{pmatrix}, \qquad X^T \mathbf{Y} = \begin{pmatrix} n\overline{Y} \\ Q(\mathbf{x}, \mathbf{Y}) \end{pmatrix},$$

$$\mathbf{B} = \begin{pmatrix} \overline{Y} \\ B \end{pmatrix}, \qquad B = \frac{Q(\mathbf{x}, \mathbf{Y})}{q(\mathbf{x}, \mathbf{x})},$$

$$\text{RSS} = \|\mathbf{Y}\|^2 - \left(n\overline{Y} \quad Q(\mathbf{x}, \mathbf{Y}) \right) \begin{pmatrix} n^{-1} & 0 \\ 0 & q(\mathbf{x}, \mathbf{x})^{-1} \end{pmatrix} \begin{pmatrix} n\overline{Y} \\ Q(\mathbf{x}, \mathbf{Y}) \end{pmatrix}$$

$$= \|\mathbf{Y}\|^2 - n\overline{Y}^2 - q(\mathbf{x}, \mathbf{x})B^2,$$

which is all as it should be.

▶ **Qb. The geometry of the general case.** Let

$$P = X(X^T X)^{-1} X^T.$$

Then it is clear that $P^T = P$ and $P^2 = P$. Hence, by Lemma 324Jb, P is perpendicular projection onto the range $\mathcal{R}(P)$ of P. Since

$$P\mathbf{v} = X\left\{(X^T X)^{-1} X^T \mathbf{v}\right\},$$

we have $\mathcal{R}(P) \subseteq \mathcal{R}(X) = X(\mathbb{R}^s)$. We want to prove that

$$\mathcal{R}(P) = \mathcal{R}(X), \quad \text{so that } P \text{ is projection onto the range of } X. \qquad \text{(Q1)}$$

If $\mathcal{R}(P)$ is a proper subspace of $\mathcal{R}(X)$, then we can find $\mathbf{a} \in \mathbb{R}^s$ such that $X\mathbf{a} \neq \mathbf{0}$ and

$$\langle X\mathbf{a}, P\mathbf{h} \rangle = 0 \quad \text{for all } \mathbf{h} \text{ in } \mathbb{R}^n.$$

But then,

$$0 = \mathbf{a}^T X^T P\mathbf{h} = \mathbf{a}^T X^T X (X^T X)^{-1} X^T \mathbf{h} = \mathbf{a}^T X^T \mathbf{h} = \langle X\mathbf{a}, \mathbf{h} \rangle = 0$$

for all $\mathbf{h} \in \mathbb{R}^n$, whence $X\mathbf{a} = 0$. This contradiction establishes 337(Q1).

We have,

> if $W := \mathcal{R}(P)$ and $B = (X^T X)^{-1} X^T \mathbf{Y}$, then
> $$X\mathbf{B} = P\mathbf{Y} = X\beta + \sigma\, \text{SN}(W),$$
> $$\text{RSS} = \|\mathbf{Y} - X\mathbf{B}\|^2 = \|P_W^{\perp}\mathbf{Y}\|^2 \sim \sigma^2 \chi_{n-s}^2,$$

and $X\mathbf{B}$ **is independent of RSS.** Note that $X\mathbf{B} = P\mathbf{Y}$ is the best fit of $\mathbf{Y}$ allowed by our model, both in ML and least-squares senses. Obviously,

$$\frac{\|X\mathbf{B} - X\beta\|^2/s}{\text{RSS}/(n-s)} \sim F_{s,n-s},$$

giving a Confidence Region for the best fit.

Qc. The case of ANOVA. Suppose we have a 2×2 two-factor ANOVA model without interaction. Then

$$\begin{pmatrix} Y_{11} \\ Y_{12} \\ Y_{21} \\ Y_{22} \end{pmatrix} = \begin{pmatrix} 1 & 1 & 0 & 1 & 0 \\ 1 & 1 & 0 & 0 & 1 \\ 1 & 0 & 1 & 1 & 0 \\ 1 & 0 & 1 & 0 & 1 \end{pmatrix} \begin{pmatrix} \mu \\ \alpha_1 \\ \alpha_2 \\ \beta_1 \\ \beta_2 \end{pmatrix} + \sigma\mathbf{G} = X\beta + \sigma\mathbf{G}.$$

However, for Linear-Algebra reasons, $X^T X$ cannot be invertible. One way to see this is that, really because of aliasing, $X\mathbf{a} = 0$ where $\mathbf{a} = (1, -1, -1, 0, 0)^T$.

Of course, there are ways round the aliasing difficulty. For example, we can use $\alpha_1 + \alpha_2 = 0$ and $\beta_1 + \beta_2 = 0$ to write instead

$$\begin{pmatrix} Y_{11} \\ Y_{12} \\ Y_{21} \\ Y_{22} \end{pmatrix} = \begin{pmatrix} 1 & 1 & 1 \\ 1 & 1 & -1 \\ 1 & -1 & 1 \\ 1 & -1 & -1 \end{pmatrix} \begin{pmatrix} \mu \\ \alpha_1 \\ \beta_1 \end{pmatrix} + \sigma\mathbf{G} = X\beta + \sigma\mathbf{G}.$$

Now $X^T X = 4I$, and $\mathbf{B}$ is what it should be.

Just think what X looks like for the $J \times K \times L$ two-factor ANOVA with interaction and replication at Subsection 331M: it's something of a mess.

Moral: if your model has structure, then make sure that your Mathematics reflects it.

8.4   Goodness of fit; robustness; hierarchical models

A. Introduction.   The previous section studied the first stages of the Mathematics of Linear-Model theory. As its title suggests, this section considers the first stages of associated Statistics.

Goodness of fit is concerned with the question: do the data suggest that our *model* is unacceptable? Suppose that we *can* assume that $Y_1, Y_2, \ldots, Y_n$ are IID with common DF F. Our model may assume that F is some particular DF G or that F belongs to some parametrized family of DFs. Is there strong evidence in the data to reject such an assumption? This is a 'distributional' test of fit.

Recall that with Y as a typical Y_k,

$$F(x) := \mathbb{P}(Y \leq x).$$

and that the Sample $\mathbf{Y} = (Y_1, Y_2, \ldots, Y_n)$ determines the so-called **Empirical Distribution Function (EDF)** $F_n(\cdot; \mathbf{Y})$, where

$$F_n(x; \mathbf{Y}) := \frac{1}{n}\sharp\{k : Y_k \leq x\}, \qquad (A1)$$

the proportion of k such that $Y_k \leq x$. One way to decide whether to abandon the assumption that $F = G$ is to consider the **Kolmogorov Statistic**:

$$D_K(n, \mathbf{Y}, G) := \sup_x |F_n(x, \mathbf{Y}) - G(x)|, \qquad (A2)$$

a distance between the functions $F_n(\cdot; \mathbf{Y})$ and G. We study this in Subsection 342C. Before that, we shall however consider a visual way, using **quantile-quantile (qq)** plots, of assessing whether there is strong evidence against the hypothesis that $F = G$. Our study of the famous **Chi-Squared** method of assessing goodness of fit begins in Subsection 345Da.

It seems fair to say that assessing goodness of distributional fit has more of the flavour of a cottage-craft industry than of a science, unless well-specified Alternative Hypotheses are available. I spend more time on qq plots than I believe the method merits, because the qq-plot method is likely to be widely used with its availability on so many packages. Each of the methods has advantages and glaring faults; and don't always expect compatibility of answers! But wait for hierarchical modelling, though that has its own difficulties.

We have talked about goodness of fit in regard to hypotheses on DFs. Brief indication of goodness of fit in regard to hypotheses of independence is given in chi-squared tests for contingency tables in Subsection 363O, in a study of correlation in Subsection 372H, and implicitly in the mention of Spearman's rank correlation coefficient at 317Ea. But there are other important aspects of testing independence which we have to skip.

Robustness features in Statistics in many ways. I have already mentioned it in connection with sensitivity to small changes in data, and in regard to testing the effect of slightly different priors and different initial values in MCMC methods. Here, however, we are thinking of the robustness of methods to changes in (the 'likelihood' part of) our model. Let's take an example. Student's 't' method of getting CIs for a mean assumes that the underlying distribution is normal. However, in many situations, the normal distribution tails off too quickly to be realistic. If the underlying distribution is different from normal, by how much are results obtained by Student's 't' method in error? Simulation often allows us to answer such questions. See Subsection 354I.

Hierarchical models, the study of which is often possible only by MCMC methods, can be more realistic than classical ones. They also do much to avoid goodness-of-fit and robustness difficulties. In particular, they can allow the machine to decide for itself amongst classical models or amongst 'mixtures' of such models. An often-used example is where we choose the underlying standardized distribution of errors to be $StandT_\nu$, where ν itself is 'random' taking values in some such set as $\{3, 4, 6, 8, 12, 16, 24, 32\}$. However, we shall see that the Gibbs sampler may run into difficulties with this.

Outliers – now *there's* a problem! An outlier is an observation for which the associated residual is significantly larger than might be anticipated from the vast majority of other observations. One must check each outlier to see whether it is genuine and not the result of an error in measurement or in recording data. Non-genuine outliers can seriously distort analysis. If outliers *are* genuine, then one must use a 'larger-tailed' model (or a hierarchical one). See Barnett and Lewis [10] and Rousseeuw and Leroy [200].

B. Quantile-quantile plots.

Suppose that $Y_1, Y_2, \ldots, Y_n$ are IID each with strictly increasing continuous DF F on $\mathbb{R}$. For $0 < p < 1$, the unique value x such that $F(x) = p$ is called the *quantile* of F corresponding to p. Recall from 50B and Fact 99G that if $U_k := F(Y_k)$, then $U_1, U_2, \ldots, U_n$ are IID, each U[0, 1]. If $U_{(1)}, U_{(2)}, \ldots, U_{(n)}$ are the Order Statistics for $U_1, U_2, \ldots, U_n$ (that is, $U_1, U_2, \ldots, U_n$ arranged in increasing order), then (see Exercise 108La)

$$\mathbb{E}\ U_{(k)} = \frac{k}{n+1},$$

so that, very roughly speaking, $U_{(k)}$ should be close to $k/(n+1)$, and $Y_{(k)}$ should be close to $F^{-1}(k/(n+1))$, the quantile of F corresponding to $k/(n+1)$. To test

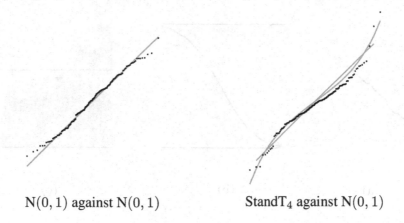

N(0, 1) against N(0, 1)   StandT$_4$ against N(0, 1)

Figure B(i): Quantile-quantile plots

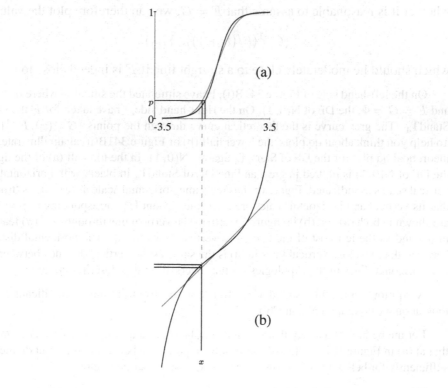

Figure B(ii): Understanding a qq plot: $p = 0.17 = \Phi(x)$

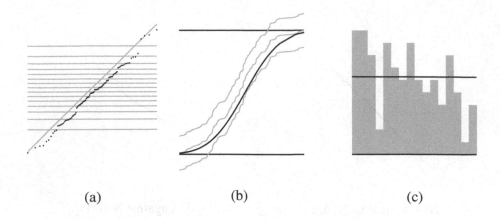

(a)   (b)   (c)

Figure B(iii): Three pictures of an example

whether it is reasonable to assume that $F = G$, we can therefore plot the values

$$\left(G^{-1}(k/(n+1)), \, Y_{(k)}\right),$$

which should lie moderately close to a straight line if F is indeed close to G.

On the left-hand side of Figure 341B(i), I have simulated the situation where $n = 200$ and $F = G = \Phi$, the DF of N$(0, 1)$. On the right-hand side, I have taken for F the DF of StandT$_4$. The grey curve is the theoretical curve through the points $\left(G^{-1}(x), F^{-1}(x)\right)$. To help you think about qq plots, the lower half (b) of Figure 341B(ii) again illustrates the theoretical qq plot for the DF of StandT$_4$ against N$(0, 1)$. In the top half (a) of the figure, the DF of N$(0, 1)$ is plotted in grey, and the DF of StandT$_4$ in black, with horizontal and vertical scales as indicated. Figure (b) has the same horizontal scale as (a), and is 'true' in that its vertical and horizontal scales are the same. Quantiles corresponding to $p = 0.17$ are shown in both (a) and (b) in Figure 341B(ii). The vertical line through $\Phi^{-1}(p)$ features in (b) and as the leftmost of the two close vertical lines in (a). The horizontal distance between the two close vertical lines in (a) is the same as the vertical distance between the two horizontal lines in (b). Apologies for the 'overkill' in this description.

A qq plot is a useful visual device. But 'how far from the line is significant'? (We look at quantifying a qq plot at Cb.)

Let me be honest and say that my first simulation of a qq plot with $F = G = \Phi$ was that at (a) in Figure B(iii). Clearly, the simulated points lie below the line. But do they lie sufficiently far below for me to suspect the random-number generator?

▶ **C. Distribution of the Kolmogorov Statistic.** Suppose that $Y_1, Y_2, \ldots, Y_n$ are IID each with strictly increasing continuous DF G on $\mathbb{R}$. Note that if m is the median of G, then, with 2 replacing 1.96 as usual, $F_n(m; \mathbf{Y})$ is approximately N$(\frac{1}{2}, \frac{1}{4}n^{-1})$ for large n, the usual $n^{-\frac{1}{2}}$ warning not to expect too great accuracy.

Kolmogorov proved the fine result, best understood in terms of the theory of Brownian motion (see Billingsley [22]) that, as $n \to \infty$,

$$\mathbb{P}\left(D_K(n, \mathbf{Y}, G) > \frac{c}{\sqrt{n}}\right) \to 2\sum_{r=1}^{\infty}(-1)^{r-1}\exp\{-2r^2c^2\}, \quad \text{(C1)}$$

where D_K is at 339(A2). The right-hand side is 5% when $c = 1.36$.

Ca. Exercise. Use the 'F inverse' Principle to check that for each fixed n, the expression on the left-hand side of (C1) does not depend on the common DF G (assumed strictly increasing continuous) of the Y_k.

Note that if $\mathbb{P}\left(D_K(n, \mathbf{Y}, G) > cn^{-\frac{1}{2}}\right) = \eta$, then

$$\mathbb{P}\left(G(x) \in \left[F_n(x, \mathbf{y}) - cn^{-\frac{1}{2}}, F_n(x, \mathbf{y}) + cn^{-\frac{1}{2}}\right]\right) = 1 - \eta,$$

so we obtain a $100(1 - \eta)\%$ Confidence Band *for the whole function G*. For the data from the qq plot at (a) in Figure 342B(iii), we would get the centre 'wavy' line in (b) as the observed EDF $F_n(\cdot, \mathbf{y}^{\text{obs}})$ and the region between the top and bottom wavy lines as Confidence Band for the true DF G. The smooth curve represents the graph of Φ which lies within this band. Indeed the 'p-value' based on the Kolmogorov statistic, the probability that D_K would be at least as large as the observed value if the common DF of the Y_k is indeed G, is about 11%. From this point of view, there is not strong evidence to reject the hypothesis that $F = G$. Of course, we are in a very difficult situation if we try to formulate an Alternative Hypothesis ... . Wait for the Chi-Squared method.

I have already remarked that the $n^{-\frac{1}{2}}$ difficulty, very clear from (C1), means that we cannot expect too accurate results. Just look at the top half (a) of Figure 341B(ii) to see that the DFs of StandT_4 and $N(0, 1)$ never differ by more than 0.053. You can see that you need a large sample to separate these DFs on the basis of the Kolmogorov Statistic.

Cb. 'Kolmogorov Statistic' quantifying of a qq plot. In the following discussion, we assume n so large that we can treat $F_n(\cdot; \mathbf{Y})$ as if it were continuous (rather than a right-continuous function with jumps). Suppose that $Y_1, Y_2, \ldots, Y_n$ are IID each with strictly increasing continuous DF G. One possible way of quantifying a qq plot is by 'inverting' the Kolmogorov result. That result states that for large n, we can with 95% confidence state that

$$\text{for all } x, \, F_n(x; \mathbf{Y}) \geq G(x) - 1.36n^{-\frac{1}{2}}, \qquad F_n(x; \mathbf{Y}) \leq G(x) + 1.36n^{-\frac{1}{2}}. \quad \text{(C2)}$$

For $x = G^{-1}(p)$, so that $G(x) = p$, we have from (C2)

$$F_n(x) := F_n(x; \mathbf{Y}) \geq p - 1.36n^{-\frac{1}{2}},$$

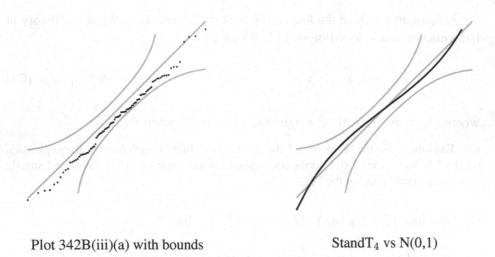

Plot 342B(iii)(a) with bounds   StandT$_4$ vs N(0,1)

Figure C(i): Quantifying qq plots for $n = 100$

so that the quantile for F_n associated with $p - 1.36n^{-\frac{1}{2}}$ is less than or equal to x. Writing c for $p - 1.36n^{-\frac{1}{2}}$, we have

$$F_n^{-1}(c) \leq G^{-1}(c + 1.36n^{-\frac{1}{2}}).$$

This leads us to say that a 95% Confidence Band for the qq plot, if the true DF is G, is the region between

an upper curve $\left\{ \left(G^{-1}(c), G^{-1}(c + 1.36n^{-\frac{1}{2}}) \right) : c < 1 - 1.36n^{-\frac{1}{2}} \right\}$

and

a lower curve $\left\{ \left(G^{-1}(c), G^{-1}(c - 1.36n^{-\frac{1}{2}}) \right) : c > 1.36n^{-\frac{1}{2}} \right\}.$

These curves for $G = \Phi$ and $n = 100$ are shown in Figure C(i). The qq plot corresponding to Figure 342B(iii)(a) is also shown in the left-hand Figure. The theoretical StandT$_4$ against N(0, 1) qq plot is shown in black on the right-hand figure. It helps emphasize that one needs large sample sizes to differentiate between certain DFs.

Cc. Exercise. Think about possible disadvantages of the method, but in a 'before the experiment' way.

▶▶ **D. The Chi-Squared Principle (ChiSqP) in Hypothesis Testing.** Neither the qq method nor the Kolmogorov method of assessing goodness of fit is based on likelihood; and it has been our contention all along that it is likelihood that matters.

We are now going to move towards the Chi-Squared method of assessing goodness of fit which *is* based on likelihood considerations and which to that extent is philosophically satisfying, but which has a lot of arbitrariness and which can fail to see obvious things. You have been warned not always to expect consistency with the Kolmogorov method.

Here is an important general principle. We saw it in the case when $s = 1$ and $m = 0$ in Subsection 230D and for the case of Linear Models in Theorem 301Ha. Revising these cases will help you understand the following theorem. Don't worry about the 'submanifold' terminology: examples will give the idea.

▶▶▶ **Da. Wilks' Theorem (Chi-Squared Principle (ChiSqP)).** Again suppose that $Y_1, Y_2, \ldots, Y_n$ are IID each with pdf/pmf $f(y \mid \theta)$. Let H_0 take the form '$\theta \in B_0$' and H_A the form '$\theta \in B_A$', where B_A is a submanifold of some $\mathbb{R}^d$ of dimension s and B_0 is a submanifold of B_A of dimension m. Then if H_0 is true and n is large,
$$\mathrm{Dev}(\mathbf{Y}) = \mathrm{Dev}(H_A, H_0; \mathbf{Y}) = 2 \ln \mathrm{LR}(\mathbf{Y}) \approx \chi^2_{s-m}.$$

For a sketched proof see Subsection 377L.

▶ **E. ChiSqP for the discrete case.** Suppose that $Y_1, Y_2, \ldots, Y_n$ are IID RVs each with values in $\{1, 2, \ldots, b\}$, such that
$$\mathbb{P}(Y_m = k) = p_k \ (1 \le k \le b, 1 \le m \le n), \qquad \sum_i p_i = 1.$$

Let N_k be the number of Y_m equal to k. Then
$$\mathrm{lhd}(\mathbf{p}; \mathbf{Y}) = p_1^{N_1} p_2^{N_2} \cdots p_b^{N_b},$$
and this is maximized by $\hat{\mathbf{p}}$ where
$$\hat{\mathbf{p}} = \frac{N_k}{n}.$$

(It is understood that $0^0 = 1$.)

Let H_0 be the Null Hypothesis that $p_k = p_{0k}$ where the p_{0k} are completely specified positive numbers summing to 1. Let H_A be the hypothesis that the p_k are arbitrary non-negative numbers summing to 1. Then (with the anticipated 'entropy' in the first equation)
$$\ln \mathrm{mlhd}(H_A; \mathbf{Y}) = \sum N_K \ln \frac{N_k}{n}, \qquad (0 \ln 0 = 0),$$
$$\ln \mathrm{mlhd}(H_0; \mathbf{Y}) = \sum N_K \ln p_{0k}.$$

Then

$$2\ln \mathrm{LR}(\mathbf{Y}) = 2\sum N_k \ln \frac{N_k}{np_{0k}}.$$

By ChiSqP, if n is large and H_0 is true, then

$$2\ln \mathrm{LR}(\mathbf{Y}) \approx \chi^2_{b-1},$$

since here, $s = b - 1$ and $m = 0$. (The condition $\sum p_k = 1$ defines a $(b-1)$-dimensional hyperplane in $\mathbb{R}^b$. The single point $(p_{01}, p_{02}, \ldots, p_{0b})$ is 0-dimensional.)

There is a famous (but not always that good) approximation to $2\ln \mathrm{LR}(\mathbf{Y})$ due to Pearson, which in fact predates LR tests. Let

$$\Delta_k := \frac{N_k - np_{0k}}{np_{0k}} = \frac{\text{'Observed - fitted'}}{\text{'fitted'}},$$

'fitted' being calculated on the basis of the Null Hypothesis, and note that $\sum np_{0k}\Delta_k = 0$ and $N_k = np_{0k}(1 + \Delta_k)$. Then

$$
\begin{aligned}
2\ln \mathrm{LR}(\mathbf{Y}) &= 2\sum np_{0k}(1 + \Delta_k)\ln(1 + \Delta_k) \\
&= 2\sum np_{0k}(1 + \Delta_k)\left(\Delta_k - \tfrac{1}{2}\Delta_k^2\right) + \text{error} \\
&= 2\sum np_{0k}\left(\Delta_k + \tfrac{1}{2}\Delta_k^2\right) + \text{error},
\end{aligned}
$$

so that

$$2\ln \mathrm{LR}(\mathbf{Y}) \approx \sum np_{0k}\Delta_k^2 =: \text{PearsonStat}.$$

We usually write PearsonStat as

$$\text{PearsonStat} = \sum \frac{(\text{Observed} - \text{fitted})^2}{\text{fitted}}.$$

Later, we shall see a nice direct explanation (with mysterious connections with quantum theory) of the fact that, if H_0 is true and n is large,

$$\text{PearsonStat is approximately } \chi^2_{b-1}.$$

Ea. Detecting 'over-enthusiasm'. An important application of this technique is to test when data fit the Null Hypothesis *better* than is believable. Fisher (see [217]) showed that data as closely in agreement with Mendel's theory of inheritance as Mendel's own 'data' would occur by Chance only about 4 times in 100000 (if Mendel's theory is true – which it *is*, of course). So there is evidence that Mendel (or someone working with him) showed a certain over-enthusiasm for Mendel's theory. I won't accuse a monk of deliberately cheating! See Wright's article in [217].

▶ **F. ChiSqP for goodness of fit.** An obvious idea for assessing whether it is unreasonable to assume that the common DF of IID $Y_1, Y_2, \ldots, Y_n$ is a given DF G is to take a 'partition'

$$-\infty = x_0 < x_1 < \cdots < x_b = \infty,$$

and apply the method of the previous section where N_k is the number of Y_m falling in $(x_{k-1}, x_k]$, with

$$p_{0k} = G(x_k) - G(x_{k-1}).$$

(Of course, if each Y_k must be positive, we can take $x_0 = 0$.) Suppose for a moment that we have decided on the value of b. Then it is best to take $x_k = G^{-1}(k/b)$, the quantile of G corresponding to k/b. There is little point in looking at cases where $n \leq 80$. There are some reasons for choosing b close to

$$15 \left(\frac{n}{100} \right)^{2/5}.$$

See, for example, Kendall, Stuart, Ord and Arnold [133].

Figure 342B(iii)(c) takes $b = 16$ and shows the observed values $n_1^{\text{obs}}, \ldots, n_b^{\text{obs}}$ of $N_1, N_2, \ldots, N_b$, the black horizontal line indicating the fitted values. The horizontal lines in Figure (a) there give 'equiprobable' spacing. One finds the following results:

```
We have for observed values N_1,N_2,...,N_b,
using 16 equal-probability intervals:
 10 10  8  2  9  7  6  9  6  5  6  4  8  5  1  4
Each fitted value is 6.25.

Pearson =  17.4400 with corresponding p-value 0.2932.
twolnLR =  20.1175 with corresponding p-value 0.1675.
```

Here then is a case when PearsonStat does not approximate $2 \ln \mathrm{LR}$ well. Recall that the p-value from the Kolmogorov statistic was 0.11.

Fa. Important Exercise. Again consider a case where a Chi-Squared test is to be performed using equi-probable intervals.
(a) Describe a situation where the Chi-Squared method would tell strongly against the $F = G$ hypothesis but the Kolmogorov statistic would not.
(b) Describe a situation where the Kolmogorov statistic would tell strongly against the $F = G$ hypothesis but the Chi-Squared method would not.
(c) Now discuss the discrepancy between the Kolmogorov and Chi-Squared results for our example.

▶ **G. Goodness of fit with parameter estimation.** Let $Y_1, Y_2, \ldots, Y_n$ yet again be IID with common DF F. The model (discussed in the last few subsections) that F is some completely specified DF G is, of course, hardly ever used. Much more common is the assumption that F belongs to some family such as the family $\{N(\mu, \sigma^2); \mu \in \mathbb{R}, \sigma > 0\}$ of normal distributions. Let us stick to this normal case for illustration.

We have to estimate μ and σ^2 from the observed sample and consider whether the 'best' $N(\mu, \sigma^2)$ should be rejected on the basis of the data. If the sample size n is large (and hence much bigger that the number, 2, of parameters being estimated), we can use the qq, Kolmogorov, or Chi-Squared method just as before without worrying too much.

If, however, the sample size is small, our method must take into account that we have estimated μ and σ^2; and the qq and Kolmogorov methods cannot do that. The only sensible question is: how is the Chi-Squared method affected by estimation of parameters?

Well, we have to use a partition

$$-\infty = x_0 < x_1 < \cdots < x_b = \infty,$$

as before, and compare how the

$$p_{0k}(\mu, \sigma^2) = G(x_k; \mu, \sigma^2) - G(x_{k-1}; \mu, \sigma^2)$$

(where $G(\cdot; \mu, \sigma^2)$ is the DF of $N(\mu, \sigma^2)$) match up to the MLE under H_A. Now, we have $\dim(B_0) = 2$ and $\dim(B_A) = b - 1$ as before, so ChiSqP relates to $b - 3$ degrees of freedom. But this use of ChiSqP is valid only if we choose (μ, σ^2) so as to maximize

$$\prod p_{0k}(\mu, \sigma^2)^{N_k},$$

and this is never done in practice. In practice, one normally uses $\overline{Y}$ and $Q(\mathbf{Y}, \mathbf{Y})/(n-1)$ as Estimators for μ and σ^2; and the limiting distribution of the deviance $2 \ln LR$ is then *not* χ^2_{b-3}. If, with these usual estimates of μ and σ^2, the χ^2_{b-1} method says that the result is significant, we go along with that; and if the χ^2_{b-3} result is not significant, we agree with that. The 'in-between' cases mean that you will have to consult the literature.

H. 'Hierarchical t' alternative to normal; and serious problems with
▶ **the Gibbs sampler.** For all its great theoretical importance, the fast tail-off of the normal density means that it may not provide that good a model for the pdf of a single observation in practice. Of course, when n is large, the CLT guarantees that $\overline{Y}$ will be approximately normal.

It is quite common these days to assume that an 'Error Variable' in Linear-Model theory has a t_ν distribution where ν itself is 'random'. This alternative to the 'normality' assumption is not without its practical difficulties however. In addition, it continues to assume a symmetric distribution of errors.

Let us concentrate for now on a simple example in which we believe that each Y_k has mean 0 and variance 1. Recall from Exercise 255Ge that the 'Standardized t_ν' distribution StandT$_\nu$ is the distribution of $\{(\nu - 2)/\nu\}^{\frac{1}{2}} T_\nu$, where $T_\nu \sim t_\nu$. We assume that the common DF of the Y_k is StandT$_\nu$, where ν is a 'random' element

$$\nu \in \mathbb{D} := \{3, 4, 6, 12, 16, 24, 32\}.$$

Of course, the StandT$_{32}$ distribution is very close to $N(0, 1)$. Indeed, the fact that all the described distributions are close needs thinking about. (Figure 341B(ii)(a) shows how close the StandT$_4$ and $N(0, 1)$ DFs are.)

Suppose that $Y_1, Y_2, \ldots, Y_n$ are IID each with the StandT$_\nu$ distribution. Even though we are not really in the Hypothesis Testing situation suggested by the following discussion, the HT language is useful here. If H_0 is '$\nu = 6$' and H_A '$\nu = 4$', then

$$\ln \text{LR} := \ln \frac{\text{lhd}(H_A; \mathbf{Y})}{\text{lhd}(H_0; \mathbf{Y})} = S := \sum_{k=1}^{n} Z_k, \quad Z_k := \ln \frac{f_4(Y_k)}{f_6(Y_k)},$$

where f_ν is the pdf of the StandT$_\nu$ distribution. Now $\alpha := \mathbb{E}_0(Z_k) \approx -0.007$, and $\beta^2 := \text{Var}(Z_k) \approx 0.0135$. (I obtained these approximate values by the cheap method of taking a million simulated values (more than once).)

Suppose now that we wish n to be large enough that we have 95% probability under H_0 of the extremely mild requirement that $\text{lhd}(H_A; \mathbf{Y}) \leq \text{lhd}(H_0; \mathbf{Y})$, that is that $\mathbb{P}(S_n \leq 0) \geq 95\%$. Now S_n is approximately $N(n\alpha, n\beta^2)$, so that

$$\mathbb{P}(S_n \leq 0) \approx \Phi(-n^{\frac{1}{2}}\alpha/\beta) \approx 95\% = \Phi(1.64)$$

so that, very roughly, we need $n \approx 1.64^2 \beta^2 / \alpha^2 \approx 740$.

Ha. Exercise. Show that a rough CLT estimate of the sample size required to have 95% probability under H_0 that LR $< 1/10$ is 1320. $\qquad\square$

The moral is that the different ν are not easily distinguished. That's something you ought to bear in mind, but I am not claiming that it invalidates the 'hierarchical' idea here.

Hb. Numerical example. I simulated a sample of size 600 from the StandT$_6$ distribution. Assuming a uniform prior pmf ($\pi(\nu) = 1/8$) on the set $\mathbb{D}$, I worked out the exact posterior pmf for $n = 5, 50, 200, 600$. The results are shown in the rows labelled 'Exact' in Table H(i). Note the considerable change in posterior pmf between $n = 200$ and $n = 600$. Now, for a sample of size 200, a likelihood ratio for $H_A : \nu = 4$ against $H_0 : \nu = 6$ at least as large as $0.7204/0.2017$ will occur about 5% of the time, and since we have chosen a worst case after seeing the data, the real p-value is surely significantly more than 5%. The exact results for $n = 600$ are more in favour of the (true!) H_0 than one normally gets at this sample size.

The Gibbs results are from a Gibbs sampler program described later in this subsection. In each case, 100000 sweeps of the sampler were used, no 'Burn-in' time being allowed for in such a huge number of sweeps. The results are good for $n = 5$ and

```
Sample size = 5
      3       4       6       8      12      16      24      32 nu
 0.2319 0.1653 0.1257 0.1112 0.0989 0.0933 0.0881 0.0855 Exact
 0.2361 0.1652 0.1244 0.1090 0.0973 0.0937 0.0881 0.0863 Gibbs
Sample size = 50
 0.0133 0.1502 0.2676 0.2291 0.1434 0.0975 0.0575 0.0412 Exact
 0.0139 0.1557 0.2690 0.2252 0.1381 0.0975 0.0580 0.0426 Gibbs
Sample size = 200
 0.0321 0.7204 0.2017 0.0408 0.0041 0.0009 0.0001 0.0000 Exact
 0.0748 0.7908 0.1094 0.0224 0.0024 0.0002 0.0000 0.0000 Gibbs
but sampler has gone rather crazy
Sample size = 600
 0.0000 0.1500 0.7514 0.0964 0.0021 0.0001 0.0000 0.0000 Exact
Gibbs sampler gives meaningless results
```

Table H(i): Results for a hierarchical t model

$n = 50$ – which persuades me that my programs (in their final versions) are correct. For $n = 200$, the Gibbs results are far worse, and in fact, as we shall see, the sampler has already gone crazy. For $n = 600$, the results are meaningless. A WinBUGS program (also described later in this subsection) coped no better. For what is going wrong, see the Discussion 353Hf later in this subsection.

Hc. Programming the exact posterior pmf. We take a prior pmf $\pi(\nu)$ $(\nu \in \mathbb{D})$, uniform on $\mathbb{D}$. We have

$$\pi(\nu \mid \mathbf{y}) \propto \pi(\nu) \prod_{k=1}^{n} f_\nu(y_k)$$

where f_ν, the pdf of the StandT$_\nu$ distribution, is given by

$$f_\nu(y) = \frac{1}{\sqrt{(\nu - 2)\pi}} \frac{\Gamma(\frac{1}{2}(\nu + 1))}{\Gamma(\frac{1}{2}\nu)} \left(1 + \frac{y^2}{\nu - 2}\right)^{-\frac{1}{2}(\nu+1)}.$$

Here are the key parts of the program:

```
int nu[9] = {0,3,4,6,8,12,16,24,32};
int main(){
  ...
  for(w=1; w<=8; w++){
    nuw = nu[w];   sum = 0.0;
    for(k=1; k<=n; k++) {
      sum = sum + log(1.0 + y[k]*y[k]/(nuw - 2.0));
    }
    lnKscaled = - 0.5*log(nuw - 2.0)
              + loggam(0.5*nuw + 0.5) - loggam(0.5*nuw);
```

```
    lnpost = n*lnKscaled - 0.5*(nuw + 1.0) * sum;
    a[w] = exp(lnpost);  asum = asum + a[w];
}
...
for(w=1; w<=8; w++) printf("%7.4f", a[w]/asum); printf("\n");
...
}
```

Hd. Programming a Gibbs sampler for this case. We use the idea from Exercise 255Ge that we can assume that

$$\rho_k \sim \text{Gamma}\big(\tfrac{1}{2}\nu, \text{rate } \tfrac{1}{2}\nu - 1\big), \qquad (Y_k \mid \rho) \sim \text{N}(0, \text{prec } \rho_k).$$

We take a uniform prior on $\mathbb{D}$ for ν. The joint pdf/pmf for $(\nu, \{(\rho_k, y_k) : 1 \le k \le n\})$ is proportional to

$$a_\nu \propto \pi(\nu) \prod_{k=1}^{n} \left[(\tfrac{1}{2}\nu - 1)^{\frac{1}{2}\nu} \rho_k^{(\frac{1}{2}\nu - 1)} \mathrm{e}^{-(\frac{1}{2}\nu - 1)\rho_k} \rho_k^{\frac{1}{2}} \mathrm{e}^{-\frac{1}{2}\rho_k y_k^2} \Gamma(\tfrac{1}{2}\nu)^{-1} \right]$$

$$= \pi(\nu)(\tfrac{1}{2}\nu - 1)^{\frac{1}{2}\nu n} \mathrm{e}^{-(\frac{1}{2}\nu - 1)\sum \rho_k} \left(\prod \rho_k\right)^{(\frac{1}{2}\nu - \frac{1}{2})} \mathrm{e}^{-\frac{1}{2}\sum \rho_k y_k^2} \Gamma(\tfrac{1}{2}\nu)^{-n}$$

We see that

$$(\rho_k \mid \nu, \rho^{(-k)}, \mathbf{y}) \sim \text{Gamma}\big(\tfrac{1}{2}\nu + \tfrac{1}{2}, \text{rate } \tfrac{1}{2}\nu - 1 + \tfrac{1}{2}y_k^2\big),$$

$$\pi(\nu \mid \rho, \mathbf{y}) \propto a_\nu.$$

where $\rho^{(-k)}$ signifies the vector of all ρ_i-values with $i \ne k$. The pmf $\pi(\nu \mid \rho, \mathbf{y})$ may be calculated from the sufficient statistics

$$\text{rhosum} = \sum \rho_k, \quad \text{sumlnrho} = \sum \ln \rho_k, \quad \text{sumrhoysq} = \sum \rho_k y_k^2. \tag{H1}$$

Here are key parts of the program:

```
int dof[9] = {0,3,4,6,8,12,16,24,32};
int code, nu,
double rhosum, sumlnrho, sumrhoysq, y[601], prob[9];
...
void ChooseRhos(){   /* Choose rhos given nu */
int k; double rho;
  rhosum = 0.0; sumlnrho = 0.0; sumrhoysq = 0.0;
  for(k=1; k<= n; k++){
    rho = rnewGrate(0.5 * (nu + 1.0),
        0.5 * (nu - 2.0 + y[k] * y[k]));
    rhosum += rho; sumlnrho += log(rho);
    sumrhoysq += rho * y[k] * y[k];
  }
```

```
}
void nuProbs(){  /* for nu = dd = dof[d] given rhos */
int d, dd;
double lna, asum = 0.0;
  for(d=1;d<=8;d++){  dd = dof[d];
    lna = 0.5 * dd * n * log(0.5*(dd-2.0)) - n * loggam(0.5*dd);
    lna = lna - 0.5*(dd - 2.0)*rhosum + 0.5*(dd - 1.0)*sumlnrho
         - 0.5*sumrhoysq;
    a[d] = exp(lna);  asum += a[d];
  }
  for(d=1;d<=8;d++) prob[d] = a[d]/asum;
}
void ChooseNu(){  /* Choose nu for the next sweep */
double U, psum = 0.0;
  code = 0; U = Unif();
  do{code++; psum += prob[code];
  }while (psum < U);
  nu = dof[code];
}
int main(){
int c, sweep, Kount[9], sumKount;  double estp[9];
  ...
  for(c=1; c<=8; c++){prob[c] = 1.0/8; Kount[c] = 0;}
  ChooseNu();
  for(sweep = 1; sweep <= Nsweeps; sweep++){
    ChooseRhos(); nuProbs(); ChooseNu();
    Kount[code]++;
  }
  for(c=1; c<=8; c++) sumKount += Kount[c];
  for(c=1; c<=8; c++) estp[c] = ((double) Kount[c])/sumKount;
  for(c=1; c<=8; c++) printf("%7.4f", estp[c]);
  ...
}
```

I *can* spell 'count'; but since it is a reserved word in some languages, I have got used to spelling it with a 'K' in programs.

He. Programming WinBUGS for this example. Here's the program:

```
model t_hier;
{
    for(a in 1:A) {
        nuProb[a] <- 1/A;
        tau[a] <- nu[a]/(nu[a] - 2);
    }
    w  ~ dcat(nuProb[1:A]);
```

```
        v <- nu[w];

    for (k in 1:n){
        y[k]~dt(0, tau[w], nu[w]); # variance = 1
    }
}
#data
list(n = 50, A = 8,   nu = c(3,4,6,8,12,16,24,32),   y=c(
    0.1519,   -0.2051,    0.3190,   -0.1753,   -2.3757,
    0.8941,   -0.0373,    1.9039,    0.7230,   -1.3515,
    ... ))
#inits
list(w = 3)
```

As stated earlier, it too runs into serious difficulties.

▶▶ **Hf. Discussion.** The problem with the Gibbs sampler is that ν tends to get stuck at the same value for long periods. The fact that it is difficult to separate ν-values from the data would incline one to think the opposite: that ν-values would fluctuate rather rapidly; but this is not the case.

n=50							
nu	3	4	6	8	12	16	24	32
3	0.8477	0.1523	0.0000	0.0000	0.0000	0.0000	0.0000	0.0000
4	0.0135	0.9131	0.0733	0.0001	0.0000	0.0000	0.0000	0.0000
6	0.0000	0.0410	0.7628	0.1940	0.0022	0.0000	0.0000	0.0000
8	0.0000	0.0003	0.2214	0.6712	0.1035	0.0036	0.0000	0.0000
12	0.0000	0.0000	0.0047	0.1606	0.6211	0.2066	0.0069	0.0001
16	0.0000	0.0000	0.0001	0.0077	0.3122	0.5523	0.1202	0.0075
24	0.0000	0.0000	0.0000	0.0000	0.0167	0.2042	0.5430	0.2361
32	0.0000	0.0000	0.0000	0.0000	0.0006	0.0206	0.3293	0.6495
n=200							
3	0.9997	0.0003	0.0000	0.0000	0.0000	0.0000	0.0000	0.0000
4	0.0000	0.9999	0.0001	0.0000	0.0000	0.0000	0.0000	0.0000

Table H(ii): Illustrating the 'sticking' phenomenon

The alarming extent of the 'sticking' phenomenon is illustrated in Table H(ii). The matrix at the top shows for sample size 50 the estimated probability that the ν-value indicated by the row is followed by that in the column. When the sample size is 200, things go completely crazy. The second row of the table for $n = 200$ indicates that for 10000 occasions when ν was 4 before a sweep, it stuck at 4 except for the one time when it switched to 6.

[[For those of you who know about Markov chains:

The transition matrix for the '$n = 50$' case in Table H(ii) is close to being tri-diagonal: it is mostly concentrated on $\{(i, j) : |j - i| \leq 1\}$. It is trivial to calculate the invariant measure for a tri-diagonal transition matrix by using 'symmetrizability' or 'time-reversibility' or whatever you want to call it. See Norris [176]. Thus, on comparing Tables 350 H(i) and H(ii) for $n = 50$, we should have

$$0.0133 \times 0.1523 \approx 0.1502 \times 0.0135,$$
$$0.1502 \times 0.0733 \approx 0.2676 \times 0.0410,$$
$$0.2676 \times 0.1940 \approx 0.2291 \times 0.2214.$$

The values of the ratio RHS/LHS (RHS being right-hand side) are respectively (to 3 places) 1.001, 0.997, 0.977. Thus, our worrying picture is coherent.]]

I am not going to delve into the sticking phenomenon here. Equation 351H1 gives some pointers.

I warned you earlier (at 275Eb) that adding as many new nodes to the Gibbs sampler as there are observations is a potential invitation to trouble if the sample size is even moderately large. Sure, the Ergodic Theorem does apply, but you may need millions of sweeps of the Gibbs sampler to derive benefit from it.

Hg. Extending the Gibbs sampler to include (μ, σ^2)**.** Suppose that our model had been that $Y_1, Y_2, \ldots, Y_n$ are IID, each with the same distribution as $\mu + \sigma Z$, where $Z \sim \text{StandT}_\nu$, and μ and σ are assigned priors. The exact method would then break down, but the Gibbs sampler is easily modified. You should do the modification as an exercise.

The Gibbs sampler is also easily modified to allow a model where

$$Y_k = \theta_\nu + \gamma_\nu T_\nu, \quad T_\nu \sim t_\nu,$$

where (say)
$$\nu \in \{1, 2, 3, 4, 6, 8, 12, 16, 24, 32, \infty\},$$

and where each (θ_ν, γ_ν) is assigned a prior.

But the 'sticking' phenomenon persists.

▶ **I. Robustness of methods: an example.** Suppose that $Y_1, Y_2, \ldots, Y_6$ is actually an IID sample from the t_4 distribution, but that we produce a '95%' CI I for the mean of the Y_k using the 'Student t' method which is exact only for samples from normal distributions. What is the true level of confidence for I? The answer is that it is just under 96%. Of course the same is true if each Y_k has the same distribution as $\mu + \gamma T_4$ where $T_4 \sim t_4$, where μ and γ are constants.

We see from this that the 't' method is robust to certain changes in the model. However, we are certainly aware by now that the t_4 and $N(0, 1)$ distributions are rather similar; and moreover, the t_4 distribution is symmetric. How would the 't' method perform if 'errors' were markedly skewed? Well, we can try the case when each Y_k has the standardized $E(1)$ distribution, the distribution of $-\ln(U) - 1$, where $U \sim U[0, 1]$. We find that the true level of confidence for I is then about 89%, so that we have now lost robustness.

Of course, we can easily estimate the true levels of robustness by simulation. Here are the main parts of a program which utilizes functions described earlier in the book:

```
double Skew(){return -log(Unif()) - 1.0;}
void Do(double F()){ long c, OK = 0; int k;
double sum,ssq, y, mean, estV, SE;
  for(c=0; c<L; c++){ sum = 0.0; ssq = 0.0;
    for(k=0; k<n; k++){y = F(); sum += y; ssq += y*y;}
    mean = sum/n; estV = (ssq - mean*sum)/(n-1);
    SE = sqrt(estV)/rtn;
    if (fabs(mean) < t*SE) OK++;
  }
  printf("\n P(mu in CI) is approx %7.5f", ((double) OK)/L);
}
int main(){ setseeds(); sethtol(1.0E-6);
  setnu(n-1); t = invT(0.025); rtn = sqrt((double) n);
  Do(aGauss); prepKM_T(4); Do(KMStudT); Do(Skew);
  return 0;
}
```

Other simple methods from Classical Statistics may be tested for robustness in similar fashion. It is obvious how to test MCMC methods for robustness by comparing CIs obtained when one modifies the model. I have earlier emphasized testing robustness to changes in priors.

▶ **J. Examining regression models.** For many subsections now, we have considered problems associated with the simplest of situations: that when $Y_1, Y_2, \ldots, Y_n$ are IID. Inevitably, when it comes to regression, etc, one has all the analogous problems and a set of new ones.

Recall from Subsection 312D that in Frequentist theory with model

$$Y_k = \mu + \beta(x_k - \overline{x}) + \varepsilon_k, \qquad \varepsilon_k\text{'s IID each } N(0, \sigma^2)$$

we use Estimators $\overline{Y}, B, \hat{\sigma}_{n-2}(\mathbf{x}, \mathbf{Y})$ for μ, β, σ, and that

$$R_i := Y_k - \overline{Y} - B(x_k - \overline{x})$$

is the ith Residual. The column vector $\mathbf{R}$ of Residuals is just

$$\mathbf{R} = P_{\text{res}} \mathbf{Y} = \sigma P_{\text{res}} \mathbf{G}.$$

and $\mathbb{E}\, R_i R_j$ is the (i, j)th component of $\mathbb{E}\,\mathbf{RR}^T$. We have, with $\mathbf{u}$ and $\mathbf{e}$ as in Subsection 312D,

$$
\begin{aligned}
\mathbb{E}\,\{\mathbf{RR}^T\} &= \sigma^2 \mathbb{E}\,\{P_{\text{res}} \mathbf{G}\,(P_{\text{res}} \mathbf{G})^T\} = \sigma^2 P_{\text{res}} \mathbb{E}\{\mathbf{GG}^T\} P_{\text{res}}^T \\
&= \sigma^2 P_{\text{res}} P_{\text{res}}^T = \sigma^2 P_{\text{res}} P_{\text{res}} = \sigma^2 P_{\text{res}} \\
&= \sigma^2 \left(I - \mathbf{uu}^T - \mathbf{ee}^T\right),
\end{aligned}
$$

since $\mathbb{E}\,\mathbf{GG}^T = I$. We have worked componentwise with expectations of matrices, and used the linearity of expectations and property 324(J2) of projections. (We are, of course, relating everything to the standard orthonormal basis $\{\mathbf{e}^{(1)}, \mathbf{e}^{(2)}, \dots, \mathbf{e}^{(n)}\}$ of $\mathbb{R}^n$.) We see that, with

$$
\delta_{ij} = \begin{cases} 1 & \text{if } i = j, \\ 0 & \text{if } i \neq j, \end{cases}
$$

as usual,

$$
\mathbb{E}\, R_i R_j = \sigma^2 \left\{ \delta_{ij} - \frac{1}{n} - \frac{(x_i - \overline{x})(x_j - \overline{x})}{q(\mathbf{x}, \mathbf{x})} \right\}.
$$

If n is large and each $(x_i - \overline{x})$ is small compared with $q(\mathbf{x}, \mathbf{x})^{\frac{1}{2}}$, that is if all the leverage coefficients are small, then the Residuals R_i behave rather like IID $N(0, \sigma^2)$ variables. (Of course, $\sum R_k = 0$ and $\sum (x_k - \overline{x}) R_k = 0$, but we can to an extent forget that. The Multivariate-Normal nature of $\mathbf{R}$ will be obvious from the next Section.)

When assessing whether a regression model is appropriate, one thing we always do is to look at a plot of the standardized residuals, the kth standardized residual being

$$
r_k^{\text{obs}} \Big/ \hat{\sigma}_{n-2}(\mathbf{x}, \mathbf{y}^{\text{obs}}), \qquad r_k^{\text{obs}} = y_k^{\text{obs}} - \overline{y}_{\text{obs}} - b^{\text{obs}}(x_k - \overline{x}).
$$

Under the conditions mentioned above (since $\hat{\sigma}_{n-2}(\mathbf{x}, \mathbf{y}^{\text{obs}}) \approx \sigma$) the standardized residuals should look rather like an IID sample from a normal distribution; and we can assess this via qq, Kolmogorov or χ^2 methods. Again, outliers should be checked for accuracy, and if genuine, we would have to consider a model 'with larger tails'. And see Rousseeuw and Leroy [200]. Look at Figure J(i). For each part $n = 50$. The top of each figure shows the regression line and the bottom the standardized residuals on a magnified scale. Figure (a) looks reasonable: we would expect on average $50 \times 0.05 = 2.5$ observations with standardized residual of size 2 or more. Figure (b) shows a situation where non-linear regression is appropriate: the method from 318F should be tried first. In (c), it is clear that the residuals have a positively skewed distribution. In (d) we have a (rather extreme) *heteroscedastic* case where the variance of the observation changes (here decreases) with the x-value.

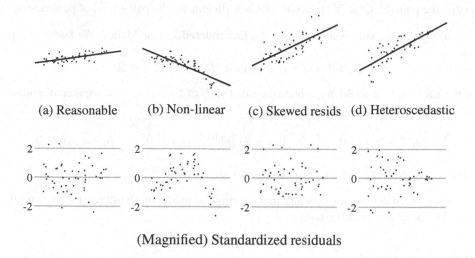

(a) Reasonable   (b) Non-linear   (c) Skewed resids   (d) Heteroscedastic

(Magnified) Standardized residuals

Figure J(i): Regression diagrams

Ja. Confounding. This can again be a problem. Compare 179Gb. Sketch a situation where the sample really divides into two samples each with a positive β-value, but where the combined sample suggests a negative β-value.

Note. 'Confounding' (which has several meanings) does not always signify something bad: deliberate confounding plays an important useful rôle in experimental designs.

▶ **K. Logistic regression; and other 'Bernoulli' models.** Suppose that $Y_1, Y_2, \ldots, Y_n$ are independent, Y_k being Bernoulli(p_k). We suppose that

$$\text{logit}(p_k) := \ln \frac{p_k}{1 - p_k} = \gamma + \beta(x_k - \overline{x}),$$

so that

$$p_k = p_k(\gamma, \beta) = \frac{e^{\gamma + \beta(x_k - \overline{x})}}{1 + e^{\gamma + \beta(x_k - \overline{x})}}.$$

We think of Y_k as the indicator function of F_k, where F_k occurs with probability p_k which depends on the level x_k of some stimulus. For example, x_k might represent age of a component and F_k the event that it will fail in the next month. Or x_k might represent an electric/chemical stimulus and F_k the event that a neuron subject to that stimulus will 'fire'.

You should check that

$$\text{lhd}(\gamma, \beta; \mathbf{Y}) = \frac{e^{\gamma n \overline{Y} + \beta Q(\mathbf{x}, \mathbf{Y})}}{\prod\{1 + e^{\gamma + \beta(x_k - \overline{x})}\}},$$

so that the pair $(\overline{Y}, Q(\mathbf{x}, \mathbf{Y}))$ of Statistics is sufficient for the pair (γ, β) of parameters.

Logistic regression is an example of a **Generalized** Linear Model. We know how to test

$$H_0 : \beta = 0, \gamma \in \mathbb{R} \text{ against } H_A : \beta \in \mathbb{R}, \gamma \in \mathbb{R}$$

by the LR test. One could only obtain the mle $(\hat{\gamma}, \hat{\beta})$ of (γ, β) on the computer, of course, by solving

$$\sum \left(y_k^{\mathrm{obs}} - \hat{p}_k\right) = 0, \qquad \sum (x_k - \overline{x}) \left(y_k^{\mathrm{obs}} - \hat{p}_k\right) = 0, \qquad \hat{p}_k := p_k(\hat{\gamma}, \hat{\beta}).$$

All standard packages will do logistic regression.

There is, of course, no difficulty in analyzing the model on `WinBUGS` for any sample size. The program would contain

```
xbar <- mean(x[]);
for(k in 1:n){
  logit(p[k]) <- gamma + beta*(x[k] - xbar);
  Y[k] ~ dbern(p[k]);
}
```

Ka. Other Bernoulli models.   With the `WinBUGS` package comes an example ('Surgical') in which for hospital j $(1 \leq j \leq J)$, r_j babies died out of n_j who had heart surgery. One way to study this is to assume that the true failure rates p_j $(1 \leq j \leq J)$ are independent (a *fixed-effects* model).

Another way is to use a *linked* (or *random-effects*) model in which it is assumed that failure rates across hospitals are in some way similar. Thus, for example, one can suppose that

$$\mathrm{logit}(p_j) = b_j, \qquad b_j \sim \mathrm{N}(\mu, \mathrm{prec}\ \rho).$$

In the prior for this hierarchical model, the parameters b_j are assumed independent $\mathrm{N}(\mu, \mathrm{prec}\ \rho)$, with the hyper-parameters μ and ρ having 'vague' priors. You can check, if you wish, that `WinBUGS` only has to sample from log-concave densities. CIs for the different p_j will be 'pulled towards one another' to some extent.

We shall look at the effects of using 'linked' models on an ANOVA case in the next subsection.

▶▶ **L. Modified ('random-effects') ANOVA models.**   How might we modify the classical model? Let's write the general additive model in the form

$$Y_{jk} = \theta_j + \beta_k + \sigma \varepsilon_{jk}, \qquad \sum \beta_k = 0,$$

where our primary concern is with the θ_j's. One thing to consider – and such things must be considered separately for every situation – is whether to adopt the idea at Ka of using a 'linked' or 'random effects' model. For example, we might use a 'Normal-linked Normal' model in which

$$\theta_1, \theta_2, \ldots, \theta_J \text{ are IID each } N(\mu, \text{prec } r),$$

where we take vague priors for μ and r. (*Note.* This model can be analyzed by classical Frequentist methods, as was done long ago by Daniels.) If one does do this on the example which we studied in Subsection 289F, then, as we shall see, one changes CIs to a fairly significant extent. For the moment, assume that an additive model (one without interaction) is appropriate.

Of course, using normal errors in any way might be less sensible than using 't' errors; and one can do this with linked or unlinked models. Consider four possible models for the numerical case at Subsection 289F.

Classical (unlinked Normal) model:

$$\theta_j, \beta_k \text{ vague}, \varepsilon\text{'s Normal}, \sigma \text{ vague}.$$

Normal-linked Normal model:

$$\theta_j \sim N(\mu, \text{prec } r), \beta_k \text{ vague}, \mu, r \text{ vague}, \varepsilon\text{'s Normal}, \sigma \text{ vague}.$$

Unlinked t_4 *model*:

$$\theta_j, \beta_k \text{ vague}, \varepsilon\text{'s } t_4, \sigma \text{ vague}.$$

t_4-*linked* t_4 *model*:

$$\theta_j = \mu + r^{-\frac{1}{2}} t_4, \beta_k \text{ vague}, \mu, r, \sigma \text{ vague}, \varepsilon\text{'s } t_4.$$

Here are results for our numerical case, the last 3 columns being produced after runs of 10000 on WinBUGS :

```
          Classical    N-linked N    unlinked t4   t4-linked t4
theta[1]  (4.73,11.27) (5.80,11.93)  (3.85,11.74)  (4.92,11.93)
theta[2]  (10.73,17.27)(8.34,15.86)  (11.14,16.83) (8.61,15.91)
theta[3]  (9.73,16.27) (8.23,14.99)  (9.89,15.87)  (8.60,14.97)
theta[4]  (5.73,12.27) (6.64,12.26)  (6.00,11.93)  (6.21,12.43)
theta[5]  (2.73, 9.27) (4.11,11.55)  (3.20, 8.88)  (3.95,11.11)
```

Because of comments in Subsection 276H, I tried several different priors, but all with similar results. That the 'N-linked N' model pulls things together compared with classical model is sensible. The results for the 't_4-linked t_4' model are not that different from those for the 'N-linked N'. The 'Unlinked t_4' results puzzled me a little, so I ran them on WinBUGS for different length runs with different seeds, all with similar results. It is interesting that this method is reflecting better than the others the fact that the first row is out of line with the other rows (see the next section), because of the slow tail-off of the t_4 density.

The degree of flexibility you have is enormous. Of course, you could use a 'hierarchical t' model, allowing different degrees of freedom, etc. In an important paper on agricultural field experiments, Besag and Higdon [20] use 'hierarchical t' models and show how to build in (amongst other things) the fact that neighbouring plots of soil will have similar fertility. For an important recent paper on these topics, see Nobile and Green [175].

La. Some comments on flexibility.   I repeat that it remains important to remember the Principle of Parsimony: use the simplest reasonable model. By all means investigate related models, but use them to convince yourself that when you talk of a 95% CI, the 95% must be taken with several pinches of salt: what you really mean is your confidence lies somewhere between 93% and 97%, or something similar.

If you have rejected various models before deciding on your final one, you should say so. And do be careful about data snooping: most pieces of data will have some strange aspects; and focusing on them after seeing the data *can* be very dangerous.

▶ **M. Some thoughts on goodness of fit for ANOVA.**   Figure M(i) is, I hope, self-explanatory. One should always draw such a figure.

What it shows for our numerical example is that the first row ($\mathcal{A}$ level 1) is 'out of line' with the other rows. It suggests the possibility either of interaction (with non-zero $(\alpha\beta)_{1k}$ terms) or increased error variance for row 1.

Of course, the fact that row 1 seems 'out of line' is reflected by the fact that its residuals are larger than those for the other rows. In fact row 1 contributes a fraction $26/48$ of the residual sum of squares. We can solve the problem: *How likely is it that some row will contribute as least as high a proportion of rss if our* **model** *is correct?* It is important to realize that if our model is correct, the described probability will be the same whether H_0 holds or not. In this case, we find that the probability is close to 10%, so we would be cautious about claiming interaction if there is no *a priori* reason to suspect that it might occur.

▶ **Ma. Exercise.** Estimate by simulation the probability that for our additive ANOVA model with 5 rows and 3 columns, some row will contribute a fraction at least $26/48$ of the total rss.

Of course, interaction should be investigated via a model with replication as in Subsection 331M. If you are presented only with a $J \times K$ table, then you

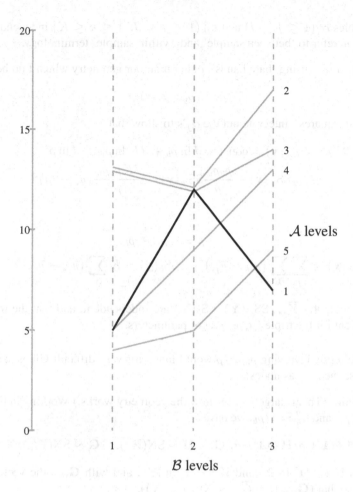

Figure M(i): Diagram showing possible interaction

must provide a diagram such as that at M(i), commenting on what possibilities of interaction it suggests. You must also do the analogue of Exercise Ma. If that exercise forces you to agree that there is interaction, your first reaction would be to beg for results with replication; but if they prove impossible to obtain, then read the discussions of interaction in Cox [48].

N. A 'paradox'. See Exercise 10.4 in Cox and Hinkley [49], based on work of Stone, Springer and Dawid. I present it mainly as an example to show how grown up you now are in that you can cope with quite complicated things and understand their geometry.

Consider the random-effects model

$$Y_{jk} = \mu + \alpha_j + \varepsilon_{jk}, \quad \alpha_j \sim N(0, \text{prec } \rho_b), \quad \varepsilon_{jk} \sim N(0, \text{prec } \rho_w),$$

with all variables $\alpha_j(1 \le j \le J)$ and $\varepsilon_{jk}(1 \le j \le J, \ 1 \le k \le K)$ independent. The suffices b and w refer to 'between sample' and 'within sample' terminology.

Na. Exercise. Find (using 'bare hands', rather than our geometry which I do below!)

$$2\ell(\mu, \rho_b, \rho_w; \mathbf{y}, \boldsymbol{\alpha}).$$

By 'completing squares', integrate out the α_j's to show that

$$2\ell(\mu, \rho_b, \rho_w; \mathbf{y}) = \text{const} + J \ln \rho_b + JK \ln \rho_w - J \ln \rho$$

$$- \rho_w \text{SS}_w(\mathbf{y}) - \frac{\rho_b \rho_w}{\rho} \text{SS}_w(\mathbf{y}) - \frac{JK\rho_b\rho_w}{\rho}(\overline{y}_{**} - \mu)^2,$$

where

$$\rho = K\rho_w + \rho_b,$$

$$\text{SS}_w(\mathbf{y}) = \sum_j \sum_k (y_{jk} - \overline{y}_{j*})^2, \quad \text{SS}_b(\mathbf{y}) = K \sum_j (\overline{y}_{j*} - \overline{y}_{**})^2.$$

Note that the variables $\overline{Y}_{**}$, $\text{SS}_w(\mathbf{Y})$, $\text{SS}_b(\mathbf{Y})$ are independent, and that the triple they form is sufficient for the triple (μ, ρ_b, ρ_w) of parameters.

Remark. The terms involving $\rho_b\rho_w/\rho$ would make for very difficult Gibbs sampling if one did not use the α_j's as nodes!

Nb. Discussion. (This is largely to see how the geometry works.) Working in $\mathbb{R}^J \otimes \mathbb{R}^K$, with $\sigma_w^2 = 1/\rho_w$ and $\sigma_b^2 = 1/\rho_b$, we have

$$\mathbf{Y} = \mu \mathbf{1} \otimes \mathbf{1} + \sigma_b \mathbf{H} \otimes \mathbf{1} + \sigma_w \mathbf{G}, \quad \mathbf{H} \sim \text{SN}(\mathbb{R}^J), \quad \mathbf{G} \sim \text{SN}(\mathbb{R}^J \otimes \mathbb{R}^K).$$

We have with $U := [\mathbf{1}]$ in $\mathbb{R}^J$ and $V := [\mathbf{1}]$ in $\mathbb{R}^K$, and with $\mathbf{G}_{\text{row}}$ the vector of row means of $\mathbf{G}$ (so that $(\mathbf{G}_{\text{row}})_j = \overline{G}_{j*} \sim \text{N}(0, \sigma_w^2/K)$),

$$P_{U \otimes V} \mathbf{Y} = \overline{Y}_{**}\mathbf{1} \otimes \mathbf{1} = \mu \mathbf{1} \otimes \mathbf{1} + \sigma_b(P_U \mathbf{H}) \otimes \mathbf{1} + \sigma_w P_{U \otimes V} \mathbf{G},$$

$$P_{U^\perp \otimes V} \mathbf{Y} = \left(P_{U^\perp}(\sigma_b \mathbf{H} + \sigma_w \mathbf{G}_{\text{row}})\right) \otimes \mathbf{1}, \qquad (\text{'between'})$$

$$\mathbb{P}_{\mathbb{R}^J \otimes V^\perp} \mathbf{Y} = \sigma_w \mathbb{P}_{\mathbb{R}^J \otimes V^\perp} \mathbf{G}. \qquad (\text{'within'})$$

We have the component formulae:

$$\left(P_{U^\perp \otimes V} \mathbf{Y}\right)_{jk} = \overline{Y}_{j*} - \overline{Y}_{**} = \sigma_b\left(\overline{H}_{j*} - \overline{H}_{**}\right) + \sigma_w\left(\overline{G}_{j*} - \overline{G}_{**}\right),$$

$$\left(\mathbb{P}_{\mathbb{R}^J \otimes V^\perp} \mathbf{Y}\right)_{jk} = Y_{jk} - \overline{Y}_{j*}.$$

We find that, with $\sigma^2 = K\sigma_b^2 + \sigma_w^2$, so that $\rho_b\rho_w/\rho = 1/\sigma^2$, we have

$$\overline{Y}_{**} \sim \text{N}(0, \sigma^2/JK),$$

$$\text{SS}_b(\mathbf{Y}) = \left\|P_{U^\perp \otimes V} \mathbf{Y}\right\|^2 \sim \sigma^2 \chi_{J-1}^2,$$

$$\text{SS}_w(\mathbf{Y}) = \left\|\mathbb{P}_{\mathbb{R}^J \otimes V^\perp} \mathbf{Y}\right\|^2 \sim \sigma_w^2 \chi_{J(K-1)}^2,$$

these variables being independent. (So, we *have* found useful consequences of the geometry!) We could find a Confidence Interval for μ via the fact that

$$\frac{(JK)^{\frac{1}{2}}\left(\overline{Y}_{**} - \mu\right)}{\{\mathrm{SS}_b(\mathbf{Y})/(J-1)\}^{\frac{1}{2}}} \sim t_{J-1}. \tag{N1}$$

This would be exactly as if we were told only the results $\overline{Y}_{j*}$ ($1 \le j \le J$). However, since in particular, $\sigma^2 \ge \sigma_w^2$, we must use all the information.

The 'paradox': If we use the prior density

$$\pi(\mu, \sigma_w, \sigma) \propto \frac{1}{\sigma\sigma_w} I_{(\sigma > \sigma_w > 0)}(\sigma_w, \sigma)$$

then the CIs for μ given all the information will be *wider*(!) than those obtained from (N1).

   Like so much else from the past, it doesn't seem terribly relevant nowadays. Surely, in any case it is (σ_b, σ_w) which should have prior $(\sigma_b\sigma_w)^{-1}$.

Nc. Exercise. Write a `WinBUGS` program to get a CI for μ, using appropriate vague priors. Use the α's as nodes, of course.

▶ **O. Contingency tables.** Since we have studied the Chi-Squared Principle, we ought to look at one of its most famous uses: that in contingency tables.

   Suppose that each individual in a large population falls into one of J categories relative to 'factor A', and into one of K categories relative to factor B. For example, some diseases are more likely to be caught by rich people than poor, some less. We might divide income into J ranges to give the A factor, and classify whether or not individuals contract a particular disease as B factor ($K = 2$).

   Let p_{jk} be the probability that an individual chosen at random belongs to category j relative to A and category k relative to B. Let $\alpha_j := \sum_k p_{jk}$ be the 'marginal' probability that an individual chosen at random belongs to category j relative to A, and define β_k analogously. Primary interest has attached to deciding whether factors A and B are independent in that the Null Hypothesis

$$H_0 : p_{jk} = \alpha_j\beta_k$$

holds. We use the 'full' Alternative Hypothesis

$$H_A : \sum\sum p_{jk} = 1.$$

This seems a situation where Hypothesis Testing makes good sense.

Suppose that a random sample of size n is taken from the population. Let N_{jk} be the number in the sample falling into category j relative to A and category k relative to B. Let $A_j := n^{-1} \sum_k N_{jk}$ and $B_k := n^{-1} \sum_j N_{jk}$. The Maximum Likelihood associated with H_0 will occur when (for all (j, k)) $\alpha_j = A_j$ and $\beta_k = B_k$. Exactly as in Subsection 345E, we find that for large n, if H_0 is true, then

$$\text{PearsonStat} := \sum \sum \frac{(N_{jk} - nA_jB_k)^2}{nA_jB_k}$$

has approximately the χ^2_ν distribution where $\nu = (J - 1)(K - 1)$, the difference in dimension between H_A (for which dim $= JK - 1$) and H_0 (for which dim $= J - 1 + K - 1$).

▶ **Oa. Exercise.** The *Star-Ship Enterprise* is approaching Planet ChiCrazy. Each individual on the planet is known to be either Green-eyed or Red-eyed. Based on certain information, Dr McCoy has set up a Null Hypothesis that for an individual chosen at random, eye-colour and sex are independent. Mr Spock, typically, has made the precise Null Hypothesis that

$$p_{\text{GM}} = p_{\text{RM}} = p_{\text{GF}} = p_{\text{RF}},$$

where p_{GM} is the proportion of Green-eyed Males.

The *Enterprise* receives a message that 400 randomly chosen individuals from the huge population will greet them. When they land, they find that these individuals form the 2×2 Table Oa (Actually, these are not quite the numbers, but I've made the calculation

		Green	Red	
--------		-------	-----	
Male		87	113	
Female		113	87	

Table O(i): ChiCrazy table

easy for you without spoiling the message.) Check that for Dr McCoy and Mr Spock, the Pearson statistic takes the value 6.76. Check that Dr McCoy has to reject his Null Hypothesis at the 1% level, but that Mr Spock does not have to reject his even at the 5% level.

"This is simply not fair;" says McCoy, "if your Null Hypothesis is true, then mine must be true. Yet I have to reject my Null Hypothesis, but you do not have to reject yours."

"Yes," says Spock, "but you see, there *is* logic in the situation ... ".

What does Mr Spock go on to say?

Ob. Exercise. Suppose that $(X_1, Y_1), (X_2, Y_2), \ldots, (X_n, Y_n)$ are n IID $\mathbb{R}^2$ variables each assumed to have the same distribution as (X, Y). Suggest a test of independence of X and Y.

8.5 Multivariate Normal (MVN) Distributions

This section brings lots of nice Mathematics to the service of Statistics. I hope that you will like it.

Let's first think about why we need some more algebra. Suppose that $X_1, X_2, \ldots, X_n$ are RVs, *all with zero mean*. Then the **variance-covariance matrix** V with

$$V_{ij} = \text{Cov}(X_i, X_j), \quad \text{so that } V = \mathbb{E}\,\mathbf{X}\mathbf{X}^T,$$

is **symmetric** $(V = V^T)$ and **nonnegative-definite** in that

$$\mathbf{w}^T V \mathbf{w} \geq 0, \quad \mathbf{w} \in \mathbb{R}^n.$$

This is because

$$0 \leq \mathbb{E}\left(\langle \mathbf{w}, \mathbf{X}\rangle^2\right) = \mathbb{E}\,\mathbf{w}^T \mathbf{X}\mathbf{X}^T \mathbf{w} = \mathbf{w}^T(\mathbb{E}\,\mathbf{X}\mathbf{X}^T)\mathbf{w} = \mathbf{w}^T V \mathbf{w}.$$

So, it is not surprising that we need to study nonnegative-definite symmetric matrices.

I am sure that you know that if B is an $n \times n$ matrix, then we can solve $B\mathbf{w} = 0$ with $\mathbf{w} \neq \mathbf{0}$ if and only if $\det(B) = 0$, that is, if and only if, B is singular.

▶▶ **A. Fact: Spectral Theorem for symmetric matrices.** *If A is a symmetric matrix, then there exist (non-trivial) orthogonal (that is, perpendicular) subspaces $W_1, W_2, \ldots, W_r$ of $\mathbb{C}^n$ with sum $\mathbb{C}^n$ such that*

$$A = \gamma_1 P_1 + \gamma_2 P_2 + \cdots + \gamma_r P_r, \tag{A1}$$
$$I = P_1 + P_2 + \cdots + P_r,$$

*where P_k denotes perpendicular projection onto W_k and $\gamma_1, \gamma_2, \ldots, \gamma_r$ are distinct **real** numbers. Then,*

$$g(A) = g(\gamma_1)P_1 + g(\gamma_2)P_2 + \cdots + g(\gamma_r)P_r \tag{A2}$$

for any polynomial function g.

Here the γ_k are the distinct eigenvalues of A, the distinct roots of

$$\det(\gamma I - A) = 0,$$

and W_k is the space of eigenvectors of A corresponding to γ_k:

$$W_k = \{\mathbf{w} \in \mathbb{C}^n : A\mathbf{w} = \gamma_k \mathbf{w}\}.$$

If $d_k = \dim(W_k)$, then

$$\det(\lambda I - A) = \prod_k (\lambda - \gamma_k)^{d_k}, \qquad \det(A) = \prod_k \gamma_k^{d_k}.$$

Proving that the roots of the equation are real is easy – see Fact 449Aa. The more difficult thing with Fact A is to prove that A has enough eigenvectors to span $\mathbb{R}^n$.

Because $P_k^2 = P_k$ and $P_k P_\ell = 0$, it follows by induction from (A1) that

$$A^m = \gamma_1^m P_1 + \gamma_2^m P_2 + \cdots + \gamma_r^m P_r,$$

so that 365(A2) follows for polynomials g.

B. Positive-definite symmetric matrices.

As we have seen, we say that a symmetric matrix A is nonnegative-definite if $\mathbf{w}^T A \mathbf{w} \geq 0$ for every $\mathbf{w} \in \mathbb{R}^n$. We say that A is **positive-definite** if $\mathbf{w}^T A \mathbf{w} > 0$ for every non-zero $\mathbf{w}$ in $\mathbb{R}^n$.

Let A have spectral decomposition as in Fact 365A. Then

- A is nonnegative-definite if and only if $\gamma_k \geq 0$ for every k,
- A is positive-definite if and only if $\gamma_k > 0$ for every k.

This is because

$$\mathbf{w}^T A \mathbf{w} = (\mathbf{w}^T P_1 + \cdots + \mathbf{w}^T P_r)(\gamma_1 P_1 \mathbf{w} + \cdots + \gamma_r P_r \mathbf{w})$$
$$= \gamma_1 \|P_1 \mathbf{w}\|^2 + \cdots + \gamma_r \|P_r \mathbf{w}\|^2,$$

while

$$\|\mathbf{w}\|^2 = \|P_1 \mathbf{w}\|^2 + \cdots + |P_r \mathbf{w}\|^2.$$

Note that if A is a nonnegative-definite symmetric matrix and we put

$$S := \sqrt{\gamma_1} P_1 + \cdots + \sqrt{\gamma_r} P_r,$$

then (S is nonnegative-definite symmetric and) $S^2 = A = SS^T$.

Ba. Exercise. Prove that if A is positive [respectively nonnegative] definite, then the matrix formed by taking just the first k rows and first k columns of A is positive [respectively nonnegative] definite. Prove that if A is positive [respectively nonnegative] definite, then $\det(A) > 0$ [respectively, $\det(A) \geq 0$].

Bb. Choleski decomposition. This decomposition is very useful for simulation. Suppose that V is positive-definite symmetric. Then we can find a lower triangular matrix S such that $SS^T = V$. Note that the matrix SS^T is symmetric, and (Why?) is positive-definite if S is non-singular. Writing out the equation $SS^T = V$, we have

$$\begin{pmatrix} s_{11} & 0 & 0 & \cdot \\ s_{21} & s_{22} & 0 & \cdot \\ s_{31} & s_{32} & s_{33} & \cdot \\ \cdot & \cdot & \cdot & \cdot \end{pmatrix} \begin{pmatrix} s_{11} & s_{21} & s_{31} & \cdot \\ 0 & s_{22} & s_{32} & \cdot \\ 0 & 0 & s_{33} & \cdot \\ \cdot & \cdot & \cdot & \cdot \end{pmatrix} = \begin{pmatrix} v_{11} & v_{12} & v_{13} & \cdot \\ v_{21} & v_{22} & v_{23} & \cdot \\ v_{31} & v_{32} & v_{33} & \cdot \\ \cdot & \cdot & \cdot & \cdot \end{pmatrix},$$

and we must have

$$s_{11}^2 = v_{11}, \quad s_{r1} = v_{1r}/s_{11} \quad (r > 1),$$
$$s_{22}^2 = v_{22} - s_{21}^2 = (v_{11}v_{22} - v_{12}v_{21})/s_{11}^2.$$

etc, etc. Note that

$$s_{11}^2 s_{22}^2 \cdots s_{kk}^2 = \det \begin{pmatrix} v_{11} & \cdot & v_{1k} \\ \cdot & \cdot & \cdot \\ v_{k1} & \cdot & v_{kk} \end{pmatrix},$$

and since the determinant is positive, it is easy to believe that the Choleski decomposition will work.

▶▶ **C. The Multivariate Normal Distribution $\mathrm{MVN}_n(\mu, V)$.** Let $\mu \in \mathbb{R}^n$ and let V be a nonnegative-definite symmetric $n \times n$ matrix. Then a Random Variable $\mathbf{Y}$ with values in $\mathbb{R}^n$ is said to have the $\mathrm{MVN}_n(\mu, V)$ distribution if, for every vector α in $\mathbb{R}^n$,

$$\mathbb{E} \exp\left(i\alpha^T \mathbf{Y}\right) = \exp\left(i\alpha^T \mu - \tfrac{1}{2}\alpha^T V \alpha\right). \tag{C1}$$

Uniqueness of the distribution is guaranteed by Characteristic-Function (Fourier) results. If S is such that $SS^T = V$ and $\mathbf{Y} = \mu + S\mathbf{G}$ where $\mathbf{G} \sim \mathrm{MVN}_n(\mathbf{0}, I)$ (so that $G_1, G_2, \ldots, G_n$ are IID each $N(0, 1)$), then

$$i\alpha^T \mathbf{Y} = i\alpha^T \mu + i\alpha^T S\mathbf{G} = i\alpha^T \mu + i(S^T \alpha)^T \mathbf{G},$$

so that

$$\mathbb{E} \exp\left(i\alpha^T \mathbf{Y}\right) = \exp\left(i\alpha^T \mu - \tfrac{1}{2}\alpha^T SIS^T \alpha\right) = \exp\left(i\alpha^T \mu - \tfrac{1}{2}\alpha^T V \alpha\right),$$

so that $\mathbf{Y}$ has the $\mathrm{MVN}_n(\mu, V)$ distribution. Note that this and the Choleski decomposition give a way of simulating Y if V is positive-definite.

Note that since $\mathbb{E}(\mathbf{G}) = \mathbf{0}$, we have $\mathbb{E}(\mathbf{Y}) = \mu$. Also,

$$\mathbb{E}(\mathbf{Y} - \mu)(\mathbf{Y} - \mu)^T = \mathbb{E}S(\mathbf{G}\mathbf{G}^T)S^T = S\mathbb{E}(\mathbf{G}\mathbf{G}^T)S^T = SS^T = V,$$

so that μ and V have the correct significance as mean vector and variance-covariance matrix for $\mathbf{Y}$.

▶ **D. Pdf for $\mathrm{MVN}_n(\mu, V)$ when $\det(V) \neq 0$.** Suppose that $\mu \in \mathbb{R}^n$ and that V is a positive-definite symmetric $n \times n$ matrix. Let $\mathbf{Y} \sim \mathrm{MVN}_n(\mu, V)$. We use the representation

$$\mathbf{Y} = \mu + S\mathbf{G}, \quad SS^T = V, \quad \text{so that } \mathbf{G} = S^{-1}(\mathbf{Y} - \mu).$$

Now, $\mathbf{G}$ has pdf on $\mathbb{R}^n$ given by

$$f_{\mathbf{G}}(\mathbf{g}) = (2\pi)^{-\frac{1}{2}n} \exp(-\tfrac{1}{2}\mathbf{g}^T \mathbf{g}).$$

If $\mathbf{g} = S^{-1}(\mathbf{y} - \boldsymbol{\mu})$, then $J\binom{\mathbf{g}}{\mathbf{y}} = \det\left((S^{-1})^T\right)$ with our conventions. Thus $|J| = 1/\sqrt{\det(V)}$ and

$$f_{\mathbf{Y}}(\mathbf{y}) = f_{\mathbf{G}}(S^{-1}(\mathbf{y} - \boldsymbol{\mu}))|J|,$$

whence, since $\mathbf{g}^T\mathbf{g} = (\mathbf{y} - \boldsymbol{\mu})^T(S^{-1})^T S^{-1}(\mathbf{y} - \boldsymbol{\mu}) = (\mathbf{y} - \boldsymbol{\mu})^T V^{-1}(\mathbf{y} - \boldsymbol{\mu})$, we have

$$f_{\mathbf{Y}}(\mathbf{y}) = \frac{1}{(2\pi)^{\frac{1}{2}n}\sqrt{\det(V)}} \exp\left\{-\tfrac{1}{2}(\mathbf{y} - \boldsymbol{\mu})^T V^{-1}(\mathbf{y} - \boldsymbol{\mu})\right\}. \qquad (\text{D1})$$

Of course, the fact that

$$\int_{\mathbb{R}^n} f_{\mathbf{Y}}(\mathbf{y})\mathrm{d}\mathbf{y} = \int_{-\infty}^{\infty} \cdots \int_{-\infty}^{\infty} f_{\mathbf{Y}}(y_1, \ldots y_n)\mathrm{d}y_1 \ldots \mathrm{d}y_n = 1 \qquad (\text{D2})$$

follows from our argument.

E. Some key properties of MVN distributions.

▶ **Ea. Linear maps preserve the MVN property.** Suppose that $\mathbf{Y} \sim \mathrm{MVN}_n(\boldsymbol{\mu}, V)$, that B is an $m \times n$ matrix, and that $\mathbf{Z} = B\mathbf{Y} \in \mathbb{R}^m$. Then, for $\boldsymbol{\alpha} \in \mathbb{R}^m$, $\boldsymbol{\alpha}^T B = (B^T\boldsymbol{\alpha})^T$, so that

$$\mathbb{E}\exp\left(\mathrm{i}\boldsymbol{\alpha}^T\mathbf{Z}\right) = \mathbb{E}\exp\left(\mathrm{i}(B^T\boldsymbol{\alpha})^T\mathbf{Y}\right) = \exp\left(\mathrm{i}\boldsymbol{\alpha}^T B\boldsymbol{\mu} - \tfrac{1}{2}\boldsymbol{\alpha}^T BVB^T\boldsymbol{\alpha}\right),$$

so that $\mathbf{Z} \sim \mathrm{MVN}_m(B\boldsymbol{\mu}, BVB^T)$. Thus, as claimed, linear transformations preserve the MVN property.

Eb. Partitioned matrices. The $(m + n) \times (r + s)$ matrix

$$\begin{pmatrix} a_{11} & \cdot & a_{1r} & b_{11} & \cdot & b_{1s} \\ \cdot & & \cdot & \cdot & & \cdot \\ a_{m1} & \cdot & a_{mr} & b_{m1} & \cdot & b_{ms} \\ c_{11} & \cdot & a_{1r} & d_{11} & \cdot & d_{1s} \\ \cdot & & \cdot & \cdot & & \cdot \\ c_{n1} & \cdot & a_{nr} & d_{n1} & \cdot & d_{ns} \end{pmatrix} \quad \text{is written} \quad \begin{pmatrix} A & B \\ C & D \end{pmatrix},$$

where A is $m \times r$, etc. Partitioned matrices 'of compatible dimensions' multiply in the usual way:

$$\begin{pmatrix} A & B \\ C & D \end{pmatrix}\begin{pmatrix} E & F \\ G & H \end{pmatrix} = \begin{pmatrix} AE + BG & AF + BH \\ CE + DG & CF + DH \end{pmatrix}.$$

▶ **Ec. 'Orthogonality implies independence within MVN'.** Suppose that $\mathbf{W} \sim \mathrm{MVN}_{r+s}(\boldsymbol{\mu}, V)$, that, with $\mathbf{Y} \in \mathbb{R}^r$, $\mathbf{Z} \in \mathbb{R}^s$, we have

$$\mathbf{W} = \begin{pmatrix} \mathbf{Y} \\ \mathbf{Z} \end{pmatrix}, \quad \mathbb{E}(\mathbf{W}) = \boldsymbol{\mu} = \begin{pmatrix} \boldsymbol{\mu}_y \\ \boldsymbol{\mu}_z \end{pmatrix},$$

$$V = \mathbb{E}(\mathbf{W} - \boldsymbol{\mu})(\mathbf{W} - \boldsymbol{\mu})^T = \begin{pmatrix} V_{yy} & 0 \\ 0 & V_{zz} \end{pmatrix}.$$

Do note the supposition about zero entries in V. Then, with $\boldsymbol{\alpha} = \left(\begin{smallmatrix} \boldsymbol{\alpha}_y \\ \boldsymbol{\alpha}_z \end{smallmatrix}\right)$, we have

$$
\begin{aligned}
\mathbb{E}\exp\left(i\boldsymbol{\alpha}_y^T\mathbf{Y} + i\boldsymbol{\alpha}_z^T\mathbf{Z}\right) &= \exp\left\{i\boldsymbol{\alpha}^T\boldsymbol{\mu} - \tfrac{1}{2}\boldsymbol{\alpha}^T V\boldsymbol{\alpha}\right\} \\
&= \exp\left\{i\boldsymbol{\alpha}_y^T\boldsymbol{\mu}_y - \tfrac{1}{2}\boldsymbol{\alpha}_y^T V_{yy}\boldsymbol{\alpha}_y\right\}\exp\left\{i\boldsymbol{\alpha}_z^T\boldsymbol{\mu}_z - \tfrac{1}{2}\boldsymbol{\alpha}_z^T V_{zz}\boldsymbol{\alpha}_z\right\} \\
&= \mathbb{E}\exp\left(i\boldsymbol{\alpha}_y^T\mathbf{Y}\right)\mathbb{E}\exp\left(i\boldsymbol{\alpha}_z^T\mathbf{Z}\right);
\end{aligned}
$$

and that this is true for all $\boldsymbol{\alpha}_y \in \mathbb{R}^r$ and all $\boldsymbol{\alpha}_z \in \mathbb{R}^s$ guarantees that $\mathbf{Y}$ and $\mathbf{Z}$ are independent. Thus,

if $\mathbf{Y}$ and $\mathbf{Z}$ have a joint MVN distribution, and if $\mathbf{Y}$ and $\mathbf{Z}$ are orthogonal (yet another sense of the term!) in that $\mathbb{E}\,\mathbf{Y}\mathbf{Z}^T = 0$, then $\mathbf{Y}$ and $\mathbf{Z}$ are independent.

Ed. Exercise. Check out result 295Ga.

▶▶ **F. Conditional distributions within an MVN scenario.** Suppose that

$$
\left(\begin{array}{c} \mathbf{X} \\ \mathbf{Y} \end{array}\right) \sim \mathrm{MVN}\left(\left(\begin{array}{c} \mathbf{0} \\ \mathbf{0} \end{array}\right), \left(\begin{array}{cc} V_{xx} & V_{xy} \\ V_{yx} & V_{yy} \end{array}\right)\right).
$$

Let $\mathbf{Z} = \mathbf{X} - A\mathbf{Y}$, where $A := V_{xy}V_{yy}^{-1}$. Then,

$$
\left(\begin{array}{c} \mathbf{Z} \\ \mathbf{Y} \end{array}\right) = \left(\begin{array}{cc} I & -A \\ 0 & I \end{array}\right)\left(\begin{array}{c} \mathbf{X} \\ \mathbf{Y} \end{array}\right)
$$

is MVN, and

$$
\mathbb{E}\mathbf{Z}\mathbf{Y}^T = \mathbb{E}\mathbf{X}\mathbf{Y}^T - A\mathbb{E}\mathbf{Y}\mathbf{Y}^T = V_{xy} - AV_{yy} = 0,
$$

so that $\mathbf{Z}$ is independent of $\mathbf{Y}$. We know that $\mathbf{Z} \sim \mathrm{MVN}(\mathbf{0}, B)$ for some B. Indeed, since $\mathbb{E}\mathbf{Z}\mathbf{Y}^T = 0$, we have

$$
B = \mathbb{E}\mathbf{Z}\mathbf{Z}^T = \mathbb{E}\mathbf{Z}\mathbf{X}^T = V_{xx} - V_{xy}V_{yy}^{-1}V_{yx}.
$$

In particular,

$$
\mathbf{X}\,|\,\mathbf{Y} \sim \mathrm{MVN}(A\mathbf{Y}, B), \quad A := V_{xy}V_{yy}^{-1}, \quad B = V_{xx} - AV_{yx}. \tag{F1}
$$

the general **theoretical regression formula** for multivariate normal variables.

▶▶ **G. The Bivariate Normal (BVN) distribution.** Suppose that X and Y are real-valued RVs such that

$$
\left(\begin{array}{c} X \\ Y \end{array}\right) \sim \mathrm{BVN}(\mu_x, \mu_y; \sigma_x^2, \sigma_y^2; \rho) := \mathrm{MVN}\left(\left(\begin{array}{c} \mu_x \\ \mu_y \end{array}\right), \left(\begin{array}{cc} \sigma_x^2 & \rho\sigma_x\sigma_y \\ \rho\sigma_y\sigma_x & \sigma_y^2 \end{array}\right)\right).
$$

Note that **the symbol ρ is now restored to its traditional 'correlation coefficient' use.**

By the result of the last subsection, we have

$$
Y\,|\,X \sim \mathrm{MVN}\left(\mu_y + \rho\frac{\sigma_y}{\sigma_x}(X - \mu_x), \sigma_y^2(1 - \rho^2)\right).
$$

Thus, as we know from Chapter 3,

$$y = \mu_y + \rho \frac{\sigma_y}{\sigma_x}(x - \mu_x)$$

is the theoretical linear regression line of Y on X. It is used to predict Y-values from X-values. Note that the variance of a prediction of a Y-value from an X-value is $\sigma_y^2(1 - \rho^2)$, so is very small if ρ is close to 1. We know from Chapter 3 that if $\rho = 1$, then we can predict Y from X exactly (with probability 1).

There are many situations in which BVN distributions have been used. One might for example have X for man's height, Y for height of his son (both at maturity). It was this example which led Francis Galton to introduce Regression Analysis. See Freedman, Pisani and Purves [83] for illuminating discussions on this type of example and especially *for the reason for Galton's term 'regression'*.

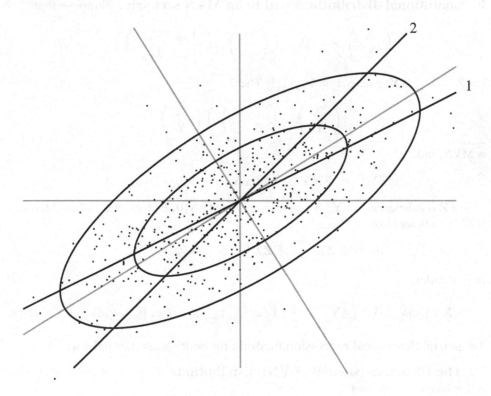

Axes of ellipses of constant likelihood in grey

Regression lines in black. 1: Y on X; 2: X on Y

Figure G(i): BVN with $(\sigma_x, \sigma_y, \rho) = (2, \sqrt{2}, 1/\sqrt{2})$

▶ **Ga. The correlation coefficient ρ as a measure of association.** If there is good evidence that $\rho \neq 0$, we say that there is evidence of an **association** between the 'X' and 'Y' 'factors'. This does **not** imply that there is any causal relation. The situation may be confounded in that there is a common cause for the two 'factors'. It is of course possible that there *is* a causal relation with (say) an increase in X causing (on average) an increase in Y; but remember that ρ is symmetric in X and Y, and (if an increase in X does on average cause an increase in Y) it is often extremely unlikely that we could also claim that an increase in Y *causes* an increase in X, even though a larger Y-value would lead us to *predict* a larger X-value as having caused it.

Gb. Contours. The logarithm of the joint pdf of X and Y is

$$
\begin{aligned}
f_{X,Y}^*&(x,y \,|\, \mu_x, \mu_y; \sigma_x, \sigma_y; \rho) \\
&= -\ln(2\pi) - \ln(\sigma_x \sigma_y) - \tfrac{1}{2}\ln(1-\rho^2) \qquad\qquad\text{(G1)} \\
&\quad - \frac{1}{2(1-\rho^2)}\left\{ \frac{(x-\mu_x)^2}{\sigma_x^2} + \frac{(y-\mu_y)^2}{\sigma_y^2} - \frac{2\rho(x-\mu_x)(y-\mu_y)}{\sigma_x \sigma_y} \right\}.
\end{aligned}
$$

The contours (level curves) on which the pdf of X and Y is constant are ellipses centred on $(\mu_x, \mu_y)^T$ and with major axis parallel to the vector

$$
\left(2\rho\sigma_x\sigma_y, \; \{(\sigma_x^2 - \sigma_y^2) + 4\rho^2\sigma_x^2\sigma_y^2\}^{\frac{1}{2}} - (\sigma_x^2 - \sigma_y^2) \right)^T.
$$

Coordinate geometry (though, in effect, the Spectral Theorem) tells us that the major-axis vector is an eigenvector of V corresponding to the larger eigenvalue of V (or, more properly, an eigenvector of V^{-1} corresponding to the smaller eigenvalue of V^{-1}).

After shifting the origin to $(\mu_x, \mu_y)^T$, a contour ellipse has the form

$$
\mathbf{z}^T V^{-1} \mathbf{z} = r^2 = (S^{-1}\mathbf{z})^T S^{-1}\mathbf{z},
$$

so that $\mathbf{z} = S\mathbf{w}$, where $\mathbf{w}$ lies on the circle of radius r. We can therefore easily draw the ellipse which contains the 'high pdf' region (the smallest region of the plane) of given probability p. We simply choose r such that

$$
p = \mathbb{P}(\mathbf{G}^T\mathbf{G} \leq r) = 1 - e^{-\frac{1}{2}r}.
$$

The theoretical regression line of Y on X bisects any vertical chord of any ellipse 'of constant likelihood', so must go through the rightmost point of any such ellipse. The theoretical regression line of X on Y bisects any horizontal chord of any ellipse 'of constant likelihood', so must go through the topmost point of any such ellipse.

Gc. Exercise. Prove the italicized result above by using the formula for the sum of the roots of a quadratic equation.

Gd. Remark. Freedman, Pisani and Purves [83] contains very much wisdom about correlation, as about so much else. Be careful though that you are not misled into thinking that their 'SD line' is the major axis of an ellipse of constant likelihood. If, for example, $\sigma_x = 1.6, \sigma_y = 1, \rho = 0.2$, then the SD line is at $32°$ while the major axis is at $11°$. The SD line has no particular theoretical significance.

▶▶ **H. Statistics associated with correlation coefficients.** Suppose that $(X_1, Y_1), (X_2, Y_2), \ldots, (X_n, Y_n)$ are IID, each pair having the $\mathrm{BVN}(\mu_x, \mu_y; \sigma_x^2, \sigma_y^2; \rho)$ distribution. Then

$$
f^*_{\mathbf{X}, \mathbf{Y}}(\mathbf{x}, \mathbf{y} \mid \mu_x, \mu_y; \sigma_x, \sigma_y; \rho)
$$
$$
= -n \ln(2\pi) - n \ln(\sigma_x \sigma_y) - \tfrac{1}{2} n \ln(1 - \rho^2)
$$
$$
- \frac{1}{2(1 - \rho^2)} \left\{ \frac{Q(\mathbf{x}, \mathbf{x}) + n(\overline{x} - \mu_x)^2}{\sigma_x^2} + \frac{Q(\mathbf{y}, \mathbf{y}) + n(\overline{y} - \mu_y)^2}{\sigma_y^2} \right.
$$
$$
\left. - 2\rho \frac{Q(\mathbf{x}, \mathbf{y}) + n(\overline{x} - \mu_x)(\overline{y} - \mu_y)}{\sigma_x \sigma_y} \right\}.
$$

For the mle $(\hat{\mu}_x, \hat{\mu}_y; \hat{\sigma}_x, \hat{\sigma}_y; \hat{\rho})$ of $(\mu_x, \mu_y; \sigma_x, \sigma_y; \rho)$, we have the hardly-surprising results:

$$
\hat{\mu}_x = \overline{x}, \quad \hat{\mu}_y = \overline{y}, \quad \hat{\sigma}_x^2 = n^{-1} Q(\mathbf{x}, \mathbf{x}), \quad \hat{\sigma}_y^2 = n^{-1} Q(\mathbf{y}, \mathbf{y}),
$$
$$
\hat{\rho} = \frac{n^{-1} Q(\mathbf{x}, \mathbf{y})}{\hat{\sigma}_x \hat{\sigma}_y}.
$$

Ha. Exercise. Check out these formulae. Have faith if it starts to look a little messy! Verify that $(\hat{\mu}_x, \hat{\mu}_y; \hat{\sigma}_x, \hat{\sigma}_y; \hat{\rho})$ is a sufficient statistic for $(\mu_x, \mu_y; \sigma_x, \sigma_y; \rho)$.

▶ **Hb. Test of independence or of 'no association'.** One of the most celebrated Hypothesis Tests in Statistics is the 'test of independence', that is, of

$$
H_0 : \rho = 0 \quad (\text{and } \mu_x, \mu_y \in \mathbb{R} \text{ and } \sigma_x, \sigma_y \in (0, \infty))
$$

against

$$
H_A : \rho \neq 0 \quad (\text{and } \mu_x, \mu_y \in \mathbb{R} \text{ and } \sigma_x, \sigma_y \in (0, \infty)).
$$

Hc. Exercise. Check that we have the very simple formula

$$
2 \ln \mathrm{LR}(\mathbf{X}, \mathbf{Y}) = -n \ln(1 - R^2),
$$

where R is the MLE of ρ, the *Sample Correlation Coefficient*:

$$
R = \frac{Q(\mathbf{X}, \mathbf{Y})}{\{Q(\mathbf{X}, \mathbf{X}) Q(\mathbf{Y}, \mathbf{Y})\}^{\frac{1}{2}}}. \qquad \square
$$

Hence, we reject H_0 at size α if $|r^{\mathrm{obs}}| > r_\alpha$, where $\mathbb{P}_{\rho=0}(|R| > r_\alpha) = \alpha$. You can see from 'shifting and scaling' that the distribution of R under H_0 does not involve any of $\mu_x, \mu_y, \sigma_x, \sigma_y$.

It was conjectured by 'Student' and proved by Fisher, that if $\rho = 0$, then

$$
T_{n-2} := \frac{R}{\sqrt{1 - R^2}} \sqrt{n - 2} \sim t_{n-2}. \tag{H1}
$$

It would be argued by some that in these days of computers, it is not necessary to know how to prove this result. However, we can prove it in a way that helps us understand our geometric picture more fully.

Proof of result (H1). Suppose that $\rho = 0$, so that we have independence of the X and Y components. Then,

$$\sigma_x^{-1}(\mathbf{X} - \overline{X}\mathbf{1}) \sim \text{SN}([1]^\perp), \quad \sigma_y^{-1}(\mathbf{Y} - \overline{Y}\mathbf{1}) \sim \text{SN}([1]^\perp).$$

Hence, T_{n-2} has the same distribution as

$$T := \frac{\langle \mathbf{U}, \mathbf{V} \rangle}{\{\langle \mathbf{U}, \mathbf{U} \rangle \langle \mathbf{V}, \mathbf{V} \rangle - \langle \mathbf{U}, \mathbf{V} \rangle^2\}^{\frac{1}{2}}} \sqrt{n-2},$$

where $\mathbf{U}, \mathbf{V}$ are independent each $\text{SN}(\mathbb{R}^{n-1})$. We can even prove that conditionally on $\mathbf{U}$, T has the t_{n-2} distribution. Suppose that $\mathbf{U} \neq \mathbf{0}$ is fixed. Then, we can write $\mathbf{V} = \alpha\mathbf{U} + \mathbf{W}$, where $\alpha\|U\| \sim \text{N}(0,1)$ and $\mathbf{W} \sim \text{SN}([\mathbf{U}]^\perp)$. We then find that

$$T = \frac{\alpha\|\mathbf{U}\|}{\|\mathbf{W}\|} \sqrt{n-2},$$

and since $\|W\|^2 \sim \chi_{n-2}^2$, the result follows from the definition of the t distribution. $\square$

Hd. Fisher's z-transform. Fisher discovered the remarkable fact that, whatever the value of ρ, for large n,

$$Z := \tfrac{1}{2}\ln\frac{1+R}{1-R} \approx \text{N}\left(\tfrac{1}{2}\ln\frac{1+\rho}{1-\rho}, \frac{1}{n-3}\right).$$

Hotelling later discovered refinements of this result. The proofs of these results are complicated, and I skip them. These days, you can even find a Confidence Interval for ρ directly on the Internet (under several sites found via 'correlation coefficient') by typing in your values of r^{obs} and n.

Note that if $\rho = 0$, then, remembering that $t_{n-2} \approx \text{N}(0,1)$ when n is large, we have

$$R^2 = \frac{T_{n-2}^2}{n-2+T_{n-2}^2} \approx \frac{\text{N}(0,1)^2}{n} = \frac{\chi_1^2}{n},$$

showing that $2\ln\text{LR} \approx \chi_1^2$, as predicted by the Chi-Squared Principle.

Of course, for large n, we could use the Chi-Squared Principle to provide a rough estimate of the significance level of the data for testing H_0 against H_A. For the value r_α for such a test when $n = 32$ and $\alpha = 5\%$, both Fisher's exact 'T' method and Fisher's approximate 'Z' method give (to 3 places) 0.349 whereas the χ^2 method gives 0.336.

And, of course, you can get individual CIs for all the parameters for this BVN situation from WinBUGS . Use a variety of sensible priors.

▶ **I. Multivariate Central Limit Theorem.**   *Suppose that* **X** *is a Random Variable with values in* $\mathbb{R}^b$, *with each* X_j *in* $\mathcal{L}^2$. *Let*

$$\boldsymbol{\mu} := \mathbb{E}\mathbf{X}, \quad V := \mathbb{E}(\mathbf{X} - \boldsymbol{\mu})(\mathbf{X} - \boldsymbol{\mu})^T.$$

Let $\mathbf{X}^{(1)}, \mathbf{X}^{(2)}, \dots$ *be independent* $\mathbb{R}^b$-*valued RVs each with the same distribution as* **X**. *Then, for large* n,

$$\mathbf{S}^{(n)} = \mathbf{X}^{(1)} + \cdots + \mathbf{X}^{(n)} \approx \mathrm{MVN}_b(n\boldsymbol{\mu}, nV)$$

in the usual sense that for any 'rectangular region' A *in* $\mathbb{R}^b$,

$$\mathbb{P}\left(\frac{\mathbf{S}^{(n)} - n\boldsymbol{\mu}}{\sqrt{n}} \in A\right) \to \mathbb{P}(Z \in A),$$

where $Z \sim \mathrm{MVN}_b(\mathbf{0}, V)$.

The proof is just a vector form of the proof of the CLT.

J. Application to the Pearson Statistic.   Return to the situation of Subsection 345E. So, suppose that $Y_1, Y_2, \dots, Y_n$ are IID RVs each with values in $\{1, 2, \dots, b\}$, such that

$$\mathbb{P}(Y_m = k) = p_k \ (1 \le k \le b, 1 \le m \le n), \quad \sum_i p_i = 1.$$

Let N_k be the number of Y_m equal to k.

Let

$$X_k^{(m)} = \begin{cases} 1 & \text{if } Y_m = k, \\ 0 & \text{if } Y_m \ne k. \end{cases}$$

Then, for fixed m, $\mathbf{X}^{(m)} \in \mathbb{R}^b$ has the same distribution as **X**, where in particular,

$$\mathbb{E}X_k = p_k, \quad \mathbb{E}X_k X_\ell = \delta_{k\ell}p_k, \quad V_{k\ell} := \mathrm{Cov}(X_k, X_\ell) = \begin{cases} p_k - p_k^2 & \text{if } k = \ell, \\ -p_k p_\ell & \text{if } k \ne \ell. \end{cases}$$

By the Multivariate CLT, for large n,

$$S^{(n)} = (N_1, N_2, \dots, N_b)^T \approx \mathrm{MVN}_b(n\mathbf{p}, nV).$$

Let

$$W_k = \frac{N_k - np_k}{\sqrt{np_k}}.$$

Then, for large n,

$$\mathbf{W} \approx \mathrm{MVN}_b(\mathbf{0}, C)$$

where (*you check*)

$$C = P_{[\mathbf{u}]}^\perp, \quad \mathbf{u} := (\sqrt{p_1}, \dots, \sqrt{p_n})^T.$$

Thus, C has one eigenvalue 0 (with associated eigenvector $\mathbf{u}$) while all other eigenvalues are 1. Thus, for $\alpha \in \mathbb{R}$, $I + 2\alpha C$ has one eigenvalue 1 and all others $1 + 2\alpha$, whence,

$$\det(I + 2\alpha C) = (1 + 2\alpha)^{\nu}, \quad \text{where } \nu = b - 1.$$

Now, if $W \sim \mathrm{MVN}(\mathbf{0}, C)$ where C is symmetric and strictly positive-definite, then with $Q = \sum W_k^2$, and using 368(D1) and 368(D2),

$$
\begin{aligned}
\mathbb{E}e^{-\alpha Q} &= (2\pi)^{-\frac{1}{2}b}(\det C)^{-\frac{1}{2}} \int_{\mathbb{R}^b} e^{-\frac{1}{2}\mathbf{w}^T(C^{-1}+2\alpha I)\mathbf{w}}\,d\mathbf{w} \\
&= \frac{\left\{\det(C^{-1} + 2\alpha I)\right\}^{-\frac{1}{2}}}{(\det C)^{\frac{1}{2}}} = \left\{\det(I + 2\alpha C)\right\}^{-\frac{1}{2}}.
\end{aligned}
$$

It is easy to argue by continuity that the equation

$$\mathbb{E}e^{-\alpha Q} = \left\{\det(I + 2\alpha C)\right\}^{-\frac{1}{2}}$$

holds if C is just nonnegative-definite, as our C is. (Just replace C by $C + \varepsilon I$ where $\varepsilon > 0$, and then let $\varepsilon \downarrow 0$.)

Since PearsonStat$= \sum_k W_k^2$ and $W \approx \mathrm{MVN}(\mathbf{0}, C)$, we now see that

$$\mathbb{E}e^{-\alpha(\text{PearsonStat})} \approx (1 + 2\alpha)^{-\frac{1}{2}\nu},$$

so that

$$\text{PearsonStat} \approx \mathrm{Gamma}(\tfrac{1}{2}\nu, \tfrac{1}{2}) = \chi_\nu^2,$$

as required.

Ja. Remarks. I describe a – loose – analogy with Quantum Theory which it seems hard to tighten into anything of real significance. As we shall see in Chapter 10, if A is the diagonal $b \times b$ matrix with (k, k)th element k and $\mathbf{v}$ is a quantum state (unit vector in $\mathbb{R}^b$), then a measurement of the 'observable' associated with A when the system is in state $\mathbf{v}$ will produce the result k with probability v_k^2, and the system will then be in state $\mathbf{e}^{(k)}$, the kth element of the standard basis. In our χ^2 context, the Null Hypothesis is – loosely – related to the statement that the quantum system is in state $\mathbf{u}$. The matrix C corresponds – loosely – to testing whether or not the Null Hypothesis fails. If the system is in state $\mathbf{v}$ and the quantum measurement associated with the observable C is made (for our particular C), then we shall obtain result 0 with probability $\langle \mathbf{u}, \mathbf{v} \rangle^2 = \cos \varphi$, where φ is the angle between $\mathbf{u}$ and $\mathbf{v}$, and result 1 with probability $\sin^2 \varphi$. If the measurement associated with C produces result 0, then the system will be in state $\mathbf{u}$ and the Null Hypothesis will be true; if 1, then the system will be in a state perpendicular to $\mathbf{u}$ which corresponds to the assertion that the Null Hypothesis is false. It is as if measuring C commits Nature to making a definite decision: is the Null Hypothesis true or not?

▶▶ **K. Asymptotic multivariate normality for MLEs.** It will not surprise you that most results from Chapter 6 have multi-parameter extensions.

Ka. Exercise. (We use r, s, t to look different from i, n.) Let Y be a real-valued RV with pdf $f(y \mid \boldsymbol{\theta})$, where $\boldsymbol{\theta} = (\theta_1, \theta_2, \ldots, \theta_s) \in \mathbb{R}^s$. Define an $\mathbb{R}^s$-valued RV $\mathbf{K}(\boldsymbol{\theta}, Y)$ and a symmetric $s \times s$ matrix $I_{f^*}(\boldsymbol{\theta})$ via

$$\{\mathbf{K}(\boldsymbol{\theta}, Y)\}_r = \frac{\partial f}{\partial \theta_r}(Y \mid \boldsymbol{\theta}), \qquad \{I_{f^*}(\boldsymbol{\theta})\}_{rt} = -\mathbb{E}\, \frac{\partial^2 f^*}{\partial \theta_r \partial \theta_t}(Y \mid \boldsymbol{\theta}).$$

Show that, with $\mathbb{E}$ denoting $\mathbb{E}(\cdot \mid \boldsymbol{\theta})$,

$$\mathbb{E}K(\boldsymbol{\theta}, Y) = 0, \qquad \mathbb{E}K(\boldsymbol{\theta}, Y)K(\boldsymbol{\theta}, Y)^T = I_{f^*}(\boldsymbol{\theta}).$$

Show that for any vector $\mathbf{w} \in \mathbb{R}^s$,

$$\mathbf{w}^T I_{f^*}(\boldsymbol{\theta})\mathbf{w} = \mathbb{E}\left\{\langle \mathbf{w}, K(\boldsymbol{\theta}, Y)\rangle^2\right\} \geq 0,$$

so that $I_{f^*}(\boldsymbol{\theta})$ is nonnegative-definite.

Now let $Y_1, Y_2, \ldots, Y_n$ be IID real-valued RVs each with pdf $f(y \mid \boldsymbol{\theta})$ which is as in the Exercise. Let $\widehat{\boldsymbol{\Theta}}$ be the MLE of $\boldsymbol{\theta}$. That the MLE is consistent in the sense of Subsection 193C may be established by the methods sketched there. So, for large n, we may assume that $\widehat{\boldsymbol{\Theta}}$ is close to $\boldsymbol{\theta}$.

From the above exercise and the Multivariate CLT, we have, for large n, with the obvious $I_\ell(\boldsymbol{\theta}) = nI_{f^*}(\boldsymbol{\theta})$,

$$\{\mathbf{U}(\boldsymbol{\theta})\}_r := \frac{\partial \ell}{\partial \theta_r}(\mathbf{Y} \mid \boldsymbol{\theta}) = \sum_{i=1}^{n}\{\mathbf{K}(\boldsymbol{\theta}, Y_i)\}_r,$$

so

$$\mathbf{U}(\boldsymbol{\theta}) \approx \mathrm{MVN}_s(0, I_\ell(\boldsymbol{\theta})).$$

Let us assume that $I_\ell(\boldsymbol{\theta})$ is *positive*-definite. By Taylor's Theorem,

$$0 = \mathbf{U}(\widehat{\boldsymbol{\Theta}}) \approx \mathbf{U}(\boldsymbol{\theta}) + I_\ell(\boldsymbol{\theta})\left(\widehat{\boldsymbol{\Theta}} - \boldsymbol{\theta}\right),$$

so $\widehat{\boldsymbol{\Theta}} - \boldsymbol{\theta} \approx -I_\ell(\boldsymbol{\theta})^{-1}\mathbf{U}(\boldsymbol{\theta})$, a Newton–Raphson result. By property 368Ea,

$$\widehat{\boldsymbol{\Theta}} - \boldsymbol{\theta} \approx \mathrm{MVN}_s\left(0, I_\ell(\boldsymbol{\theta})^{-1}\right),$$

and that's the Frequentist story.

Kb. Exercise. Mimic Subsection 204F to give the Bayesian version.

The fact that, for large n, Frequentists and Bayesians can behave as if

$$\boldsymbol{\theta} - \widehat{\boldsymbol{\Theta}} \approx \mathrm{MVN}_s\left(0, I_\ell(\widehat{\boldsymbol{\Theta}})^{-1}\right)$$

was the mainstay of important computer programs such as GLIM and GENSTAT. You can see how one could do (for example) the large-sample analysis for logistic regression at 357K. But really, what is the point when WinBUGS can treat any sample size?

▶ **L. Heuristic explanation of Wilks' Chi-Squared Principle.**   We now set out to give a heuristic explanation of Wilks' Chi-Squared Principle at Subsection 345Da. We need a preliminary Lemma.

La. Lemma. *If* $\mathbf{Y} \sim \mathrm{MVN}_r(\mathbf{0}, V)$ *where* V *is non-singular, then*

$$\mathbf{Y}^T V^{-1} \mathbf{Y} \sim \chi_r^2.$$

Proof. With $V = SS^T$, we have

$$\mathbf{Y}^T V^{-1} \mathbf{Y} = \mathbf{G}^T S^T V^{-1} S \mathbf{G} = \mathbf{G}^T \mathbf{G} \sim \chi_r^2. \qquad \square$$

Differential-geometry considerations (see the Subsection JP3) mean that in seeking to understand the Chi-Squared Principle, we can reduce to the case when $\boldsymbol{\theta} = \binom{\boldsymbol{\varphi}}{\boldsymbol{\nu}} \in \mathbb{R}^s$, $\boldsymbol{\nu}$ being a nuisance parameter in $\mathbb{R}^m$ and $\boldsymbol{\varphi}$ having values in $\mathbb{R}^r$, where $r = s - m$ in the notation of Subsection 345Da, and where we have

$$H_0 : \boldsymbol{\varphi} = \boldsymbol{\varphi}_0, \, \boldsymbol{\nu} \in \mathbb{R}^m, \qquad H_A : \boldsymbol{\varphi} \in \mathbb{R}^r, \boldsymbol{\nu} \in \mathbb{R}^m,$$

$\boldsymbol{\varphi}_0$ being a fixed vector in $\mathbb{R}^r$. We suppose that the Maximum-Likelihood Estimator associated with H_0 is $\hat{\boldsymbol{\theta}}_0 = \binom{\boldsymbol{\varphi}_0}{\hat{\boldsymbol{\nu}}_0}$ and the MLE associated with H_A is $\hat{\boldsymbol{\theta}}_A = \binom{\hat{\boldsymbol{\varphi}}_A}{\hat{\boldsymbol{\nu}}_A}$.

Define R to be the $s \times s$ matrix with (i, j)th component

$$R_{ij} := -\frac{\partial^2 \ell}{\partial \theta_i \partial \theta_j} \text{ evaluated at } \hat{\boldsymbol{\theta}}_A$$

and partition R as

$$R = \begin{pmatrix} R_{\varphi\varphi} & R_{\varphi\nu} \\ R_{\nu\varphi} & R_{\nu\nu} \end{pmatrix}.$$

With $\partial \ell / \partial \boldsymbol{\nu}$ denoting the column vector with kth component $\partial \ell / \partial \nu_k$, we have

$$\mathbf{0} - \mathbf{0} = \frac{\partial \ell}{\partial \boldsymbol{\nu}} (\mathbf{y} \mid \boldsymbol{\varphi}_0, \hat{\boldsymbol{\nu}}_0) - \frac{\partial \ell}{\partial \boldsymbol{\nu}} (\mathbf{y} \mid \hat{\boldsymbol{\varphi}}_A, \hat{\boldsymbol{\nu}}_A)$$

$$\approx R_{\nu\varphi} (\hat{\boldsymbol{\varphi}}_A - \boldsymbol{\varphi}_0) + R_{\nu\nu} (\hat{\boldsymbol{\nu}}_A - \hat{\boldsymbol{\nu}}_0).$$

Thus,

$$\hat{\boldsymbol{\nu}}_A - \hat{\boldsymbol{\nu}}_0 \approx -R_{\nu\nu}^{-1} R_{\nu\varphi} (\hat{\boldsymbol{\varphi}}_A - \boldsymbol{\varphi}_0). \tag{L1}$$

By Taylor's Theorem around $\hat{\boldsymbol{\theta}}_A$ (remember the minus sign in the definition of R),

$$-2 \ln \mathrm{LR} = 2\ell (\mathbf{y} \mid \boldsymbol{\varphi}_0, \hat{\boldsymbol{\nu}}_0) - 2\ell (\mathbf{y} \mid \hat{\boldsymbol{\varphi}}_A, \hat{\boldsymbol{\nu}}_A)$$

$$\approx - \begin{pmatrix} \boldsymbol{\varphi}_0 - \hat{\boldsymbol{\varphi}}_A \\ \hat{\boldsymbol{\nu}}_0 - \hat{\boldsymbol{\nu}}_A \end{pmatrix}^T R \begin{pmatrix} \boldsymbol{\varphi}_0 - \hat{\boldsymbol{\varphi}}_A \\ \hat{\boldsymbol{\nu}}_0 - \hat{\boldsymbol{\nu}}_A \end{pmatrix},$$

and so, substituting from (L1), we have

$$2 \ln \text{LR}$$

$$\approx (\hat{\varphi}_A - \varphi_0)^T \left(I, \, -R_{\varphi\nu}R_{\nu\nu}^{-1}\right) \begin{pmatrix} R_{\varphi\varphi} & R_{\varphi\nu} \\ R_{\nu\varphi} & R_{\nu\nu} \end{pmatrix} \begin{pmatrix} I \\ -R_{\nu\nu}^{-1}R_{\nu\varphi} \end{pmatrix} (\hat{\varphi}_A - \varphi_0)$$

$$= (\hat{\varphi}_A - \varphi_0)^T \left(R_{\varphi\varphi} - R_{\varphi\nu}R_{\nu\nu}^{-1}R_{\nu\varphi}\right) (\hat{\varphi}_A - \varphi_0).$$

Now *suppose that H_0 is true*, so that the true value of θ is $\binom{\varphi_0}{\nu}$ for some $\nu \in \mathbb{R}^m$. We have seen on several occasions that, by consistency of the MLE, the value of R at θ_A is essentially equal to its value at the true θ. Moreover, we have seen that the Strong Law (or better, the CLT) effectively allows us to replace R by $I_\ell(\theta)$, the two differing by a factor $1 + O(n^{-\frac{1}{2}})$ in a rough sense.

By the asymptotic-normality results of the previous subsection,

$$\begin{pmatrix} \hat{\varphi}_A - \varphi_0 \\ \hat{\nu}_A - \nu \end{pmatrix}^T \approx \text{MVN}_s \left(\begin{pmatrix} 0 \\ 0 \end{pmatrix}, \begin{pmatrix} R_{\varphi\varphi} & R_{\varphi\nu} \\ R_{\nu\varphi} & R_{\nu\nu} \end{pmatrix}^{-1} \right),$$

R now being identified with $I_\ell(\theta)$. All that remains is to prove that

$$\begin{pmatrix} R_{\varphi\varphi} & R_{\varphi\nu} \\ R_{\nu\varphi} & R_{\nu\nu} \end{pmatrix}^{-1}$$

has 'top-left entry' $\left(R_{\varphi\varphi} - R_{\varphi\nu}R_{\nu\nu}^{-1}R_{\nu\varphi}\right)^{-1}$, because the desired result then follows from Lemma 377La. The following exercise settles the algebra.   □

Lb. Exercise. Prove that if A and D are invertible square matrices, then

$$\begin{pmatrix} A & B \\ C & D \end{pmatrix}^{-1} = \begin{pmatrix} (A - BD^{-1}C)^{-1} & -A^{-1}B(D - CA^{-1}B)^{-1} \\ -D^{-1}C(A - BD^{-1}C)^{-1} & (D - CA^{-1}B)^{-1} \end{pmatrix},$$

provided the left-hand side exists.

▶▶ **M. Model choice, Frequentist and Bayesian.**   What we have been seeing in the previous two subsections is that for large sample size, results from Linear Model theory transfer to the general situation. I am not going to chase it through, but we would find that the Akaike Information Criterion may be motivated for the general context much as it was in Subsections 236J and 303J. Again see Burnham and Anderson [35].

In Bayesian theory as in Frequentist, one wants to balance parsimony against 'accuracy' and so to minimize a suitable expression of the form

$$2 \ln(\text{maximum likelihood under model})$$

$$- \text{function}(\text{sample size, number of parameters in model})$$

The recent paper by Spiegelhalter, Best and Carlin [215] presents a very interesting development of this idea which covers hierarchical models, etc. The 'number of parameters' itself needs definition for hierarchical models. Some hints which might help you understand the paper are scattered throughout this book. See also Exercise Ma below. Since this book is already getting too much like an encyclopaedia, I leave discussion to the experts.

But do note that the S–B–C paper employs many topics originating in Frequentist theory. To be sure, these topics could all be presented within a Bayesian approach, but I think that the situation helps exemplify the need for a broad background in Statistics culture.

Ma. Exercise. The *trace*, $\text{tr}(M)$, of a square matrix M is the sum of its diagonal elements. Let A be an $m \times n$ matrix, and B an $n \times m$ matrix. Prove that $\text{tr}(AB) = \text{tr}(BA)$. Deduce that if $\mathbf{Y} \sim \text{MVN}(\mathbf{0}, V)$, then

$$\mathbb{E}(\mathbf{Y}^T B \mathbf{Y}) = \text{tr}(BV).$$

Prove that the trace of a perpendicular projection is equal to the dimension of its range. *Hint.* Show that if $\mathbf{u}$ is a unit vector, then the trace of $P_{[\mathbf{u}]}$ $(= \mathbf{u}\mathbf{u}^T)$ is 1. (Alternatively, use the fact that the trace of a square matrix M is the sum of the eigenvalues of M.)

N. Reference priors and Differential Geometry.

We now consider the multiparameter analogue of Subsection 204G. Suppose that $\boldsymbol{\theta}$ is a multidimensional parameter: $\boldsymbol{\theta} \in \mathbb{R}^s$. Under a one-one change of variables $\boldsymbol{\theta} = \mathbf{v}(\boldsymbol{\varphi})$, we would like our assignment of priors to satisfy

$$\pi(\boldsymbol{\varphi}) = \pi(\mathbf{v}(\boldsymbol{\theta})) \left| J\left(\frac{\boldsymbol{\theta}}{\boldsymbol{\varphi}}\right) \right|.$$

Let $h(y \mid \boldsymbol{\varphi}) = f(y \mid \mathbf{v}(\boldsymbol{\varphi}))$, so that

$$\frac{\partial h^*}{\partial \varphi_j} = \sum_i \frac{\partial f^*}{\partial \theta_i} \frac{\partial \theta_i}{\partial \varphi_j}, \qquad \tilde{D}h = (Df)M,$$

where $\tilde{D}h$ is the row vector with jth component $\partial h^*/\partial \varphi_j$, Df is the obvious analogue, and M is the matrix with (i, j)th element $\partial \theta_i/\partial \varphi_j$, so that $\det(M) = J$. Verify that

$$\tilde{I}_{ij}(\boldsymbol{\varphi}) = \int_y \frac{\partial h^*}{\partial \varphi_i} \frac{\partial h^*}{\partial \varphi_j} h(y \mid \boldsymbol{\varphi}) \, \mathrm{d}y = \sum_k \sum_\ell M_{ki} I_{k\ell}(\boldsymbol{\theta}) M_{\ell j}, \qquad (\text{N1})$$

so that

$$\tilde{I}(\boldsymbol{\varphi}) = M^T I(\boldsymbol{\theta}) M, \qquad \det(\tilde{I}(\boldsymbol{\varphi})) = J^2 \det(I(\boldsymbol{\theta})).$$

Hence, the Jeffreys prior $\det(I(\boldsymbol{\theta}))^{\frac{1}{2}}$ is invariant.

►► **Na. Differential Geometry and Statistics.** The results just described can be seen as part of a 'differential geometry of Statistics', something in which there has been, and is, much interest. See, for example, Amari [4], Barndorff-Nielsen and Cox [9], Kass and Vos [126], and Murray and Rice [168]. Marriott and Salmon, in the introductory chapter of [158], provide a concise and readable introduction to some of the ideas. It is clear that *Differential Geometry provides a useful language for some statistical considerations, and that it can focus attention on new issues of importance*. It is however true that the marriage of the two subjects is not as perfect as one would hope: as we shall see below, 'mathematical correctness' from the point of view of geometry need not correspond to statistical good sense. Yet, and even though the fusion of the subjects seems unlikely to produce results which approach in depth those in other areas of Differential Geometry, it remains a *very* useful venture.

In this discussion, we shall take $\mathbb{R}^s$ (rather than a general manifold) as our parameter space $\mathcal{P}$. The idea is to make this parameter space into a Riemannian manifold by using the *Fisher-information metric* to assign *appropriate* lengths to paths in $\mathcal{P}$, lengths which will generally not be the Euclidean lengths of the paths. We shall assume that $I_\ell(\boldsymbol{\theta})$ is everywhere positive-definite. (Recall that it has to be nonnegative-definite.)

Suppose that we have a map $t \mapsto \boldsymbol{\theta}(t)$ from $[0, \infty)$ into parameter space $\mathbb{R}^s$. Think of $\boldsymbol{\theta}(t)$ as the position of a particle at time t. We wish to speak of the (Fisher metric) distance $S(t)$ travelled by the particle during time-interval $[0, t]$. The idea is that the speed $S'(t) \geq 0$ is given by a 'modified Pythagorean' formula

$$S'(t)^2 \;=\; \boldsymbol{\theta}'(t)^T I(\boldsymbol{\theta}(t))\boldsymbol{\theta}'(t) \;=\; \mathbb{E}\left\{\left(\frac{\partial}{\partial t}f^*(\mathbf{Y} \mid \boldsymbol{\theta}(t))\right)^2\right\}. \qquad \text{(N2)}$$

The arc-length $S(t)$ is of course $\int_{[0,t]} S'(u)\mathrm{d}u$. The formula 379(N1), describing the 'covariant-tensor' property of I shows that arc length is unaffected by a change in parameterization. Indeed, with obvious notation, $\boldsymbol{\theta}'(t) = M\boldsymbol{\varphi}'(t)$, whence

$$\boldsymbol{\varphi}'(t)^T \tilde{I}(\boldsymbol{\varphi}))\boldsymbol{\varphi}'(t) \;=\; \boldsymbol{\theta}'(t)^T I(\boldsymbol{\theta}(t))\boldsymbol{\theta}'(t).$$

The appropriate definition of the volume element for the Riemannian manifold is then $(\det(I(\boldsymbol{\theta}))^{\frac{1}{2}}\mathrm{d}\boldsymbol{\theta}$, exactly in agreement with the Jeffreys prescription. We know that this definition is coordinate-free, and it can be motivated via the use of 'normal coordinates' at $\boldsymbol{\theta}$.

Some of the above ideas are necessary for the 'differential geometry' reduction utilized at the start of the proof of Wilks' Theorem in Subsection 377L. I skip the details.

Of course, the Fisher metric and Jeffreys prior belong only at the very beginning of Differential Geometry in Statistics. The next stages inevitably involve *connections* and *curvature*. Read the literature mentioned above.

On all mathematical grounds, the Jeffreys prior, the Riemannian volume element, is 'correct'. But as Statistics, the Jeffreys prior does not necessarily make much sense, as the following example (due to Neyman and Scott) illustrates.

▶▶ **Nb. Exercise.** Suppose that $(Y_{jk} : 1 \leq j \leq J, 1 \leq k \leq K)$ are independent RVs with

$$Y_{jk} \sim \mathrm{N}(\mu_j, \ \mathrm{prec} \ \rho).$$

Prove that the Jeffreys reference prior is

$$\pi(\mu_1, \mu_2, \ldots, \mu_J, \rho) \ \propto \ \rho^{\frac{1}{2}J-1}.$$

You may well agree with me that the inappropriateness of the Jeffreys prior as Statistics is already evident: why should the prior depend on J? Suppose that we do use this prior. Prove that then

$$(\rho \,|\, \mathbf{Y}) \ \sim \ \mathrm{Gamma}(\tfrac{1}{2}JK, \ \mathrm{rate} \ \tfrac{1}{2}W(\mathbf{Y})),$$

where

$$W(\mathbf{Y}) := \sum\sum(Y_{jk} - \overline{Y}_{j*})^2,$$

and that

$$W(\mathbf{Y}) \sim \frac{1}{\rho}\chi^2_{J(K-1)}.$$

Consider letting $J \to \infty$ while keeping K fixed. Then, by the Strong Law,

$$\frac{W(\mathbf{Y})}{JK} \to \frac{1}{\rho}\frac{K-1}{K},$$

so that the distribution of $(\rho \,|\, \mathbf{Y})$ will concentrate around $K\rho/(K-1)$, not around ρ. If K is small, the difference really matters.

Jeffreys himself modified the idea of a reference prior to cope with this type of example. See Bernardo and Berger [15] (with the amusing comments from McCullogh). I think McCullogh is right in that a definitive theory of reference priors seems impossible.

People will rightly continue to use *proper* 'vague' priors in their computer work, and as long as they check robustness to change in prior, change in model, change in initial values for Gibbs sampler, things should generally be OK.

Remark. I do apologize that I often tell you that something is a good idea only to have to say later that it does not work in many important cases. That is the way that Statistics is: full of surprises and never dull (to take the charitable view!)

O. A final thought on Statistics – for now! These few chapters (and parts of Chapter 9) have been the only things I have ever written on Statistics, though I have always had a keen interest in the subject. I have greatly enjoyed writing this account. (Please note that there are further Statistical things in Chapter 9: explanation of de Finetti's Theorem, discussion of Sequential Sampling, some discussion of reference priors, etc.)

Mathematicians approaching Statistics might begin by feeling very frustrated about the lack of any definitive solution to various problems; and it has to be agreed that there is nothing of the depth of the Riemann Hypothesis around. But the more one thinks about Statistics, the more one finds its sheer untameability attractive: it is, after all, trying to cope with the real world – and with very important things in that world.

Amongst the main motivations for writing this book are my own sadness that I have not been more actively involved in Statistics throughout my career and my wish to persuade others not to follow my example in that regard.

SOME FURTHER PROBABILITY

In this chapter, we look briefly at three further topics from Probability: Conditional Expectations (CEs); Martingales; Poisson Processes. The first of these leads directly on to the second. The third is somewhat different.

First, we look at the conditional expectation $\mathbb{E}(X \mid A)$ of a Random Variable X given that event A occurs, and at the decomposition

$$\mathbb{E}(X) = \mathbb{P}(A)\mathbb{E}(X \mid A) + \mathbb{P}(A^c)\mathbb{E}(X \mid A^c).$$

We used these ideas in solving the 'average wait for patterns' problems in Chapter 1.

We then examine one of the central breakthroughs in Probability, the definition (due in its most general form largely to Kolmogorov) of $\mathbb{E}(X \mid Y)$, the Random Variable which is the conditional expectation of X given the Random Variable Y. This is a rather subtle concept, for which we shall later see several motivating examples. To get the idea: if X and Y are the scores on two successive throws of a fair die, then $\mathbb{E}(X + Y \mid Y) = 3\frac{1}{2} + Y$. Here, $\mathbb{E}(X \mid Y) = \mathbb{E}(X) = 3\frac{1}{2}$ because, since X is independent of Y, knowing the value of Y does not change our view of the expectation of X. Of course, $\mathbb{E}(Y \mid Y) = Y$ because 'if we are given Y, then we know what Y is', something which needs considerable clarification!

Oa. Exercise. What is $\mathbb{E}(Y \mid X + Y)$ for this example? (We return to this shortly, but common sense should allow you to write down the answer even without knowing how things are defined.)

It is to be understood whenever the definition of conditional expectation is the more fundamental one, a corresponding conditional probability is defined as illustrated by the examples

$$\mathbb{P}(F \mid Y) := \mathbb{E}(I_F \mid Y), \quad \mathbb{P}(F \mid \mathcal{G}) := \mathbb{E}(I_F \mid \mathcal{G}).$$

One of the most useful results in Probability is the Conditional Mean Formula (one of the 'Golden Rules' of Conditional Expectation)

$$\mathbb{E}(X) = \mathbb{E}\mathbb{E}(X \mid Y).$$

The 'Two-Envelope Paradox' makes a very instructive example, so study it carefully.

Doob's Martingale Theory is one of Probability's great success stories. As we shall see, a martingale is a mathematical abstraction of the evolution in time of a gambler's fortune in a fair game. Amongst the main results is the Stopping-Time Principle that 'one cannot cheat the system'.

It is really amazing that so apparently simple a concept as that of 'martingale' should have had such wide application. When I mentioned much earlier that Probability has been able to solve – in a number of key cases, for the first time – problems from Complex Analysis, Partial Differential Equation Theory, Potential Theory, etc, it is really Martingale Theory which has allowed this success. It is no accident that the definitive account of Martingale Theory is in the book *Probability and Potential* by Dellacherie and Meyer. (That account is *very* much more advanced than this book.) I shall sometimes refer to my book, [W] for proofs.

In our brief look at martingales, we shall see how they allow us to solve the 'average wait for patterns' problem, to solve the hitting-time problems for Random Walks, to prove the Ballot Theorem, etc. Likelihood-ratio martingales are important, and the application of the Stopping-Time Principle to them is associated with (Wald's) method of Sequential Sampling, an important technique in Medical Statistics and for Quality Control. (Sequential Sampling, however, goes beyond the stage where Martingale Theory helps. See Jennison and Turnbull [119].) I promised a martingale explanation of de Finetti's Theorem and of the Strong Law, and I stick to that, though I cannot provide all details of the proofs here. We take a brief look at the very fashionable (in my opinion, too fashionable!) topic of martingales in finance, proving the simplest case of the Black–Scholes Theorem.

The Poisson process models (under conditions which you should later think about in relation to the real world) the arrival of customers at a queue, or of calls to a telephone exchange, or of 'radioactive' particles at a Geiger counter. There are 'spatial' versions, too. Poisson processes form a good area in which to develop your intuition.

9.1 Conditional Expectation

▶ **A. Conditional Expectation $\mathbb{E}(X \mid A)$ of a Random Variable X given that event A occurs.** Let A be an event for which $\mathbb{P}(A) > 0$, and let X either be in $\mathcal{L}^1$ or be non-negative (or both).

We have
$$\mathbb{P}(B \mid A) := \frac{\mathbb{P}(A \cap B)}{\mathbb{P}(A)} = \frac{\mathbb{E}(I_A I_B)}{\mathbb{P}(A)}.$$

If we interpret $\mathbb{E}(I_B \mid A)$ as $\mathbb{P}(B \mid A)$, this suggests defining

$$\mathbb{E}(X \mid A) := \frac{\mathbb{E}(I_A X)}{\mathbb{P}(A)} = \mathbb{P}\text{-average of } X \text{ over } A. \qquad \text{(A1)}$$

To understand the last equality, note that if Ω is discrete with all its subsets as events, then

$$\mathbb{E}(X \mid A) = \frac{\sum_{\omega \in A} X(\omega)\mathbb{P}(\omega)}{\mathbb{P}(A)} = \sum_A X(\omega)\mathbb{P}(\{\omega\} \mid A). \qquad \text{(A2)}$$

For a discrete RV X (in $\mathcal{L}^1$),

$$\mathbb{E}(X) = \sum_x x\mathbb{P}(X = x), \qquad \mathbb{E}(X \mid A) = \sum_x x\mathbb{P}(X = x \mid A),$$

which you can easily check, and which is reassuring.

It is very convenient to write

$$\mathbb{E}(X; A) := \mathbb{E}(I_A X), \qquad \text{(A3)}$$

the (Lebesgue) integral of X over A. Then $\mathbb{E}(X; A)$ is read as 'the expectation of X *and* A'; and then the expectation $\mathbb{E}(X \mid A)$, the expectation of X *given* A, satisfies $\mathbb{E}(X \mid A) = \mathbb{E}(X; A)/\mathbb{P}(A)$.

B. Rules for $\mathbb{E}(X \mid A)$.

▶ **Ba.** *If X is independent of A (that is, X and I_A are independent RVs), then* $\mathbb{E}(X \mid A) = \mathbb{E}X$. Intuition: 'If we are told that A has occurred, then that does not change our estimate of the mean of X'. The result is obvious because independence guarantees that $\mathbb{E}(I_A X) = \mathbb{E}(I_A)\mathbb{E}(X) = \mathbb{P}(A)\mathbb{E}(X)$.

▶ **Bb.** We have

$$\mathbb{E}(X) = \mathbb{P}(A)\mathbb{E}(X \mid A) + \mathbb{P}(A^c)\mathbb{E}(X \mid A^c). \tag{B1}$$

Proof. This is obvious because $X = (I_A + I_{A^c})X$.   □

When $X = I_B$, equation (B1) reduces to the familiar

$$\mathbb{P}(B) = \mathbb{P}(A)\mathbb{P}(B \mid A) + \mathbb{P}(A^c)\mathbb{P}(B \mid A^c).$$

We use (B1) in 'commonsense' (rather than formal mathematical) fashion.

▶ **Bc. Linearity.** We have, for $X_1, X_2 \in \mathcal{L}^1$ and $\alpha_1, \alpha_2 \in \mathbb{R}$,

$$\mathbb{E}(\alpha_1 X_1 + \alpha_2 X_2 \mid A) = \alpha_1 \mathbb{E}(X_1 \mid A) + \alpha_2 \mathbb{E}(X_2 \mid A).$$

This is because by linearity of the 'absolute expectation',

$$\mathbb{E}[(\alpha_1 X_1 + \alpha_2 X_2)I_A] = \alpha_1 \mathbb{E}[X_1 I_A] + \alpha_2 \mathbb{E}[X_2 I_A].$$

▶ **Bd. Important Note.** If X, X_1, X_2 are non-negative and α_1, α_2 are non-negative, then the previous three rules apply even if some or all of $\mathbb{E}(X), \mathbb{E}(X_1), \mathbb{E}(X_2)$ are infinite.

C. Duration of gambling game.

Consider the gambling game at Subsection 117B. The gambler starts with a fortune of a where $0 \le a \le b$. Her fortune thereafter behaves as SRW(p) up until the first time $T \ge 0$ when it reaches either 0 or b. We wish to calculate $y_a := \mathbb{E}_a(T)$. Let A be the event that the gambler wins the first game. Then it is intuitively obvious that $\mathbb{E}_a(T \mid A) = 1 + \mathbb{E}_{a+1}(T)$ for $1 \le a \le b - 1$: the '1+' counts the first game, and then the process 'starts afresh from $a+1$'. It is a very common mistake to forget the '1+'. See Discussion below for how this intuition can be made rigorous. We have, from (B1),

$$y_a = p(1 + y_{a+1}) + q(1 + y_{a-1})$$

or, as we more usually write it,

$$y_a = 1 + py_{a+1} + qy_a, \qquad (1 \le a \le b-1).$$

All we have to do is to solve this difference equation with boundary conditions $y_0 = y_b = 0$. See the next subsection for how to do this.

Ca. Discussion. WHY can we say that $\mathbb{E}_a(T \mid A) = 1 + \mathbb{E}_{a+1}(T)$, if A is the event that the gambler wins the first game? This is one of those intuitively obvious things which are tricky to pin down rigorously. The following argument is perhaps best skipped on a first reading of the book! But DO the next Exercise.

Ⓜ If we think of the walk $\mathrm{SRW}_a(p)$ (where $0 < a < b$) as

$$W_n = a + X_1 + X_2 + \cdots + X_n,$$

then $T(\omega)$ is some (measurable) function

$$T(\omega) = h\big(a, X_1(\omega), X_2(\omega), \ldots\big)$$

of $a, X_1(\omega), X_2(\omega), \ldots$. *On the set* $A = \{\omega : X_1(\omega) = +1\}$, we have the obvious relation $T(\omega) = 1 + h\big(a + 1, X_2(\omega), X_3(\omega), \ldots\big)$. The variable $h\big(a + 1, X_2(\omega), X_3(\omega), \ldots\big)$ is independent of X_1 (see Fact 99G which relied on the π-system Lemma 45M) and (the π-system Lemma is needed again here) has the same DF as $h\big(a + 1, X_1(\omega), X_2(\omega), \ldots\big)$, whence

$$
\begin{aligned}
\mathbb{E}_a(T \mid A) &= 1 + \mathbb{E}\left(h\big(a+1, X_2(\omega), X_3(\omega), \ldots\big) \,\Big|\, A \right) \\
&= 1 + \mathbb{E}\, h\big(a+1, X_2(\omega), X_3(\omega), \ldots\big) \\
&= 1 + \mathbb{E}\, h\big(a+1, X_1(\omega), X_2(\omega), \ldots\big) \\
&= 1 + \mathbb{E}_{a+1}(T).
\end{aligned}
$$

▶ **Cb. Important Exercise.** Let W be $\mathrm{SRW}(p)$ started at 1. Let $T := \inf\{n : W_n = 0\}$, and let $y_k := \mathbb{E}_k(T)$ for $k = 1, 2, \ldots$. Prove that

$$y_2 = 2y_1, \qquad y_1 = 1 + py_2 + 0.$$

(The first of these really appeals to the Strong Markov Theorem. Compare Subsection 130D.) Solve for y_1 when $p = \frac{1}{2}$ and when $p > \frac{1}{2}$. Write down the obviously correct answer when $p < \frac{1}{2}$. Show that if $y_k(n) := \mathbb{E}_k(T \wedge n)$, where $T \wedge n := \min(T, n)$ as usual, then

$$y_1(n) = 1 + py_2(n-1), \quad y_2(n) \le 2y_1(n).$$

Compare the discussion at 119Da.

D. Solving difference equations, 2. This continues from 118C. The General Solution of the *differential* equation

$$ax''(t) + bx'(t) + cx(t) = Ke^{\gamma t}, \qquad (a \ne 0)$$

is (with α, β the roots of $am^2 + bm + c = 0$)

$$x(t) = \begin{cases} \frac{Ke^{\gamma t}}{a(\gamma-\alpha)(\gamma-\beta)} + Ae^{\alpha t} + Be^{\beta t} & \text{if } \alpha, \beta, \gamma \text{ are distinct,} \\ \frac{Kte^{\alpha t}}{a(\alpha-b)} + Ae^{\alpha t} + Be^{\beta t} & \text{if } \gamma = \alpha \neq \beta, \\ \frac{Kt^2 e^{\alpha t}}{2a} + (At + B)e^{\alpha t} & \text{if } \gamma = \alpha = \beta. \end{cases}$$

The pattern can be appreciated via $\partial_\alpha e^{\alpha t} = te^{\alpha t}$ and $\partial_\alpha^2 e^{\alpha t} = t^2 e^{\alpha t}$, where $\partial_\alpha = \partial/\partial\alpha$. You will have done the 'Particular Integral plus Complementary Function' idea.

These ideas transfer immediately to difference equations. The General Solution of the *difference* equation

$$ax(n + 2) + bx(n + 1) + bx(n) = \gamma^n$$

is given by

$$x(n) = \begin{cases} \frac{K\gamma^n}{a(\gamma-\alpha)(\gamma-\beta)} + A\alpha^n + B\beta^n & \text{if } \alpha, \beta, \gamma \text{ are distinct,} \\ \frac{Kn\alpha^{n-1}}{a(\alpha-\beta)} + A\alpha^n + B\beta^n & \text{if } \gamma = \alpha \neq \beta, \\ \frac{Kn^2\alpha^{n-2}}{2a} + (An + B)\alpha^n & \text{if } \gamma = \alpha = \beta. \end{cases}$$

You would see the pattern better if $n^2\alpha^{n-2}$ were replaced by $n(n-1)\alpha^{n-2}$ and $n\alpha^n$ by $n\alpha^{n-1}$ in the last expression.

▶▶ **E. The conditional expectation $\mathbb{E}(X \mid Y)$, Y discrete.** Suppose that X and Y are RVs, with Y discrete and X either non-negative or in $\mathcal{L}^1$ or both. We then define the *Random Variable* $Z = \mathbb{E}(X \mid Y)$ via

$$Z := c(Y), \quad \text{that is,} \quad Z(\omega) = c(Y(\omega))$$

where

$$c(y) := \mathbb{E}(X \mid Y = y).$$

[Of course, there is no RV '$(X \mid Y)$' of which $\mathbb{E}(X \mid Y)$ is the 'expectation' any more than there is an event '$A \mid B$' of which $\mathbb{P}(A \mid B)$ is the 'probability'.]

When you consider $Z(\omega) = \mathbb{E}(X \mid Y)(\omega)$, all that you know about ω is the value $Y(\omega)$. Let's consider an example.

Ea. Example. Suppose that a tetrahedral die with score $(1, 2, 3$ or 4 on each throw) is thrown twice. Suppose that

$$\mathbb{P}(\text{Score is } i \text{ on any throw}) = p_i, \quad (i = 1, 2, 3, 4).$$

Let

$$X = \text{sum of scores}, \quad Y = \text{product of scores}.$$

$$\Omega: \begin{array}{cccc} (1,1) & (1,2) & (1,3) & (1,4) \\ (2,1) & (2,2) & (2,3) & (2,4) \\ (3,1) & (3,2) & (3,3) & (3,4) \\ (4,1) & (4,2) & (4,3) & (4,4) \end{array}$$

$$X: \begin{array}{cccc} 2 & 3 & 4 & 5 \\ 3 & 4 & 5 & 6 \\ 4 & 5 & 6 & 7 \\ 5 & 6 & 7 & 8 \end{array} \qquad Y: \begin{array}{cccc} 1 & 2 & 3 & 4 \\ 2 & 4 & 6 & 8 \\ 3 & 6 & 9 & 12 \\ 4 & 8 & 12 & 16 \end{array}$$

$$Z: \begin{array}{cccc} 2 & 3 & 4 & z_4 \\ 3 & z_4 & 5 & 6 \\ 4 & 5 & 6 & 7 \\ z_4 & 6 & 7 & 8 \end{array}$$

$$P(\omega) = p_i p_j \text{ if } \omega = (i, j)$$

Table E(i): X, Y and $Z = \mathbb{E}(X \mid Y)$ for an example

In Table E(i), values of X, Y and $Z = \mathbb{E}(X \mid Y)$ are listed. If the actual outcome is $(1,4)$, then $Y = 4$. But if you are told only that $Y = 4$, then the outcome could be any of $(1,4), (2,2), (4,1)$, with corresponding X-values $5, 4, 5$ and absolute probabilities $p_1 p_4$, $p_2 p_2$, $p_4 p_1$. Hence, we have

$$z_4 = \frac{5 p_1 p_4 + 4 p_2 p_2 + 5 p_4 p_1}{p_1 p_4 + p_2 p_2 + p_4 p_1}.$$

Check out all the other (easy) entries.

Eb. Exercise. The following experiment (considered at Exercise 52Da) is performed. A fair coin is tossed twice and the number N of Heads is recorded. The coin is then tossed N more times. Let X be the total number of Heads obtained, Y the total number of tails obtained, and let $Z = \mathbb{E}(X \mid Y)$. Continue the table from the earlier exercise to the form beginning

ω	$\mathbb{P}(\omega)$	$X(\omega)$	$Y(\omega)$	$Z(\omega)$
HHHH	$\frac{1}{16}$	4	0	4

▶ **Ec. Essential Exercise.** Prove that if Y is discrete and if h is a function on the range of Y such that $h(Y)$ is either non-negative or in $\mathcal{L}^1$, then

$$\mathbb{E}(h(Y) \mid Y) = h(Y).$$

Prove more generally that *if either X and $h(Y)$ are both non-negative or $h(Y)X$ is in $\mathcal{L}^1$, then we have the* **'taking out what is known' Golden Rule**

$$\mathbb{E}(h(Y)X \mid Y) = h(Y)\mathbb{E}(X \mid Y). \tag{E1}$$

▶▶ **F. A fundamental property.**   We continue with the assumptions of the previous paragraph.

Things have been fixed so that the $\mathbb{P}$-average of X over a set $\{\omega : Y(\omega) = y\}$ is equal to the $\mathbb{P}$-average of Z over this set. (Indeed, over this set, Z is a constant equal to that $\mathbb{P}$-average of X over the set.) More significantly, *over any set of the form $Y \in A$, Z has the same $\mathbb{P}$-average as does X.* Let's spell this out.

If $Z = \mathbb{E}(X \mid Y)$, then, for any set A,

$$\mathbb{E}(X; Y \in A) = \mathbb{E}(Z; Y \in A), \qquad \mathbb{E}(X \mid Y \in A) = \mathbb{E}(Z \mid Y \in A). \quad \text{(F1)}$$

Indeed, (for our discrete Y) $Z = \mathbb{E}(X \mid Y)$ is characterized by the fact that Z is a function of Y and $\mathbb{E}(X; Y \in A) = \mathbb{E}(Z; Y \in A)$ for every set A.

Of course, $\mathbb{E}(X; Y \in A)$ means $\mathbb{E}(X; B)$ where B is the event $\{\omega : Y(\omega) \in B\}$. Subsection 399M on the Two-Envelopes Problem, which you can read now if you wish, explains clearly the benefits of this way of thinking.

Proof of (F1).   With Measure Theory ('Dominated and Monotone Convergence Theorems') providing justification if A is infinite, we have the intuitively obvious argument:

$$\mathbb{E}(X; Y \in A) = \sum_{y \in A} \mathbb{E}(X; Y = y) = \sum_{y \in A} \mathbb{P}(Y = y)\mathbb{E}(X|Y = y)$$

$$= \sum_{y \in A} \mathbb{P}(Y = y)c(y) = \mathbb{E}(c(Y); Y \in A) = \mathbb{E}(Z; Y \in A).$$

Why is the 'characterization' part obvious?   □

Note that the **Conditional Mean Formula** which states that

$$\mathbb{E}(X) = \mathbb{E}\mathbb{E}(X \mid Y), \quad\quad\quad\quad \text{(F2)}$$

is the case $A = \Omega$.

▶ **G. Linearity.** *Suppose that X_1 and X_2 are in $\mathcal{L}^1$, that Y is discrete, and that λ_1 and λ_2 are real numbers. Then,*

$$\mathbb{E}(\lambda_1 X_1 + \lambda_2 X_2 \mid Y) = \lambda_1 \mathbb{E}(X_1 \mid Y) + \lambda_2 \mathbb{E}(X_2 \mid Y).$$

The result is also true if $X_1, X_2, \lambda_1, \lambda_2$ are non-negative and Y is discrete.

Ga. Exercise. Prove this.

Gb. Exercise. The purpose of this exercise is to give a clearer explanation of the result at Exercise 71Kb. Again, μ is the mean, and σ^2 the variance, of a person chosen at random from Bath. We write t for the total height of all people in Bath.

Suppose that two different people are chosen at random from the n Bath residents. Let X be the height of the first and Y the height of the second. Express $\mathbb{E}(X \mid Y)$ in terms of t, Y and n. What is $\mathbb{E}(XY \mid Y)$? Use your answer to find $\mathbb{E}(XY)$ in terms of μ, σ^2 and n; and finally deduce the result from Exercise 71Kb:

$$\text{Cov}(X, Y) = -\frac{\sigma^2}{n-1}.$$

▶ **Gc. Exercise.** Let X and Y be the scores on two consecutive throws of a tetrahedral die. I've changed to tetrahedral so that you can look at Table 389 E(i). Consider the map $(i, j) \mapsto (j, i)$ from Ω to itself. Does this map change probabilities? In the light of the map, how are

$$\mathbb{E}(X \mid X + Y) \text{ and } \mathbb{E}(Y \mid X + Y)$$

related? What is the sum of these two variables? What is $\mathbb{E}(X \mid X + Y)$?

▶ **Gd. Important Exercise.** This exercise is an essential step in the martingale proofs of the Strong Law and of de Finetti's Theorem.

Let $X_1, X_2, \ldots, X_n$ be discrete RVs with the property that for every permutation τ of $\{1, 2, \ldots, n\}$,

$$\mathbb{P}(X_{\tau(1)} = x_1, \ldots, X_{\tau(n)} = x_n) = \mathbb{P}(X_1 = x_1, \ldots, X_n = x_n)$$

for all choices of $x_1, x_2, \ldots, x_n$. It is easily seen that $X_1, X_2, \ldots, X_n$ are identically distributed (Why?), but they need not be independent. Prove that if

$$S_n = X_1 + X_2 + \cdots + X_n \quad (\text{whence } S_n = X_{\tau(1)} + \cdots + X_{\tau(n)})$$

then $\mathbb{E}(X_1 \mid S_n) = S_n/n$. After the previous exercise, you should not need any hints.

▶▶ **H. Conditional Variance.** Recall the Parallel-Axis result that if $X \in \mathcal{L}^2$ and $\mathbb{E}(X) = \mu$, then

$$\text{Var}(X) := \mathbb{E}[(X - \mu)^2] = \mathbb{E}(X^2) - \mu^2.$$

If $X \in \mathcal{L}^2$, Y is discrete, and $Z := \mathbb{E}(X \mid Y)$, then

$$0 \le \text{Var}(X \mid Y) := \mathbb{E}[(X - Z)^2 \mid Y] = \mathbb{E}(X^2 \mid Y) - Z^2.$$

I am not going to fuss over the fact that $Z \in \mathcal{L}^2$. The algebra of the result is clear because, since Z is a function $c(Y)$ of Y,

$$\mathbb{E}(ZX \mid Y) = \mathbb{E}(c(Y)X \mid Y) = c(Y)\mathbb{E}(X \mid Y) = Z^2.$$

Note that (assuming that $Z \in \mathcal{L}^2$) we have

$$\mathbb{E}(Z^2) \le \mathbb{E}\mathbb{E}(X^2 \mid Y) = \mathbb{E}(X^2).$$

That Z *is* in $\mathcal{L}^2$ is proved by a 'truncation' (or 'staircase') argument.

For $X \in \mathcal{L}^2$ and Y discrete, we have the **Conditional Variance Formula**

$$\mathrm{Var}(X) = \mathbb{E}[\mathrm{Var}(X \mid Y)] + \mathrm{Var}[\mathbb{E}(X \mid Y)], \qquad (H1)$$

another of the 'Golden Rules'.

▶ **Ha. Exercise.** Prove property (H1).

▶ **Hb. Exercise.** Let $X \in \mathcal{L}^2$ and let Y be discrete. Let $c(Y) = \mathbb{E}(X|Y)$. Let f be any function on the range of Y. Show that

$$\mathbb{E}\left\{[X - f(Y)]^2\right\} = \mathbb{E}\left(X^2\right) - \mathbb{E}\left\{c(Y)^2\right\} + \mathbb{E}\left\{[f(Y) - c(Y)]^2\right\}.$$

Hence, $\mathbb{E}(X \mid Y)$ **is the least-squares best predictor of** X **given** Y **amongst 'linear' and 'non-linear' predictors.**

▶ **I. Variance of a 'Random Sum'.**   Suppose that N is an RV taking values in $\{0, 1, 2, \ldots\}$, and that $X_1, X_2, \ldots$ are identically distributed RVs each with the same distribution as an RV X. Suppose that

$$N, X_1, X_2, \ldots \text{ are independent.}$$

In particular, $X_1, X_2, \ldots$ are IID. We suppose that N and X are in $\mathcal{L}^2$ and write

$$\mu_X := \mathbb{E}(X), \quad \sigma_X^2 := \mathrm{Var}(X), \quad \mu_N := \mathbb{E}(N), \quad \sigma_N^2 := \mathrm{Var}(N).$$

Let S_N be the 'Random Sum' (the number of *terms* is random):

$$S_N := X_1 + X_2 + \cdots + X_N \qquad (:= 0 \text{ if } N = 0).$$

We want to find the mean and variance of S_N. We write

$$S_n := X_1 + X_2 + \cdots + X_n$$

as usual.

You should **note the steps in the following argument carefully** because it is an argument which is used very frequently. For each fixed n,

$$
\begin{aligned}
c(n) &:= \mathbb{E}(S_N \mid N = n) & \\
&= \mathbb{E}(S_n \mid N = n) & \text{(logic applied to the definition!)} \\
&= \mathbb{E}(S_n) & (X_1 + X_2 + \cdots + X_n \text{ is independent of } N) \\
&= n\mu_X & \text{(standard).}
\end{aligned}
$$

Here we have used the fact that on the set $\{\omega : N(\omega) = n\}$, we have $(S_N)(\omega) := S_{N(\omega)}(\omega) = S_n(\omega)$, and the fact that for a fixed integer n, S_n is independent of N.

Hence,

$$\mathbb{E}(S_N \mid N) = c(N) = N\mu_X,$$

and, a result which is hardly a surprise,

$$\mathbb{E}(S_N) = \mathbb{E}\mathbb{E}(S_N \mid N) = \mu_N\mu_X.$$

▶ **Ia. Exercise.** Prove that

$$\text{Var}(S_N) = \sigma_N^2 \mu_X^2 + \mu_N \sigma_X^2.$$

Explain the good intuitive sense of this.

Assume that the number of coaches which visit Bath on a winter's day is a Poisson RV with parameter 20 (and hence mean 20 and variance 20). Assume that each coach has 32 seats and that, 'independently of everything else', the probability that any particular seat is taken is 3/4. Let T be the total number of people who visit Bath by coach on some winter's day. Find $\mathbb{E}(T)$ and $\text{Var}(T)$.

▶ **J. Generating Functions for Random Sums.** Again suppose that N is an RV taking values in $\{0, 1, 2, \ldots\}$, and that $X_1, X_2, \ldots$ are identically distributed RVs each with the same distribution as an RV X.

▶▶ **Ja. Important Exercise.** *This exercise is crucial for Branching-Process Theory.* Suppose that each X_k takes values in $\{0, 1, 2, \ldots\}$, and let $g_X(\alpha)$ denote the probability generating function of X. Let S_n be as in the previous subsection. Prove that

$$g_{S_N}(\alpha) = g_N(g_X(\alpha)).$$

Hint. Your argument should mimic that in the previous subsection:

$$c(n) := \mathbb{E}\left(\alpha^{S_N} \mid N = n\right) = \cdots.$$

Jb. Exercise: A CLT for a random sum. Suppose that N is Poisson(n), and that each X is exponential E(1). Prove that S_N has mean n and variance $2n$. Show that (for $\alpha < 1$)

$$M_{S_N}(\alpha) = \mathbb{E}\exp(\alpha S_N) = g_N(M_X(\alpha)) = \exp\left(\frac{n\alpha}{1-\alpha}\right).$$

Prove that the CGF $C_{Z_n}(\alpha)$ of the standardized form

$$Z_n := (S_N - n)/\sqrt{2n}$$

of S_N satisfies (for $\alpha < \sqrt{2n}$)

$$C_{Z_n}(\alpha) = \frac{\frac{1}{2}\alpha^2}{1 - (2n)^{-\frac{1}{2}}\alpha}.$$

Deduce that, as $n \to \infty$, $\mathbb{P}(Z_n \leq x) \to \Phi(x)$ for every $x \in \mathbb{R}$.

►► **K. The Galton–Watson Branching Process.** The original problem might now be seen as sexist. A certain Reverend Watson, the last of the male line descended from some Watson in the past, posed the question: what is the probability that the male line descended from a newborn boy will die out? Francis Galton, whom we have already met in connection with regression, discussed the problem (and got it wrong!).

Let's make it non-sexist by having asexual reproduction for some type of animal, so we have 'animal and children' rather than 'man and sons'. Note however that many interpretations are possible. In constructing the atomic bomb or nuclear power station, the 'Watson' question had to be asked where 'animal' equals atom and where a 'child of that atom' is another atom split by a neutron emitted from the first. For another important application of the idea, see the next subsection.

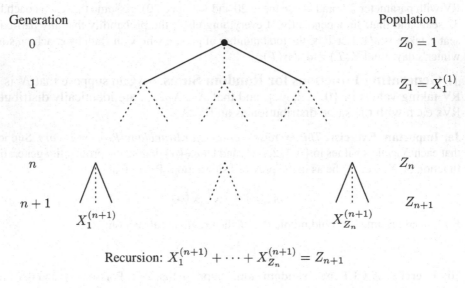

Recursion: $X_1^{(n+1)} + \cdots + X_{Z_n}^{(n+1)} = Z_{n+1}$

Figure K(i): A Galton–Watson branching process

Figure K(i) illustrates the type of family tree (starting with one individual) in which we are interested. We write $X_r^{(n+1)}$ for the number of children (who will be in the $(n+1)$th generation) of the rth animal in the nth generation. Let Z_n denote the number of animals in the nth generation. I hope that the diagram explains why the fundamental recurrence relation is

$$Z_{n+1} = X_1^{(n+1)} + X_2^{(n+1)} + \cdots + X_{Z_n}^{(n+1)}. \tag{K1}$$

The key assumption is that

$$X_1^{(1)}, \quad X_2^{(1)}, \quad X_3^{(1)}, \quad \ldots$$
$$X_1^{(2)}, \quad X_2^{(2)}, \quad X_3^{(2)}, \quad \ldots$$
$$X_1^{(3)}, \quad X_2^{(3)}, \quad X_3^{(3)}, \quad \ldots$$
$$\ldots, \quad \ldots, \quad \ldots, \quad \ldots$$

are IID RVs with values in $\{0, 1, 2, \ldots\}$, each with the same distribution as an RV X. We write μ for μ_X, and we assume that it is finite. We want to calculate

$$\pi := \mathbb{P}(\text{extinction}) = \mathbb{P}(Z_n = 0 \text{ for some } n).$$

We assume throughout that $\mathbb{P}(X = 0) > 0$.

From Subsection 392I, we know that $\mathbb{E}(Z_{n+1}) = \mathbb{E}(Z_n)\mu$, so that $\mathbb{E}(Z_n) = \mu^n$, as intuition suggests. It is therefore plausible that

$$\pi = 1 \text{ if } \mu < 1, \text{ and } \pi < 1 \text{ if } \mu > 1.$$

We now define

$$\pi_n := \mathbb{P}(\text{extinction by time } n) = \mathbb{P}(Z_n = 0).$$

Measure Theory in the form of 43D(a) guarantees the plausible result that

$$\pi_n \uparrow \pi \text{ as } n \uparrow \infty,$$

and we shall take this for granted. Now,

$$\pi_{n+1} = \mathbb{P}(Z_{n+1} = 0)$$
$$= \sum_{k=0}^{\infty} \mathbb{P}(Z_1 = k; Z_{n+1} = 0)$$
$$= \sum_{k=0}^{\infty} \mathbb{P}(Z_1 = k)\mathbb{P}(Z_{n+1} = 0 \,|\, Z_1 = k).$$

But $\mathbb{P}(Z_{n+1} = 0 \,|\, Z_1 = k)$ is the probability that k independent family trees each starting from one individual will die out by time n, that is π_n^k. Thus,

$$\pi_{n+1} = \sum_{k=0}^{\infty} \mathbb{P}(X = k)\pi_n^k = g(\pi_n),$$

g being the pgf of X. We have

$$\pi_{n+1} = g(\pi_n), \qquad \pi_n \uparrow \pi, \qquad \pi = g(\pi),$$

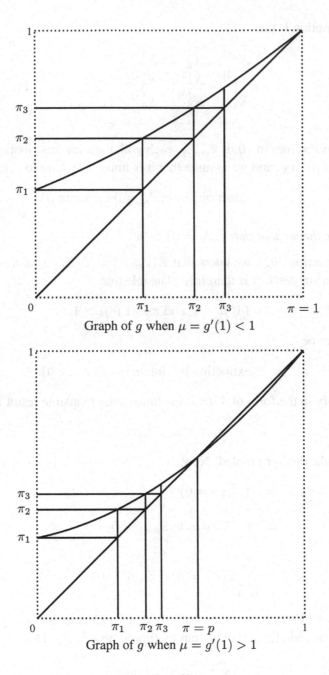

Graph of g when $\mu = g'(1) < 1$

Graph of g when $\mu = g'(1) > 1$

Figure K(ii): The famous Branching-process pictures

the last by continuity. Since $g''(t) \geq 0$ for $t \geq 0$, g is convex: its slope is never decreasing.

Recall that $g'(1) = \mathbb{E}(X) = \mu$.

Figure K(ii) shows the two situations which can arise. In the top figure, $g'(1) = \mu < 1$, so the slope of the g-curve is always less than 1 within $[0, 1]$. In this case, there is only one root, namely 1, of $g(\pi) = \pi$ in $[0, 1]$. The case when $\mu = 1$ is exactly similar, except that the $45°$ line touches the curve at $(1, 1)$. Thus,

$$\pi = 1 \text{ when } \mu \leq 1.$$

The case when $\mu = g'(1) > 1$ is illustrated in the lower figure. The curve is steeper than the line at $(1, 1)$, and so will cut the line at precisely one point in $(0, 1)$. From the picture, $\pi_n \uparrow \pi_\infty$, where π_∞ is the unique root of $g(\pi) = \pi$ in $(0, 1)$; so, $\pi = \pi_\infty$.

We often argue as follows. First,

$$\pi_1 = \mathbb{P}(Z_1 = 0) = g(0) \leq \pi_\infty,$$

and, since g is a non-decreasing function,

$$\pi_n \leq \pi_\infty \text{ implies that } \pi_{n+1} = g(\pi_n) \leq g(\pi_\infty) = \pi_\infty,$$

whence, by induction, $\pi_n \leq \pi_\infty$ for all n, and so $\pi = \pi_\infty$.

Summary. *If $\mu \leq 1$ then $\pi = 1$, and if $\mu > 1$, then π is the unique root of $g(\pi) = \pi$ in $(0, 1)$.*

Ka. Exercise. Suppose that $\mathbb{P}(X = k) = pq^k$ for $k = 0, 1, 2, \ldots$, where $0 < p = 1 - q < 1$. Show that $g(\alpha) = p/(1 - q\alpha)$, $\mu = g'(1) = q/p$, and $\pi = p/q$ if $q > p$ and 1 if $q \leq p$.

▶ **Kb. Use of pgfs.** The fundamental recurrence relation 394(K1) combined with the 'Random Sums' result from Exercise 393Ja shows that

$$g_{n+1}(\alpha) := \mathbb{E}\left(\alpha^{Z_{n+1}}\right) = g_n(g_X(\alpha)) = g_n(g(\alpha)).$$

Hence,

$$g_2 = g \circ g, \quad g_3 = g \circ g \circ g, \quad g_n = g \circ g \circ \cdots \circ g \text{ (n-fold iteration)}.$$

Only in a few simple cases is it possible to give a closed-form expression for the nth iterate. Note that $g_{n+1} = g \circ g_n$, so that

$$\pi_{n+1} = \mathbb{P}(Z_{n+1} = 0) = g_{n+1}(0) = g(g_n(0)) = g(\pi_n).$$

▶ **Kc. Exercise.** Show that if

$$G = \begin{pmatrix} g_{11} & g_{12} \\ g_{21} & g_{22} \end{pmatrix},$$

is a non-singular 2×2 matrix and we define the associated fractional linear transformation via

$$G(\alpha) = \frac{g_{11}\alpha + g_{12}}{g_{21}\alpha + g_{22}},$$

then, if H is another such matrix, we have

$$G(H(\alpha)) = (GH)(\alpha),$$

GH being the matrix product. Composition of linear fractional transformations therefore corresponds to multiplication of matrices, and since there are standard 'eigenvalue' methods of computing powers of matrices, we can handle the case described in Exercise 397Ka. Indeed, for that case, as you can verify by induction, for $p \neq q$,

$$g_n(\alpha) = \frac{p\mu^n(1-\alpha) + q\alpha - p}{q\mu^n(1-\alpha) + q\alpha - p}.$$

Check that, for $p \neq q$,

$$\pi_n = \frac{p(q^n - p^n)}{q^{n+1} - p^{n+1}}.$$

For more on this, see, for example, [W].

▶ **L. A 'typed' branching process.** Suppose that our population consists of animals of two types, A and B. Suppose that an A type

either, and with probability $\frac{1}{2}$, has 1 A-type child,

or, and with probability $\frac{1}{4}$, has 1 B-type child,

or, and with probability $\frac{1}{4}$, has no children.

Suppose that a B-type

either, and with probability $\frac{1}{2}$, has 2 B-type children,

or, and with probability $\frac{1}{2}$, has 1 A-type child.

Let a [respectively, b] be the probability of extinction starting from a population of just 1 A-type animal [respectively, just 1 B-type animal].

▶ **La. Exercise.** Calculate a and b, *proving* that both are less than 1.

▶ **Lb. Challenging Exercise*.** Continue with the preceding exercise, but suppose in addition that each animal at birth has a chance p of being transferred right out of the population under consideration. Find the critical value p_0 such that the animal population started with 1 animal has a positive probability of surviving for ever if, and only if, $p < p_0$. *Answer:* $p_0 = (2/\sqrt{3}) - 1$.

Consider the population consisting of all people with AIDS. Divide it into types: heterosexual men, bisexual men, homosexual men, heterosexual women, bisexual women, lesbians. Let a 'child' of an individual be a healthy person who catches AIDS from that individual (could be a newborn baby). Transferring out of population could be (we hope for the future) by cure, or else by death or moral decision. I hope that this indicates that this type of question is not just for fun, even if it hardly provides a sophisticated model in the case of AIDS.

For key further reading on branching processes, see Kendall [134, 135], Harris [109], Athreya and Ney [6].

▶▶ **M. 'Paradoxes' of 'Two-Envelopes' type.** Please re-read Subsection 16Q. You may remember that I said that I would try to present this discussion in a way much of which is comprehensible if you have only read up to page 16. But the problems *are* subtle, and it is *necessary* to go into the matter in the detail given here.

Accept for the moment that we can find a constant $c > 1$ and pair (X, Y) of positive-integer-valued Random Variables such that

$$\mathbb{P}(X = x;\ Y = y) = \mathbb{P}(Y = x;\ X = y) \text{ for all positive integers } x, y, \tag{M1}$$

and

$$\mathbb{E}(Y \mid X = x) > cx \text{ for every value } x \text{ which } X \text{ can take.} \tag{M2}$$

(In case you have jumped on to this subsection, I explain that $\mathbb{E}(Y \mid X = x)$ is the (correct probabilistic) average value of Y over those outcomes for which $X = x$.) Then, of course,

$$\mathbb{E}(X \mid Y = y) > cy \text{ for every value } y \text{ which } Y \text{ can take.} \tag{M3}$$

In a certain game, you are told the probabilities at (M1). An independent sequence

$$(X_1, Y_1), (X_2, Y_2), (X_3, Y_3), \ldots$$

of pairs of integer-valued Random Variables is then chosen, each pair having the same probability distribution as (X, Y). For each n, you will be given an envelope containing X_n dollars, while your 'opponent' will be given an envelope containing Y_n dollars.

Suppose first that at each game, *you* are given the choice of whether you wish to swap envelopes with your opponent. (In some sense, you are favoured over your opponent at every game.) Based on (M2), you decide to swap at every game.

(a) Let's be very clear what equation (M2) has to say about this situation. Let A be any non-empty finite subset of the integers such that X has positive probability of being in A. Let T_A be the set of those values m for which X_m is in the set A, and let $T_A(n)$ be the set of values $m \leq n$ for which m is in T_A (that is, for which X_m is in A). Then the content of (M2) is that (with probability 1) for all large values of n, the sum of Y_m values for which m is in $T_A(n)$ is at least c times the sum of X_m values for which m is in $T_A(n)$. In this sense, *your swapping strategy works for those games for which the value of X_m is in A.*

(b) Let us plod laboriously through the analogous situation where it is your opponent who is given the chance to swap, and does swap at every game. Let B be any non-empty finite subset of the integers such that Y has positive probability of being in B. Let W_B be the set of those values m for which Y_m is in the set B, and let $W_B(n)$ be the set of values $m \leq n$ for which m is in W_B (that is, for which Y_m is in B). Then the content of (M3) is that (with probability 1) for all large values of n, the sum of X_m values for which m is in $W_B(n)$ is at least c times the sum of Y_m values for which m is in $W_B(n)$. In this sense, *your opponent's swapping strategy works for those games for which the value of Y_m is in B.*

(c) **There is no contradiction here.** You cannot find A, B such that $T_A = W_B$. Thus, *(a) and (b) refer to different subsets of the set of all games.* This is why there is no contradiction. The difference may seem 'small', but we can only have 399(M1) and 399(M2) holding simultaneously if both X and Y have infinite expectation: thus, what happens for outcomes of very small probability can have a huge impact on averages.

(d) It is indeed the case that we can look at the set of all games. However (as already remarked and as is proved in the discussion at Mb below) it is a consequence of our assumptions about X and Y that $\mathbb{E}(X) = \mathbb{E}(Y) = \infty$; and, of course, '$c$ times infinity is still infinity'. However, for the final resolution of the paradox, we would really need the fact that, with probability 1,

$$\liminf \frac{\sum_{m \leq n} X_m}{\sum_{m \leq n} Y_m} \leq c^{-1}; \quad \limsup \frac{\sum_{m \leq n} X_m}{\sum_{m \leq n} Y_m} \geq c. \tag{M4}$$

It is moderately easy to prove this if we further assume that Y/X is bounded: that, with probability 1, there is a finite deterministic constant K such that $Y \leq KX$. This is the case in all examples which concern us. I give a proof of a stronger result for this case at Appendix A9, p501.

(e) Playing the swapping strategy when $X \leq k$ will gain for those games when $X \leq k$, but what happens on those games will be completely dominated by what happens on the other games.

(f) This will all be very clearly visible in the following example. I repeat that all versions of the 'Two-Envelopes Paradox' will have the features of this example, even though these features are harder to see in other versions.

Ma. Example. Suppose that X and Y have joint probability mass function as shown in Table M(i). Thus, for $k = 0, 1, 2, \ldots,$

$$\mathbb{P}(X = 4^k; \ Y = 4^{k+1}) = \mathbb{P}(Y = 4^k; \ X = 4^{k+1}) = 2^{-(k+2)}.$$

Check that

$$\mathbb{E}(Y \mid X = x) = \begin{cases} 4x & \text{if } x = 1, \\ 1.5x & \text{if } x = 4^k \ (k \geq 1). \end{cases}$$

If we sum down to row D (inclusive), then the contributions to the means $\mathbb{E}(X)$ of X and $\mathbb{E}(Y)$ of Y are equal. However, if we consider the swapping policy 'Swap if X-value is less than or equal to 4^k', then this takes us down to row E, by which time the contribution to $\mathbb{E}(Y)$ is more than 1.5 times the contribution to $\mathbb{E}(X)$.

If you sum down to row B you have the sum corresponding to $X \leq 4^{k-1}$. To get the sum corresponding to $Y \leq 4^{k-1}$, you have to replace row B by row D, and just look at the impact of that change.

					Contribs to			
	Prob	X	Y	S	$\mathbb{E}(X)$	$\mathbb{E}(Y)$	C	$\mathbb{E}(Y\mid X)$
	2^{-2}	1	4		$\frac{1}{4}$	1		1×4
	2^{-2}	4	1		1	$\frac{1}{4}$		4×1.5
	2^{-3}	4	4^2		$\frac{1}{2}$	2		4×1.5
	2^{-3}	4^2	4		2	$\frac{1}{2}$		$4^2 \times 1.5$
	2^{-4}	4^2	4^3		1	4		$4^2 \times 1.5$
	.	.	.	.	.	.	.	.
B	$2^{-(k+1)}$	4^{k-1}	4^k		2^{k-3}	2^{k-1}		$4^{k-1} \times 1.5$
D	$2^{-(k+1)}$	4^k	4^{k-1}		2^{k-1}	2^{k-3}		$4^k \times 1.5$
E	$2^{-(k+2)}$	4^k	4^{k+1}		2^{k-2}	2^k		$4^k \times 1.5$
	.	.	.	.	.	.	.	.

'S' refers to symmetric grouping,
'C' to 'conditional on X' grouping

Table M(i): Joint pmf and other information about (X, Y)

The sheer 'lack of robustness' is well displayed by this example. Rows far further down the table have an even greater impact on averages than does row E, even though they have still smaller probabilities.

Notice that the fact that X and Y are symmetric does not prevent the fact that it is obviously better for you to swap if $X = 1$ because you then know that $Y = 4$.

▶ **Mb. The rôle of infinite expectations.** *It is impossible to find Random Variables U and V both in $\mathcal{L}^1$ such that*

$$\mathbb{P}(\text{both } E(U \mid V) > V \text{ and } \mathbb{E}(V \mid U) > U) = 1.$$

for if this were so, then the Conditional Mean Formula would lead to the contradiction

$$\mathbb{E}(V) = \mathbb{E}\mathbb{E}(V \mid U) > \mathbb{E}(U) = \mathbb{E}\mathbb{E}(U \mid V) > \mathbb{E}(V).$$

People have correctly identified that the presence of infinite expectations is necessary for 'Two-Envelope Paradoxes'. However, I do not regard the assertion that certain expectations are infinite as any kind of *explanation* of such 'paradoxes'. (It is however true that, as David Chalmers has pointed out, the naive intuitive idea that 'on average I gain by swapping' in the Two-Envelopes problem makes little sense if the expected amount I receive is infinite whether or not I swap.)

Be extremely careful if using conditional expectations of RVs which are not in $\mathcal{L}^1$. In particular, in contexts with improper Bayes priors, always try to avoid using any conditional-probability statements; otherwise apparent paradoxes such as the 'Two-Envelopes Paradox' and that in Subsection 281K will always be lurking in the background. As we have seen, this can happen with proper priors if variables have infinite expectation.

The Two-Envelopes Problem is taken up again at 404Ob and Appendix A9,p501.

▶▶ **N. Further important properties of CEs.**   Results such as those in Subsection 390F form a general pattern summarized by the **Tower Property:**

$$\mathbb{E}\left(X \mid \text{information}\right) = \mathbb{E}\left(\mathbb{E}\left(X \mid \text{more information}\right) \Big| \text{information}\right) \qquad \text{(N1)}$$

of which the rigorous form is given at equation 405(P1) below. Examples are

$$\mathbb{E}(X) = \mathbb{E}\mathbb{E}(X \mid Y, U),$$
$$\mathbb{E}(X \mid Y) = \mathbb{E}\big(\mathbb{E}(X \mid Y, U) \mid Y\big). \qquad \text{(N2)}$$

Discussion of equation (N2). If Y and U are discrete, then

$$T := \mathbb{E}(X \mid Y, U) = c(y, u) := \mathbb{E}(X \mid Y = y; U = u)$$
$$\text{on } \{\omega : Y(\omega) = y; U(\omega) = u\}.$$

But then, for a subset H of the range of (Y, U), and with double summations over $\{(y, u) \in H\}$,

$$\mathbb{E}(X; (Y, U) \in H) = \sum\sum \mathbb{E}(X; Y = y; U = u)$$
$$= \sum\sum \mathbb{P}(Y = y; U = u) c(y, u) = \mathbb{E}(T; (Y, U) \in H).$$

Especially, taking $H = A \times \mathbb{R}$, we have

$$\mathbb{E}(X; Y \in A) = \mathbb{E}(T; Y \in A),$$

whence $\mathbb{E}(X \mid Y) = \mathbb{E}(T \mid Y)$, which is result (N2).   □

▶ **Na. An independence result for CEs.**   *If X is non-negative or in $\mathcal{L}^1$, and $Y, U, \ldots$ are finitely many discrete RVs, and if X is independent of $(Y, U, \ldots)$, then*

$$\mathbb{E}(X \mid Y, U, \cdots) = \mathbb{E}(X).$$

You prove this.

▶▶ **Nb. A more subtle independence result for CEs.** *If X is non-negative or in $\mathcal{L}^1$, U and Y are discrete and U is independent of the pair (X, Y), then*

$$\mathbb{E}(X \mid Y, U) = \mathbb{E}(X \mid Y). \qquad \text{(N3)}$$

Proof. Let A be a subset of the range of Y, and B be a subset of the range of U. Let $Z = \mathbb{E}(X \mid Y)$, a function $c(Y)$ of Y. Since Y is independent of U, whence $ZI_A(Y) = c(Y)I_A(Y)$ is independent of $I_B(U)$,

$$
\begin{aligned}
\mathbb{E}(Z; Y \in A; U \in B) &= \mathbb{E}(\{ZI_A(Y)\}I_B(U)) = \mathbb{E}(ZI_A(Y))\mathbb{E}(I_B(U)) \\
&= \mathbb{E}(Z; Y \in A)\mathbb{P}(U \in B) = \mathbb{E}(X; Y \in A)\mathbb{P}(U \in B),
\end{aligned}
$$

using the Golden Rule 390(F1) at the last step. But, since (X, Y) is independent of U, whence $XI_A(Y)$ is independent of $I_B(U)$, we have

$$
\begin{aligned}
\mathbb{E}(X; Y \in A; U \in B) &= \mathbb{E}(\{XI_A(Y)\}I_B(U)) = \mathbb{E}(XI_A(Y))\mathbb{E}(I_B(U)) \\
&= \mathbb{E}(X; Y \in A)\mathbb{P}(U \in B) = \mathbb{E}(Z; Y \in A; U \in B),
\end{aligned}
$$

using the previous equation at the last step. Thus, $Z = \mathbb{E}(X \mid Y, U)$, as required. That for any subset H of the range of (Y, U),

$$
\mathbb{E}[Z; (Y, U) \in H] = \mathbb{E}[X; (Y, U) \in H]
$$

is clear (why?). $\qquad\square$

Nc. Exercise. Give an example to show that if we only know that X and U are independent, then 402(N3) need not hold. What is the intuition behind this?

▶ **O. Conditional expectation for the 'continuous' case.** When Y is not discrete, $\mathbb{E}(X \mid Y)$ is defined only modulo subsets of Ω of probability 0. Recall that 'almost surely' or 'a.s.' for short means 'with probability 1'.

A Random Variable Z is called a version of $\mathbb{E}(X \mid Y)$, and we write

$$
Z = \mathbb{E}(X \mid Y), \quad \text{a.s.,}
$$

if Z is a nice (Borel) function $c(Y)$ of Y and for every nice (Borel) subset A of $\mathbb{R}$,

$$
\mathbb{E}(Z; Y \in A) = \mathbb{E}(X; Y \in A). \tag{O1}
$$

Two versions Z and $\tilde{Z}$ of $\mathbb{E}(X \mid Y)$ will agree almost surely: $\mathbb{P}(Z = \tilde{Z}) = 1$.

The π-system Lemma implies that if (O1) holds for every set A of the form $(-\infty, a]$, then it will hold for all Borel sets A.

You will guess what happens: if X and Y have a joint pdf $f_{X,Y}(x, y)$, then

$$
Z = c(Y), \quad \text{a.s.,} \tag{O2}
$$

where

$$
c(y) = \mathbb{E}(X \mid Y \in dy) = \int x f_{X|Y}(x \mid y) \, dx.
$$

▶ **Oa. Exercise.** Check that (O2) works when (X, Y) is 'continuous', ignoring nonsenses about sets of measure zero. Check that 389(E1), 390(F2), and Result 390G transfer to the situation of arbitrary Y.   □

If $(X, Y) \sim \text{BVN}(\mu_x, \mu_y, \sigma_x^2, \sigma_y^2, \rho)$, then, as we have essentially seen already,

$$\mathbb{E}(X \mid Y) = \mu_x + \rho \frac{\sigma_x}{\sigma_y}(Y - \mu_y).$$

(This property is shared by any elliptically-contoured distribution centred on (μ_x, μ_y) – Yes?!) We know (for the BVN case) that

$$\mathbb{E}\,\text{Var}(X \mid Y) = \mathbb{E}\sigma_x^2(1 - \rho^2) = \sigma_x^2(1 - \rho^2).$$

Moreover,

$$\text{Var}\big(\mathbb{E}(X \mid Y)\big) = \frac{\rho^2\sigma_x^2}{\sigma_y^2}\sigma_y^2 = \rho^2\sigma_x^2,$$

and we see that this tallies with the Conditional Variance Formula 392(H1).

Ob. The Two-Envelopes Problem – again.   *You puzzle out the following case. Give your intuition free rein.*

Suppose that someone chooses a real number Z at random, puts Z units in one envelope and $2Z$ units in another. You choose one of the envelopes at random, and are then asked whether you wish to swap it for the other.

Take the situation where $Z\ (> 0)$ has *proper* pdf

$$f_Z(z) = \frac{1}{(1+z)^2} \quad (z > 0). \tag{O3}$$

Let X be the amount in the first envelope you pick, Y the amount in the other. Then (you check!)

$$(dx)^{-1}\mathbb{P}\{X \in (x, x + dx); Y = 2X\} = \tfrac{1}{2}f_Z(x) = \frac{1}{2(1+x)^2} =: r_1(x),$$

$$(dx)^{-1}\mathbb{P}\{X \in (x, x + dx); Y = \tfrac{1}{2}X\} = (dx)^{-1}\tfrac{1}{2}\mathbb{P}\{Z \in (\tfrac{1}{2}x, \tfrac{1}{2}x + \tfrac{1}{2}dx)\}$$

$$= \tfrac{1}{4}f_Z(\tfrac{1}{2}x) = \frac{1}{4(1+\tfrac{1}{2}x)^2} =: r_2(x).$$

So, with $r(x) := r_1(x) + r_2(x)$, then conditionally on $X \in dx$,

$$Y = 2x \text{ with probability } p_1(x) := r_1(x)/r(x),$$
$$Y = \tfrac{1}{2}x \text{ with probability } p_2(x) := r_2(x)/r(x).$$

Conditionally on $X \in dx$, the expectation of Y is

$$2xp_1(x) + \tfrac{1}{2}xp_2(x),$$

and because $8(1 + \frac{1}{2}x)^2 > 2(1 + x)^2$ for $x > 0$, this is greater than x for $x > 0$. We therefore have

$$\text{both } \mathbb{E}\,(Y \mid X) > X \text{ and } \mathbb{E}\,(X \mid Y) > Y.$$

Convince yourself that there is no contradiction. (The condition that A is a non-empty finite set in our previous arguments should be replaced by the assumption that $\mathbb{E}\,(X; A)$ is finite and positive.)

Of course, in this problem, (X, Y) is not a 'continuous' variable in $\mathbb{R}^2$: X and Y do not have a joint pdf. That's why I said "Give your intuition free rein".

▶▶ **P. Remarks on the general theory of CEs.** Perhaps the greatest of Kolmogorov's many great contributions to Probability was the realization that the right general concept is that of (a version of) the conditional expectation $\mathbb{E}\,(X \mid \mathcal{G})$ of an RV X given a sub-σ-algebra $\mathcal{G}$ of $\mathcal{F}$.

The crucial sets of the form $Y^{-1}(A) := \{\omega : Y(\omega) \in A\}$, A a Borel subset of $\mathbb{R}$, over which we integrated within the definition of $\mathbb{E}\,(X \mid Y)$, form a σ-algebra of subsets of Ω, the σ-algebra $\sigma(Y)$ generated by Y. The conditional expectation $\mathbb{E}\,(X \mid Y)$ is really $\mathbb{E}\,(X \mid \sigma(Y))$. Analogously, $\mathbb{E}\,(X \mid Y, U)$ is really $\mathbb{E}\,(X \mid \sigma(Y, U))$ where $\sigma(Y, U)$ is the σ-algebra generated by (Y, U), that is, the collection of subsets of Ω of the form $\{\omega : (Y(\omega), U(\omega)) \in A\}$, where A is a Borel subset of $\mathbb{R}^2$.

In general, if Z is a version of $\mathbb{E}\,(X \mid \mathcal{G})$, then

$$\mathbb{E}\,(Z; G) = \mathbb{E}\,(X; G) \quad \text{for every } G \text{ in } \mathcal{G}.$$

The idea that Z must be a function of Y is replaced by the idea that Z must be $\mathcal{G}$-measurable in that $\sigma(Z) \subseteq \mathcal{G}$. A version is then determined almost surely. The benefits of such a generalization are considerable. Amongst other things, it equips Probability with a quite different level of subtlety, something of which we shall get a hint in our discussion of de Finetti's Theorem.

The **general Tower Property** says that if $\mathcal{G}$ and $\mathcal{H}$ are sub-σ-algebras of the σ-algebra $\mathcal{F}$ of all events, and $\mathcal{G} \subseteq \mathcal{H}$, then

$$\mathbb{E}\,(X \mid \mathcal{G}) = \mathbb{E}\,\big(\mathbb{E}\,(X \mid \mathcal{H}) \mid \mathcal{G}\big), \qquad \text{a.s.;} \tag{P1}$$

and this is just definition chasing. Every property which we have seen for conditioning relative to a discrete Random Variable Y holds in general, provided we add 'a.s.' here and there.

With regret, I have decided that conditional expectations relative to σ-algebras are too technical for this book. See, for example, [W] and Rogers and Williams [199] for detailed accounts, both of which base the theory on the least-squares best predictor idea at 392Hb. See Bingham and Kiesel [23] for a well-motivated account.

9.2   Martingales

This is only a brief introduction to a great topic. It relies on the theory of conditional expectation, and we shall use results for that theory which we have proved only when the conditioning RV is discrete. In non-mathematical English, a martingale is a kind of gambling strategy or else part of a horse's harness which stops the horse from throwing its head up. Doob's Martingale Theory does relate to gambling systems, and it puts constraints ('Doob's Upcrossing Lemma', not studied here) on oscillatory behaviour.

Many common structures support enough martingales to allow us to derive any desired properties by using Doob's theory. (And many more-subtle ones don't.)

▶▶ **A. Definition of martingale.**   A (discrete-time) **stochastic process** (or, simply, **process**) is a finite or infinite sequence $\mathbf{X} = (X_1, X_2, \ldots)$ of random variables.

We study martingales *associated with an underlying process* $\mathbf{X} = (X_1, X_2, \ldots)$. The idea is that the values $X_1(\omega^{\text{act}}), X_2(\omega^{\text{act}}), \ldots, X_n(\omega^{\text{act}})$ are known to the observer at time n.

A martingale **M** relative to $\mathbf{X}$ is a process $M = (M_0, M_1, M_2, \ldots)$ such that M_0 is a deterministic constant and the following three properties hold:

(M1) $M_n = f_n(X_1, X_2, \ldots, X_n), \qquad f_n : \mathbb{R}^n \to \mathbb{R},$

(f_n being a deterministic (and Borel) function), so that the value $M_n(\omega^{\text{act}})$ is known to the observer at time n;

(M2) $M_n \in \mathcal{L}^1$ for every n,

(M3) $\mathbb{E}(M_{n+1} \mid X_1, X_2, \ldots, X_n) = M_n$ (a.s.) for every $n \geq 0$.

When $n = 0$, (M3) states that $\mathbb{E}(M_1) = \mathbb{E}(M_0)$.

I usually write $\mathbb{E} M_0$ rather than M_0 because the theory extends to cases where M_0 is not a deterministic constant.

We can regard M_n as the hypothetical fortune at time n of a hypothetical gambler in a hypothetical fair game: if the present time is n, given the past and present history $X_1, X_2, \ldots, X_n$, the gambler's fortune M_{n+1} one time step into the future is on average what it is now, M_n. (Even if we are studying a 'real' gambling game, we may wish to build from it a hypothetical game in which (for example) the hypothetical fortune at time n of a hypothetical gambler is some function of our real gambler's fortune and the value n. We shall exploit this idea in later examples.)

Aa. Exercise. Prove that if $\mathbf{M}$ is a martingale relative to $\mathbf{X}$, then $\mathbb{E}(M_n) = \mathbb{E}(M_0)$ for every n. Prove that if $m \leq n$, then

$$\mathbb{E}(M_n \mid X_1, X_2, \ldots, X_m) = M_m, \qquad \text{a.s..}$$

It is convenient to write $\mathcal{F}_n$ for the history up to time n, so that the sequence $(\mathcal{F}_n)$ represents the 'evolution'. Compare the situation in Section 127. Strictly speaking (see that section), $\mathcal{F}_n$ is the σ-algebra of those events of the form

$$\{\omega : (X_1(\omega), X_2(\omega), \ldots X_n(\omega)) \in U\} \tag{A1}$$

where U is a nice (Borel) subset of $\mathbb{R}^n$. Then property (M1) states that M_n is $\mathcal{F}_n$-measurable. The definition of $\mathcal{F}_0$ as the trivial σ-algebra $\mathcal{F}_0 = \{\emptyset, \Omega\}$ carrying no information, tallies with our assumption that M_0 is a deterministic constant. Then the expectation on the left of (M3) is fundamentally $\mathbb{E}(M_{n+1} \mid \mathcal{F}_n)$, and the key martingale property may be written in the neater form

$$\mathbb{E}(M_{n+1} \mid \mathcal{F}_n) = M_n, \qquad \text{a.s..}$$

The precise significance of (M3) is that if F_n is any set of the form (A1), then

$$\mathbb{E}(M_{n+1}; F_n) = \mathbb{E}(M_n; F_n). \tag{A2}$$

In practice, the martingale property is often easy to verify, as examples will show.

▶ **Ab. Creating martingales by 'projection'.** Here is one important way in which martingales arise. Suppose that we have an evolution $(\mathcal{F}_n)$, and that $\xi \in \mathcal{L}^1$. Then, if we define
$$M_n := \mathbb{E}(\xi \mid \mathcal{F}_n),$$
then M is a martingale. This is because, by the Tower property,

$$M_n = \mathbb{E}(\xi \mid \mathcal{F}_n) = \mathbb{E}(\mathbb{E}(\xi \mid \mathcal{F}_{n+1}) \mid \mathcal{F}_n) = \mathbb{E}(M_{n+1} \mid \mathcal{F}_n).$$

B. 'Sum', 'product' and 'Likelihood-Ratio' martingales.

▶ **Ba. 'Sum' martingales.**   Suppose that $X_1, X_2, \ldots, X_n$ are independent RVs, each in $\mathcal{L}^1$ and each of mean 0. Let $a \in \mathbb{R}$, and define

$$M_n := a + X_1 + X_2 + \cdots + X_n, \qquad M_0 := a.$$

Then (M1) and (M2) automatically hold, and, as regards (M3), we have

$$\mathbb{E}(M_{n+1} \mid X_1, X_2, \ldots, X_n) = \mathbb{E}(M_n + X_{n+1} \mid X_1, X_2, \ldots, X_n)$$
$$= \mathbb{E}(M_n \mid X_1, X_2, \ldots, X_n) + \mathbb{E}(X_{n+1} \mid X_1, X_2, \ldots, X_n)$$
$$= M_n + \mathbb{E}X_{n+1} = M_n, \qquad \text{a.s..}$$

Here, we used the facts that M_n is known when $X_1, X_2, \ldots, X_n$ are known, that X_{n+1} is independent of $(X_1, X_2, \ldots, X_n)$, and that $\mathbb{E}X_{n+1} = 0$. Hence, **M** is a martingale relative to **X**.

Of course, we could have used the shorthand

$$\mathbb{E}(M_{n+1} \mid \mathcal{F}_n) = \mathbb{E}(M_n + X_{n+1} \mid \mathcal{F}_n) = \mathbb{E}(M_n \mid \mathcal{F}_n) + \mathbb{E}(X_{n+1} \mid \mathcal{F}_n)$$
$$= M_n + \mathbb{E}X_{n+1} = M_n, \qquad \text{a.s..}$$

▶ **Bb. Exercise: 'Product' martingales.**   Suppose that the Random Variables $Z_1, Z_2, \ldots, Z_n$ are independent, each nonnegative and in $\mathcal{L}^1$ and of mean 1. Let

$$M_n := Z_1 Z_2 \ldots Z_n, \qquad M_0 := 1.$$

Show that **M** is a martingale relative to **Z**. Note that what we really need is that

$$\mathbb{E}(Z_{n+1} \mid \mathcal{F}_n) = 1,$$

which does not require independence.

▶ **Bc. Exercise on Random Walks.** Let $X_1, X_2, \ldots$ be IID Random Variables with

$$\mathbb{P}(X_n = +1) = p, \qquad \mathbb{P}(X_n = -1) = q := 1 - p,$$

where $0 < p < 1$. Let $a \in \mathbb{Z}$ and define

$$W_n := a + X_1 + X_2 + \cdots + X_n,$$

so that **W** is $\mathrm{SRW}_a(p)$. Prove that if

$$M_n := W_n - (p - q)n, \qquad V_n = \left(\frac{q}{p}\right)^{W_n},$$

then **M** and **V** are martingales relative to **X**. *Hint* for the '**V**' part: Use the previous exercise.

▶ **Bd. 'Likelihood-Ratio' martingales.** Suppose that f and g are two pdfs, positive on the whole of $\mathbb{R}$. Let $Y_1, Y_2, \ldots$ be a finite or infinite sequence of IID Random Variables each with pdf f. Define the Likelihood-Ratio for time n:

$$R_n := \frac{g(Y_1)g(Y_2)\ldots g(Y_n)}{f(Y_1)f(Y_2)\ldots f(Y_n)}$$

with $R_0 := 1$. By using Exercise 408Bb, prove that **R** is a martingale relative to **Y**.

▶ **C. 'Markovian' martingales.** These form an important class.

Ca. The proportion martingale for Pólya's urn. Let B_n denote the number of Black balls in Pólya's urn of Exercise 76D at time n, and let

$$M_n := \frac{B_n}{n+2},$$

the proportion of Black balls in the urn at time n. We prove that **M** is a martingale relative to **B**. There is an obvious 'Markovian' ('lack-of-memory') property

$$\mathbb{E}(B_{n+1} \mid B_1, B_2, \ldots B_n) = \mathbb{E}(B_{n+1} \mid B_n)$$

(*the system does not remember how it arrived at the B_n situation*), and it is also clear that

$$c(b) := \mathbb{E}(B_{n+1} \mid B_n = b) = \frac{b}{n+2}(b+1) + \frac{n+2-b}{n+2}(b) = \frac{b(n+3)}{n+2}.$$

Hence,

$$\mathbb{E}(B_{n+1} \mid B_1, B_2, \ldots B_n) = c(B_n) = \frac{B_n(n+3)}{n+2},$$

and the desired result is now obvious.

▶ **Cb. Exercise: A 'game of cards' martingale.** A pack consisting of b Black and r Red cards is shuffled and placed face down on the table. The cards are then turned over one by one. Show that if M_n is the proportion of Black cards left before the nth card is revealed, then **M** is a martingale relative to **B**-process, where B_n is the number of Black cards left just before the nth card is revealed. Of course, the time-parameter set for **M** is $\{1, 2, \ldots, b+r\}$.

▶ **Cc. Exercise: A 'quadratic' martingale.** Suppose that $X_1, X_2, \ldots$ are IID RVs in $\mathcal{L}^2$ with $\mathbb{E}(X_k) = 0$ and $\mathrm{Var}(X_k) = \sigma^2$. Let **M** denote the martingale with $M_n = a + X_1 + X_2 + \cdots + X_n$. Show that if

$$V_n := M_n^2 - n\sigma^2,$$

then **V** is a martingale relative to **X**.

Cd. Exercise: Other martingales associated with Pólya's urn. We change the notation for Pólya's urn by letting the number of Black balls in the urn at time n be $B_n + 1$, so that B_n **now denotes the number of 'new' Black balls in the urn at time** n (that is, not counting the original Black ball). Prove that for $\theta \in (0, 1)$ (and with R_n the number of new Red balls, of course),

$$U_n := \frac{(n+1)!}{B_n!(n - B_n)!}(1 - \theta)^{B_n}\theta^{R_n}$$

defines a martingale $\mathbf{U}$ relative to $\mathbf{B}$. *Hint.* Consider the two values which $Z_{n+1} := U_{n+1}/U_n$ can take for a given value b of B_n, and calculate $\mathbb{E}(Z_{n+1} \mid B_n)$.

Discussion. *How could we anticipate that* $\mathbf{U}$ *is a martingale?* The answer is provided by the isomorphism in Subsection 79H. Then, in the notation of Subsection 78G, we have from 79(G4),

$$(\mathrm{d}\theta)^{-1}\mathbb{P}(H_{\mathrm{d}\theta} \mid \mathcal{F}_n) = U_n,$$

where $\mathcal{F}_n$ is the history up to time n for the urn process (or, under the isomorphism, for the coin-tossing process). The answer now follows heuristically from result 407Ab.

Ce. Kingman's 'OK Corral' martingales. In a paper *The OK Corral and the power of the law*, Paul McIlroy and I studied the following problem.

Two lines of gunmen face each other, there being initially m on one side, n on the other. Each person involved is a hopeless shot, but keeps firing at the enemy until either he himself is killed or there is no-one left on the other side. Let $\mu(m, n)$ be the expected number of survivors. Clearly, we have boundary conditions:

$$\mu(m, 0) = m, \quad \mu(0, n) = n. \tag{C1}$$

We also have the equation

$$\mu(m, n) = \frac{m}{m + n}\mu(m, n - 1) + \frac{n}{m + n}\mu(m - 1, n) \quad (m, n \geq 1). \tag{C2}$$

This is because the probability that the first successful shot is made by the side with m gunmen is $m/(m + n)$.

m	$\mu(m, m)$	$2Km^{\frac{3}{4}}$
2048	319.556354	319.857107
4096	537.627362	537.933399
8192	904.382093	904.692518

Table C(i): Three-quarters power law for the 'OK Corral' problem

McIlroy and I considered the case when $m = n$, for which the computer produces Table C(i), in which

$$K := 3^{-\frac{1}{4}}\pi^{-\frac{1}{2}}\Gamma(\tfrac{3}{4}) = 0.52532558.$$

We had arrived at the fact that $\mu(m, m) \sim 2Km^{\frac{3}{4}}$ heuristically by using a diffusion approximation. Kingman, in a fine paper [137] with a well-chosen title, found lots of martingales for this problem, and used them to *prove* our conjecture and much more.

▶▶ **D. Stopping times.** Please re-read the beginning of Section 127 up to the end of Subsection 128A. We allow our stopping times now to take values in $\{0, 1, 2, \ldots; \infty\}$, making the obvious modification from those with values in $\{1, 2, 3, \ldots; \infty\}$.

▶ **E. The 'Constant-Risk' Principle (CRP).** It is essential to have good ways of showing (when it is true) that a stopping time T is finite with probability 1, equivalently, that $\mathbb{P}(T = \infty) = 0$.

Here is a simple, very useful, criterion.

(CRP): *Suppose that T is a stopping time relative to the evolution $\{\mathcal{F}_n\}$. Suppose that for some $k \geq 1$ and some $c > 0$, we have for every n,*

$$\mathbb{P}(T \leq n + k \,|\, \mathcal{F}_n) \geq c \quad \text{for every } \omega$$

(oh, OK, for almost every ω). Then $\mathbb{P}(T = \infty) = 0$.

Ea. Application to SRW. Let's see an application of this before proving it. So consider Simple Random Walk $\text{SRW}_a(p)$, $\mathbf{W}$, started at a, where $0 < a < b$ and $0 < p < 1$. Let

$$T := \inf\{n : W_n = 0 \text{ or } W_n = b\}. \tag{E1}$$

How do we *prove* the (hardly surprising) fact that T is almost surely finite?

Consider $\mathbb{P}(T \leq n + b \,|\, \mathcal{F}_n)$. (Note that we have taken '$k = b$', as it were.) We have $\mathbb{P}(T \leq n + b \,|\, \mathcal{F}_n) = 1$ on the set $T \leq n$ which is in $\mathcal{F}_n$. So suppose that ω is such that $T(\omega) > n$; then W_n lies between 0 and b, and b 'Heads' (+1 X-values) in a row would take W above b, forcing T to have occurred before time $n + b$. In other words, for every ω,

$$\mathbb{P}(T \leq n + b \,|\, \mathcal{F}_n) \geq p^b \text{ for every } n,$$

so that the conditions of (CRP) hold.

Proof of (CRP). Using the obvious fact that $\{T > n + k\} = \{T > n\} \cap \{T > n + k\}$, we have

$$\begin{aligned}
\mathbb{P}(T > n + k) &= \mathbb{E} I_{\{T > n+k\}} = \mathbb{E} I_{\{T > n\}} I_{\{T > n+k\}} \\
&= \mathbb{E}\mathbb{E}\left(I_{\{T>n\}} I_{\{T>n+k\}} \,|\, \mathcal{F}_n\right) = \mathbb{E}\left[I_{\{T>n\}} \mathbb{E}\left(I_{\{T>n+k\}} \,|\, \mathcal{F}_n\right)\right] \\
&\leq \mathbb{E}\left[I_{\{T>n\}}(1 - c)\right] = (1 - c)\mathbb{P}(T \geq n).
\end{aligned}$$

Thus,
$$\mathbb{P}(T > n + k) \le (1 - c)\mathbb{P}(T > n),$$
and so, as you can easily see,
$$\mathbb{P}(T > mk) \le (1 - c)^{m-1},$$
and $\mathbb{P}(T = \infty) = 0$.   □

▶▶ **F. Doob's Stopping-Time Principle (STP).** This is one of the most interesting subsections in the book, describing one of the most important results in the subject and a few first applications. The result says that *under suitable conditions,* we can use the following idea. A martingale is a fair game and a stopping time is a time after which our hypothetical gambler can decide to stop playing. His hypothetical fortune would then be M_T, and since he cannot cheat the system, we should have $\mathbb{E}\,M_T = \mathbb{E}\,M_0$.

Now, *we clearly need conditions.* For let $\mathbf{W}$ be SRW$(\frac{1}{2})$ started at 0, and let $T := \inf\{n : W_n = 1\}$. We know from Subsection 118D that $\mathbb{P}(T = \infty) = 0$. However, in this case, $\mathbb{E}\,W_T = 1 \ne 0 = \mathbb{E}\,W_0$. Here, you *can* cheat the system, and we shall examine why at 414Fk below.

▶▶▶ **Fa. Theorem: Doob's Stopping-Time Principle (STP).** Let $\mathbf{X}$ be a process, $\mathbf{M}$ a martingale relative to $\mathbf{X}$, and T a stopping time with $\mathbb{P}(T = \infty) = 0$. Assume that we can write **either**

$$M_n = B_n + I_n \text{ for all } n \le T$$

or

$$M_n = B_n - I_n \text{ for all } n \le T$$

where for some deterministic constant K,

$$|B_n(\omega)| \le K \text{ for all } \omega \text{ and all } n \le T(\omega),$$

and

$$0 \le I_{n-1}(\omega) \le I_n(\omega) \text{ for all } \omega \text{ and all } n \le T(\omega).$$

Then $\mathbb{E}\,M_T = \mathbb{E}\,M_0$.

The idea of the notation is that up to time T, the process $\mathbf{B}$ is Bounded and the process $\mathbf{I}$ is Increasing (or at least non-decreasing).

Before seeing how to prove this result, we look at a number of applications.

▶ **Fb. Probability of gambler's success for $p = \frac{1}{2}$.** Consider $\mathrm{SRW}_a(\frac{1}{2})$, where $0 < a < b$. Define T as at 411(E1). Then we already know that $\mathbb{P}(T = \infty) = 0$. Here, we can take $\mathbf{B} = \mathbf{W}$, with $\mathbf{I}$ the process always equal to 0, because $\mathbf{B}$ is then bounded between 0 and b up to time T. Hence $\mathbb{E}(W_T) = \mathbb{E}(W_0) = a$. But

$$\mathbb{E}(W_T) = b\mathbb{P}(W_T = b) + 0\mathbb{P}(W_T = 0) = b\mathbb{P}(W_T = b),$$

so that $\mathbb{P}_a(W_T = b) = a/b$.

▶ **Fc. Exercise: Probability of gambler's success for $p \neq \frac{1}{2}$.** Consider the same problem where $0 < p < 1$ and $p \neq \frac{1}{2}$. Find $\mathbb{P}_a(W_T = b)$ by using the martingale $\mathbf{V}$ at Exercise 408Bc.

▶ **Fd. Average duration of gambler's game when $p = \frac{1}{2}$.** Suppose that $p = \frac{1}{2}$, and that T continues to be defined by 411(E1). We know from Exercise 409Cc that $A_n := W_n^2 - n$ defines a martingale. But W_n^2 is bounded by b^2 for $n \leq T$ and the process $\mathbf{I}$ with $I_n(\omega) := n$ is increasing. Hence we can apply STP to conclude that

$$\mathbb{E}(W_T^2 - T) = \mathbb{E}(W_0^2 - 0) = a^2.$$

But $\mathbb{E}(W_T^2) = b^2 \mathbb{P}(W_T = b) = b^2 \times (a/b) = ab$, and hence $\mathbb{E}(T) = a(b - a)$.

▶ **Fe. Exercise: Average duration of gambler's game when $p \neq \frac{1}{2}$.** Solve the same problem as in the previous exercise for $p \neq \frac{1}{2}$. You have already met the right martingales.

▶ **Ff. Exercise: Distribution of duration of gambler's game.** Let $0 < p < 1$. For the game where the initial fortune is a, we have

$$W_n = a + X_1 + X_2 + \cdots + X_n,$$

as usual. Prove that for $0 < \theta < 1$, then for some real number α we have $\mathbb{E}(\alpha^X) = 1/\theta$, and then

$$\alpha^b \mathbb{E}\{\theta^T ; W_T = b\} + \alpha^0 \mathbb{E}\{\theta^T ; W_T = 0\} = \alpha^a.$$

Find $\mathbb{E}(\theta^T)$, which, in principle gives the distribution of T. This helps illustrate the point that there *are* often enough martingales to solve problems in which we are interested.

▶ **Fg. Waiting times for patterns – an example.** This very nice way of solving the 'Waiting for Patterns' problem is due to S.Y.R. Li. Let me describe the intuitive idea.

> "How long on average do you have to wait for HH if you keep tossing a fair coin?
>
> Consider the situation where a fair coin tossed in a fair casino falls
>
> $$\text{T} \quad \text{H} \quad \text{T} \quad \text{H} \quad \text{H.}$$
>
> Suppose that just before each toss, a new gambler with a total fortune of £1 arrives, and bets his £1 that the first toss he sees will produce Heads; if he wins, he bets all he then has, £2, that the next toss will also produce Heads. When two Heads in succession occur, the casino closes down.

Our 1st gambler leaves with nothing. So do the 2nd and 3rd gamblers. However, our 4th gambler leaves with £4, and the 5th with £2. Whatever way we get to HH, the total fortune of the gamblers at the end is £6. Because everything is fair, the average fortune brought in by the gamblers is £6, but this is exactly the average number of tosses to get HH. By the same argument, the average number of tosses to get a Head followed by a Tail is 4."

Fh. Exercise on waiting times for patterns. Give some thought as to how the argument just given may be justified. (Consider the total amount won by the gamblers up to time n minus the total amount brought in by the gamblers up to time n.) Then solve the following problem by the same method. Every second a monkey types a capital letter chosen from the 26 letters of the alphabet. How long on average is it before he types the pattern ABRACADABRA?

▶ **Fi. Exercise: A card game.** A pack consisting of b Black and r Red cards is shuffled and placed face down on the table. The cards are then turned over one by one. Just before some card is turned over, you must say 'The next card is Black'. If you are right, you win; otherwise, you lose. Show that whatever strategy you adopt, your probability of winning is $b/(b+r)$.

▶ **Fj. Exercise: a martingale proof of the Ballot Theorem.** Prove the Ballot Theorem 126L by looking at the previous exercise – perhaps in a suitable 'mirror'.

▶ **Fk. Why can you sometimes cheat the system?** Again, let $\mathbf{W}$ be SRW$(\frac{1}{2})$ started at 0, and let $T := \inf\{n : W_n = 1\}$. We know that $\mathbb{P}(T = \infty) = 0$ and that, in this case, $\mathbb{E} W_T = 1 \neq 0 = \mathbb{E} W_0$. However, let D be the gambler's minimum fortune before time T:

$$D := \inf\{W_n : n \leq T\}.$$

We know from 413Fb that for $k \in \mathbb{N}$, we have

$$\mathbb{P}(D \leq -k) = \frac{1}{k+1},$$

and the trouble derives in part from the fact that $\mathbb{E}(|D|) = \infty$.

If $T_k := \inf\{n : W_n = 1 \text{ or } W_n = -k\}$, then $T_k \to T$ (a.s.) as $k \to \infty$ and $W_{T_k} \to W_T$. However, $0 = \mathbb{E} W_{T_k} \not\to \mathbb{E} W_T = 1$. Here, W_{T_k} is equal to $-k$ with probability $1/(k+1)$ resulting in a contribution of $-k/(k+1) \approx -1$ to $\mathbb{E} W_{T_k}$ from a set of very small probability. What is wrong here is what is called in Measure Theory 'lack of uniform integrability'. The conditions imposed within the STP preclude this kind of bad behaviour.

▶ **Fl. Proof of the Stopping-Time Principle.** Suppose first that for some positive integer

r, we have, for all ω, $T(\omega) \leq r$. Then

$$\mathbb{E}(M_T) = \sum_{n=0}^{r} \mathbb{E}(M_T; T = n) \qquad\qquad \text{(Linearity)}$$

$$= \sum_{n=0}^{r} \mathbb{E}(M_n; T = n) \qquad\qquad \text{(Logic)}$$

$$= \sum \mathbb{E}(M_r; T = n) \qquad \text{(by 407(A2), since } \{T = n\} \in \mathcal{F}_n)$$

$$= \mathbb{E}(M_r) \qquad\qquad\qquad\qquad \text{(Linearity)}$$

$$= \mathbb{E}(M_0) \qquad\qquad\qquad\qquad \text{(by Exercise 407Aa).}$$

Now, let us suppose that T is any stopping time with $\mathbb{P}(T = \infty) = 0$. Then, for each r in $\mathbb{N}$, $T \wedge r := \min(T, r)$ is a stopping time (*you* check this) which is at most r, so, by what we have just proved,

$$\mathbb{E} M_{T \wedge r} = \mathbb{E} M_0.$$

To finish the proof, we *have* to use Measure Theory. Let's stick to the '+' case of the 'either/or' in the theorem. Then,

$$M_{T \wedge r} = B_{T \wedge r} + I_{T \wedge r}, \qquad \mathbb{E} M_{T \wedge r} = \mathbb{E} B_{T \wedge r} + \mathbb{E} I_{T \wedge r}.$$

As $r \to \infty$, $B_{T \wedge r} \to B_T$ (a.s.) and, since $|B_{T \wedge r}| \leq K$, the Bounded-Convergence Theorem 65L shows that we have $\mathbb{E} B_{T \wedge r} \to \mathbb{E} B_T$. Next by the Monotone-Convergence Theorem 60D, $\mathbb{E} I_{T \wedge r} \to \mathbb{E} I_T$. Finally,

$$\mathbb{E} M_0 = \mathbb{E} M_{T \wedge r} \to \mathbb{E} B_T + \mathbb{E} I_T = \mathbb{E} M_T,$$

and the proof is complete. (There is no possibility that $\mathbb{E} I_T$ can be infinite: it must be $\mathbb{E} M_0 - \mathbb{E} B_T$.) The '−' case is exactly similar of course; or we can use the fact that $-\mathbf{M}$ is a martingale!. $\qquad\square$

G. Optional Sampling. Here is an important extension of the Stopping-Time Principle.

▶▶▶ **Ga. Theorem: Doob's Optional-Sampling Theorem (OST).** *Make the assumptions of Doob's Stopping-Time Principle, and let S be a stopping time with $S \leq T$. Let $\mathcal{F}_S$ be the σ-algebra describing the evolution of $\mathbf{X}$ up to time S, as explained in Subsection 129B. Then (M_S is $\mathcal{F}_S$-measurable and)*

$$\mathbb{E}(M_T \mid \mathcal{F}_S) = M_S, \quad (a.s.).$$

This is a nice extension of the martingale property to stopping times, and it plays a huge part in more advanced theory. The proof really mirrors exactly that of the STP, the first step being to prove that if T satisfies $T \leq r$ for some positive integer r, then

$\mathbb{E}(M_r \mid \mathcal{F}_T) = M_T$, a.s.. The desired result when $T \leq r$ is then a consequence of the Tower Property; and one uses the Bounded-Convergence and Monotone-Convergence Theorems as before.

►► **H. Wald's Sequential Hypothesis Test.**   This is a key topic for Quality Control and for Clinical Trials.

► **Ha. Lemma.** *If M is a nonnegative martingale and T is a stopping time such that* $\mathbb{P}(T = \infty) = 0$, *then*

$$\mathbb{E}\,M_T \leq \mathbb{E}\,M_0.$$

In the light of the proof of the Stopping-Time Principle, this follows immediately because 'Fatou's Lemma' in Measure Theory implies that

$$\mathbb{E}\,M_T = \mathbb{E}\left(\lim_{r \to \infty} M_{T \wedge r}\right) \leq \liminf \mathbb{E}\,M_{T \wedge r} \leq \mathbb{E}\,M_0.$$

Hb. The testing problem.   Suppose that we are able to make observations of IID RVs $Y_1, Y_2, \ldots$ and have two hypotheses:

$$H_0 : \text{the common pdf/pmf of } Y_1, Y_2, \ldots \text{ is } f,$$
$$H_A : \text{the common pdf/pmf of } Y_1, Y_2, \ldots \text{ is } g,$$

with f and g being strictly positive with the same domain. Suppose that we have to choose whether to accept H_0 or to accept H_A. (I know that we do not usually talk of accepting hypotheses, but stick with it.) We want Type I and Type II errors controlled by

$$\mathbb{P}(\text{reject } H_0 \mid H_0) \leq \alpha, \qquad \mathbb{P}(\text{reject } H_A \mid H_A) \leq \beta.$$

We wish to make our decision utilizing on average the smallest sample size.

► **Hc. Quality-Control Example.** We might have the situation where on a production line

$$Y_k = \begin{cases} 1 & \text{if the } k\text{th item produced is defective,} \\ 0 & \text{if the } k\text{th item produced is OK,} \end{cases}$$

where $Y_1, Y_2, \ldots$ are IID with

$$\mathbb{P}(Y_k = 1) = p = 1 - \mathbb{P}(Y_k = 0),$$

where $0 < p < 1$. We wish to test $H_0 : p = p_0$ against $H_A : p = p_1$ (or, to be more sensible, $H_0 : p \leq p_0$ against $H_A : p \geq p_1$, H_0 corresponding to an acceptable situation, H_A to a situation where something has gone wrong).

►► **Hd. Wald's Strategy.** Actually, I present a slightly different (and, I think, neater) strategy than the usual one.

Let R be the Likelihood-Ratio martingale under H_0 of Exercise 409Bd. Consider the strategy:

$n = 0$;

Repeat

    $n = n + 1$; *Observe* Y_n;

    Calculate R_n;

Until $R_n \geq \alpha^{-1}$ *or* $R_n \leq \beta$;

If $R_n \geq \alpha^{-1}$ *accept* H_A;

Otherwise accept H_0.

Of course,

$$T := \inf\{n : R_n \geq \alpha^{-1} \text{ or } R_n \leq \beta\}$$

is a stopping time relative to $\mathbf{Y}$. With $\mathbb{P}_0$ relating to H_0, we have $P_0(T = \infty) = 0$. We know this from the consistency result 193(C1), though inevitably this result has a martingale proof due to Doob (for which see [W]).

By 416Ha, we have

$$1 = \mathbb{E}_0 R_0 \geq \mathbb{E}_0 R_T \geq \alpha^{-1} \mathbb{P}_0 \left(R_T \geq \alpha^{-1} \right) = \alpha^{-1} \mathbb{P}(\text{reject } H_0 \mid H_0),$$

so that $\mathbb{P}(\text{reject } H_0 \mid H_0) \leq \alpha$, as required. By interchanging the rôles of H_0 and H_A, we find that $\mathbb{P}(\text{reject } H_A \mid H_A) \leq \beta$.

The whole point is that *on average, fewer observations are needed for this sequential test than for a fixed-sample-size test with the same error bounds.*

He. Example. Consider the case when $Y_k \sim N(\theta, 1)$ with $H_0 : \theta = 0$ and $H_A : \theta = \mu = 0.4$. We want $\alpha \leq 0.05$ and $\beta \leq 0.10$. Now, under H_0,

$$\ln R_n = \sum \mu(Y_k - \tfrac{1}{2}\mu) \sim N(-\tfrac{1}{2}\mu^2 n, \mu^2 n).$$

Very crudely, the expected time to get $R_n \leq \beta$ is therefore $(\ln \beta)/(-\tfrac{1}{2}\mu^2) \approx 29$, so $\mathbb{E}_0 T$ should be at most a value near this. By contrast, the fixed-sample-size test at Subsection 229Ce required a sample size of 54.

You can see that, as common sense suggests, the Average Sample Number for the sequential test will be greatest when the *true* mean θ is $\tfrac{1}{2}\mu$. Then there is no drift in $\ln R$, and we have to rely entirely on random fluctuations to bring the test to a conclusion.

Suppose now that the true value of θ is $\tfrac{1}{2}\mu$. Then it is easy to see from the CRP that $\mathbb{P}(T = \infty) = 0$ – you can take $k = 1$. Now, $\ln R_n$ is a martingale, so that roughly (ignoring 'overshoot')

$$0 = \mathbb{E} \ln R_0 = \ln(1/\alpha)\mathbb{P}(\text{accept } H_A) + \ln(\beta)\mathbb{P}(\text{accept } H_0),$$

whence

$$\mathbb{P}(\text{accept } H_A) = \frac{b}{a + b}, \quad \text{where } a = \ln(1/\alpha), b = \ln(1/\beta),$$

in analogy with the discussion at 413Fb. But, by Exercise 409Cc,

$$(\ln R_n)^2 - n\mu^2$$

is a martingale, whence we might expect that

$$\mu^2 \mathbb{E}(T) \approx \frac{b}{a+b}a^2 + \frac{a}{a+b}b^2 = ab,$$

in analogy with the discussion at 413Fd. This suggests that in the worst case, the ASN is about $ab\mu^{-2} = 43$. □

That's all I'm going to say about sequential testing. For a serious study of its most important application, see Jennison and Turnbull [119].

▶ **I. Doob's Martingale–Convergence Theorem.** This theorem is one of the most important results in Mathematics, not just in Probability. For Doob's marvellous 'upcrossing' proof, see [W].

▶▶▶ **Ia. Fact: Martingale-Convergence Theorem (MCT).** *Let* M *be a martingale relative to* X. *Suppose either that* M *is bounded in* $\mathcal{L}^1$ *(for some finite K, we have* $\mathbb{E}(|M_n|) \leq K$ *for all n) or that each* M_n *is nonnegative. Then*

$$M_\infty := \lim M_n \quad \text{exists almost surely.}$$

It is important that the convergence will not necessarily be in $\mathcal{L}^1$: *it will be in* $\mathcal{L}^1$ *if and only if* $M_n = \mathbb{E}(M_\infty \mid \mathcal{F}_n)$ *(a.s.) for every n. However, if* M *is bounded in* $\mathcal{L}^2$ *(for some finite K, we have* $\mathbb{E}(|M_n|^2) \leq K$ *for all n), then* $M_n \to M_\infty$ *in* $\mathcal{L}^2$ *(whence also in* $\mathcal{L}^1$).

Ib. Application to Pólya's urn. We see that the proportion of Black balls in Pólya's urn will almost surely converge to a limit Θ, something we also know from the isomorphism explained in Subsection 79H.

▶ **Ic. Application to branching processes.** If Z_n is the population size at time n for a Galton–Watson branching process for which $\mu := \mathbb{E}(X) < \infty$, then

$$\mathbb{E}(Z_{n+1} \mid Z_1, Z_2, \ldots, Z_n) = \mathbb{E}(Z_{n+1} \mid Z_n) = \mu Z_n,$$

so that $M_n := Z_n/\mu^n$ defines a nonnegative Z-martingale. We know that if $\mu \leq 1$, then (assuming as always that $\mathbb{P}(X = 1) \neq 1$), the population will eventually die out, so that $M_\infty = 0$ and we do not get $\mathcal{L}^1$ convergence. Kesten proved that if $\mu > 1$, then we get $\mathcal{L}^1$ convergence if and only if $\mathbb{E}[X \ln(1 + X)] < \infty$.

See [W] for a little more on this branching-process martingale and lots more on the MCT.

▶ **J. Lévy's Theorem for 'Reverse' Martingales.** At first sight, this seems a rather crazy idea. However, it leads to a very nice proof both of de Finetti's Theorem (which, you may remember, many Bayesians see as one of the main justifications of their philosophy) and of the cornerstone of the whole subject, the Strong Law.

Let $\mathbf{S} = (S_1, S_2, ...)$ be a process. A reverse martingale relative to $\mathbf{S}$ is a process $\mathbf{R}$ such that $R_1 \in \mathcal{L}^1$ and

$$R_m = \mathbb{E}(R_1 \mid S_m, S_{m+1}, \ldots) \quad \text{a.s..}$$

[*Clarification.* If we consider for $n \geq m$, the RV

$$R_{m,n} := \mathbb{E}(R_1 \mid S_m, S_{m+1}, \ldots S_{m+n}) \quad \text{a.s.,}$$

then by result described at 407Ab, $R_{m,n}$ will be a standard martingale as n varies, and in fact

$$R_m = \lim_n R_{m,n} \quad \text{(a.s.) and in } \mathcal{L}^1.$$

Accept this.]

▶▶ **Ja. Fact (Lévy's Theorem).** *If* $\mathbf{R}$ *is a reverse martingale relative to* $\mathbf{S}$, *then*

$$R_\infty := \lim R_m \quad \text{exists almost surely and in } \mathcal{L}^1$$

and $\mathbb{E}(R_1 \mid R_\infty) = R_\infty$, *almost surely.*

For proof, see [W].

▶ **Jb. Martingale proof of de Finetti's Theorem.** Please re-read Subsection 220P.

Proof of Theorem 221Pa. Let

$$S_n := Y_1 + Y_2 + \cdots + Y_n.$$

Define $R_1 := Y_1$ and (for $n \geq m$)

$$R_m := \mathbb{E}(R_1 \mid S_m, S_{m+1}, \ldots), \qquad R_{m,n} := \mathbb{E}(R_1 \mid S_m, S_{m+1}, \ldots S_{m+n}).$$

Now, by logic,

$$R_{m,n} = \mathbb{E}(R_1 \mid S_m, X_{m+1}, \ldots X_{m+n}).$$

But the n-tuple $(X_{m+1}, \ldots X_{m+n})$ is independent of the pair (R_1, S_m), so that by an obvious extension of Result 402Nb,

$$R_{m,n} = \mathbb{E}(R_1 \mid S_m) = \mathbb{E}(Y_1 \mid S_m).$$

Let τ be a permutation of $\mathbb{N}$ which leaves each number greater than m fixed. This permutation will preserve the probabilistic structure of the whole process, and will leave S_m exactly the same (for each ω). Hence, we see that

$$\mathbb{E}(Y_1 \mid S_m) = \mathbb{E}(Y_2 \mid S_m) = \cdots = \mathbb{E}(Y_m \mid S_m)$$
$$= \text{their average} = \mathbb{E}\left(\frac{S_m}{m} \,\Big|\, S_m\right) = \frac{S_m}{m}.$$

We now see that $R_m = S_m/m$ (a.s.), and it follows from Lévy's Theorem that $\Theta :=$ $\lim S_m/m$ exists almost surely.

We now use the fact that

$$\sum Y_{i_1} Y_{i_2} \ldots Y_{i_r} = S_m(S_m - 1) \ldots (S_m - r + 1), \qquad (\text{J1})$$

the sum being over all $m(m-1)\ldots(m-r+1)$ r-tuples $i_1, i_2, \ldots, i_r$ of *distinct* numbers chosen within $\{1, 2, \ldots, m\}$. Assume this for the moment. By the obvious extension of the 'invariance under permutation' argument, we see that for distinct $i_1, i_2, \ldots, i_r$ within $\{1, 2, \ldots, m\}$,

$$\mathbb{E}(Y_{i_1} Y_{i_2} \ldots Y_{i_r} \mid S_m, S_{m+1}, \ldots)$$
$$= \frac{S_m(S_m - 1) \ldots (S_m - r + 1)}{m(m - 1) \ldots (m - r + 1)}.$$

But, as $m \to \infty$, this tends (a.s.) to Θ^r, just because S_m/m tends to Θ. Hence for distinct $i_1, i_2, \ldots, i_r$,

$$\mathbb{E}(Y_{i_1} Y_{i_2} \ldots Y_{i_r} \mid \Theta^r) = \Theta^r \quad \text{a.s.},$$

that is,

$$\mathbb{P}(Y_{i_1} = 1; Y_{i_2} = 1; \ldots; Y_{i_r} = 1 \mid \Theta) = \Theta^r \quad (\text{a.s.}),$$

and that is the desired result: given Θ, the variables Y_i are IID each Bernoulli(Θ). $\square$

Jc. Exercise. Prove the identity (J1) using the fact that $Y_i^2 = Y_i$. This is very simple; don't let the complicated appearance of the result put you off. Try it for $r = 2$ first, using the fact that with i, j restricted to $\{1, 2, \ldots, m\}$,

$$\sum_{j \neq i} Y_j = S_m - Y_i.$$

▶▶ **Jd. Martingale proof of the Strong Law.** Let $X_1, X_2, \ldots$ be IID Random Variables in $\mathcal{L}^1$ with $\mu = \mathbb{E}(X_k)$. You can see that the argument used for de Finetti's Theorem shows that

$$L := \lim S_n/n \text{ exists almost surely, where } S_n := X_1 + X_2 + \cdots + X_n.$$

What we need to do is to prove that $L = \mu$ almost surely. The $\mathcal{L}^1$ convergence guarantees that $\mathbb{E}L = \mu$.

The crucial point is that for each m, L (a.s.) depends only on the values of $X_m, X_{m+1}, \ldots$; indeed,

$$L = \lim \frac{X_{m+1} + \cdots + X_{m+n}}{n} \quad (\text{a.s.}),$$

just by Analysis. Hence L is independent of $m^{-1}S_m$, so that, for $\alpha, \beta \in \mathbb{R}$,

$$\mathbb{E} \exp(i\alpha L + i\beta m^{-1}S_m) = \mathbb{E} \exp(i\alpha L) \, \mathbb{E} \exp(i\beta m^{-1}S_m).$$

Let $m \to \infty$ and use the Bounded-Convergence Theorem to see that the Characteristic Function $\varphi_L := \mathbb{E}\,\mathrm{e}^{\mathrm{i}\alpha L}$ of L satisfies

$$\varphi(\alpha + \beta) = \varphi(\alpha)\varphi(\beta),$$

so that $\varphi(\alpha) = \mathrm{e}^{K\alpha}$ for some K. However, $\varphi(-\alpha) = \mathbb{E}\,\mathrm{e}^{-\mathrm{i}\alpha L}$ must be the complex conjugate of $\varphi(\alpha)$, so that K is pure imaginary: $K = \mathrm{i}\gamma$ for some real γ. But then $\mathbb{E}\,\mathrm{e}^{\mathrm{i}\alpha L} = \mathrm{e}^{\mathrm{i}\alpha\gamma}$, and so L has the same distribution as the deterministic variable equal to γ on the whole of Ω. Clearly, $\gamma = \mu$, and the proof of the Strong Law is complete. $\qquad\square$

▶▶ **K. The Black–Scholes option-pricing formula.** Option pricing, the Black–Scholes formula, etc, are all based on martingales and stochastic calculus; and at one stage, even being able to say the word 'martingale' tended to get you a highly paid job in finance. Nowadays, everyone in finance knows something about martingales, so you would need to, too. Because you might be interested, I try hard here to improve on the account of the Mathematics in [W], though I also express my unease about the topic more explicitly than I did there.

I do think that the Black–Scholes formula is nice if one accepts the 'hedging strategy' philosophy, and *am* well aware that finance has posed some problems on martingales of real intrinsic interest. But I must also confess that I cannot summon up too great an enthusiasm for anything which has to do with money, and that in many ways I regret the prodigious talent that has gone into mathematical finance (when people could have been trying much harder problems in Statistical Mechanics, for example). The site (by Uwe Wystup)

$$\mathtt{http://www.mathfinance.de}$$

is informative about the literature on finance. (Of course, Bingham and Kiesel [23], 'already' in the bibliography, is very relevant. I also recommend Hunt and Kennedy [114] and Karatzas and Shreve [124].)

Idealized situation. We assume that there is only one type of stock you can own, and one type of bond. The value B_n of one unit of bond at time n is deterministic: it is $B_n = B_0(1 + r)^n$, with fixed *known* interest rate r. The value of one unit of stock jumps at time n to a new value S_n according to 'random' instantaneous interest R_n. We have

$$S_n = (1 + R_n)S_{n-1}, \quad B_n = (1 + r)B_{n-1}.$$

We assume for our idealized world that it is known that R_n can only ever take one of the two *known* values a and b where $a < r < b$:

$$R_n \in \{a, b\}, \quad -1 < a < r < b.$$

The value S_0 is a known constant.

European(K, N) **option.** In this option, you pay an amount x at time 0 to have the option to buy at time N one unit of stock for an amount (strike price) K. If your $S_N > K$, you will exercise that right at time N, gaining value $S_N - K$; otherwise, you will not exercise that option. Thus, your gain at time N will be $(S_N - K)^+$, the positive part of $S_N - K$.

For now, we regard N and K as given constants.

The problem: *What is the fair price you should pay at time* 0 *for that option to buy one unit of stock at cost K at time N?*

Perhaps it might seem that the answer should be $\mathbb{E}\left\{(S_N - K)^+\right\}$, the mean value of your gain at time N. However, the Black–Scholes formulation does not require any probabilistic structure to be imposed on the sequence $R_1, R_2, \ldots$.

Previsible portfolio strategy.   Let me help by saying straight away that time 0 is rather anomalous in the formulation I now give: there is no real distinction between A_1 and A_0. Or to put it another way, A_0 and V_0 play no rôle in what follows except 'to fill out a pattern'. Suppose that at time 0 you own A_0 units of stock and V_0 units of bond, so your fortune is $X_0 = A_0 S_0 + V_0 B_0$. Before time 1, you rearrange your portfolio reinvesting the amount X_0 as A_1 units of stock and V_1 units of bond. So, $X_0 = A_1 S_0 + V_1 B_0$. Thus (A_1, V_1) is your 'stake' on the 'game' at time 1 after which your fortune jumps to $X_1 = A_1 S_1 + V_1 B_1$. Generally, for $n \geq 1$,

- your fortune just before time n is $X_{n-1} = A_n S_{n-1} + V_n B_{n-1}$,

- your fortune just after time n is $X_n = A_n S_n + V_n B_n$, which you then reinvest as

$$X_n = A_{n+1} S_n + V_{n+1} B_n \quad (= A_n S_n + V_n B_n).$$

Note that we must be able to calculate A_n from the values $R_1, R_2, \ldots, R_{n-1}$ (equivalently from $S_1, S_2, \ldots, S_{n-1}$): we do not know the value of R_n at the time when we have to decide on A_n. The process **A** is called **previsible** relative to the process **R**: 'it is known one time unit before it looks as if it should be'. This *is* the right way to formulate things: previsible processes are the things we integrate in stochastic calculus.

Hedging strategy for our European option.   By a hedging strategy for our European(K, N) option, with initial value x, we mean a previsible portfolio strategy such that $X_0 = x$ and that, *for every possible outcome, every sequence of values that* $R_1, R_2, \ldots$ *can take*, we have

$$X_N = (S_N - K)^+.$$

Thus the strategy exactly matches the option price at time N.

The Black–Scholes concept of a Fair Price. We say that x is a Black–Scholes fair price at time 0 for the option if there exists a hedging strategy for the option, with initial value x.

▶▶ **THEOREM (Black–Scholes).** *There exists a unique Black–Scholes fair price x for our European(K, N) option, and a unique associated hedging strategy. Moreover,*

$$x = (1 + r)^{-N} \mathbb{E}_{BS} \left\{ (S_N - K)^+ \right\} \tag{K1}$$

where $\mathbb{E}_{BS}$ is expectation associated with the probability measure $\mathbb{P}_{BS}$ which makes $R_1, R_2, \ldots$ independently identically distributed, each with the unique distribution on $\{a, b\}$ with mean r, so that

$$\mathbb{P}_{BS}(R_k = b) = p := \frac{r - a}{b - a}, \quad \mathbb{P}_{BS}(R_k = a) = q := \frac{b - r}{b - a}.$$

I emphasize that nothing is being assumed in the theorem about the true 'real-world' probabilistic structure (in which the values of $R_1, R_2, \ldots$ might well be highly correlated). The law $\mathbb{P}_{BS}$ is a contrived law, created merely to allow neat expression of some elementary algebra.

Remark. I have heard several statisticians express some unease about the formula because it does not involve *real* probabilities, unease which I share. Of course, we are aware that *the people who offer the option used their own beliefs about the real probabilities in deciding on the appropriate value for K, and we would use our beliefs about those probabilities in deciding whether or not actually to buy the option at the Black–Scholes fair price.* But still ... .

The case when $N = 1$. Let's consider the case when $N = 1$ without utilizing any probabilistic structure. First assume that a fair price and associated hedging strategy exist. Let us prove that they are unique.

We must have, for every possible outcome,

$$A_1 S_0 + V_1 B_0 = x, \quad A_1 S_1 + V_1 B_1 = (S_1 - K)^+.$$

Since $S_1 = (1 + R_1) S_0$ and $B_1 = (1 + r) B_0$, we have

$$(1 + r)x + A_1 S_0 (R_1 - r) = (S_0 + R_1 S_0 - K)^+.$$

Since $R_1 = b$ or $R_1 = a$,

$$(1+r)x + A_1 S_0(b-r) = (S_0 + bS_0 - K)^+,$$
$$(1+r)x + A_1 S_0(a-r) = (S_0 + aS_0 - K)^+.$$

whence,

$$(b-a)A_1 S_0 = (S_0 + bS_0 - K)^+ - (S_0 + aS_0 - K)^+.$$

If $(1+a)S_0 > K$, then

$$A_1 = 1, \quad (1+r)x = S_0(1+r) - K, \quad V_1 = -\frac{K}{(1+r)B_0}.$$

If $(1+b)S_0 > K > (1+a)S_0$, then

$$(b-a)A_1 S_0 = S_0(1+b) - K,$$
$$(1+r)x = \frac{r-a}{b-a}\{S_0(1+b) - K\},$$
$$V_1 = -\frac{a+1}{r+1} \cdot \frac{S_0(1+b) - K}{b-a}.$$

If $K > (1+b)S_0$, then $A_1 = V_1 = x = 0$.

Thus in each case there is at most one Black–Scholes fair price x at time 0. Conversely, since we now know what the hedging strategy must be, we can confirm that x is a Black–Scholes fair price.

Ka. Exercise. Prove that the values of x just found agree with 423(K1) with $N = 1$.

That V_1 is generally negative is another disturbing feature of the Black–Scholes philosophy. It means that in the hedging strategy, we have to borrow an amount $V_1 B_0$ of money against the bond rate of interest in order to buy our stock. (Would we be able to do this? I doubt it very much.)

But, on with the Mathematics.

Proof of uniqueness of x for general N.   Suppose that a hedging strategy $(A_1, V_1, A_2, V_2, \ldots)$ exists, with initial value x, and that we adopt it. Recall the situation

$$X_n = A_n S_n + V_n B_n, \quad X_{n-1} = A_n S_{n-1} + V_n B_{n-1},$$
$$S_n = S_{n-1} + S_{n-1} R_n, \quad B_n = B_{n-1} + r B_{n-1}.$$

Hence,

$$X_n - X_{n-1} = A_n S_{n-1} R_n + V_n B_{n-1} r$$
$$= r X_{n-1} + A_n S_{n-1}(R_n - r).$$

If we set

$$Y_n = (1+r)^{-n} X_n,$$

the discounted value of our fortune at time n, then

$$Y_n - Y_{n-1} = (1+r)^{-n} A_n S_{n-1} (R_n - r).$$

Now, as we have observed previously, A_n is a function of $R_1, R_2, \ldots, R_{n-1}$, as is S_{n-1}. Hence, for our contrived measure $\mathbb{P}_{\text{BS}}$,

$$
\begin{aligned}
\mathbb{E}_{\text{BS}} & (Y_n - Y_{n-1} \mid R_1, R_2, \ldots, R_{n-1}) \\
&= (1+r)^{-n} A_n S_{n-1} \mathbb{E}_{\text{BS}} (R_n - r \mid R_1, R_2, \ldots, R_{n-1}) \\
&= (1+r)^{-n} A_n S_{n-1} \mathbb{E}_{\text{BS}} (R_n - r) \quad \text{(by independence)} \\
&= 0 \quad \text{(since } \mathbb{E}_{\text{BS}} R_n = r\text{)}.
\end{aligned}
$$

Thus, $\mathbf{Y}$ is a martingale relative to $\mathbf{R}$. But, by the definition of a hedging strategy, $X_N = (S_N - K)^+$, so, by the martingale property,

$$x = X_0 = \mathbb{E}_{\text{BS}}(Y_N) = (1+r)^{-N} \mathbb{E}_{\text{BS}} \left\{ (S_N - K)^+ \right\}.$$

Thus, there is at most one Black–Scholes fair price, and if a fair price does exist, it is given by formula 423(K1). $\qquad\square$

Proof of existence of a hedging strategy with initial value x as given by 423(K1). The obvious thing is to start afresh and to *define*

$$Y_n := \mathbb{E}_{\text{BS}} \left\{ (1+r)^{-N} (S_N - K)^+ \mid R_1, R_2, \ldots, R_n \right\}. \tag{K2}$$

Then, by the result at 407Ab, $\mathbf{Y}$ is a martingale (on time-parameter set $\{0, 1, 2, \ldots, N\}$) relative to $\mathbf{R}$. For some function f_n on $\{a, b\}^n$, we have

$$Y_n = f_n(R_1, R_2, \ldots, R_n).$$

But now, the martingale property

$$\mathbb{E}_{\text{BS}} (Y_n \mid R_1, R_2, \ldots, R_{n-1}) = Y_{n-1}$$

yields

$$
\begin{aligned}
&= p f_n(R_1, R_2, \ldots, R_{n-1}, b) + q f_n(R_1, R_2, \ldots, R_{n-1}, a) \\
&= Y_{n-1} = f_{n-1}(R_1, R_2, \ldots, R_{n-1}). \tag{K3}
\end{aligned}
$$

Hence,

$$
\begin{aligned}
&\frac{f_n(R_1, R_2, \ldots, R_{n-1}, b) - f_{n-1}(R_1, R_2, \ldots, R_{n-1})}{q} \\
&= \frac{f_{n-1}(R_1, R_2, \ldots, R_{n-1}) - f_n(R_1, R_2, \ldots, R_{n-1}, a)}{p} \\
&= H_n \text{ (say)}.
\end{aligned}
$$

Note that H_n does not involve R_n, so **H** is clearly previsible relative to **R**. We have

$$
\begin{aligned}
Y_n &- Y_{n-1} \\
&= f_n(R_1, R_2, \ldots, R_{n-1}, R_n) - f_{n-1}(R_1, R_2, \ldots, R_{n-1}) \\
&= I_b(R_n)\{f_n(R_1, R_2, \ldots, R_{n-1}, b) - f_{n-1}(R_1, R_2, \ldots, R_{n-1})\} \\
&+ I_a(R_n)\{f_n(R_1, R_2, \ldots, R_{n-1}, a) - f_{n-1}(R_1, R_2, \ldots, R_{n-1})\} \\
&= \{qI_b(R_n) - pI_a(R_n)\} H_n.
\end{aligned}
$$

Now,

$$
\begin{aligned}
R_n - r &= I_b(R_n)(b - r) + I_a(R_n)(a - r) \\
&= (b - a)\{qI_b(R_n) - pI_a(R_n)\},
\end{aligned}
$$

and so, finally,

$$
Y_n - Y_{n-1} = (b - a)^{-1} H_n (R_n - r),
$$

where **H** is previsible.

So, we can define a previsible process **A** to satisfy

$$
(1 + r)^{-n} A_n S_{n-1} = (b - a)^{-1} H_n.
$$

Then, of course, we define

$$
X_n := (1 + r)^n Y_n, \quad V_n := (X_n - A_n S_{n-1})/B_{n-1},
$$

and we have all the elements of the hedging strategy, with initial value Y_0 given as x in the Black–Scholes formula 423(K1). The Black–Scholes Theorem is proved. □

That each H_n is positive, whence each A_n is positive, is clear from 425(K3) and the *obvious* fact that

$$
f_n(R_1, R_2, \ldots, R_{n-1}, b) > f_n(R_1, R_2, \ldots, R_{n-1}, a).
$$

The fact that each A_n is positive means that there is no 'short-selling' of stock.

Important Remarks. In the discrete-time case which we have been considering, the introduction of the probability measure $\mathbb{P}_{BS}$ was merely a device to allow neat expression of some elementary algebra. However, *in the continuous-time version*, where the algebra is no longer meaningful, *martingale theory provides the essential language for doing the appropriate analysis.*

9.3   Poisson Processes (PPs)

The Poisson Process is used to model (under conditions on which you should decide later) *arrivals*

- of customers at a queue,

- of calls to a telephone exchange,

- of particles at a Geiger counter,

etc, etc. For now, we work with the time-parameter set $[0, \infty)$, our arrivals falling in this half-line. We let $N(A)$ denote the number of arrivals in a time interval A, and write, for example, $N(t, t + u]$ for $N(A)$ when $A = (t, t + u]$.

▶▶ **A. Two descriptions of a PP(λ).**   Let me begin by giving the two most

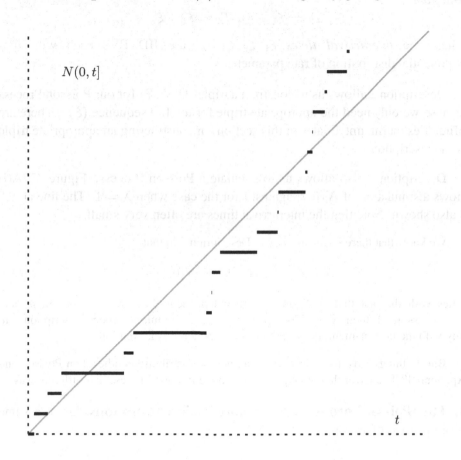

Figure A(i): Simulation of Poisson $N(0, t]$ when $\lambda = 1$

direct ways of describing a Poisson Process, the first as a case of the more general concept of a Poisson Point Process (PPP).

Description 1. A Poisson Process of rate λ, denoted by PP(λ), is one whose arrival times form a PPP of constant intensity λ on $[0, \infty)$; which means that

- the numbers of arrivals $N(A_1), N(A_2), \ldots, N(A_n)$ in a collection $A_1, A_2, \ldots, A_n$ of *disjoint* time-intervals are *independent* RVs;

- the number arriving in any interval of length t has the Poisson(λt) distribution.

Description 2. A PP(λ) can also be fully described by the fact that the arrival times are
$$T_1 := \xi_1, \; T_2 := \xi_1 + \xi_2, \; T_3 := \xi_1 + \xi_2 + \xi_3, \; \ldots,$$
where the *interarrival times* $\xi_1, \xi_2, \xi_3, \ldots$ are IID RVs each with the exponential distribution of rate parameter λ.

Description 2 allows us to construct a triple $(\Omega, \mathcal{F}, \mathbb{P})$ for our Poisson Process because we only need the appropriate triple for the IID sequence (ξ_n). There are difficulties (surmounted later in this section) in constructing an appropriate triple from Description 1.

Description 2 also allows us to simulate a Poisson Process. Figure 427A(i) shows a simulation of $N(0, t]$ against t for the case when $\lambda = 1$. The line at $45°$ is also shown. Note that the interarrival times are often very small.

We know that there is consistency in Description 1 in that

$$N(0, t] + N(t, t + h] \; = \; N(0, t + h] \tag{A1}$$

tallies with the fact that if X and Y are independent RVs, $X \sim$ Poisson(μ) and $Y \sim$ Poisson(ν), then $X + Y \sim$ Poisson($\mu + \nu$). We might expect Description 2 to tally with the lack-of-memory property of the exponential distribution.

But do things *have* to be this way: can we use distributions other than Poisson and exponential? To answer this, it helps to develop the Poisson Process in a different way.

▶▶ **B. The 'Poisson' property from more basic assumptions.** Let us assume of our process of arrivals of rate λ that

- the numbers of arrivals $N(A_1), N(A_2), \ldots, N(A_n)$ in a collection $A_1, A_2, \ldots, A_n$ of *disjoint* time-intervals are *independent* RVs;

- the distribution of the number of arrivals in an interval depends only on the length of that interval;

- for the number $N(0, h]$ of arrivals during time-interval $(0, h]$, we assume for small h that

$$\mathbb{P}\{N(0, h] = 1\} = \lambda h + o(h),$$
$$\mathbb{P}\{N(0, h] > 1\} = o(h).$$

Here $g(h) = o(h)$ means that $g(h)/h \to 0$ as $h \downarrow 0$. See Appendix A3, p496.

We clearly have

$$\mathbb{P}\{N(0, h] = 0\} = 1 - \lambda h + o(h).$$

▶ **Ba. Theorem.** *Under the above assumptions,*

$$N(0, t] \sim \text{Poisson}(\lambda t).$$

Proof. Fix α with $0 \le \alpha \le 1$. From our study of probability generating functions, we know that it is enough to prove that

$$g_t(\alpha) := \mathbb{E}\alpha^{N(0,t]} = e^{\lambda t(\alpha - 1)}.$$

Let $t, h > 0$. Since the two RVs $N(0, t]$ and $N(t, t + h]$ are independent and $N(t, t + h]$ has the same distribution as $N(0, h]$, we find that

$$g_{t+h}(\alpha) = g_t(\alpha)g_h(\alpha),$$

so the exponential form of g is not surprising.

For small h, we have, heuristically,

$$\begin{aligned}
g_h(\alpha) &= \mathbb{E}\alpha^{N(0,h]} \\
&= \alpha^0 \mathbb{P}\{N(0, h] = 0\} + \alpha^1 \mathbb{P}\{N(0, h] = 1\} + \cdots \\
&= \alpha^0\{1 - \lambda h + o(h)\} + \alpha^1\{\lambda h + o(h)\} + o(h) \\
&= 1 + \lambda(\alpha - 1)h + o(h).
\end{aligned}$$

Hence,

$$g_{t+h}(\alpha) = g_t(\alpha)[1 + \lambda(\alpha - 1)h + o(h)]$$

or

$$\frac{g_{t+h}(\alpha) - g_t(\alpha)}{h} = \lambda(\alpha - 1)g_t(\alpha) + \frac{g_t(\alpha)o(h)}{h}.$$

Let $h \downarrow 0$ to get

$$\frac{\partial}{\partial t}g_t(\alpha) = \lambda(\alpha - 1)g_t(\alpha), \qquad g_0(\alpha) = \mathbb{E}\alpha^0 = 1. \tag{B1}$$

But if $y'(t) = by(t)$ and $y(0) = 1$, then $y(t) = e^{bt}$. Hence,

$$g_t(\alpha) = e^{\lambda(\alpha-1)t} = e^{\lambda t(\alpha-1)},$$

as required. (*Note.* If are keen on rigour, check that (B1) holds where the derivative is two-sided, not just a derivative to the right.)   □

I hope that this helps you understand why Description 1 is as it is.

▶ **C. Tying in Descriptions 1 and 2.**   Make the assumptions of the previous subsection, equivalently, those of Description 1.

Let T_n be the time at which the nth customer arrives. We have

$$\mathbb{P}(T_1 > t) = \mathbb{P}\{N(0,t] = 0\} = e^{-\lambda t}.$$

(Of course, the probability that a customer arrives at time 0 is 0.)

By combining this idea with independence of numbers of arrivals in disjoint time intervals, one can prove that Description 2 is valid. I skip doing this – though see remarks in Subsection 433E below.

Let us work in the other direction to see that Description 2 implies Description 1. So, assume that

$$T_1 := \xi_1, \ T_2 := \xi_1 + \xi_2, \ T_3 := \xi_1 + \xi_2 + \xi_3, \ \ldots,$$

where the interarrival times $\xi_1, \xi_2, \xi_3, \ldots$ are IID RVs each with the exponential distribution of rate parameter λ. Postulate that the nth customer arrives at time T_n. In this formulation,

$$N(0,t] := \max\{n : T_n \leq t\}.$$

Let us show that $N(0,t] \sim \text{Poisson}(\lambda t)$. We know that T_n has the Gamma$(n, \text{ rate } \lambda)$ distribution with pdf

$$f_{T_n}(t) = \frac{\lambda^n t^{n-1} e^{-\lambda t}}{(n-1)!}.$$

It is intuitively clear that we can use the independence of ξ_{n+1} and T_n in a calculation

$$\mathbb{P}\{N(0,t] = n\} = \mathbb{P}\{T_n \leq t < T_{n+1}\}$$

$$= \int_{s=0}^{t} \mathbb{P}\{T_n \in ds; \ \xi_{n+1} > t - s\} = \int_0^t \mathbb{P}\{T_n \in ds\}\mathbb{P}\{\xi_{n+1} > t - s\}$$

$$= \int_0^t f_{T_n}(s) e^{-\lambda(t-s)} ds = \int_s^t \frac{\lambda^n t^{n-1} e^{-\lambda t}}{(n-1)!} e^{-\lambda(t-s)} ds$$

$$= \frac{\lambda^n e^{-\lambda t}}{(n-1)!} \int_0^t s^{n-1} ds = \frac{\lambda^n e^{-\lambda t}}{(n-1)!} \frac{t^n}{n}$$

$$= \frac{(\lambda t)^n e^{-\lambda t}}{n!},$$

as required.

We see how closely interrelated Poisson and exponential distributions are. The martingale explanation is mentioned at Subsection 433E.

Ca. Exercise. Calculate

$$\mathbb{P}\Big(N(0,1] = 0, \ N(1,2] = 1 \Big),$$

first (easily!) using Description 1, and then using Description 2. Do the same for

$$\mathbb{P}\Big(N(0,1] = 1, \ N(1,2] = 1 \Big).$$

▶ **Cb. Exercise.** This re-does Exercise 22Gb in a way which is now more meaningful and more informative.

Let S be the first time when there has been a gap of length c during which no customers have arrived. Let L_S be the Laplace transform (see Subsection 155Q)

$$L_S(\alpha) := \mathbb{E}\left(e^{-\alpha S}\right), \qquad (\alpha > 0).$$

Let ξ denote the arrival time of the first customer. Give a clear explanation of how the IID property of the interarrival times leads one to the conclusion that

$$\mathbb{E}\left(e^{-\alpha S} \mid \xi\right) = \begin{cases} e^{-\alpha c} & \text{if } \xi > c, \\ e^{-\alpha \xi} L_S(\alpha) & \text{if } \xi < c. \end{cases}$$

Deduce from the Conditional Mean Formula that

$$L_S(\alpha) = e^{-\alpha c} e^{-\lambda c} + \int_0^c \lambda e^{-\lambda s} e^{-\alpha s} L_S(\alpha) \mathrm{d}s,$$

and that

$$L_S(\alpha) = \frac{\lambda + \alpha}{\lambda + \alpha e^{(\lambda+\alpha)c}}.$$

Recall that λ is fixed. By evaluating

$$\frac{d}{d\alpha} L_S(\alpha) \quad \text{and setting } \alpha = 0,$$

show that

$$\mathbb{E}(S) = \frac{e^{\lambda c} - 1}{\lambda}.$$

Cc. Exercise. Let W denote the first time when the gap between the arrival of two consecutive customers is *less* than c. Find $L_W(\alpha)$ and $\mathbb{E}(W)$.

Cd. Exercise. Deduce from the equivalence of Descriptions 1 and 2 that, for $n \geq 1$,

$$\int_0^t \frac{\lambda^n s^{n-1} e^{-\lambda s}}{(n-1)!} \mathrm{d}s = \sum_{k=n}^{\infty} e^{-\lambda t} \frac{(\lambda t)^k}{k!}.$$

Hint. Interpret in terms of T_n and N_t.

▶ **D. The 'Boys and Girls' Principle.** Recall the result of Subsection 100I. It said that

if

the Number of births is Poisson(γ) and
each birth ('independently of everything else') produces
a boy with probability p, a girl with probability q,

then

Number of boys $\sim$ Poisson(γp),
Number of girls $\sim$ Poisson(γq),
the Number of boys is independent of the Number of girls.

Here is an important extension of that principle.

▶▶▶ **Da. Fact: 'Boys and Girls' Principle for Poisson processes.** The following two Models have exactly the same probabilistic structure.

Model 1:

Men arrive at a queue in a PP(λ) process,
Women arrive at the queue in a PP(μ) process,
and these two Poisson Processes are independent.

Model 2:

People arrive at the queue in a PP($\lambda + \mu$) process,
Each person (independently of everything else) is
a man with probability p, a woman with probability q,
where

$$p = \frac{\lambda}{\lambda + \mu}, \quad q = \frac{\mu}{\lambda + \mu}.$$

Assume this. It should not be too surprising now.

▶ **Db. Exercise.** Men arrive at a shop in a Poisson Process of rate α, and Women arrive at the shop in an independent Poisson Process of rate β.

Let $M(t, u]$, $W(t, u]$, $N(t, u]$ be the numbers of Men, Women, People arriving during time-interval $(t, u]$. If the time-interval is $(0, t]$, write $M(t)$, $W(t)$, $N(t)$ for short.

Calculate

(a) the probability that the first person to arrive is a Woman;
(b) the probability that the first two to arrive are both Women;
(c) $\mathbb{P}[M(t) = m \mid N(t) = n]$, where $m \leq n$,
(d) $\mathbb{P}[N(t) = n \mid M(t) = m]$, where $n \geq m$.

▶ **Dc. 'Coupon-collector problem': a Poisson version.** Of course I eat three *Barleyabix*^{TM} per day when I've got some. Every time I buy a packet, I get one of the collection of n dinosaurs.

I buy packets at random, at the times of a Poisson Point Process of rate 1 per week – from a 24-hour, 7-days-a-week shop. Explain why the set of times when I get a *T. Rex* is a PPP of rate $1/n$ independent of the PPP of times when I get a *Velociraptor*. Let S be the time at which I get my first *T. Rex*, and let T be the time in weeks that I have to wait to complete my collection. Prove that

$$\mathbb{P}(S \leq t) = 1 - \exp\left(-\frac{t}{n}\right), \qquad \mathbb{P}(T \leq t) = \left\{1 - \exp\left(-\frac{t}{n}\right)\right\}^n,$$

On average, how long do I have to wait for my first dinosaur? When I have k *different* dinosaurs ($k < n$), how much longer on average do I have to wait for my next *new* dinosaur? Explain (don't try by integration!) why

$$\mathbb{E}(T) = n\left(\frac{1}{n} + \frac{1}{n-1} + \cdots + \frac{1}{2} + 1\right),$$

so that, by well-known mathematics, $\mathbb{E}(T) \sim n \ln n$ in that the ratio of the two sides tends to 1 as $n \to \infty$. How does $\mathrm{Var}(T)$ behave as $n \to \infty$? Prove that, for any $\varepsilon > 0$,

$$\mathbb{P}(|T_n/(n \ln n) - 1| > \varepsilon) \to 0 \quad \text{as } n \to \infty.$$

▶ **E. Martingale characterizations of Poisson Processes.** Let $N(t) := N(0, t]$, where we use Description 1. Then
    (a) $N(0) = 0$;
    (b) $t \mapsto N_t$ is right-continuous and 'constant except for jumps by 1',
    (c) $M_t := N_t - \lambda t$ defines a martingale relative to **N**.
In regard to (c) we mean that

$$\mathbb{E}\{N_{t+h} - \lambda(t+h) \mid N_s : s \leq t\} = N_t - \lambda t.$$

This rearranges as

$$\mathbb{E}\{N_{t+h} - N_t \mid N_s : s \leq t\} = \lambda h;$$

and this is intuitively obvious since $N_{t+h} - N_t = N(t, t+h]$ is independent of $\{N_s : s \leq t\}$ and has mean λh. We saw similar things in our study of additive martingales.

The amazing thing, which shows the power of the martingale concept, is the **martingale characterization result** that *properties (a), (b) and (c) imply that* **N** *is a PP(λ)*: in particular that $N_{t+h} - N_t$ is independent of $\{N_s : s \leq t\}$ with the Poisson(λh) distribution.

To prove this, one first uses **stochastic calculus** (see, for example, Volume 2 of Rogers and Williams [199]) to show that, for $0 < \alpha < 1$,

$$\alpha^{N_t} e^{-\lambda t(\alpha - 1)} \text{ is a martingale relative to } \mathbf{N}.$$

In particular, therefore, $\mathbb{E}\alpha^{N_t} = e^{\lambda t(\alpha-1)}$, so that $N_t \sim \text{Poisson}(\lambda t)$. Next, one uses the fact that if T is a stopping time (for example, the time of the first arrival, the first jump of **N**), then by the appropriate Optional Sampling Theorem,

$$\mathbb{E}\left(N_{T+t+h} - N_{T+t} \mid \mathcal{F}_{T+t}\right) = h,$$

$\mathcal{F}_{T+t}$ being the σ-algebra describing the history up to the stopping time $T + t$. Putting these ideas together carefully, one finds that

$$N_{T+t} - N_T \text{ is a PP}(\lambda) \text{ independent of } \mathcal{F}_T.$$

You can start to see why martingale theory settles elegantly the fact that Description 1 implies Description 2. Let T_1 be the time of first jump of **N**. For $\theta > 0$,

$$\left(\frac{\lambda + \alpha}{\lambda}\right)^{N_t} e^{-\alpha t}$$

defines a martingale relative to **N** which is bounded on $[0, T]$. Thus,

$$\mathbb{E}\left\{\frac{\lambda + \alpha}{\lambda} e^{-\alpha T_1}\right\} = 1, \qquad \mathbb{E}e^{-\alpha T_1} = \frac{\lambda}{\lambda + \alpha},$$

whence T_1 is exponential rate λ. Analogously, T_n is Gamma$(n, \text{rate } \lambda)$. And so on.

▶▶ **F. Poisson Point Processes (PPPs).** We have already met a PPP of constant intensity λ on $[0, \infty)$. Generalizations are immediate. For example, a PPP of constant intensity λ, PPP(λ), on $\mathbb{R}^2$ is a random set $\mathcal{P}$ of 'points' in $\mathbb{R}^2$ such that

- the numbers of points $N(A_1), N(A_2), \ldots, N(A_n)$ in a collection $A_1, A_2, \ldots, A_n$ of *disjoint* (Borel) subsets of $\mathbb{R}^2$ are *independent* RVs;

- the number of points in any (Borel) region of area a has the Poisson(λa) distribution.

We might use this to model the positions of faults randomly produced in some surface.

Fa. Exercise. Suppose that we have a PPP(λ) $\mathcal{P}$ on $\mathbb{R}^2$. Fix a point in $\mathbb{R}^2$. What is the pdf of the distance from the nearest point of $\mathcal{P}$ to that fixed point?

▶ **G. The 'Waiting-time Paradox'.** Let $\mathcal{P}$ be a PPP(λ) on $\mathbb{R}$, and let a be any point of $\mathbb{R}$. Let T_k $(k = 1, 2, 3, \ldots)$ be the kth point of $\mathcal{P}$ to the right of a, and let T_{-k} $(k = 1, 2, 3, \ldots)$ be the kth point of $\mathcal{P}$ to the left of a. Then

$$\ldots, T_{-2} - T_{-3},\ T_{-1} - T_{-2},\ a - T_{-1},\ T_1 - a,\ T_2 - T_1,\ T_3 - T_2,\ \ldots,$$

are IID each exponential rate λ with pdf $\lambda e^{-\lambda t}$ on $(0, \infty)$. People do not like the fact that the gap around a is the sum of two IID exponentials and has pdf $\lambda^2 t e^{-\lambda t}$. In a sense, '$a$ is more likely to fall in a bigger gap, and

$$\mathbb{P}\{(\text{gap around } a) \in dt\} \propto \mathbb{P}\{(\text{typical gap}) \in dt\} \times t,$$

where we have boosted the probability by the length of the gap'. You arrived at the bus stop at a bad time!

Ga. Exercise. Find ways of making the 'boosting by length' idea rigorous. You might for example wish to consider a whole sequence of observers arriving at times $\ldots, -2\delta, -\delta, 0, \delta, 2\delta, \ldots$.

▶▶ **H. Conditional Uniformity.** This is another version of the 'Boys and Girls' Principle.

Suppose that we have a PPP(λ) $\mathcal{P}$ on $\mathbb{R}^n$. *Given* that a (Borel) region A contains n points of $\mathcal{P}$, these points are distributed in A exactly as if we made n independent choices uniformly within A. Suppose, for example, that B is a (Borel) subset of A. Then,

$$\mathbb{P}\{N(B) = k \mid N(A) = n\} = \binom{n}{k} p^k q^{n-k}, \tag{H1}$$

where $p = |B|/|A|$, $|A|$ denoting the (hyper-)volume of A.

Ha. Exercise. Prove (H1). How would you prove the general result?

▶ **Hb. Exercise.** Continue with Exercise 432Db. Calculate
(e) $\mathbb{P}[M(s) = k \mid M(t) = m]$, where $s \leq t$ and $k \leq m$.
(f) $\mathbb{P}(M(1,3] = 2 \mid M(0,2] = 2)$.
Part (f) is a bit tricky.

Hc. Exercise. Consider a Poisson process of rate λ and suppose that we use the improper 'Frequentist' prior $\pi(\lambda) = \lambda^{-1}$ for λ. Let A denote the event 'K arrivals during $[0, \alpha]$' and B the event '$K + n$ arrivals in $[0, \alpha + 1]$'. Show that (*formally*)

$$\mathbb{P}_\pi(A) = \frac{1}{K}, \quad \mathbb{P}_\pi(B) = \frac{1}{Kn}, \quad \mathbb{P}_\pi(A \mid B) = \binom{n + K}{n}\left(\frac{\alpha}{\alpha + 1}\right)^K,$$

and that $\mathbb{P}(B \mid A)$ therefore agrees with the answer to Exercise 213Ke.

That $P_\pi(A) = K^{-1}$ irrespective of α shows just how daft improper priors are!

▶ **Hd. Conditional uniformity and sufficiency.** Suppose that we know that an observed set of points within a region A of $\mathbb{R}^2$ is a realization of a PPP(λ) on A, and that we wish to estimate λ. Then, from the point of view of sufficiency, all that is relevant to inference about λ is the total number of points $n^{\text{obs}}(A)$ in A. This is because after Zeus chose $n^{\mathrm{obs}}(A)$ and reported its value to Tyche, Tyche (without knowing the value of λ) then chose the $n^{\mathrm{obs}}(A)$ points in A independently, each according to the uniform distribution on A.

▶ **He. Rigorous construction of a PPP.** Suppose that for each (Borel) subset A of $\mathbb{R}^n$ there is defined a RV $N(A)$ with the Poisson($\lambda|A|$) distribution, that for disjoint subsets $B_1, B_2, \ldots, B_n$ with union B the RVs $N(B_1), N(B_2), \ldots, N(B_n)$ are independent and

$$\mathbb{P}\{N(B) = N(B_1) + N(B_2) + \cdots + N(B_n)\} = 1.$$

Can we say that

$$\mathbb{P}\{N(C \cup D) = N(C) + N(D) \text{ whenever } C \text{ and } D \text{ are disjoint}\} = 1? \qquad \text{(H2)}$$

Here we meet a serious difficulty. There are uncountably many Borel subsets C of $\mathbb{R}^2$, and there is no reason why the set F (say) within {} in (H2) should be measurable; and if it is not measurable, it is impossible to assign a probability to it.

However, the conditional-uniformity result does allow us to construct a model for a PPP(λ) $\mathcal{P}$ where F is measurable with measure 1. We do this by concentrating more on the *points* in $\mathcal{P}$ than on the $N(\cdot)$ values. Divide up $\mathbb{R}^n$ into a disjoint union of sets $A_1, A_2, \ldots$ each of volume 1. Choose IID Random Variables $N(A_1), N(A_2), \ldots$ each with the Poisson(λ) distribution. Within each A_i, scatter randomly $N(A_i)$ points; and there you have your PPP. For any Borel subset B of $\mathbb{R}^n$, $N(B)$ is now defined to be the number of points of $\mathcal{P}$ within B.

▶ **I. Reference priors: the controversial case of rare accidents.** All of the nuclear-powered electricity-generating stations on Planet Altair-3 were installed by Altair Nuclear Industries (ANI) exactly 1 of their years ago; and no more are to be built. It is known that nuclear accidents in the power stations happen at the points of a PP(λ) process with an Altair-3 year as unit of time. The first accident has just occurred. The resulting repair will mean that the PP(λ) property will be preserved. A public inquiry wants to answer the question: how many accidents will occur on average in the next Altair-3 year?

Witness 1, a Frequentist, says that the number of accidents in the next year will be Poisson(λ) and that the best estimate of λ from just about any point of view is 1, so he answers '1'. ANI object that it is not fair that the situation is being considered just after a nuclear accident. Witness 1 quotes the conditional-uniformity result and says that sufficiency means that only the total number of events up to now matters, not when they occurred. He backs this up by saying that if T is the Time of the first accident, then

$$f_T(1 \mid \lambda) = \lambda e^{-\lambda},$$

and that if N is the number of accidents in the first year, then

$$p_N(1 \mid \lambda) = \lambda e^{-\lambda},$$

so that the likelihood as a function of λ is exactly the same whether or not you are told that the first accident happened at time 1 or that 1 accident occurred during the first year.

Witness 2 is a Bayesian who believes that your prior represents your prior degree of belief about λ and cannot depend on the way that data is collected. She takes a prior λ^{-1} for both the Poisson and exponential descriptions, and basically agrees with the Frequentist.

ANI are not happy; but then they hear that Bayesians who believe in reference priors will assign a lower mean to the exponential situation.

Witness 3 is called. He is a Bayesian who believes strongly in reference priors. He explains to the inquiry team about invariance's being essential for consistent behaviour. He says that he always considers the whole picture: prior knowledge together with how the data are collected. He says that one should use prior λ^{-1} for the approach based on T leading to a mean number of accidents for the next year as 1, but prior $\lambda^{-\frac{1}{2}}$ for the approach based on N leading to an estimated mean of $1\frac{1}{2}$. So though he reflects the *difference* between the models in the way that ANI want, the *values* of his estimates seem only to make things worse for them.

Witness 4 is You. What do you say in the light of the above?

▶ **J. Poisson Line Processes (PLPs).** How should we model the random

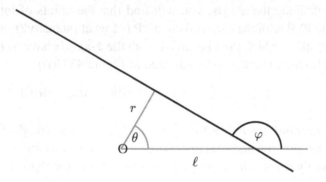

Figure J(i): The line $x \cos \theta + y \sin \theta = r$

scattering of lines on the plane $\mathbb{R}^2$? Here's the answer.

Let $\mathcal{P}$ be a PPP(λ) on the rectangle $(0, \infty) \times [0, 2\pi)$. Each point of $\mathcal{P}$ is a pair (r, θ), and that determines the line

$$x \cos \theta + y \sin \theta = r$$

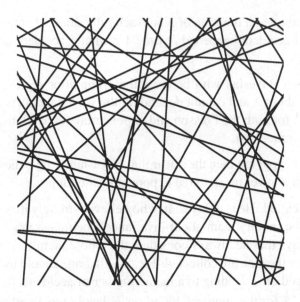

Figure J(ii): A window on a simulation of a PLP

shown in Figure J(ii). These lines make up your PLP. The key thing is that $\lambda \mathrm{d}r\mathrm{d}\theta$ is preserved under translations, rotations and reflections. Check this.

If you walk along the x-axis, you will find that the points of intersection of the lines of the PLP with the x-axis form a PP (of what intensity?) on $\mathbb{R}$, but that the angles $\varphi \in [0, \pi)$ which the lines make with the x-axis behave as if having pdf $\sin \varphi$. This is because (for the case indicated in Figure 437J(i))

$$r = \ell \sin \varphi, \quad \theta = \varphi - \tfrac{1}{2}\pi, \quad \lambda \mathrm{d}r\mathrm{d}\theta = \lambda(\sin \varphi)\mathrm{d}\ell \mathrm{d}\varphi.$$

Bear this in mind when you take your first drive in a straight line over the surface of Planet Poisson and find that more of the mysterious lines on its surface are approximately perpendicular to your direction of motion than close to it.

For the fascinating subject of Stochastic Geometry, see Stoyan, Kendall and Mecke [222].

▶▶ **K. Just when things were about to get really interesting ...** I stop my account of classical Probability and move on to Quantum Probability. I've said many times that it disturbs me that Quantum Probability is based on quite different principles and that I feel I ought therefore to include an introduction to it.

Even so, Quantum Probability is no more interesting than Classical Probability when the latter explodes into new directions which we have not met in this book. [W] and Rogers and Williams [199] are natural follow-ups to this book, but see Appendix D for other books on stochastic calculus.

To get an idea of the wider scope, do look at Diaconis [62], Davis [54], Pemantle [182], for fascinating extensions of the Pólya-urn idea, at Hall [106], at Aldous [3] for enough problems to keep you going, etc, etc. If you are keen on seeing standard Group Theory applied to Probability and Statistics, see Diaconis [63]. In regard to more advanced Group Theory, see below.

Statistical Mechanics is full of problems which I consider to be at least as interesting as the Riemann Hypothesis (and some of which may well be related to it), and contains some of the most profound Probability ever done. The papers on dimer coverings by Kastelyn [127] and by Temperley and (M.E.) Fisher [224] make a great lead-in. See also Baxter [12] and Thompson [226]. In connection with the Ising model (now popular with statisticians for image processing), look at Onsager's wonderful (Nobel-Prize-winning) paper [180], at Kaufman's also-wonderful simpler proof ([128]), and at Baxter and Enting [13] for a glorious piece of ingenuity. Now that I think of it, the papers by Onsager and by Kaufman use some of the Pauli-matrix theory in the next chapter. See Itzykson and Drouffe [116]. For the remarkable polaron problem, see Varadhan [231] and Deuschel and Stroock [61].

Do at all costs just peep at the paper [145] on Cardy's brilliant work for which Grimmett's book [102] provides excellent background. When I said in the Preface that at very advanced level, Group Theory and Probability are linked in an amazing way, I was thinking about how Virasoro algebras and Conformal Field Theory link work such as Cardy's to the work of Borcherds and a galaxy of other stars on the Monster Group and *'moonshine'*. See Borcherds [27], Goddard [98], di Francesco, Mathieu and Sénéchal [64] for some of the best Maths there is. On some of the amazing story surrounding the fact, quoted by Borcherds, that

$$\exp\left(\pi\sqrt{163}\right) = 262537412640768743.99999999999925\cdots,$$

see Cox [47], a book rightly described by American Mathematical Monthly as 'unique and sensational'.

So it's not entirely a case of including a chapter on Quantum Probability to help provide desired education for a well-rounded probabilist: Onsager, Kaufman, Cardy and others have shown that Quantum Probability is apparently necessary for some of the most challenging problems in Classical Probability. (And understanding phase transitions is relevant to some MCMC studies ... .)

Sure, most of the books and papers just mentioned are miles beyond the level of this book; but that's the great thing.

QUANTUM PROBABILITY

and QUANTUM COMPUTING

The true logic of this world lies in the calculus of probabilities
James Clerk Maxwell

Absolutely right, James Clerk, as always. But *which* calculus of probabilities? At present, we have two, totally different, calculi: Classical and Quantum. For discussion of the relation between the two calculi, see Subsection 461M.

Please see my thanks to Feynman, Gruska, Isham, Nielsen and Chuang, Penrose, Preskill, ..., in Appendix D.

In computer jargon, this chapter is a 'several-pass' account: we begin with vague ideas, and tighten these up in stages.

No previous knowledge of Quantum Theory is assumed. Except for a brief discussion of the Uncertainty Principle, we deal exclusively with comparatively easy finite-dimensional situations. We do behave with somewhat greater freedom in our use of Linear Algebra.

I ought to stress the following: *Quantum Computing is by no means the most wonderful part of Quantum Theory.* If Nature had not chosen to exploit in the design of spin-$\frac{1}{2}$ particles the fact that (as we shall see later) 'SU(2) is the double cover of SO(3)', then there would be no Chemistry, no Biology, no *you!*. *To a large extent, we are using Quantum Computing as the simplest context in which to learn some Quantum Probability.*

My hope is that by the time you finish this chapter, you will be better equipped to read critically the huge amount of 'popular Science' (some very good, a lot not so good) on the Uncertainty Principle, entanglement, etc, and that, more importantly, you will be persuaded to learn more from the real experts.

Note. At the time of writing this, the distinguished experimenter Humphrey Maris claims that at very low temperatures, it is possible to split an electron into 'electrinos'. This, if true, would have serious impact on some aspects of Quantum Theory, but the great success of the subject in so many areas remains of course.

10.1 Quantum Computing: a first look

I assume units chosen so that $\hbar = 1$, where $\hbar = h/(2\pi)$, h being the so-called Planck's constant. I use c for $2^{-\frac{1}{2}}$, not for the velocity of light! (In Section 10.5, $\hbar$ and c are restored to their conventional usage in Physics.)

A. Introduction.

Interference. The quantum world is extremely strange. We know that if we have a small light source, a screen, and a sheet in which there is a small hole between the light and the screen, then the light will illuminate a small region of the screen. If we now make another hole very near the first, then some of the region previously illuminated will now become dark: there is interference. [The Mach-Zender interferometer illustrates the effect much more spectacularly.] We understood this classically in terms of the trough of one wave being superimposed on a crest of another. For brief discussion of Quantum Theory and 'realism', see Subsection 491A.

But when the light is let through one photon at a time (as can be done in the laboratory), the same phenomenon occurs. The probability that we will find a photon in some region is the square of the modulus of the value of a wave function, and, since the sum of two wave functions can be zero at a point where neither is itself zero, probabilities can 'cancel each other out' in a way totally unexpected from classical Probability. The photon cannot be considered to pass through one hole rather than another: it is somehow aware of the existence of both holes; it somehow investigates the whole system.

Utilizing interference. In quantum computers, we can utilize interference to cook things so that even though all answers are possible, the correct answer is much more likely than others. We rely on the fact that in some sense a quantum computer can simultaneously investigate many possibilities, something often referred to as quantum parallelism.

Utilizing entanglement. Quantum parallelism also refers to the weird phenomenon of entanglement. Entanglement, within quantum computers and in its manifestation in the Aspect experiment, is discussed in later sections.

States. Quantum states are vectors, which in this book we usually restrict to lie in a finite-dimensional space. States evolve in a fully-understood way described by *Schrödinger's equation*. As a linear equation, this is in some ways simpler than equations in classical mechanics. In a quantum computer, we must operate on states by cooking up the right Schrödinger equation and running it for precisely the right amount of time to make the changes we want.

U-mode quantum mechanics. The evolution of quantum states determined by Schrödinger's equation is what is called *unitary*, which is linked to the fact that probabilities have to sum (or integrate) to 1. No controversy surrounds the way in which a quantum system behaves when we are not observing it.

R-mode quantum mechanics. If we make a measurement of some 'observable' (position, momentum, spin, ...) when the quantum system is in state $\mathbf{v}$, then, according to the 'orthodox' (or 'mathematical' or 'operator') theory, the measurement obtained is randomly chosen in a way depending upon $\mathbf{v}$. All predictions made on the basis of the Bohr interpretation have been confirmed in the most spectacular way by experiment. However, many serious thinkers doubt that this randomness (R-mode quantum mechanics, but the R is for 'reduction') will be the last word on the topic. (See for example Penrose [183], where the 'U' and 'R' terminology is used. Penrose makes a convincing case that taking account of gravity will provide the eventual explanation of the R-mode. It may well be the case that when Witten has finished off everything else, he will apply M-theory to this too. See [100].) All that matters for us is that the mathematical formalism gives the right final answers for predicting the results of experiments.

So, in a quantum computer, every 'logic gate' will correspond to a unitary evolution; and when we make a measurement at the end, the answer will be random.

What can/could quantum computers do? The first demonstration of what a quantum computer could do was that of *David Deutsch* who showed that it was possible to tell if a function from the two-point set $\{0, 1\}$ to itself satisfies $f(0) = f(1)$ with only *one* call to the (quantum-encoded) function. A classical computer would need two calls. But then a quantum computer can in some sense do lots of things simultaneously.

Grover's algorithm showed that a quantum computer could find the correct name corresponding to a telephone number in a (quantum-encoded) telephone directory with N numbers in about $\sqrt{N}(\ln N)^3$ steps, whereas any conventional program would take on average $(N + 1)/2$ steps. This algorithm has actually been implemented on a quantum computer for $N = 8$, and probably by the time you read this, for larger N.

Grover, and then others, also described how certain statistics (mean, variance, etc) of a sample could be computed more quickly. This leads on to faster numerical computation of multivariate integrals (for given average error). These things also lead to a reduction 'from N to $\sqrt{N}$' in computing time.

Schor's famous factorization algorithm showed that a quantum computer could factorize a number into prime factors exponentially faster than a conventional computer can. For an L-bit number, Schor's algorithm needs on average about $L^2 \ln L \ln \ln L$ operations, whereas the best existing algorithm for a conventional computer needs about $\exp(KL^{\frac{1}{3}}(\ln L)^{\frac{2}{3}})$ operations, where K is a known constant. (*Note.* Much security – on the Internet, in financial transactions, in secret-service doings – is based on the apparent impossibility of factorizing numbers which are the product of two large primes. This idea is the basis for the RSA algorithm which made a fortune for Rivest, Shamir and Adleman, but which had been independently invented earlier by British cryptographers, Ellis, Cocks and Williamson. Results such as Shor's put such security under threat. But now, seemingly uncrackable quantum codes exist. See Preskill [187], Nielsen and Chuang [174].)

Quantum computers could do certain types of simulation (that of spin-glass systems, and presumably of other interacting systems of 'Ising' type widely used by statisticians in image analysis). They could probably speed up (other) MCMC methods for complicated situations.

That quantum computers will out-perform parallel classical ones at some of the most common computations is far from clear. It is 'bad news' that a quantum computer cannot beat a classical one at one of the most important operations in numerical mathematics: iteration of functions. See (`quant-ph 9712051`).

Technologies. Several completely different technologies are being investigated. It seems fair to say that none of these stands much chance of leading to a quantum computer of useful size. Here's where new ideas are most desperately needed.

At the moment of writing this, the leading technology in the sense that it can deal with the most qubits, doing some calculations with 7 qubits, is based on NMR techniques. However, it will surely soon be overtaken by quantum-dot or ion-trap technology, and possibly (see Knill, Laflamme and Milburn [138]) by optical-computer technology. NMR computing requires unbelievably sophisticated technology, essentially used for many years by physicists and chemists. Very roughly, the idea is to use spins of nuclei, etc, within molecules to store 0 (spin up) and 1 (spin down); but we must remember that the quantum state will be a superposition of these! The natural frequencies of these spins, and the way in which 'spins can precess around each other', are very accurately known; and this allows manipulation of spins by electromagnetic pulses of very precisely controlled frequency and duration. The precession provides the coupling necessary for things such as the CNOT gates described later. The manipulation of a single spin relies on resonance with its natural frequency. For a little more on NMR technology, see Subsection 470F.

The Nielsen–Chuang book gives an excellent account of the state of the art.

Doubts. A number of physicists doubt that quantum computers of useful size will ever be made. In many of the technologies, it is immensely difficult to stop the computer's being influenced by the environment. NMR technology has the advantages that there is a high degree of isolation from the environment, and that one has about 10^{18} separate computers, one for each molecule. However, only a minute excess of those 'computers' are doing what you really want. Even its most enthusiastic advocates doubt that an NMR computer will be able to handle a 'practically useful' number of qubits. Some people seem to question whether NMR computers are truly 'quantum', but it seems to me that since they can do Grover's algorithm, they utilize entangled states at a certain stage of the calculation.

Localized computing. The geometry of real-world implementations makes it difficult ever to have full quantum parallelism, so what could be achieved with only local parallelism needs thorough investigation. This process has been begun by Seth Lloyd and others.

B. 'A second pass'. We now look at things a little more closely.

▶ **Qubits.** A 1-bit classical computer deals just with the numbers 0 and 1. A 1-qubit quantum computer deals with complex superpositions of 0 and 1. The (normalized) *state* of such a computer is a vector in $\mathbb{C}^2$, that is, a vector

$$\mathbf{v} = \begin{pmatrix} v_0 \\ v_1 \end{pmatrix} = v_0 \begin{pmatrix} 1 \\ 0 \end{pmatrix} + v_1 \begin{pmatrix} 0 \\ 1 \end{pmatrix} = v_0 \mathbf{z} + v_1 \mathbf{u},$$

where v_0 and v_1 are *complex* numbers satisfying $|v_0|^2 + |v_1|^2 = 1$. Thus if $v_r = x_r + iy_r$, the point (x_0, y_0, x_1, y_1) lies on the 3-dimensional sphere S^3 of radius 1 in 4-dimensional real space $\mathbb{R}^4$. (Recall that in Mathematics, 'sphere' equals spherical (hyper-)surface; that a sphere together with its interior constitute a ball.)

▶▶ **Self-adjoint matrices; observables.** For a square matrix A with complex entries, *let $A^\dagger$ be the complex conjugate of the transpose of A. A matrix is said to be Hermitian, or self-adjoint if $A^\dagger = A$. A mathematical observable is a self-adjoint matrix.*

Every real-world observable derives from a self-adjoint operator (perhaps on an infinite-dimensional space); but it may not be possible to make real measurements corresponding to some mathematical observables.

▶ **Measuring our qubit.** For our qubit, we have the special observable M represented by the matrix

$$M = \begin{pmatrix} 0 & 0 \\ 0 & 1 \end{pmatrix}.$$

If we observe M when the system is in normalized state $\mathbf{v}$, then
• with probability $|v_0|^2$, we get the measurement 0, and the state jumps to $\mathbf{z}$,
• with probability $|v_1|^2$, we get the measurement 1, and the state jumps to $\mathbf{u}$.
This is quantum mechanics in R-mode.

Thus state $\mathbf{z}$ (for 'zero') corresponds to the deterministic measurement 0, and state $\mathbf{u}$ (for 'unity') corresponds to the deterministic measurement 1. In Dirac's 'bra and ket' notation, $\mathbf{z}$ would be written as $|0\rangle$ and $\mathbf{u}$ as $|1\rangle$. Thus

$$|0\rangle := \mathbf{z} := \begin{pmatrix} 1 \\ 0 \end{pmatrix}, \qquad |1\rangle := \mathbf{u} := \begin{pmatrix} 0 \\ 1 \end{pmatrix}. \tag{B1}$$

Please note that I now **define** *'deterministic' to mean 'with probability 1'.*

▶ **Superposition.** In a generic state $\mathbf{v}$, the system is in some mysterious superposition of 0 and 1. 'Superposition' is a rather dangerous term. *You must* **never** *think of state v as signifying that 'with probability $|v_0|^2$, the state is $\mathbf{z}$ and with probability $|v_1|^2$ the state is $\mathbf{u}$'.* The conflict between superposition and probabilistic mixture will be emphasized several times. We have seen that it *is* true that if a measurement of the special observable M is made when the system is in state $\mathbf{v}$, then *after* the measurement of that *particular* observable, the system will be in state $\mathbf{z}$ with probability $|v_0|^2$, and in state $\mathbf{u}$ with probability $|v_1|^2$. The states $\mathbf{z}$ and $\mathbf{u}$ are specially related to M: they are eigenvectors of M. (More on this shortly.)

▶ **Schrödinger's equation** for the unitary evolution of the state $\mathbf{v}(t)$ at time t is expressed in terms of a special self-adjoint operator $\mathcal{H}$, the energy or **Hamiltonian** operator $\mathcal{H}$. (I am using $\mathcal{H}$ for the Hamiltonian so as to avoid confusion with the Hadamard gate H utilized below.) Schrödinger's equation reads

$$\frac{\mathrm{d}\mathbf{v}(t)}{\mathrm{d}t} = -\mathrm{i}\mathcal{H}\mathbf{v}(t) \tag{B2}$$

with solution (clarified later, but not surprising to our intuition)

$$\mathbf{v}(t) = U_t\mathbf{v}(0), \quad U_t = \mathrm{e}^{-\mathrm{i}t\mathcal{H}} := I + \frac{(-\mathrm{i}t\mathcal{H})}{1!} + \frac{(-\mathrm{i}t\mathcal{H})^2}{2!} + \cdots .$$

▶ **Unitary matrices.** Recall that a matrix U is called *unitary* if $UU^\dagger = U^\dagger U = I$, that is, if $U^{-1} = U^\dagger$. We shall see later that $U_t = \mathrm{e}^{-\mathrm{i}t\mathcal{H}}$ is unitary for every $t > 0$ if and only if $\mathcal{H}$ is self-adjoint; and that if U is unitary, then $\|U\mathbf{v}\|^2 = \|\mathbf{v}\|^2$, the correct law for mapping one state to another.

▶ **Logic gates: a NOT gate.** A NOT logic gate on a classical computer interchanges 0 and 1. On a quantum computer, it has to be represented by a unitary matrix N, and it must interchange $\mathbf{z}$ and $\mathbf{u}$. Thus,

$$N = \begin{pmatrix} 0 & 1 \\ 1 & 0 \end{pmatrix}.$$

[[**N.B.** We are skipping for a moment the problems associated with 'phase'.]]

To implement a NOT gate, we therefore need to find a self-adjoint operator $\mathcal{H}$ such that $N = \mathrm{e}^{-\mathrm{i}t\mathcal{H}}$ for some t, make sure that $\mathcal{H}$ serves as the *Hamiltonian* operator for our physical implementation, run the system for time t so that N *is* the desired unitary evolution, and we have our NOT gate.

We could take $N = U_t$ where

$$t = \tfrac{1}{2}\pi, \quad \mathcal{H} = \begin{pmatrix} 1 & -1 \\ -1 & 1 \end{pmatrix}, \quad U_t = \mathrm{e}^{-\mathrm{i}t}\begin{pmatrix} \cos t & \mathrm{i}\sin t \\ \mathrm{i}\sin t & \cos t \end{pmatrix}.$$

You should check that $(\mathrm{d}/\mathrm{d}t)U_t = -\mathrm{i}\mathcal{H}U_t$.

▶ **Phase.** If instead we take

$$\mathcal{H} = \begin{pmatrix} 0 & -\mathrm{i} \\ \mathrm{i} & 0 \end{pmatrix}$$

then

$$U_t := \mathrm{e}^{-\mathrm{i}t\mathcal{H}} = \begin{pmatrix} \cos t & -\sin t \\ \sin t & \cos t \end{pmatrix},$$

and, if we choose $t = \tfrac{1}{2}\pi$, then $U_t\mathbf{z} = \mathbf{u}$ and $U_t\mathbf{u} = -\mathbf{z}$. However, $-\mathbf{z}$ yields a deterministic M-measurement 0 just as $\mathbf{z}$ does, because of the $|\cdot|^2$ in the rules determining probabilities. But, of course, $|\mathrm{e}^{\mathrm{i}\theta}|^2 = 1$ for any real θ. Thus, the normalized states $\mathbf{v}$ and $\mathrm{e}^{\mathrm{i}\theta}\mathbf{v}$ lead to the same probabilities: they are usually considered as corresponding to the same normalized state. However, we need to watch phases carefully. Still, we could use our new U_t, where $t = \frac{1}{2}\pi$, to implement a NOT gate. But this has all been Mathematics. *To implement such things in the real world, $\mathcal{H}$ has to be restricted to be an observable which describes the* **energy** *of the implementation system.*

Solving Deutsch's problem. Here's a plan for a 1-qubit computer to solve Deutsch's problem. Encode the function f via the unitary map U_f where

$$(U_f \mathbf{v})_k := (-1)^{f(k)} v_k, \qquad k = 0, 1.$$

Clearly U_f corresponds to the diagonal matrix with kth diagonal entry $(-1)^{f(k)} = \pm 1$. Let ρ be the unitary transformation

$$\rho := \frac{1}{\sqrt{2}} \begin{pmatrix} 1 & -1 \\ 1 & 1 \end{pmatrix} = e^{-it\mathcal{H}}, \quad \text{where } \mathcal{H} := \begin{pmatrix} 0 & -i \\ i & 0 \end{pmatrix}.$$

Then, you can check that if you 'input' $\mathbf{z}$, then apply ρ, then U_f, then ρ^{-1}, and finally use M to measure, you obtain deterministic results

$$\text{measured value of } M = \begin{cases} 0 & \text{if } f(0) = f(1), \\ 1 & \text{if } f(0) \neq f(1). \end{cases}$$

The encoding U_f of the function f is called an **oracle**, a 'black box' which you can query or consult to find desired information about f.

Think of f as the ordered pair $[f(0), f(1)]$. Then in 1 call to the f-oracle, the quantum computer can distinguish between the 2 sets $\{[0, 0], [1, 1]\}$ and $\{[0, 1], [1, 0]\}$. A classical computer needs 2 calls, but can then distinguish between the 4 possible functions.

The point is that a good quantum algorithm can sometimes get some desired result faster than a classical algorithm. Compare 'good versus bad' for classical algorithms.

▶ **C. Grover's algorithm.**   This algorithm originates in Lov Grover [104]. It is usually presented as finding the name corresponding to a given number in a telephone directory of K numbers. To do this in any classical way would, on average, involve $\frac{1}{2}(K + 1)$ operations. [Actually, if we are so unlucky as to miss the correct number on the first $K - 1$ consultations of the directory, we know that the Kth number must be the one we are looking for. So the mean number of consultations can reduced by $1/K$.] For a quantum computer, the phone book is encoded as an **oracle**, and you need only consult this oracle a constant multiple of $\sqrt{K}$ times (on average). We now turn to the mathematical statement of the problem, and give the geometric interpretation which was discovered by several people after they read Grover's more algebraic account.

Suppose that a function f on $\{1, 2, \ldots, K\}$ is given such that there is precisely one value k_0 of k such that $f(k_0) = 1$ and that $f(k) = 0$ for $k \neq k_0$. We want to start with some well-chosen quantum state and perform a sequence of steps which will take that vector nearer and nearer to the **target** vector

$$\mathbf{t} := |k_0\rangle := (0, 0, , \ldots, 0, 1, 0, \ldots, 0),$$

the 1 being in the k_0th place. Let

$$\mathbf{e} := K^{-\frac{1}{2}}(1, 1, \ldots, 1),$$

with equal components. Then we know that, whatever the value of k_0, the inner product $\langle \mathbf{t}, \mathbf{e} \rangle = K^{-\frac{1}{2}}$, and this knowledge makes $\mathbf{e}$ a good choice of initial vector. We work entirely in the plane $[\mathbf{t}, \mathbf{e}]$ spanned by the real vectors $\mathbf{t}$ and $\mathbf{e}$. Write $R_{[\mathbf{e}]}$ for reflection in the line through $\mathbf{0}$ and $\mathbf{e}$; and define $R_{[\mathbf{t}]}$ analogously. Then (**accept that for this discussion that**) $R_{[\mathbf{t}]}$ is the (**not necessarily 'well-encoding'**) oracle which encodes the phone book. Define the Grover transformation

$$G := -R_{[\mathbf{e}]}R_{[\mathbf{t}]}. \tag{C1}$$

It is geometrically obvious that this extends to a unitary map on $\mathbb{C}^K$. We return to this point later when we consider implementation.

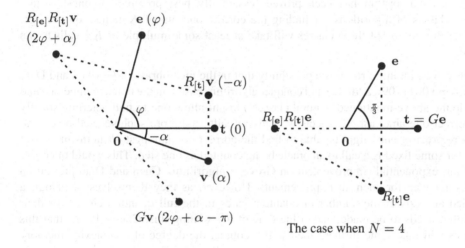

Figure C(i): The effect of Grover's G on states in $[\mathbf{t}, \mathbf{e}]$

The left-hand picture at Figure C(i) (in which α is negative!) shows (within the plane spanned by $\mathbf{t}$ and $\mathbf{e}$) the effect of the operation G on a vector $\mathbf{v}$. All vectors from $\mathbf{0}$ will be of unit length. Consider the angles they make with $\mathbf{t}$, calling φ the angle from $\mathbf{t}$ to $\mathbf{e}$ with φ in $(0, \frac{1}{2}\pi)$. We already know that $\cos\varphi = K^{-\frac{1}{2}}$. We see that if $\mathbf{v}$ is at an angle α, then $R_{[\mathbf{t}]}\mathbf{v}$ is at an angle $-\alpha$, then $R_{[\mathbf{e}]}R_{[\mathbf{t}]}\mathbf{v}$ is at an angle $(2\varphi + \alpha)$, and finally, $G\mathbf{v}$ is at an angle $(2\varphi + \alpha - \pi)$. Hence, the effect of G is to rotate $\mathbf{v}$ through an angle $2\varphi - \pi$. Set $\theta := \pi/2 - \varphi$, so that $\sin\theta = K^{-\frac{1}{2}}$. Then $G^r\mathbf{e}$ will be at an angle $\varphi - 2r\theta = \frac{1}{2}\pi - (2r+1)\theta$. We therefore choose r to make $(2r+1)\theta$ as close as possible to $\frac{1}{2}\pi$. Since $\sin\theta = K^{-\frac{1}{2}}$, then for large K, $\theta \approx K^{-\frac{1}{2}}$. so that $r \approx \frac{1}{4}\pi\sqrt{K}$. If K is large and we start with quantum state $\mathbf{e}$ and evolve with G^r for this r, then we have high probability $\sin^2(2r+1)\theta$ that a measurement of the state with $M = \mathrm{diag}(1, 2, \ldots, K)$

will yield the correct value k_0. If it doesn't, try again. Boyer, Brassard, Høyer and Tapp (quant-ph 9605034) show how Grover's algorithm may be speeded up somewhat on average by using a smaller r (and trying again if the scheme fails).

The right-hand picture at Figure 447C(i) shows (at smaller scale) that if $K = 4$, then one obtains (deterministically) the right answer after just one Grover iteration!

Ca. Exercise. Show that for Grover's algorithm with $K = 4$ and with $\mathbf{t} = (0, 0, 1, 0)^T$, we have

$$\mathbf{e} = (\tfrac{1}{2}, \tfrac{1}{2}, \tfrac{1}{2}, \tfrac{1}{2})^T, \quad R_{[\mathbf{t}]}\mathbf{e} = (-\tfrac{1}{2}, -\tfrac{1}{2}, \tfrac{1}{2}, -\tfrac{1}{2})^T, \quad R_{[\mathbf{e}]}R_{[\mathbf{t}]}\mathbf{e} = (0, 0, -1, 0)^T,$$

confirming that $G\mathbf{e} = \mathbf{t}$.

D. Fascinating thought.   Is it possible to exploit the striking property of $K = 4$ just described?

Grover's algorithm has been proved 'essentially best possible' amongst certain natural classes of algorithms for finding the correct one out of K items: on average, an algorithm in any of these classes will take at least some multiple of $K^{\frac{1}{2}}$ calls to the oracle.

However, intuition raises the possibility that in the nice phrase of G. Chen and Diao (quant-ph 0011109), a 'divide and conquer algorithm based on the magic number 4' (and outside the above-mentioned 'natural classes') might allow one to find deterministically the correct one amongst 4^n objects with just n calls to a more sophisticated oracle: to be more precise, one might be able to find the correct one amongst 4^n items in $O(n^r)$ steps for some fixed r, a call to an oracle being counted as one step. This would of course mark an 'exponential' improvement on Grover's algorithm. Chen and Diao present an interesting idea for such an improvement. However, as they themselves point out, a detailed analysis of the number of quantum gates in the full quantum circuit for their algorithm needs to be made; and at the time of my writing this, Chen tells me that this analysis will take some further months. (Of course, the degree of complexity, memory requirements, etc, for implementation of an oracle, would be an important factor in practice, particularly for sophisticated oracles.) I look forward with interest to the Chen-Diao results which will surely be posted on the quant-ph site.

10.2   Foundations of Quantum Probability

How would one start to think about implementing Grover's algorithm in the real world? Before we can answer that, we need a 'third pass' through the subject, this time finalizing several things, including the Fundamental Postulate of Quantum Probability 451C.

A. Some basic algebra.   Much of this is material covered in Chapter 8 with slight modification because we now work with $\mathbb{C}^n$ rather than $\mathbb{R}^n$ and our 'field of scalars' is $\mathbb{C}$, not $\mathbb{R}$. Thus we regard $\mathbb{C}^n$ as a vector space of dimension n over $\mathbb{C}$, never as a vector space of dimension $2n$ over $\mathbb{R}$.

With superscript T for transpose as before, for a (column) vector $(v_1, v_2, \ldots, v_n)^T$ in $\mathbb{C}^n$, we define $\mathbf{v}^\dagger := (\overline{v}_1, \overline{v}_2, \ldots, \overline{v}_n)$, with 'bar' denoting complex conjugate. We define (for $\mathbf{v}, \mathbf{w} \in \mathbb{C}^n$),

$$\langle \mathbf{v}, \mathbf{w} \rangle := \mathbf{v}^\dagger \mathbf{w} = \sum \overline{v}_k w_k, \qquad \|\mathbf{v}\|^2 := \langle \mathbf{v}^\dagger, \mathbf{v} \rangle = \sum |v_k|^2.$$

Note that $\langle \mathbf{w}, \mathbf{v} \rangle = \overline{\langle \mathbf{v}, \mathbf{w} \rangle}$. Two vectors $\mathbf{v}, \mathbf{w}$ are called orthogonal if $\langle \mathbf{v}, \mathbf{w} \rangle = 0$. I now revert to 'orthogonal' rather than perpendicular, because 'perpendicularity' is not that clear a concept in complex spaces. The 'Extension to Basis' Principle in its 'general' and orthogonal forms carries over. The definition of basis remains the same. If $\{ \mathbf{w}^{(1)}, \mathbf{w}^{(2)}, \ldots, \mathbf{w}^{(r)} \}$ is an orthonormal basis for a subspace W of $\mathbb{C}^n$, then orthogonal projection P_W onto W is obtained via

$$P_W \mathbf{v} = \sum \langle \mathbf{w}^{(k)}, \mathbf{v} \rangle \mathbf{w}^{(k)},$$

so that P_W has the matrix representation

$$P_W = \sum \mathbf{w}^{(k)} (\mathbf{w}^{(k)})^\dagger.$$

A matrix P corresponds to an orthogonal projection onto a subspace if and only if (compare Lemma 324Jb)

$$P^\dagger = P = P^2.$$

We know that a complex $n \times n$ matrix A is called *self-adjoint* if $A = A^\dagger$, equivalently (as you can check) if

$$\langle A\mathbf{v}, \mathbf{w} \rangle = \langle \mathbf{v}, A\mathbf{w} \rangle$$

for all $\mathbf{v}, \mathbf{w}$. Especially, if A is self-adjoint, then, for any $\mathbf{v} \in \mathbb{C}^n$,

$$\mathbf{v}^\dagger A\mathbf{v} = \langle \mathbf{v}, A\mathbf{v} \rangle = \langle A\mathbf{v}, \mathbf{v} \rangle = \overline{\langle \mathbf{v}, A\mathbf{v} \rangle},$$

so that $\mathbf{v}^\dagger A\mathbf{v}$ is real. We shall see that, when $\mathbf{v}$ is a **state**, that is, when $\|\mathbf{v}\| = 1$, $\mathbf{v}^\dagger A\mathbf{v}$ represents the mean or expectation $\mathbb{E}_\mathbf{v}(A)$ of A when the system is in state $\mathbf{v}$.

The following result is extremely important.

▶▶ **Aa. Fact: Spectral Theorem for self-adjoint matrices.** *If A is a self-adjoint matrix, then there exist orthogonal subspaces $V_1, V_2, \ldots, V_r$ of $\mathbb{C}^n$ with sum $\mathbb{C}^n$ such that*

$$A = \alpha_1 P_1 + \alpha_2 P_2 + \cdots + \alpha_r P_r, \tag{A1}$$

*where P_k denotes orthogonal projection onto V_k and $\alpha_1, \alpha_2, \ldots, \alpha_r$ are distinct **real** numbers. Then,*

$$g(A) = g(\alpha_1)P_1 + g(\alpha_2)P_2 + \cdots + g(\alpha_r)P_r \tag{A2}$$

for polynomial or exponential functions g. Each P_k is a polynomial in A with real coefficients, and therefore commutes with any operator which commutes with A.

Here the α_k are the distinct eigenvalues of A, the distinct roots of

$$\det(\alpha I - A) = 0,$$

and V_k is the space of eigenvectors of A corresponding to α_k:

$$V_k = \{\mathbf{v} \in \mathbb{C}^n : A\mathbf{v} = \alpha_k \mathbf{v}\}.$$

Note that if $\mathbf{v} \neq \mathbf{0}$ and $A\mathbf{v} = \alpha \mathbf{v}$, then

$$\overline{\alpha}\|\mathbf{v}\|^2 = \langle \alpha\mathbf{v}, \mathbf{v}\rangle = \langle A\mathbf{v}, \mathbf{v}\rangle = \langle \mathbf{v}, A\mathbf{v}\rangle = \langle \mathbf{v}, \alpha\mathbf{v}\rangle = \alpha\|\mathbf{v}\|^2,$$

so that $\overline{\alpha} = \alpha$ and α is real. Of course, the tricky thing with Fact 449Aa is to prove that A has enough eigenvectors to span $\mathbb{C}^n$.

Because $P_k^2 = P_k$ and $P_k P_m = 0$ hen $k \neq m$, it follows by induction from 449(A1) that

$$A^n = \alpha_1^n P_1 + \alpha_2^n P_2 + \cdots + \alpha_r^n P_r,$$

so that 449(A2) follows for polynomials g.

For any $n \times n$ matrix B, we define

$$e^B := \exp(B) := I + B + \frac{B^2}{2!} + \cdots,$$

the series converging componentwise. Fact: we have for any $n \times n$ matrix $\mathcal{H}$,

$$\frac{\mathrm{d}}{\mathrm{d}t} e^{-it\mathcal{H}} = -i\mathcal{H}e^{-it\mathcal{H}} = e^{-it\mathcal{H}}(-i\mathcal{H}),$$

interpreting the derivative componentwise. Compare Schrödinger's equation. Fact: we have

$$e^{A+B} = e^A e^B \quad \text{when} \quad AB = BA.$$

B. Notes on unitary matrices.

We already know that a matrix U is called unitary if $UU^\dagger = U^\dagger U = I$. If U is unitary then (you prove this) $\|U\mathbf{v}\|^2 = \|\mathbf{v}\|^2$ for every vector $\mathbf{v}$. We need evolutions to be unitary to preserve probabilities.

Ba. Exercise. Show that if $\|U\mathbf{v}\|^2 = \|\mathbf{v}\|^2$ for every vector $\mathbf{v}$ in $\mathbb{C}^n$, then U is unitary. *Hint.* Show that if $A := UU^\dagger - I$, then for all vectors $\mathbf{v}$ and $\mathbf{w}$ in $\mathbb{C}^2$, $\Re(\mathbf{v}^\dagger A\mathbf{w}) = 0$. Why does the desired result now follow immediately?

Bb. Exercise. Show that for a square matrix A, $(e^A)^\dagger = \exp(A^\dagger)$. Deduce that if $\mathcal{H}$ is self-adjoint, then $e^{-it\mathcal{H}}$ is unitary for all $t \geq 0$. Prove the converse result that if $e^{-it\mathcal{H}}$ is unitary for all $t \geq 0$, then $\mathcal{H}$ is self-adjoint.

▶▶▶ **C. The Fundamental Postulate of (finite-dimensional) Quantum Probability.** Suppose that $\mathbb{C}^n$ is the vector space in which the (normalized) quantum states are vectors of unit length, and that A is an associated observable with spectral decomposition as at 449(A1). Then, if a measurement of A is made when the system is in the state $\mathbf{v}$ (where $\|\mathbf{v}\| = 1$), then Nature

- chooses a number in $\{1, 2, \ldots, r\}$, choosing k with probability

$$p_k(\mathbf{v}) = \|P_k\mathbf{v}\|^2,$$

- reveals α_k as the measured value of A, and

- makes the state jump to $P_k\mathbf{v}/\|P_k\mathbf{v}\|$.

We write $\mathbb{P}_\mathbf{v}$ for the probability law associated with state $\mathbf{v}$, and so will write

$$\mathbb{P}_\mathbf{v}(A \text{ is measured as } \alpha_k) = p_k(\mathbf{v}) = \|P_k\mathbf{v}\|^2.$$

As I have already mentioned, several leading physicists suspect that the postulate may not be the last word. What matters to us is that all calculations made on the basis of the (general form of the) postulate have been verified experimentally.

Let's take $n = 2$. We have already met the matrix $\begin{pmatrix} 0 & 1 \\ 1 & 0 \end{pmatrix}$ in its unitary guise, as a NOT gate. But the matrix is also self-adjoint, and as such it is an important observable, σ_x, one of the Pauli spin matrices. Let us again write M for $\mathrm{diag}(0, 1)$ and $c = 2^{-\frac{1}{2}}$.

If our system is in state $\mathbf{v} = c(\mathbf{z} + \mathbf{u})$, then a measurement of M will produce the result 0 or 1 with probability $\frac{1}{2}$ each, and the state will jump to $\mathbf{z}$ or $\mathbf{u}$ accordingly.

Let $A = \sigma_x$ for the moment. If A is measured either in state $\mathbf{z}$ or in state $\mathbf{u}$, we obtain $+1$ or -1, each with probability $\frac{1}{2}$. But if A is measured in state $\mathbf{v}$ which is a 'superposition' of $\mathbf{z}$ and $\mathbf{u}$, then, since $A\mathbf{v} = \mathbf{v}$, we obtain the deterministic result 1. You can see why we need to be careful about superposition.

Ca. Eigenvectors of the Hamiltonian. If $\mathcal{H}$ is the observable which is the Hamiltonian for Schrödinger's equation and if $\mathbf{v}_0$, the state of the system at time 0, is an eigenstate (eigenvector) of $\mathcal{H}$ corresponding to (real) eigenvalue λ, then the state at time t will be $\mathbf{v}_t = \mathrm{e}^{-\mathrm{i}\lambda t}\mathbf{v}_0$, still an eigenvector of $\mathcal{H}$ with eigenvalue λ. As mentioned before, $\mathcal{H}$ describes the energy of the quantum system, and the fact just described is a 'conservation of energy' law. (We have to be very careful though that, for example, we do not become trapped into thinking that an electron will stay for ever in an excited state: we have to remember that simple idealized models will be subject to all sorts of perturbations and to 'second-order' forces. Even the vacuum in Quantum Theory is not a classical vacuum.)

▶ **D. Density matrices and expectations.** For the mean of observable A in state $\mathbf{v}$, we have

$$\mathbb{E}_{\mathbf{v}}(A) = \sum \alpha_k \|P_k \mathbf{v}\|^2 = \sum \alpha_k (P_k \mathbf{v})^\dagger P_i \mathbf{v}$$
$$= \mathbf{v}^\dagger A \mathbf{v} = \mathrm{trace}(\mathbf{v}^\dagger A \mathbf{v}) = \mathrm{trace}(\mathbf{v}\mathbf{v}^\dagger A) = \mathrm{trace}(\rho_{\mathbf{v}} A), \qquad \text{(D1)}$$

where $\rho_{\mathbf{v}}$ is the density matrix $\rho_{\mathbf{v}} = \mathbf{v}\mathbf{v}^\dagger$ associated with $\mathbf{v}$. Recall that the trace of a square matrix is the sum of its diagonal elements, that $\mathbf{v}^\dagger A \mathbf{v}$ is a number (in fact a real one) and is therefore equal to its trace, and that if F is an $r \times s$ matrix and G is an $s \times r$ matrix, then $\mathrm{trace}(FG) = \mathrm{trace}(GF)$.

▶ **E. Superposition versus probabilistic mixture.** Let $\mathbf{v}$ and $\mathbf{w}$ be orthogonal states, and let $\mathbf{x}$ be the superposition $c(\mathbf{v} + \mathbf{w})$ where $c = 2^{-\frac{1}{2}}$. The density matrix

$$\tfrac{1}{2}(\rho_{\mathbf{v}} + \rho_{\mathbf{w}})$$

corresponds to a probabilistic mixture of states: 'With probability $\frac{1}{2}$ prepare state $\mathbf{v}$; otherwise, prepare state $\mathbf{w}$'. However,

$$\rho_{\mathbf{x}} = \tfrac{1}{2}(\mathbf{v} + \mathbf{w})(\mathbf{v} + \mathbf{w})^\dagger = \tfrac{1}{2}(\rho_{\mathbf{v}} + \rho_{\mathbf{w}}) + \tfrac{1}{2}(\mathbf{v}\mathbf{w}^\dagger + \mathbf{w}\mathbf{v}^\dagger),$$

showing a clear distinction between superposition and probabilistic mixtures.

Probabilistic mixtures do play an important part in the theory, especially in the study of subsystems of a full quantum system. We shall see the density matrix $\frac{1}{2}(\rho_{\mathbf{v}} + \rho_{\mathbf{w}})$ arise in this way in Subsection 467C.

▶▶ **F. Commuting observables.** Suppose that A and B are commuting observables such that $AB = BA$. Then (a fact which you can probably see how to prove) there exists an observable C and polynomials g and h with real coefficients such that $A = g(C)$, $B = h(C)$. We can measure A and B simultaneously, because the experiment of measuring them simultaneously is equivalent to that of measuring C to give a result γ (in which case A yields measurement $g(\gamma)$ and B measurement $h(\gamma)$).

A much better way to think about this is that *if $AB = BA$, then the 'probability that A is measured as α and B is measured as β' is the same if A is measured before B as if B is measured before A: commuting observables A and B have a proper joint distribution.*

Suppose that A and B are *any* two operators, with spectral decompositions

$$A = \alpha_1 P_1 + \alpha_2 P_2 + \cdots + \alpha_r P_r,$$
$$B = \beta_1 Q_1 + \beta_2 Q_2 + \cdots + \beta_s Q_s.$$

Then, if A is measured before B,

$$\mathbb{P}_{\mathbf{v}}(A \text{ is measured as } \alpha_k \text{ then } B \text{ is measured as } \beta_\ell)$$
$$= \|P_k \mathbf{v}\|^2 \frac{\|Q_\ell P_k \mathbf{v}\|^2}{\|P_k \mathbf{v}\|^2} = \|Q_\ell P_k \mathbf{v}\|^2.$$

As explained in the statement of the Spectral Theorem 449Aa, if A and B commute, then Q_ℓ commutes with P_k, and so the above probability does equal

$$\mathbb{P}_\mathbf{v}(B \text{ is measured as } \beta_\ell \text{ then } A \text{ is measured as } \alpha_k)$$

when B is measured before A.

Fa. Exercise on non-commuting observables. Let

$$A = \begin{pmatrix} 1 & 0 \\ 0 & -1 \end{pmatrix}, \quad B = \begin{pmatrix} 0 & 1 \\ 1 & 0 \end{pmatrix}, \quad \mathbf{v} = \begin{pmatrix} 1 \\ 0 \end{pmatrix}.$$

Show that if the state is $\mathbf{v}$ and A is measured before B, then

$$\mathbb{P}_\mathbf{v}(A \text{ is measured as } 1 \text{ then } B \text{ is measured as } 1) = \tfrac{1}{2},$$

whereas if B is measured before A, then

$$\mathbb{P}_\mathbf{v}(B \text{ is measured as } 1 \text{ then } A \text{ is measured as } 1) = \tfrac{1}{4}.$$

▶▶ **G. The Heisenberg Uncertainty Principle as Mathematics.** Suppose that A and B are observables. Let

$$\mu_\mathbf{v}(A) := \mathbb{E}_\mathbf{v}(A) = \langle \mathbf{v}, A\mathbf{v} \rangle$$

be the mean of A when the state is $\mathbf{v}$, let

$$\tilde{A}_\mathbf{v} := A - \mu_\mathbf{v}(A)I$$

(analogously to $\tilde{X} = X - \mu_X$), and define

$$\mathrm{Var}_\mathbf{v}(A) := \mathbb{E}_\mathbf{v}(\tilde{A}^2) = \langle \mathbf{v}, \tilde{A}^2\mathbf{v} \rangle = \langle \tilde{A}\mathbf{v}, \tilde{A}\mathbf{v} \rangle = \|\tilde{A}\mathbf{v}\|^2.$$

Of course, $\mathrm{Var}_\mathbf{v}(A) = \mathbb{E}_\mathbf{v}(A^2) - [\mu_\mathbf{v}(A)]^2$. We define

$$\mathrm{SD}_\mathbf{v}(A) := [\mathrm{Var}_\mathbf{v}(A)]^{\frac{1}{2}} = \|\tilde{A}\mathbf{v}\|.$$

Now, $\tilde{A}\tilde{B}$ is generally not self-adjoint (it *is* if A and B commute). However, we can write

$$\tilde{A}\tilde{B} = S_+ + iS_-,$$

where

$$S_+ = \frac{\tilde{A}\tilde{B} + \tilde{B}\tilde{A}}{2}, \quad S_- = \frac{\tilde{A}\tilde{B} - \tilde{B}\tilde{A}}{2i} = \frac{AB - BA}{2i},$$

and S_+ and S_- are self-adjoint. Thus, since $\langle \mathbf{v}, S_\pm\mathbf{v} \rangle$ is real,

$$|\langle \mathbf{v}, \tilde{A}\tilde{B}\mathbf{v} \rangle|^2 = |\langle \mathbf{v}, S_+\mathbf{v} \rangle|^2 + |\langle \mathbf{v}, S_-\mathbf{v} \rangle|^2.$$

Combining this with the complex Cauchy–Schwarz inequality, we obtain

$$|\langle \mathbf{v}, S_-\mathbf{v} \rangle| \leq |\langle \mathbf{v}, \tilde{A}\tilde{B}\mathbf{v} \rangle| = |\langle \tilde{A}\mathbf{v}, \tilde{B}\mathbf{v} \rangle| \leq \|\tilde{A}\mathbf{v}\| \, \|\tilde{B}\mathbf{v}\|,$$

and we have the Heisenberg Uncertainty Principle:

$$\text{SD}_\mathbf{v}(A)\text{SD}_\mathbf{v}(B) \geq |\langle \mathbf{v}, S_-\mathbf{v} \rangle|, \quad \text{where } S_- = \frac{AB - BA}{2\mathrm{i}}. \tag{G1}$$

Note that for the spinor representation of spins,

$$\text{SD}_\mathbf{v}(\sigma_x)\text{SD}_\mathbf{v}(\sigma_y) \geq |\langle \mathbf{v}, \sigma_z\mathbf{v} \rangle|.$$

▶▶ **H. The Heisenberg Uncertainty Principle as Physics.** We have to be very careful about interpreting the Uncertainty Principle as Physics. I stick to the orthodox 'operator' picture.

The Uncertainty Principle gives a lower bound for the product of $\text{SD}_\mathbf{v}(A)$ and $\text{SD}_\mathbf{v}(B)$ both associated with the *same* state $\mathbf{v}$. The quantity $\text{SD}_\mathbf{v}(A)\text{SD}_\mathbf{v}(B)$ is therefore relevant only to situations in which state $\mathbf{v}$ is prepared on a large set of occasions, where A is measured on a large subset of those occasions and B is measured on a large *disjoint* subset of those occasions. (One of the ways in which we could know that the system is in state $\mathbf{v}$ is if $[\mathbf{v}]$ is the 1-dimensional eigenspace of some operator C corresponding to some eigenvalue λ and if C has been measured with result λ.)

The quantity $\text{SD}_\mathbf{v}(A)\text{SD}_\mathbf{v}(B)$ *has no relevance to simultaneous measurement.* If, for example, we measure B before A, then the measurement of B generally changes the state to a new state $\mathbf{w}$, and then the value $\text{SD}_\mathbf{v}(A)$ has no relevance.

Ha. Example. Take the example of Exercise 453Fa. Here, $\text{SD}_\mathbf{v}(A)$ is obviously 0 because A is deterministically 1 in state $\mathbf{v}$. However, if B is measured before A, then that measurement changes $\mathbf{v}$ to a new state $\mathbf{w}$, and (you check) whatever the result of the measurement on B, A will then be measured as $+1$ or -1 with probability $\frac{1}{2}$ each, so that $\text{SD}_\mathbf{w}(A) = 1$.

Note. In this example, $\mathbb{E}_\mathbf{v}(A) = 1$, but for each of the two possible values b of the measurement of B, we have "$\mathbb{E}(A \,|\, B = b) = 0$". So there is no 'tower property of conditional expectations'. We wouldn't expect there to be because the measurement of B has changed the state. Of course, I have misused conditional expectations, which is why there are 'double quotes'.

▶▶ **I. Classical Heisenberg Uncertainty Principle for position and momentum.** I give a fairly detailed discussion because this is one of the most widely discussed topics in Science. It is part of infinite-dimensional Quantum Theory, and we can only treat it heuristically here because (amongst other things) self-adjointness now becomes a rather subtle concept needing precise specifications of domains of operators. From time to time, however, I make some remarks on the way to make things rigorous. (I want to persuade you to read Reed and Simon [191].)

The state of a particle moving on the real line is a map

$$\mathbf{v} : \mathbb{R} \to \mathbb{C}, \quad x \mapsto v(x),$$

such that $\mathbf{v} \in \mathcal{L}^2(\mathbb{R})$ in that

$$\|\mathbf{v}\|^2 := \int |v(x)|^2 \mathrm{d}x = 1.$$

There is no difference between v and $\mathbf{v}$; it's just that I often use $\mathbf{v}$ to relate to the finite-dimensional situation we have studied. We define the inner product of two states $\mathbf{v}$ and $\mathbf{w}$ in $\mathcal{L}^2(\mathbb{R})$ as

$$\langle \mathbf{v}, \mathbf{w} \rangle := \int \overline{v(x)}\, w(x) \, \mathrm{d}x.$$

The position of our particle is represented by the observable A, where

$$(A\mathbf{v})(x) := xv(x), \quad \text{whence } (g(A)\mathbf{v})(x) = g(x)v(x).$$

In particular, for a subinterval Γ of $\mathbb{R}$,

$$\begin{aligned}
\mathbb{P}_\mathbf{v}\,(\text{position in } \Gamma) &= \mathbb{E}_v I_\Gamma(A) = \langle \mathbf{v}, I_\Gamma(A)\mathbf{v} \rangle \\
&= \int \overline{v(x)} I_\Gamma(x)v(x) \, \mathrm{d}x = \int_\Gamma |v(x)|^2 \mathrm{d}x,
\end{aligned}$$

so that $|v(x)|^2$ is the pdf for position when the state is $\mathbf{v}$.

Momentum is described by the operator

$$(B\mathbf{v})(x) := -iv'(x), \quad B = -i\frac{\mathrm{d}}{\mathrm{d}x} = -i\partial_x.$$

Note that B is (formally) self-adjoint because integration by parts 'with things at infinity vanishing' shows that

$$\langle \mathbf{v}, B\mathbf{w} \rangle = \int \overline{v}\{-iw'\}\mathrm{d}x = \int i\overline{v}'w \, \mathrm{d}x = \int \overline{\{-iv'\}}w \, \mathrm{d}x = \langle B\mathbf{v}, \mathbf{w} \rangle.$$

Now we have

$$(AB - BA)v = x\{-iv'(x)\} + i\partial_x\{xv(x)\} = iv(x),$$

so that we have, formally, the famous **Canonical Commutation Relation (CCR)**

$$AB - BA = iI,$$

and $S_- = \frac{1}{2}I$. Thus in this case we have the classical Uncertainty Principle (in units of $\hbar$) for position A and momentum B:

$$\mathrm{SD}_\mathbf{v}(A)\mathrm{SD}_\mathbf{v}(B) \geq \tfrac{1}{2} \quad \text{for all states } v. \tag{I1}$$

Some steps towards a rigorous formulation of the CCR may be found at 458Ie below.

Ia. Exercise. Show that if we have the 'Gaussian' state

$$v(x) \; = \; (2\pi\sigma^2)^{-\frac{1}{4}} \exp\left(-\frac{x^2}{4\sigma^2}\right),$$

then the distribution of the position of the particle in state **v** is $N(0, \sigma^2)$ and we have

$$\mathbb{E}_{\mathbf{v}}(A) = 0, \quad \mathrm{Var}_{\mathbf{v}}(A) = \sigma^2, \quad \mathbb{E}_{\mathbf{v}}(B) = 0, \quad \mathrm{Var}_{\mathbf{v}}(B) = \frac{1}{4\sigma^2},$$

so that in this case, we have equality at (I1).

Ib. Having fun. Let's have some (heuristic) fun by continuing with the context of the Exercise. We have

$$
\begin{aligned}
\{\exp(\mathrm{i}\beta B)\mathbf{v}\}(x) \; &= \; \{\exp(\beta\partial_x)\mathbf{v}\}(x) \\
&= \; v(x) + \beta v'(x) + \frac{\beta^2}{2!}v''(x) + \cdots \\
&= \; v(x + \beta) \quad \text{(by Taylor's Theorem)} \\
&= \; (2\pi\sigma^2)^{-\frac{1}{4}} \exp\left\{-\frac{(x^2 + 2\beta x + \beta^2)}{4\sigma^2}\right\}.
\end{aligned}
$$

Hence we can evaluate the Characteristic Function of momentum B in state **v** as

$$\mathbb{E}_{\mathbf{v}}\exp(\mathrm{i}\beta B) \; = \; \exp\left(-\frac{\beta^2}{4\sigma^2}\right) \int \frac{1}{(2\pi\sigma^2)^{\frac{1}{2}}} \exp\left(\frac{\beta x}{2\sigma^2}\right) \exp\left(-\frac{x^2}{2\sigma^2}\right) \, \mathrm{d}x$$

and the integral is the value evaluated at $\beta/(2\sigma^2)$ of the MGF of $N(0, \sigma^2)$, namely,

$$\exp\left(\frac{1}{2}\sigma^2\frac{\beta^2}{4\sigma^4}\right).$$

So, the Characteristic Function of B in state **v** is

$$\mathbb{E}_{\mathbf{v}}\exp(\mathrm{i}\beta B) \; = \; \exp\left(-\frac{1}{2}\beta^2\frac{1}{4\sigma^2}\right),$$

and hence, in state **v**, momentum B has the $N(0, 1/(4\sigma^2))$ distribution.

You may be uneasy because $\exp(\mathrm{i}\beta B)$ is not self-adjoint (it *is* unitary, not that this is relevant now) and so I should not be using $\mathbb{E}_{\mathbf{v}}\exp(\mathrm{i}\beta B)$. However, I am sure that you will allow me to extend $\mathbb{E}_{\mathbf{v}}$ by linearity and continuity. I could alternatively work with moments of B, with the same results.

Of course, it is meaningless in this quantum setting to talk of a joint pdf of (A, B) in state v. We know that the observable $(AB - BA)/(2\mathrm{i})$ takes the value $\frac{1}{2}$, and this fact is rather difficult to square with the idea of a joint distribution of (A, B)!!

▶ **Ic. Fourier duality.** Let's extend the idea just explained to calculate the distribution of momentum for an arbitrary state $\mathbf{v}$ in $\mathcal{L}^2$. For a nice function $v : \mathbb{R} \to \mathbb{C}$, define the **Fourier transform** $v^\wedge$ of v via

$$v^\wedge(b) := \frac{1}{\sqrt{2\pi}} \int e^{-iba} v(a)\, da,$$

all integrals around here being over $\mathbb{R}$. Then we have the **Fourier inversion formula**

$$v(a) = \frac{1}{\sqrt{2\pi}} \int e^{iab} v^\wedge(b)\, db.$$

Compare 167(A2).

Those with a keen eye for rigour are advised that we can extend the Fourier transform and its inverse to continuous maps from $L^2(\mathbb{R})$ to $L^2(\mathbb{R})$, where $L^2(\mathbb{R})$ is $\mathcal{L}^2(\mathbb{R})$ with functions equal except on sets of Lebesgue measure zero being identified.

Indeed, **Parseval's formula** holds for $\mathbf{v}, \mathbf{w} \in L^2(\mathbb{R})$:

$$\langle \mathbf{v}, \mathbf{w} \rangle = \langle \mathbf{v}^\wedge, \mathbf{w}^\wedge \rangle,$$

so that Fourier transform is a unitary map. Now, formally,

$$(Bv)(a) = -iv'(a) = \frac{1}{\sqrt{2\pi}} \int e^{iab} b v^\wedge(b)\, db.$$

so that we must have

$$(Bv)^\wedge(b) = bv^\wedge(b).$$

Next, with one forgivable misuse of notation,

$$\mathbb{E}_{\mathbf{v}}\left(e^{i\theta B}\right) = \langle \mathbf{v}, e^{i\theta B} \mathbf{v} \rangle = \langle \mathbf{v}^\wedge, (e^{i\theta B}\mathbf{v})^\wedge \rangle$$

$$= \langle \mathbf{v}^\wedge, (e^{i\theta b}\mathbf{v})^\wedge \rangle = \int e^{i\theta b} |v^\wedge(b)|^2 db,$$

so that

$$\mathbb{P}_{\mathbf{v}}(B \in db) = |v^\wedge(b)|^2 db,$$

a formula which makes sense for all $\mathbf{v} \in L^2(\mathbb{R})$.

Note that

$$i(v^\wedge)'(b) = \frac{1}{\sqrt{2\pi}} \int e^{-iba} a v(a)\, da.$$

We therefore have the nice **duality** between the 'position' and 'momentum' representations of the CCR:

$$\begin{aligned}
(Av)(a) &= av(a), & (Bv)(a) &= -iv'(a), \\
(Av)^\wedge(b) &= i(v^\wedge)'(b), & (Bv)^\wedge(b) &= bv^\wedge(b),
\end{aligned}$$

so that Fourier transforms interchange the rôles of position and momentum modulo the sign change necessary to preserve the CCR.

▶ **Id. Discussion: more quantum weirdness.** You may wonder about such questions as the following. Suppose that I set up an experiment which returns 1 if the particle is observed as being in the interval (α, β), 0 if not. The corresponding observable is C, where

$$(Cv)(x) = I_{(\alpha,\beta)}(x)v(x).$$

If the system is in state **v**, and C is measured as 1, then the new state **w** will satisfy

$$w(x) = \gamma^{-1}I_{(\alpha,\beta)}(x)v(x), \quad \text{where } \gamma := \left\{ \int_\alpha^\beta |v(y)|^2 dy \right\}^{\frac{1}{2}}.$$

What can we say about the particle's momentum when it is in state **w**?

You can immediately see a strange feature. The value $Bw(x) = -iw'(x)$ will generally not exist when $x = \alpha$ or β, except as a multiple of a delta function. We therefore expect the value $\mathbb{E}_\mathbf{w}(B^2)$ to be infinite. Indeed, if one knows anything about delta functions, one expects the pdf of B in state **w** to tail off essentially as $1/b^2$, so that B will have neither variance nor proper mean.

Let us work out the easy case when $\alpha = -\delta$ and $\beta = \delta$, where $0 < \delta < \frac{1}{2}$ and where v is a nice infinitely differentiable function on $\mathbb{R}$ with $\int |v(x)|^2 = 1$ and such that $v \equiv 1$ on $(-\delta, \delta)$. Then the pdf of B in state **w** is (you check!)

$$|w^\wedge(b)|^2 = \frac{\sin^2(\delta\beta)}{\pi\delta b^2},$$

tallying exactly with our guess. From the point of view of classical Probability, the 'pseudo-periodic' behaviour of the pdf of B is bizarre, but if you take Fourier transforms when there are delta functions lurking in the background, such things are bound to happen.

Ie. Interesting mathematical digression. The CCR is not that easy to formulate rigorously because A and B are defined only on subspaces of $L^2(\mathbb{R})$. Hermann Weyl's solution was to rewrite the CCR as

$$T_t U_s = e^{ist}U_s T_t, \quad \text{where } U_s = e^{-isA}, \; T_t = e^{-itB}. \tag{I2}$$

We have for the position representation, for every $\mathbf{v} \in L^2(\mathbb{R})$,

$$(U_s v)(x) = e^{-isx}v(x), \quad (T_t v)(x) = v(x+t), \tag{I3}$$

the second being *formally* the Taylor-series expansion we have seen before. However, the correct theory (of 'strongly continuous one-parameter semigroups', or, in this context, of 'strongly continuous one-parameter unitary groups') means that the T_t equation at (I3) is correct for all v in $\mathcal{L}^2(\mathbb{R})$, even for nowhere-differentiable v and for those infinitely differentiable v for which the Taylor series sometimes converges to the wrong value. When A and B are (truly) self-adjoint operators, we can define T_t and U_s as at (I2).

If the commutation relation at (I2) holds, then the CCR follows formally on taking $\partial^2/\partial s\partial t$ and then setting $s = t = 0$. Conversely, if the CCR holds, then the commutation relation at (I2) follows formally from the '*Baker–Campbell–Hausdorff formula*'. Let's take a heuristic look. Suppose that operators F and G satisfy

$$FC = CF, \quad GC = CG, \text{ where } C := FG - GF.$$

You show by induction that $F^n G - G^n F = nF^{n-1}C$, whence

$$e^{sF}G - Ge^{sF} = se^{sF}C, \quad e^{sF}G = (G + sC)e^{sF}.$$

Hence, $e^{sF}G^n = (G + sC)^n e^{sF}$, and

$$e^{sF}e^{tG} = e^{tG}e^{stC}e^{sF} = e^{tG}e^{sF}e^{stC}.$$

Formally, 458(I2) is the case when $F = -iA$, $G = -iB$, $C = -iI$.

J. The Uncertainty Principle and Fisher information.

You may have thought while reading the above treatment of the classical Uncertainty Principle, "Haven't we seen something like this before when we studied the Cramér–Rao MVB inequality?".

Return to the classical Heisenberg principle and suppose that $\mathbf{v}$ is such that

$$\mathbb{E}_{\mathbf{v}}(A) = 0 = \mathbb{E}_{\mathbf{v}}(B).$$

Let $h(y)$ be the pdf $|v(y)|^2 = \overline{v(y)}v(y)$ for position. Consider now the location-parameter situation of Subsection 198H. We have, with the notation there,

$$I(Y) = \int \frac{h'(y)^2}{h(y)} dy.$$

But

$$|h'(y)| = 2\Re\{\overline{v(y)}v'(y)\} \leq 2|v(y)||v'(y)|,$$

so that

$$I(Y) \leq 4\int |v'(y)|^2 dy = 4\|Bv\|^2 = 4\text{Var}_{\mathbf{v}}(B).$$

And now, since the Cramér–Rao inequality says that $I(Y) \geq 1/\text{Var}_{\mathbf{v}}(A)$, the classical Heisenberg principle follows.

I do not read too much into this, since both the Cramér–Rao Theorem and the Uncertainty Principle are almost-immediate consequences of the Cauchy–Schwarz inequality.

Frieden's claim ([86]) to derive much Physics from Fisher information is, unsurprisingly, controversial. His interpretation of the Uncertainty Principle appears to be different from ours. (On this, as with much else in his interesting book, I would have welcomed much fuller discussion.)

K. Tensor products – again.

I reiterate points made in Chapter 8, but with a slightly different (but strictly equivalent) way of thinking about tensor products. Recall that the $(m+n) \times (r+s)$ matrix

$$\begin{pmatrix} a_{11} & \cdot & a_{1r} & b_{11} & \cdot & b_{1s} \\ \cdot & \cdot & \cdot & \cdot & & \cdot \\ a_{m1} & \cdot & a_{mr} & b_{m1} & \cdot & b_{ms} \\ c_{11} & \cdot & c_{1r} & d_{11} & \cdot & d_{1s} \\ \cdot & \cdot & \cdot & \cdot & & \cdot \\ c_{n1} & \cdot & c_{nr} & d_{n1} & \cdot & d_{ns} \end{pmatrix} \quad \text{is written} \quad \begin{pmatrix} A & B \\ C & D \end{pmatrix},$$

where A is $m \times r$, etc. Partitioned matrices 'of compatible dimensions' multiply in the usual way:

$$\begin{pmatrix} A & B \\ C & D \end{pmatrix} \begin{pmatrix} E & F \\ G & H \end{pmatrix} = \begin{pmatrix} AE+BG & AF+BH \\ CE+DG & CF+DH \end{pmatrix}.$$

The tensor product $\mathbf{v} \otimes \mathbf{w}$ of a vector $\mathbf{v}$ in $\mathbb{C}^n$ with $\mathbf{w}$ in $\mathbb{C}^s$ is the vector in $\mathbb{C}^{ns}$:

$$\mathbf{v} \otimes \mathbf{w} = \begin{pmatrix} v_1 \mathbf{w} \\ \cdot \\ v_n \mathbf{w} \end{pmatrix}.$$

If we are given the tensor product $\mathbf{v} \otimes \mathbf{w}$ of two vectors of length 1, then (you check) the pair $(\mathbf{v}, \mathbf{w})$ is defined uniquely only modulo phase transformations

$$\mathbf{v} \mapsto e^{i\alpha}\mathbf{v}, \quad \mathbf{w} \mapsto e^{-i\alpha}\mathbf{w}. \tag{K1}$$

Note that

$$\langle \mathbf{v} \otimes \mathbf{w}, \mathbf{s} \otimes \mathbf{t} \rangle = \langle \mathbf{v}, \mathbf{s} \rangle \langle \mathbf{w}, \mathbf{t} \rangle, \quad \| \mathbf{v} \otimes \mathbf{w} \| = \| \mathbf{v} \| \, \| \mathbf{w} \|.$$

The tensor product $A \otimes B$ of an $n \times n$ matrix A and an $s \times s$ matrix B is the $ns \times ns$ matrix

$$A \otimes B = \begin{pmatrix} a_{11}B & \cdot & a_{1n}B \\ \cdot & \cdot & \cdot \\ a_{n1}B & \cdot & a_{nn}B \end{pmatrix},$$

and then, for $\mathbf{v} \in \mathbb{C}^n$ and $\mathbf{w} \in \mathbb{C}^s$,

$$(A \otimes B)(\mathbf{v} \otimes \mathbf{w}) = \begin{pmatrix} a_{11}B & \cdot & a_{1n}B \\ \cdot & \cdot & \cdot \\ a_{n1}B & \cdot & a_{nn}B \end{pmatrix} \begin{pmatrix} v_1 \mathbf{w} \\ \cdot \\ v_n \mathbf{w} \end{pmatrix} = (A\mathbf{v}) \otimes (B\mathbf{w}).$$

You proved in Chapter 8 that not every vector in $\mathbb{C}^n \otimes \mathbb{C}^s = \mathbb{C}^{ns}$ is of the form $\mathbf{v} \otimes \mathbf{w}$ (most are 'entangled'). You noticed that if $n > 1$ and $s > 1$ and

$$\mathbf{t} = (t_{11}, \dots, t_{1s}, \dots, t_{ns})^T = \mathbf{v} \otimes \mathbf{w} \in \mathbb{C}^n \otimes \mathbb{C}^s,$$

then a number of algebraic relations follow: for example, $t_{jj}t_{kk} = t_{jk}t_{kj}$, since $t_{jk} = v_j w_k$, etc. However, a linear map on $\mathbb{C}^n \otimes \mathbb{C}^s$ *is* determined by its action on $\mathbf{v} \otimes \mathbf{w}$ vectors.

We have

$$(A \otimes B)^\dagger = A^\dagger \otimes B^\dagger.$$

▶ **L. Pure-product states; 'Multiply means Independence'.** Suppose that A is as in Fundamental Postulate 451C and that $\mathbf{v} \in \mathbb{C}^n$ is a state of the quantum system ('first system') associated with A. Suppose that B is a mathematical observable on $\mathbb{C}^s$, with spectral decomposition

$$B = \beta_1 Q_1 + \beta_2 Q_2 + \cdots + \beta_t Q_t,$$

and that $\mathbf{w} \in \mathbb{C}^s$ is a state of the quantum system ('second system') associated with B. The pure-product state $\mathbf{v} \otimes \mathbf{w}$ represents 'first system in state $\mathbf{v}$ and second in state $\mathbf{w}$', but there is an important 'Multiply means Independence' element in all this.

If we study the joint system, then 'the observable A for the first system' becomes $A \otimes I$, and its spectral decomposition is

$$A \otimes I = \sum \alpha_i (P_i \otimes I).$$

If a measurement of $A \otimes I$ is performed when the system is in pure-product state $\mathbf{v} \otimes \mathbf{w}$, then the result is α_i with probability $\|P_i \mathbf{v}\|^2$, and the new state of the total system is

$$(\|P_i \mathbf{v}\|)^{-1} (P_i \mathbf{v}) \otimes \mathbf{w}.$$

This is of the product form $\tilde{\mathbf{v}} \otimes \mathbf{w}$, so the second system is still in state $\mathbf{w}$. You can check that in the **pure-product** state $\mathbf{v} \otimes \mathbf{w}$, A and B behave as **independent** Random Variables in that $A \otimes I$ and $I \otimes B$ commute and

$$\mathbb{P}_{\mathbf{v} \otimes \mathbf{w}} (A \otimes I \text{ is measured as } \alpha_i; \text{ then } I \otimes B \text{ is measured as } \beta_j)$$
$$= \|P_i \mathbf{v}\|^2 \|Q_j \mathbf{w}\|^2$$
$$= \mathbb{P}_{\mathbf{v} \otimes \mathbf{w}} (A \otimes I \text{ is measured as } \alpha_i) \mathbb{P}_{\mathbf{v} \otimes \mathbf{w}} (I \otimes B \text{ is measured as } \beta_j)$$
$$= \mathbb{P}_{\mathbf{v}} (A \text{ is measured as } \alpha_i) \mathbb{P}_{\mathbf{w}} (B \text{ is measured as } \beta_j)$$

I think that the independence assumption which is built into the use of pure-product states should often be more emphasized in the literature. Sometimes, I get niggling doubts because of it.

For entangled states which are not pure products, things become much more interesting.

▶▶ **M. Classical Probability and Quantum Probability.** Michael Stein has, rightly, pressed me to discuss the question: *Are Classical Probability and Quantum Probability in conflict?* Here is an attempt at a brief answer.

Classical Probability is a special case, the so-called 'commutative case', of Quantum Probability: it can *calculate* probabilities associated with a model if and only if that model is a 'commutative' one in the set of all Quantum Probability models. Where both theories can calculate, their answers are therefore in agreement, but Classical Probability cannot be used to *calculate* probabilities for 'non-commutative' situations. Of course, when probabilities associated with a 'non-commutative' model have been calculated by Quantum Probability, the long-term relative-frequency interpretation of those probabilities relies on the classical Strong Law. The calculation of probabilities connected with the Aspect experiment could not be further from Classical Probability; but when the significance of the results of many experiments is assessed, it is via the classical use of standard deviations, Quantum Probability having told Classical Statistics what hypotheses to test.

Classical Probability as commutative Quantum Probability. Let me explain for the simple experiment of tossing a fair coin twice how the Classical Probability model may be viewed as a commutative quantum model. Let E_k denote 'Heads on the kth toss', and let X_k denote the indicator function of E_k. We have the familiar picture

ω	HH	HT	TH	TT
$X_1(\omega)$	1	1	0	0
$X_2(\omega)$	1	0	1	0
$\mathbb{P}(\omega)$	$\frac{1}{4}$	$\frac{1}{4}$	$\frac{1}{4}$	$\frac{1}{4}$

For the quantum setting, the space $\mathbb{V}$ of states is the 4-dimensional space (over $\mathbb{C}$) of functions mapping Ω to $\mathbb{C}$ (equivalently of complex 4-vectors parametrized by Ω). Let $\mathcal{X}_k$ be the operator (diagonal matrix) consisting of multiplication by X_k, so that $(\mathcal{X}_k \mathbf{v})(\omega) = X_k(\omega)v(\omega)$. Let ρ be the (diagonal) quantum density matrix corresponding to multiplication by $\mathbb{P}$. Then

$$\mathbb{E}_\rho(\mathcal{X}_k) = \text{trace}(\rho X_k) = \mathbb{E}_{\text{class}}(X_k).$$

Note that all our operators are diagonal and therefore commute with one another.

In the quantum setting, $\mathcal{I}_{E_k}$, multiplication by I_{E_k}, is orthogonal projection onto the subspace $\mathcal{E}_k := \{ \mathbf{z} \in \mathbb{V} : z \equiv 0 \text{ off } E_k \}$. This subspace stands for the event in Quantum Probability. Note that $\mathcal{E}^\perp$ corresponds to E^c.

What about the intersection $E \cap G$ of two events E and G? For the classical picture, $I_{E \cap G} = I_E I_G$. For our quantum picture, which in this case has to agree with the classical, $\mathcal{I}_{\mathcal{E} \cap \mathcal{G}} = \mathcal{I}_\mathcal{E} \mathcal{I}_\mathcal{G}$.

When Quantum Probability differs. Regard an 'event' as a subspace U of the space of states, and its 'indicator' as the orthogonal projection P_U onto U. Then, except in special circumstances, $P_U P_V$, $P_V P_U$ and $P_{U \cap V}$ are all different. (Take U to be the x-axis, V the line at $45°$, in $\mathbb{R}^2$, and note the effect of the three operators on $(1, 2)$. Draw the picture.)

The logic of sub**spaces** (quantum events) is completely different from the logic of sub**sets** (classical events). In particular, the correct analogue of 'disjoint events' is '*orthogonal* subspaces'. Thus the classical result

$$I_{E \cup F} = I_E + I_F \iff E \cap F = \emptyset \iff I_E I_F = 0,$$

corresponds to the quantum result (see Exercise 300Gb)

$$P_{U+V} = P_U + P_V \iff U \perp V \iff P_U P_V = 0.$$

Quantum logic does, of course, have some similarities with the theory of Linear Models.

Ma. Exercise. Let U and V be subsets of $\mathbb{C}^n$ (or of $\mathbb{R}^n$). Show that if any two of $P_U P_V$, $P_V P_U$ and $P_{U \cap V}$ are equal, then they are equal to the third; and that a necessary and sufficient condition for equality of the three is that

$$(U \cap Y^\perp) \perp (V \cap Y^\perp) \quad \text{where } Y := U \cap V.$$

[[In Classical Probability, for events U, V, it is automatic that

$$(U \cap Y^c) \cap (V \cap Y^c) = \emptyset \quad \text{where } Y := U \cap V.]]$$

10.3   Quantum computing: a closer look

This is our final pass through quantum computing.

A. Quantum gates and circuits. A quantum gate is just a unitary map U taking an input state in the space spanned by tensor products of qubits into an output state in the same space. Some simple cases, now to be explained, are shown in Figure A(i). For each picture, the input state (often a tensor product of qubits, but sometimes an entangled state) is on the left, the output state on the right. A quantum circuit is just a series of gates.

▶ **Aa. The 1-qubit NOT gate.** Figure A(i)(a) shows the NOT gate which we have already studied. It is usually represented as shown in the left-hand or top-middle picture. The top-middle figure indicates that an input **z** [respectively, **u**] will produce an output **u** [respectively, **z**].

▶ **Ab. The Hadamard gate.** An important 1-qubit gate is the Hadamard gate (Figure A(i)(b)) associated with the unitary matrix

$$H = c \begin{pmatrix} 1 & 1 \\ 1 & -1 \end{pmatrix} \quad \text{where } c = 2^{-\frac{1}{2}}.$$

Of course, this has no classical analogue.

▶ **Ac. The CNOT gate.** The most important 2-qubit gate is the CNOT (controlled-NOT) gate shown in Figure A(i)(c). In a classical computer, this changes ('NOTs') the second bit (the lower one in the picture) if the first (control) bit is 1, and leaves the second bit unchanged if the first bit is 0. The first bit is left unchanged.

Each of the two diagrams in Figure A(i)(c) is really four diagrams, one for each (i, j)-component. The right one signifies that the unitary map must take

$$\mathbf{z} \otimes \mathbf{z} \mapsto \mathbf{z} \otimes \mathbf{z}, \quad \mathbf{z} \otimes \mathbf{u} \mapsto \mathbf{z} \otimes \mathbf{u}, \quad \mathbf{u} \otimes \mathbf{z} \mapsto \mathbf{u} \otimes \mathbf{u}, \quad \mathbf{u} \otimes \mathbf{u} \mapsto \mathbf{u} \otimes \mathbf{z},$$

the first of these corresponding to reading the first component in each 'matrix'. The unitary transformation associated with this CNOT gate is

$$U = P_0 \otimes I + P_1 \otimes N, \qquad P_0 = P_{[\mathbf{z}]}, \quad P_1 = P_{[\mathbf{u}]}.$$

which conveys well the sense of what U does. You will have seen that we are taking the top qubit first in tensor products.

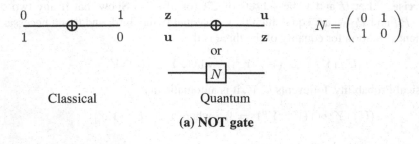

Classical   Quantum

(a) NOT gate

$$N = \begin{pmatrix} 0 & 1 \\ 1 & 0 \end{pmatrix}$$

$$H = c \begin{pmatrix} 1 & 1 \\ 1 & -1 \end{pmatrix} \text{ where } c = 2^{-\frac{1}{2}}$$

(b) Hadamard gate

$$U = P_0 \otimes I + P_1 \otimes N$$

(c) CNOT gate

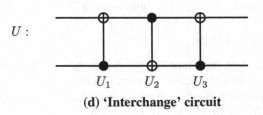

(d) 'Interchange' circuit

Figure A(i): Some simple quantum gates

Ad. The 'Interchange' circuit. The 'circuit' in Figure A(i)(d) is a three-gate circuit which interchanges the two input qubits (modulo the phase ambiguity at 460(K1)): we have

$$U(\mathbf{v} \otimes \mathbf{w}) = \mathbf{w} \otimes \mathbf{v}.$$

Let's check out the 'unitary' picture. We have $U = U_3 U_2 U_1$ (note the order!), where

$$U_1 = U_3 = I \otimes P_0 + N \otimes P_1, \qquad U_2 = P_0 \otimes I + P_1 \otimes N.$$

We have

$$P_0 = \begin{pmatrix} 1 & 0 \\ 0 & 0 \end{pmatrix}, \quad N = \begin{pmatrix} 0 & 1 \\ 1 & 0 \end{pmatrix}, \quad P_1 = \begin{pmatrix} 0 & 0 \\ 0 & 1 \end{pmatrix},$$

whence

$$P_0 N = \begin{pmatrix} 0 & 1 \\ 0 & 0 \end{pmatrix} = N P_1, \quad P_1 N = \begin{pmatrix} 0 & 0 \\ 1 & 0 \end{pmatrix} = N P_0.$$

We shall always work out products in the order typified by

$$(a + b)(c + d + \cdots) = ac + ad + \cdots + bc + bd + \cdots,$$

but will omit terms which are obviously zero because of such identities as

$$0 = P_0 P_1 = P_0 N P_0 = P_1 N P_1.$$

We have

$$
\begin{aligned}
U_2 U_1 &= P_0 \otimes P_0 + P_0 N \otimes P_1 + P_1 \otimes N P_0 + P_1 N \otimes N P_1, \\
U_3 U_2 U_1 &= P_0 \otimes P_0 + P_1 N \otimes P_0 N P_1 + N P_0 N \otimes P_1 + N P_1 \otimes P_1 N \\
&= P_0 \otimes P_0 + N P_0 \otimes N P_1 + P_1 \otimes P_1 + N P_1 \otimes N P_0 \\
&= (P_0 \otimes P_0 + P_1 \otimes P_1)(I \otimes I) + (N \otimes N)(P_0 \otimes P_1 + P_1 \otimes P_0).
\end{aligned}
$$

That U does the right thing to $\mathbf{z} \otimes \mathbf{z}, \mathbf{z} \otimes \mathbf{u}, \mathbf{u} \otimes \mathbf{z}, \mathbf{u} \otimes \mathbf{u}$ is now clear.

▶ **B. Entangling circuits.** Consider the circuit at Figure B(i)(a), where the input state is $\mathbf{z} \otimes \mathbf{z}$. You check that the output state $\mathcal{U}$ is given by

$$\mathcal{U} = (P_0 \otimes I + P_1 \otimes N)(H \otimes I)(\mathbf{z} \otimes \mathbf{z}) = c(\mathbf{z} \otimes \mathbf{z} + \mathbf{u} \otimes \mathbf{u}),$$

where $c = 2^{-\frac{1}{2}}$.

It is not possible to express $\mathcal{U} = c(1, 0, 0, 1)^T$ as a tensor product $\mathbf{v} \otimes \mathbf{w}$. *For no pair* $(\mathbf{v}, \mathbf{w})$ *of qubits is it possible to regard* $\mathcal{U}$ *as corresponding to 'top qubit is* $\mathbf{v}$ *and bottom qubit is* $\mathbf{w}$': *the state* $\mathcal{U}$ *is* entangled *and cannot be separated out.* In a sense now to be explained, 'a measurement on the top qubit will affect the bottom qubit' so we have a different situation from that in Subsection 461L.

In Figure B(i)(b), we consider the effect of measurement. Measurement of the top qubit corresponds to using the observable $M \otimes I$, and of the bottom qubit to the observable $I \otimes M$. Note that these two observables commute.

Consider what happens when the top measurement is made first when the state is $\mathcal{U}$. Check that we have the spectral decomposition

$$M \otimes I = 0 P_{W_0} + 1 P_{W_1}, \quad W_0 = [\mathbf{z}] \otimes \mathbb{C}^2, \quad W_1 = [\mathbf{u}] \otimes \mathbb{C}^2.$$

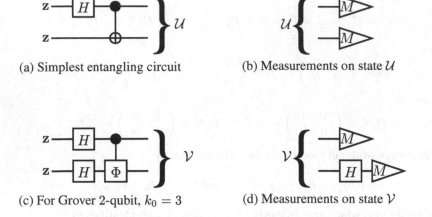

(a) Simplest entangling circuit   (b) Measurements on state $\mathcal{U}$

(c) For Grover 2-qubit, $k_0 = 3$ (d) Measurements on state $\mathcal{V}$

Figure B(i): Entanglement in quantum circuits

Note that $W_0 = [\mathbf{z} \otimes \mathbf{z}, \mathbf{z} \otimes \mathbf{u}]$ and W_0 *is* the space of eigenvectors of M corresponding to eigenvalue 0. Now, $c\mathbf{z} \otimes \mathbf{z} \in W_0$ and

$$\mathcal{U} - c\mathbf{z} \otimes \mathbf{z} = c\mathbf{u} \otimes \mathbf{u},$$

so that

$$\langle \mathcal{U} - c\mathbf{z} \otimes \mathbf{z}, \mathbf{z} \otimes \mathbf{z} \rangle = 0 = \langle \mathcal{U} - c\mathbf{z} \otimes \mathbf{z}, \mathbf{z} \otimes \mathbf{u} \rangle.$$

Hence $\mathcal{U} - c\mathbf{z} \otimes \mathbf{z} \perp W_0$, and so

$$P_{W_0}\mathcal{U} = c\mathbf{z} \otimes \mathbf{z}, \quad P_{W_1}\mathcal{U} = c\mathbf{u} \otimes \mathbf{u}.$$

A measurement of $M \otimes I$ when the system is in state $\mathcal{U}$ will produce a result 0 or 1 with probability $\frac{1}{2}$ each. If the result is 0, then the new state is $\mathbf{z} \otimes \mathbf{z}$, so that (modulo the type of phase transformation at 460(K1)) each of the top and bottom qubits is in well-defined state $\mathbf{z}$; and a measurement of the bottom qubit must give 0. We see that the two measurements must give the same results. The system is involved in a strange conspiracy. There is more on entanglement later.

Of course, we can have entangled triplets, etc. Entangled triplets feature in the amazing phenomenon of **quantum teleportation** about which you can find far too much on the Net. But Bouwmeester, Ekert and Zeilinger [28] have the correct story.

▶ **Ba. Exercise.** Figure B(i)(c) shows part of the circuit for Grover's algorithm for the case when $K = 4$. (Compare the next subsection.) The controlled Φ gate is

$$P_0 \otimes I + P_1 \otimes \Phi, \quad \Phi = \begin{pmatrix} 1 & 0 \\ 0 & -1 \end{pmatrix}.$$

Show that the output state $\mathcal{V}$ is $\frac{1}{2}(1, 1, 1, -1)^T$ and that it is entangled. Show that if $\mathcal{V}$ is inputted into the circuit in Figure B(i)(d), then the two measurements produce the same results, this independently of the order in which the $M \otimes I$ measurement and $I \otimes H$ operation are performed.

▶ **C. Probabilistic mixtures for subsystems.** Consider the 2-qubit system in state $\mathcal{U}$ as in the previous subsection. Let A be an observable associated with the first qubit, so A is 'really' $A \otimes I$. We have

$$\begin{aligned} \mathbb{E}_{\mathcal{U}}(A) &= \langle \mathcal{U}, (A \otimes I)\mathcal{U} \rangle \\ &= \tfrac{1}{2}\langle \mathbf{z} \otimes \mathbf{z}, (A \otimes I)\mathbf{z} \otimes \mathbf{z} \rangle + \tfrac{1}{2}\langle \mathbf{z} \otimes \mathbf{z}, (A \otimes I)\mathbf{u} \otimes \mathbf{u} \rangle \\ &\quad + \tfrac{1}{2}\langle \mathbf{u} \otimes \mathbf{u}, (A \otimes I)\mathbf{z} \otimes \mathbf{z} \rangle + \tfrac{1}{2}\langle \mathbf{u} \otimes \mathbf{u}, (A \otimes I)\mathbf{u} \otimes \mathbf{u} \rangle \\ &= \tfrac{1}{2}\langle \mathbf{z}, A\mathbf{z} \rangle + \tfrac{1}{2}\langle \mathbf{u}, A\mathbf{u} \rangle \\ &= \operatorname{trace}(\rho A) \quad \text{where } \rho = \tfrac{1}{2}(\mathbf{z}\mathbf{z}^\dagger + \mathbf{u}\mathbf{u}^\dagger) = \tfrac{1}{2}I. \end{aligned}$$

So, we can use the 1-qubit density matrix ρ to work out expectations within the 'first qubit' system.

If A has distinct eigenvectors α, β with corresponding unit eigenvectors $\mathbf{x}, \mathbf{y}$, then a measurement of A, that is, of $A \otimes I$, is made when the 2-qubit system is in state $\mathcal{U}$, then

$$\mathbb{P}(A \text{ is measured as } \alpha) = \mathbb{P}(A \text{ is measured as } \beta) = \tfrac{1}{2}.$$

If A is measured as α, then the new state of the 2-qubit system is the normalized version of $\mathbf{x} \otimes \mathbf{w}$, where

$$\mathbf{w} = \langle \mathcal{U}, \mathbf{x} \otimes \mathbf{x} \rangle \mathbf{x} + \langle \mathcal{U}, \mathbf{x} \otimes \mathbf{y} \rangle \mathbf{y},$$

so that each qubit is now in a definite state (modulo the usual phase ambiguity).

D. Circuitry for Grover's algorithm. Before continuing, please check that tensor products are associative in that

$$(\mathbf{t} \otimes \mathbf{v}) \otimes \mathbf{w} = \mathbf{t} \otimes (\mathbf{v} \otimes \mathbf{w}), \qquad (A \otimes B) \otimes C = A \otimes (B \otimes C).$$

We suppose that the number K of 'telephone entries' in Grover's algorithm is 8, the numbers being $0, 1, 2, \ldots, 7$. We use $\mathbb{C}^8$ as $\mathbb{C}^2 \otimes \mathbb{C}^2 \otimes \mathbb{C}^2$ as our underlying space, and regard

$$|0\rangle = \mathbf{z} \otimes \mathbf{z} \otimes \mathbf{z}, \quad |1\rangle = \mathbf{z} \otimes \mathbf{z} \otimes \mathbf{u}, \quad \ldots \quad |7\rangle = \mathbf{u} \otimes \mathbf{u} \otimes \mathbf{u},$$

as the computational basis corresponding to the measurement $M = \operatorname{diag}(0, 1, 2, \ldots, 7)$. Suppose that the target vector $\mathbf{t}$ is $|7\rangle$, that is, $\mathbf{u} \otimes \mathbf{u} \otimes \mathbf{u}$. Then simple geometry shows that reflection $R_{[\mathbf{t}]}$ in the line joining $\mathbf{0}$ to $\mathbf{t}$ is given by

$$R_{[\mathbf{t}]} = 2P_{[\mathbf{t}]} - I.$$

[[Note that if $P^2 = P = P^\dagger$, and $R := 2P - I$, then $RR^\dagger = R^2 = 4P^2 - 4P + I = I$.]] In particular, $-R_{[\mathbf{t}]}$ flips the phase of $\mathbf{t}$ (taking it to $-\mathbf{t}$) but preserves the phase of the other elements of the computational basis. So, how do we build a circuit for the gate $-R_{[\mathbf{t}]}$?

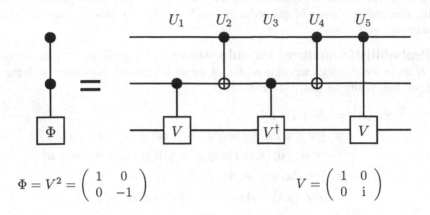

$$\Phi = V^2 = \begin{pmatrix} 1 & 0 \\ 0 & -1 \end{pmatrix} \qquad\qquad V = \begin{pmatrix} 1 & 0 \\ 0 & i \end{pmatrix}$$

Figure D(i): A phase-flip Toffoli-type gate for Grover's algorithm

In 'unitary' terms, we want our circuit to act as U where (I have written the sum in the order which your later calculation will yield)

$$U = P_0 \otimes P_0 \otimes I + P_0 \otimes P_1 \otimes I + P_1 \otimes P_1 \otimes \Phi + P_1 \otimes P_0 \otimes I,$$

where

$$\Phi = \begin{pmatrix} 1 & 0 \\ 0 & -1 \end{pmatrix}.$$

On the right-hand side of Figure D(i) is an implementation in terms of 2-qubit gates, where

$$V = \begin{pmatrix} 1 & 0 \\ 0 & i \end{pmatrix}, \quad VV^\dagger = I, \quad V^2 = \Phi.$$

By considering the effect of the right-hand circuit on each element of the computational basis, check that it has the desired effect. In terms of unitary transformations, we have

$$U_5 = P_0 \otimes I \otimes I + P_1 \otimes I \otimes V, \qquad U_4 = P_0 \otimes I \otimes I + P_1 \otimes N \otimes I,$$

etc. *You* can if you wish, for practice with tensor products, prove that $U_5 U_4 U_3 U_2 U_1$ is the desired U. As a check on an earlier stage in your calculation, you should have

$$U_3 U_2 U_1 = P_0 \otimes P_0 \otimes I + P_1 \otimes N P_1 \otimes V + P_0 \otimes P_1 \otimes I + P_1 \otimes N P_0 \otimes V^\dagger.$$

[[Note that common sense and a look at Figure D(i) immediately suggest that

$$U_3 U_2 U_1 = (I \otimes I \otimes I)(P_0 \otimes P_0 \otimes I) + (I \otimes I \otimes I)(P_0 \otimes P_1 \otimes I)$$
$$+ (I \otimes N \otimes V^\dagger)(P_1 \otimes P_0 \otimes I) + (I \otimes N \otimes V)(P_1 \otimes P_1 \otimes I). \quad]]$$

So, we have our circuit for the oracle $-R_{[\mathbf{t}]}$.

But if S is a unitary map such that $S\mathbf{v} = \mathbf{w}$, where $\mathbf{v}$ and $\mathbf{w}$ are unit vectors, then

$$R_{[\mathbf{w}]} = 2\mathbf{w}\mathbf{w}^\dagger - I = S(2\mathbf{v}\mathbf{v}^\dagger - I)S^\dagger = SR_{[\mathbf{v}]}S^\dagger.$$

Now, with $\mathbf{t} = \mathbf{u} \otimes \mathbf{u} \otimes \mathbf{u}$, we have $(N \otimes N \otimes N)\mathbf{t} = \mathbf{z} \otimes \mathbf{z} \otimes \mathbf{z}$. Moreover,

$$\mathbf{e} = 8^{-\frac{1}{2}}(1, 1, \dots, 1)^T = (H \otimes H \otimes H)(\mathbf{z} \otimes \mathbf{z} \otimes \mathbf{z}).$$

Of course, Grover's algorithm does not actually need the minus sign in formula 447(C1). It should now be clear that we can implement Grover's algorithm if we can construct

- all 1-qubit gates of type N or H,
- all 2-qubit gates of type CNOT, controlled V, or controlled $V^\dagger$.

The 'all' refers to the fact that we wish to be able to apply N or H to each individual qubit, and CNOT, controlled V and controlled $V^\dagger$ to each ordered pair of qubits. Of course, if we have all CNOT gates, then we have all Interchange gates available. You will realize that we can build the required Grover oracle, whatever the value of k_0 in $\{0, 1, 2, \dots, 7\}$.

▶ **E. A universality result.** It is a fact, important for theory if not so much for practice, that *we can achieve any n-qubit unitary transformation provided that we have*

- *all unitary 1-qubit gates,*
- *all 2-qubit gates of type CNOT.*

See, for example Chapter 4 of Nielsen and Chuang [174].

Crucial to this is the Linear-Algebra fact that, given any 2×2 unitary matrix U, we can find unitary 2×2 matrices A, B, C and a real number α such that

$$ABC = I, \quad U = e^{i\alpha}ANBNC, \quad N = \begin{pmatrix} 0 & 1 \\ 1 & 0 \end{pmatrix}.$$

Then we can implement a controlled-U gate as shown in Figure E(i).

The point is that, as should now be obvious to you,

$$U_5U_4U_3U_2U_1 = P_0 \otimes (ABC) + P_1 \otimes (ANBNC)$$
$$= P_0 \otimes I + e^{-i\alpha}P_1 \otimes U.$$

Since $U_6 = (P_0 + e^{i\alpha}P_1) \otimes I$, we have

$$U_6U_5U_4U_3U_2U_1 = P_0 \otimes I + P_1 \otimes U,$$

as required.

Check that for $U = \begin{pmatrix} 1 & 0 \\ 0 & i \end{pmatrix}$, so U equals the V of Figure 468D(i), we can take

$$A = \begin{pmatrix} e^{-i\pi/4} & 0 \\ 0 & e^{i\pi/4} \end{pmatrix}, \quad B = C = \begin{pmatrix} e^{i\pi/8} & 0 \\ 0 & e^{-i\pi/8} \end{pmatrix}, \quad \alpha = \pi/4.$$

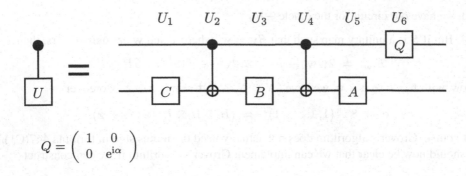

$$Q = \begin{pmatrix} 1 & 0 \\ 0 & e^{i\alpha} \end{pmatrix}$$

Figure E(i): Implementation of a controlled-U gate

F. An NMR implementation of Grover's algorithm.

The ideas here are borrowed from Gershenfeld and Chuang [92]. My account is very much a 'first pass' at this topic. The lead possessed at the time of writing by NMR computing is already seriously under threat.

In NMR computing, each molecule in a liquid is a separate quantum computer, so one has about 10^{18} separate quantum computers. However, one can only arrange that a small excess of molecules have close to the desired initial state; and all experimental measurements detect only the 'majority view' held by a very small percentage majority of the molecules.

Gershenfeld and Chuang successfully carried out Grover's algorithm for $K = 4$, using chloroform CHCℓ_3, but with the carbon-13 isotope of carbon rather than the standard carbon-12 so as to impart a spin to the carbon nucleus in the molecule. The spins of the C and H nuclei are the observables which give the 'bits'. As do Gershenfeld and Chuang in their paper, we now take a naive *non-quantum* view of things to get a very rough sense of the flavour. For this purpose, we think of the spin of a nucleus as an angular-momentum vector which describes a tiny bar magnet. So take a very classical deterministic view for the moment.

A constant, vertically up, magnetic field is imposed on the liquid. This means that an excess of spins of the C and H nuclei will point upwards (giving us an excess of the desired $\mathbf{z} \otimes \mathbf{z}$ state). (Think of $\mathbf{z}$ as 'spin up', $\mathbf{u}$ as 'spin down'.) The spins of H nuclei can be 'rotated' by electro-magnetic waves of the correct 'resonant' frequency; similarly for the carbon spins. This fact provides 1-qubit gates.

Figure F(i) is my crude picture of how a CNOT gate may be implemented. We consider the spin of the carbon nucleus which, as stated earlier, can be rotated about the x or y axes by RFPs (radio-frequency pulses). The carbon spin may be allowed to precess about the vertical z-direction in a sense determined by whether the spin of the H nucleus is down or up. The 'spherical' picture in our figure shows possible successive

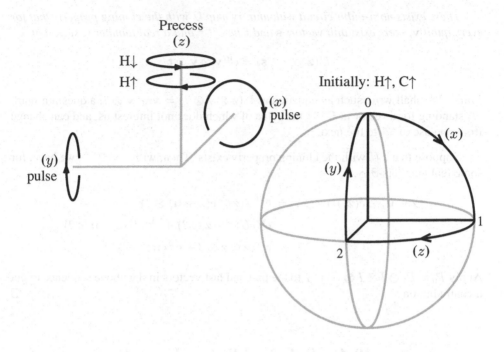

Figure F(i): Intuitive picture of NMR operations on carbon spin

positions $0, 1, 2, 3$ of the carbon spin after the '3-gate' operation:

$$90° \, \text{RFP}(x), \quad \text{precess}(z) \text{ through} \begin{cases} +90° & \text{if H spin is down,} \\ -90° & \text{if H spin is up,} \end{cases} \quad 90° \, \text{RFP}(y).$$

Check out that for this operation, if the H spin is down, then the carbon spin is reversed between up and down, while if the H spin is up, the sense of the carbon spin is preserved, exactly what is needed for a CNOT gate. You can see that RFPs of the correct nature have to be applied for precisely the correct time, and that precession time has to be precisely controlled too.

Of course, spin is a much more complex matter – see the next section. One's rotations are unitary evolutions of states which determine probabilities, etc. What one has to do is to use the known explicit Hamiltonians which describe the '1-qubit' response to RFPs and the '2-qubit' interaction.

G. The 'No-cloning theorem'. This result says that it is impossible to build a circuit which, for every qubit **v**, makes a copy of **v** onto a 'wire' which was previously occupied by **z** (say). In the following formulation, the **t** and **s** should be thought of as occupying 'work-space'.

There exists no n-qubit circuit with unitary map U with the cloning property that for every qubit $\mathbf{v}$, there exist unit vectors $\mathbf{s}$ and $\mathbf{t}$ in $\mathbb{C}^{n-2}$ and a real number α such that

$$U(\mathbf{z} \otimes \mathbf{v} \otimes \mathbf{s}) = e^{i\alpha} \mathbf{v} \otimes \mathbf{v} \otimes \mathbf{t}.$$

Proof. We shall write such an equation as $U(\mathbf{z} \otimes \mathbf{v} \otimes ?) = \mathbf{v} \otimes \mathbf{v} \otimes ?$, a question-mark '?' standing for a vector in $\mathbb{C}^{n-2}$ the value of which does not interest us, and can change from one use of '?' to the next.

Suppose that a U with the cloning property exists. Then, with $c = 2^{-\frac{1}{2}}$, we have, for some real numbers β, γ, δ,

$$
\begin{aligned}
c(\mathbf{z}+\mathbf{u}) \otimes c(\mathbf{z}+\mathbf{u}) \otimes ? &= e^{i\beta} U\big(\mathbf{z} \otimes c(\mathbf{z}+\mathbf{u}) \otimes ?\big) \\
&= e^{i\beta} cU(\mathbf{z} \otimes \mathbf{z} \otimes ?) + e^{i\beta} cU(\mathbf{z} \otimes \mathbf{u} \otimes ?) \\
&= e^{i\gamma} c\mathbf{z} \otimes \mathbf{z} \otimes ? + e^{i\delta} c\mathbf{u} \otimes \mathbf{u} \otimes ?.
\end{aligned}
$$

Apply $P_0 \otimes P_1 \otimes I \otimes I \otimes \cdots \otimes I$ to the first and last vectors in the above sequence to get a contradiction. $\qquad\qquad\square$

10.4   Spin and Entanglement

Entanglement is one of the most mind-bending things known to Science. Before discussing entanglement, we need to discuss spin.

My purpose in this section is to provide background which will allow you to read the many other discussions of 'Bell-Aspect' material, that is, of Bell's inequality (though we concentrate on the Bell–Clauser–Horne–Shimony–Holt (Bell–CHSH) version) and the famous experiments (on the Bell–CHSH inequality) done by Alain Aspect and since by many others.

I try to motivate the spinor representation quickly with what is very much a *mathematical* version of spin. The way that spin features as intrinsic angular momentum in Physics is described in the next section.

A. Notation and introductory comments.   We shall have

Vectors $\mathbf{V}, \mathbf{A}$, etc, in $\mathbb{R}^3$,

Quaternions (explained below) $\mathcal{Q}, \mathcal{V}, \mathcal{A}$, etc;

Quantum states: vectors $\mathbf{v}, \mathbf{w}$, etc, in $\mathbb{C}^2$ and tensor products of these;

SO(3), the group of familiar rotations of $\mathbb{R}^3$, the group of real 3×3 matrices S which are orthogonal ($S^{-1} = S^T$) and of determinant $+1$;

SU(2), the group of complex 2×2 matrices U which are unitary ($U^{-1} = U^\dagger$) and have determinant $+1$.

The first results of interest (explained fully below) are as follows:

- SU(2) may be identified with S^3, the unit sphere in $\mathbb{R}^4$;

- SO(3) may be considered as S^3 with opposite points identified: SU(2) is the 'simply-connected double cover' of SO(3).

The SU(2) description of rotations is used in computer graphics and in robotics, this because it gives the best way of doing calculations with SO(3). See, for example, Exercise 476Da below.

Aa. Comments. Nature views the spins of spin-$\frac{1}{2}$ particles (electrons, protons, neutrons, etc) in terms of the SU(2) picture, rather than the 'obvious' SO(3) picture. As mentioned earlier: but for this extraordinary 'piece of luck', we would not be here. It is a fact, discovered by **Pauli**, that *spin-$\frac{1}{2}$ particles must obey the* **Exclusion Principle:** *no two spin-$\frac{1}{2}$ particles can occupy the same quantum state. It is essential to realize that, here, quantum state of a particle signifies the* **full** *quantum state (which may involve energy, kinds of momentum, spin, ...) of that particle,* **not** *just the 'spin state' on which we shall be focusing in this section.* The Exclusion Principle lies very deep, requiring Relativity Theory, Quantum Theory and other deep Mathematics. Not even Feynman (see III-4-1 in [78]) could find any simple explanation of it. For a fine introduction, see Olive [179], and for a full study, see Streater and Wightman [220], but be warned that the latter is *very* much more advanced than this present book. It is the Pauli Exclusion Principle which leads to determination of electron shells in atoms, explains which elements are possible and how atoms combine to form molecules, and thus forms the basis of 'modern' Chemistry and Biology. See Feynman [78].

I mentioned that the Pauli Exclusion Principle relies on Relativity. However, as we shall see later, it is the case that Nature's utilization of SU(2) rather than SO(3) is itself very much tied up with Relativity.

▶ **B. The Special Unitary Group SU(2).** Suppose that a matrix $U = \begin{pmatrix} a & b \\ c & d \end{pmatrix}$ is unitary and of determinant 1. Then, since

$$UU^\dagger = \begin{pmatrix} a & b \\ c & d \end{pmatrix} \begin{pmatrix} \bar{a} & \bar{c} \\ \bar{b} & \bar{d} \end{pmatrix} = \begin{pmatrix} |a|^2 + |b|^2 & a\bar{c} + b\bar{d} \\ \bar{a}c + \bar{b}d & |c|^2 + |d|^2 \end{pmatrix}$$

we must have

$$|a|^2 + |b|^2 = 1 = |c|^2 + |d|^2, \quad \bar{a}c + \bar{b}d = 0, \quad ad - bc = \det(U) = 1.$$

Hence,

$$\bar{a} = \bar{a}(ad - bc) = |a|^2 d - b\bar{a}c = (|a|^2 + |b|^2)d = d.$$

Similarly, $c = -\bar{b}$. Thus

$$U = \begin{pmatrix} a & b \\ -\bar{b} & \bar{a} \end{pmatrix}, \qquad \text{where } |a|^2 + |b|^2 = 1.$$

Thus, if $a = a_0 + ia_1$, $b = b_0 + ib_1$, then $a_0^2 + a_1^2 + b_0^2 + b_1^2 = 1$, and we have the fact that $SU(2) = S^3$.

▶ **C. Quaternions.**   The great Irish mathematician Sir William Rowan Hamilton introduced (in addition to Hamiltonians, parts of Hamilton–Jacobi theory, etc) the number system of quaternions. We can think of a quaternion as being of the form

$$\mathcal{Q} = q_0 + \mathcal{V}, \quad \mathcal{V} = q_1\mathcal{I} + q_2\mathcal{J} + q_3\mathcal{K},$$

where q_0, q_1, q_2, q_3 are *real* numbers, and where

$$\mathcal{I}^2 = \mathcal{J}^2 = \mathcal{K}^2 = -1,$$
$$\mathcal{JK} = \mathcal{I} = -\mathcal{KJ}, \quad \mathcal{KI} = \mathcal{J} = -\mathcal{IK}, \quad \mathcal{IJ} = \mathcal{K} = -\mathcal{JI}. \tag{C1}$$

Thus, with **V** as the vector $(q_1, q_2, q_3)^T$ corresponding to $\mathcal{V}$,

$$(q_0 + \mathcal{V})(r_0 + \mathcal{W}) = [q_0 r_0 - \mathcal{V}.\mathcal{W}] + \mathcal{V} \times \mathcal{W},$$

where, on the right, the term in $[\,]$ is real, $\mathcal{V}.\mathcal{W}$ denotes the scalar product of the vectors **V** and **W**, and $\mathcal{V} \times \mathcal{W}$ is their vector product. The quaternions form a skew field: we have all the usual rules for sums, products, inverses, etc, except that multiplication is not commutative. We have, if $\mathcal{Q} \neq 0 \ (= \text{`}0 + \mathbf{0}\text{'})$,

$$\mathcal{Q}\mathcal{Q}^{-1} = 1 = \mathcal{Q}^{-1}\mathcal{Q}, \quad \text{where } \mathcal{Q}^{-1} = (q_0 - \mathcal{V})/\|\mathcal{Q}\|^2,$$

where

$$\|\mathcal{Q}\|^2 = q_0^2 + q_1^2 + q_2^2 + q_3^2.$$

The map from the set of quaternions to a subset of complex 2×2 matrices defined by

$$a_0 + a_1\mathcal{I} + b_0\mathcal{J} + b_1\mathcal{K} \mapsto \begin{pmatrix} a & b \\ -\bar{b} & \bar{a} \end{pmatrix}$$

is an isomorphism onto its range in that it preserves sums, products, etc. Note that the unit quaternions, those of norm 1, those on S^3, map onto $SU(2)$. Now,

$$\begin{pmatrix} a & b \\ -\bar{b} & \bar{a} \end{pmatrix} = a_0 + i(a_1\sigma_z + b_0\sigma_y + b_1\sigma_x),$$

where $\sigma_z, \sigma_y, \sigma_x$ are the *Pauli matrices:*

$$\sigma_z = \begin{pmatrix} 1 & 0 \\ 0 & -1 \end{pmatrix}, \quad \sigma_y = \begin{pmatrix} 0 & -i \\ i & 0 \end{pmatrix}, \quad \sigma_x = \begin{pmatrix} 0 & 1 \\ 1 & 0 \end{pmatrix}.$$

This is Pauli's time-honoured notation.

Let $\mathcal{A} = a_1 \mathcal{I} + a_2 \mathcal{J} + a_3 \mathcal{K}$ be an imaginary quaternion ($a_0 = 0$) of norm 1. Then,

$$\mathcal{A}^2 = -(a_1^2 + a_2^2 + a_3^2) = -1.$$

[[*Nota bene*. We will have in the product $\mathcal{A}\mathcal{A}$ the expression $a_1 a_2 \mathcal{I}\mathcal{J} + a_2 a_1 \mathcal{J}\mathcal{I}$ which will be zero because $\mathcal{I}\mathcal{J} = -\mathcal{J}\mathcal{I}$.]] Hence, as in de Moivre's Theorem for complex numbers,

$$\mathcal{R}(\tfrac{1}{2}\theta, \mathcal{A}) := e^{\frac{1}{2}\theta\mathcal{A}} = [\cos \tfrac{1}{2}\theta] + (\sin \tfrac{1}{2}\theta)\mathcal{A}. \tag{C2}$$

This is a unit quaternion, and so essentially an element of SU(2).

▶ **D. The Special Orthogonal group SO(3) of rotations.** Consider *right-handed rotation* through an angle θ about the axis specified by the unit vector **A**. It is

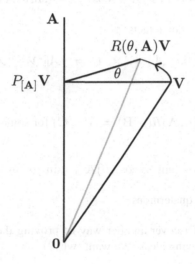

Figure D(i): Rotation $R(\theta, \mathbf{A})$ through angle θ about axis **A**

clear from Figure D(i) that, for any vector **V**, we have

$$\begin{aligned} R(\theta, \mathbf{A})\mathbf{V} &= P_{[\mathbf{A}]}\mathbf{V} + (\cos\theta)(\mathbf{V} - P_{[\mathbf{A}]}\mathbf{V}) + (\sin\theta)\mathbf{A} \times \mathbf{V} \\ &= (1 - \cos\theta)(\mathbf{A}.\mathbf{V})\mathbf{A} + (\cos\theta)\mathbf{V} + (\sin\theta)\mathbf{A} \times \mathbf{V}. \end{aligned}$$

The key result we now prove is the quaternion representation

$$R(\theta, \mathbf{A})\mathbf{V} = \mathcal{R}(\tfrac{1}{2}\theta, \mathcal{A})\mathcal{V}\mathcal{R}(-\tfrac{1}{2}\theta, \mathcal{A}). \tag{D1}$$

Here, $\mathcal{A}$ and $\mathcal{V}$ are **A** and **V** considered as imaginary quaternions, and $\mathcal{R}(\tfrac{1}{2}\theta, \mathcal{A})$ is at equation (C2).

Proof. In the following calculation, we write $c = \cos\tfrac{1}{2}\theta$, $s = \sin\tfrac{1}{2}\theta$, and write $\mathcal{V}.\mathcal{A}$ for **V**.**A** and $\mathcal{V} \times \mathcal{A} = \mathbf{V} \times \mathbf{A}$, identifying vectors with the associated imaginary quaternions. We enclose real numbers in square brackets when this helps clarify things. We find that

$$
\begin{aligned}
\mathcal{R}(\tfrac{1}{2}\theta, \mathcal{A})\mathcal{V}\mathcal{R}(-\tfrac{1}{2}\theta, \mathcal{A}) &= ([c] + s\mathcal{A})\mathcal{V}([c] - s\mathcal{A}) \\
&= ([c] + s\mathcal{A})\{c\mathcal{V} + [s\mathcal{V}.\mathcal{A}] - s\mathcal{V} \times \mathcal{A}\} \\
&= c^2\mathcal{V} + [cs\mathcal{V}.\mathcal{A}] - cs\mathcal{V} \times \mathcal{A} \\
&\quad - [sc\mathcal{A}.\mathcal{V}] + sc\mathcal{A} \times \mathcal{V} + s^2(\mathcal{V}.\mathcal{A})\mathcal{A} \\
&\quad + s^2\mathcal{A}.(\mathcal{V} \times \mathcal{A}) - s^2\mathcal{A} \times (\mathcal{V} \times \mathcal{A}) \\
&= c^2\mathcal{V} + 2sc\mathcal{A} \times \mathcal{V} + s^2(\mathcal{V}.\mathcal{A})\mathcal{A} - s^2\mathcal{V} + s^2(\mathcal{A}.\mathcal{V})\mathcal{A} \\
&= 2s^2(\mathcal{A}.\mathcal{V})\mathcal{A} + (c^2 - s^2)\mathcal{V} + 2sc\mathcal{A} \times \mathcal{V} \\
&= (1 - \cos\theta)(\mathcal{A}.\mathcal{V})\mathcal{A} + (\cos\theta)\mathcal{V} + (\sin\theta)\mathcal{A} \times \mathcal{V},
\end{aligned}
$$

as required. We used the well-known facts

$$
\mathcal{A}.(\mathcal{V} \times \mathcal{A}) = 0, \quad \mathcal{A} \times (\mathcal{V} \times \mathcal{A}) = \|\mathcal{A}\|^2\mathcal{V} - (\mathcal{A}.\mathcal{V})\mathcal{A},
$$

in the above calculation.   □

Da. Exercise. Prove that $R(\alpha, \mathbf{A})R(\beta, \mathbf{B}) = R(\gamma, \mathbf{C})$ for some γ, where $\mathbf{C}$ is a scalar multiple of

$$
(\cos\tfrac{1}{2}\alpha \sin\tfrac{1}{2}\beta)\mathbf{B} + (\sin\tfrac{1}{2}\alpha \cos\tfrac{1}{2}\beta)\mathbf{A} + (\sin\tfrac{1}{2}\alpha \sin\tfrac{1}{2}\beta)\mathbf{A} \times \mathbf{B}.
$$

Now try to prove this without quaternions.

Db. Exercise (optional). Discover another way of proving the results earlier in this subsection, based on the following ideas. We want (why?)

$$
\frac{\mathrm{d}}{\mathrm{d}\theta}R(\theta, \mathbf{A})\mathbf{V} = A \times R(\theta, \mathbf{A})\mathbf{V},
$$

whence, for $\mathbf{V} \perp \mathbf{A}$,

$$
\frac{\mathrm{d}^2}{\mathrm{d}\theta^2}R(\theta, \mathbf{A})\mathbf{V} = -R(\theta, \mathbf{A})\mathbf{V},
$$

an easily solved 'simple harmonic' equation. Next, suppose that we know that for any imaginary quaternion $\mathcal{V}$,

$$
\mathcal{S}(\mathcal{V}) := \mathcal{R}(\tfrac{1}{2}\theta, \mathcal{A})\mathcal{V}\mathcal{R}(-\tfrac{1}{2}\theta, \mathcal{A})
$$

is another imaginary quaternion. Then, using

$$
\frac{\mathrm{d}}{\mathrm{d}\theta}\mathcal{R}(\tfrac{1}{2}\theta, \mathcal{A}) = \tfrac{1}{2}\mathcal{A}\mathcal{R}(\tfrac{1}{2}\theta, \mathcal{A}), \qquad \frac{\mathrm{d}}{\mathrm{d}\theta}\mathcal{R}(-\tfrac{1}{2}\theta, \mathcal{A}) = -\tfrac{1}{2}\mathcal{R}(\tfrac{1}{2}\theta, \mathcal{A})\mathcal{A},
$$

show that $(\mathrm{d}/\mathrm{d}\theta)\mathcal{S}(\mathcal{V}) = \mathcal{A}\mathcal{S}(\mathcal{V})$.

Dc. The double-cover homomorphism. The map $\mathcal{R}(\frac{1}{2}\theta, \mathcal{A}) \mapsto R(\theta, \mathbf{A})$ defined via 475(D1) is a group homomorphism (it preserves products) from SU(2) to SO(3). However, the mapping is two-to-one: $\mathcal{R}(\frac{1}{2}(\theta + 2\pi), \mathcal{A}) = -\mathcal{R}(\frac{1}{2}\theta, \mathcal{A})$ leads to the same value of $R(\theta, \mathbf{A})$ as does $\mathcal{R}(\frac{1}{2}\theta, \mathcal{A})$. (In Group-Theory language, 'the kernel of the homomorphism is $\{+1, -1\}$'.)

Of course, we have 'non-uniqueness of representation of rotation' in that

$$R(\theta + 2n\pi, \mathbf{A}) = R(\theta, \mathbf{A}), \quad R(\theta, \mathbf{A}) = R(-\theta, \mathbf{A}).$$

However the two-to-one property of the map $\mathcal{R}(\frac{1}{2}\theta, \mathcal{A}) \mapsto R(\theta, \mathbf{A})$ does hold because

$$\mathcal{R}(\frac{1}{2}(\theta + 2n\pi), \mathcal{A}) = (-1)^n \mathcal{R}(\frac{1}{2}\theta, \mathcal{A}), \quad \mathcal{R}(\frac{1}{2}\theta, \mathcal{A}) = \mathcal{R}(-\frac{1}{2}\theta, -\mathcal{A}).$$

Precisely two elements of SU(2) map onto each element of SO(3).

▶ **Dd. Nature's view of spin-$\frac{1}{2}$ particles.** In effect, Nature sees what we think of as $R(\theta, \mathbf{A})$ as $\mathcal{R}(\frac{1}{2}\theta, \mathcal{A})$: She sees things before we take the homomorphism. So Nature views the rotation as the element

$$\mathcal{R}(\frac{1}{2}\theta, \mathcal{A}) = (\cos \frac{1}{2}\theta)I + (\sin \frac{1}{2}\theta)\mathrm{i}(A_1\sigma_z + A_2\sigma_y + A_3\sigma_x)$$

of SU(2). Again,

$$\mathcal{R}(\frac{1}{2}t, \mathcal{A}) = \mathrm{e}^{-\mathrm{i}t(\frac{1}{2}\mathrm{i}\mathcal{A})}$$

and so *infinitesimal rotation* – which is **spin** – about axis $\mathcal{A}$ is considered to be the observable (the 'Hamiltonian for the $\mathcal{R}(\frac{1}{2}t, \mathcal{A})$ evolution')

$$\tfrac{1}{2}\mathrm{i}\mathcal{A} = -\tfrac{1}{2}(A_1\sigma_z + A_2\sigma_y + A_3\sigma_x).$$

But

$$q_0 + q_1\mathcal{I} + q_2\mathcal{J} + q_3\mathcal{K} \mapsto q_0 - q_3\mathcal{I} - q_2\mathcal{J} - q_1\mathcal{K}$$

is an isomorphism ('automorphism') of the skew field of quaternions, really because $(-\mathcal{K}, -\mathcal{J}, -\mathcal{I})$ is a right-handed system like $(\mathcal{I}, \mathcal{J}, \mathcal{K})$. This observation brings the Hamilton and Pauli notations into line.

De. Topology of the double cover (optional). This topic is not strictly necessary for understanding the rest.

The two-to-one mapping taking $\mathcal{R}(\frac{1}{2}\theta, \mathcal{A})$ to $R(\theta, \mathbf{A})$ is continuous from SU(2) to SO(3). If we take a (continuous) path $t \mapsto S(t)$ ($t \in [0, t_0]$) on SO(3) and choose a point $\mathcal{R}(0)$ of SU(2) which projects onto $S(0)$, there is a unique continuous way of 'lifting the S-path' onto a continuous path $\mathcal{R}(t)$ ($t \in [0, t_0]$) on SU(2) such that for each t, $\mathcal{R}(t)$ projects onto $S(t)$. We cannot have sudden jumps to antipodal points.

Now fix a unit vector $\mathbf{A}$, and consider the path $\mathcal{R}(\frac{1}{2}t, \mathcal{A})$ $t \in [0, 2\pi]$ in $S^3 = \text{SU}(2)$. This path goes from I to its antipodal point $-I$. Its projection on SO(3) is a closed path going from I to I. It is clear that this path on SO(3) cannot be deformed continuously into a path consisting of a single point; for at each stage of the deformation, the lifted path on S^3 would have to connect a point to its antipodal point. By contrast, S^3 is simply connected in that any path may be deformed continuously into a path consisting of a single point by pulling towards a suitable 'pole' of the sphere S^3. Thus, SU(2) is much nicer topologically than SO(3).

Df. Rotations as elements of SO(3) (optional). We have not proved that *a real* 3×3 *matrix corresponds to a rotation if and only if*

$$S^T = S^{-1} \quad and \quad \det(S) = 1. \tag{D2}$$

We do not actually need this fact, which takes quite a little time to prove in full.

Sketched proof of 'if' part. Suppose that S satisfies (D2). First, S extends to a map $S : \mathbb{C}^3 \to \mathbb{C}^3$. If $S\mathbf{W} = \lambda\mathbf{W}$ where $\lambda \in \mathbb{C}$ and $\mathbf{W} \in \mathbb{C}^3 \setminus \{\mathbf{0}\}$, then

$$|\lambda|^2\|\mathbf{W}\|^2 = \bar{\lambda}\mathbf{W}^\dagger\lambda\mathbf{W} = \mathbf{W}^\dagger S^\dagger S\mathbf{W} = \mathbf{W}^\dagger S^T S\mathbf{W} = \mathbf{W}^\dagger\mathbf{W} = \|\mathbf{W}\|^2,$$

so that $|\lambda| = 1$. Now the characteristic equation $\det(\lambda I - S) = 0$ has real coefficients, and the product of its roots is $\det(S) = 1$. Complex roots occur in a conjugate pair. You can see that one root must be $+1$. Hence we can find a unit vector $\mathbf{A}$ such that $S\mathbf{A} = \mathbf{A}$. (Of course, $\mathbf{A}$ or $-\mathbf{A}$ will determine the axis of rotation of S.) Next, we use

$$S^{-1} = C(S)^T / \det(S), \quad C(S) \text{ the matrix of cofactors of } S, \text{ so } C(S) = S,$$

in proving that $S(\mathbf{V} \times \mathbf{W}) = (S\mathbf{V}) \times (S\mathbf{W})$. If $\mathbf{V} \perp \mathbf{A}$ then $(S\mathbf{V})^T(S\mathbf{A}) = \mathbf{V}^T S^T S\mathbf{A} = \mathbf{V}^T\mathbf{A} = 0$, so that $S\mathbf{V} \perp S\mathbf{A}$. Hence, for a unit vector $\mathbf{V} \in [\mathbf{A}]^\perp$, we can write

$$S\mathbf{V} = \lambda_\mathbf{V}\mathbf{V} + \mu_\mathbf{V}(\mathbf{A} \times \mathbf{V}), \quad \text{where} \quad \lambda_\mathbf{V}^2 + \mu_\mathbf{V}^2 = \|S\mathbf{V}\|^2 = \|\mathbf{V}\|^2 = 1.$$

You prove, using the fact that $S(\mathbf{A} \times \mathbf{V}) = A \times S\mathbf{V}$, that

$$\lambda_{\mathbf{A}\times\mathbf{V}} = \lambda_\mathbf{V} \quad \text{and} \quad \mu_{\mathbf{A}\times\mathbf{V}} = \mu_\mathbf{V}.$$

Next, if $\mathbf{W} \in [\mathbf{A}]^\perp$, then $W = \alpha\mathbf{V} + \beta\mathbf{A} \times \mathbf{V}$ for some α and β. Prove that $\lambda_\mathbf{W} = \lambda_\mathbf{V}$, $\mu_\mathbf{W} = \mu_\mathbf{V}$, and deduce the required result. $\qquad\square$

Sketched proof of the 'only if' part. Suppose that S is a rotation: $S = R(\theta, \mathbf{A})$ for some $(\theta, \mathbf{A})$. Then S preserves lengths, whence, for $\mathbf{V}, \mathbf{W} \in \mathbb{R}^3$,

$$2\langle S\mathbf{V}, S\mathbf{W}\rangle = \|S(\mathbf{V} + \mathbf{W})\|^2 - \|S\mathbf{V}\|^2 - \|S\mathbf{W}\|^2 = 2\langle\mathbf{V}, \mathbf{W}\rangle,$$

whence $\mathbf{V}^T(S^T S - I)\mathbf{W} = 0$ and $S^T S = I$. Hence $\det(S)^2 = \det(S^T)\det(S) = 1$. The continuous function $\varphi \mapsto \det(R(\varphi, A))$ for $\varphi \in [0, \theta]$, takes values in $\{-1, 1\}$ and so must only take the value 1 (which it takes when $\varphi = 0$). $\qquad\square$

▶ **E. The spinor representation of spin.** Recall that

$$\sigma_x = \begin{pmatrix} 0 & 1 \\ 1 & 0 \end{pmatrix}, \quad \sigma_y = \begin{pmatrix} 0 & -i \\ i & 0 \end{pmatrix}, \quad \sigma_z = \begin{pmatrix} 1 & 0 \\ 0 & -1 \end{pmatrix}. \tag{E1}$$

Henceforth, *we shall regard*

$$\tfrac{1}{2}\sigma_\mathbf{A} := \tfrac{1}{2}(A_1\sigma_x + A_2\sigma_y + A_3\sigma_z)$$

as Nature's observable for the spin of our spin-$\frac{1}{2}$ particle about **A**. This observable is a self-adjoint matrix acting on $\mathbb{C}^2$. This is the 'spinor' representation for a spin-$\frac{1}{2}$ particle. We immediately derive from Hamilton's relations at 474(C1) the properties

$$\sigma_x^2 = \sigma_y^2 = \sigma_z^2 = I,$$

$$\sigma_y \sigma_z = -\sigma_z \sigma_y, \quad \sigma_z \sigma_x = -\sigma_x \sigma_z, \quad \sigma_x \sigma_y = -\sigma_y \sigma_x.$$

We write as

$$\{\sigma_r, \sigma_r\} = 2\delta_{rs} I, \quad r, s \in \{x, y, z\},$$

where $\{F, G\}$ is the *anticommutator*

$$\{F, G\} := FG + GF.$$

Hence, whatever the unit vector **A**, we have

$$\sigma_A^2 = I.$$

If γ and **w** are such that $\sigma_A \mathbf{w} = \gamma \mathbf{w}$, where $\mathbf{w} \neq \mathbf{0}$, then

$$\mathbf{w} = \sigma_A^2 \mathbf{w} = \gamma \sigma_A \mathbf{w} = \gamma^2 \mathbf{w},$$

so that $\gamma = \pm 1$. Hence, *whatever the value of* **A**, *a measurement of the spin* $\frac{1}{2}\sigma_A$ *of our particle in the direction* **A** *will produce a value which is either* $\frac{1}{2}$ *or* $-\frac{1}{2}$.

Note that the eigenvector **w** corresponding to spin measurement $+\frac{1}{2}$ for spin $\frac{1}{2}\sigma_y$ about the y-axis is a multiple of $(1, i)^T$, and this will represent the state of the system after the measurement. *It is not wise to consider what the significance of the eigenvectors is.* Remember this when the answers for real eigenvectors in the later calculations seem strange.

▶ **Ea. The 'opposite-spins' state** $\mathcal{W}$ **in** $\mathbb{C}^2 \otimes \mathbb{C}^2$. Please note that the symbol $\mathcal{W}$ will now stand for a (very special) vector in $\mathbb{C}^2 \otimes \mathbb{C}^2$, not for a quaternion. We shall not see quaternions again in this book. In coordinates,

$$\mathcal{W} = \frac{1}{\sqrt{2}} \begin{pmatrix} 0 \\ 1 \\ -1 \\ 0 \end{pmatrix} = 2^{-\frac{1}{2}} (\mathbf{z} \otimes \mathbf{u} - \mathbf{u} \otimes \mathbf{z}). \tag{E2}$$

Note that if **v** and **w** are vectors in $\mathbb{C}^2$, so that $\mathbf{v} = \alpha\mathbf{z} + \beta\mathbf{u}$, $\mathbf{w} = \gamma\mathbf{z} + \delta\mathbf{u}$ for some $\alpha, \beta \gamma, \delta \in \mathbb{C}$, then (you check!)

$$\mathbf{v} \wedge \mathbf{w} := \mathbf{v} \otimes \mathbf{w} - \mathbf{w} \otimes \mathbf{v} = (\alpha\delta - \beta\gamma)\sqrt{2}\mathcal{W}.$$

It is no coincidence that the determinant $(\alpha\delta - \beta\gamma)$ appears in this Grassmann (or exterior-algebra) product. This is part of the correct algebra for Jacobians.

Let **A** be any unit vector. Let $\mathbf{w}_A^+$ and $\mathbf{w}_A^-$ be unit eigenvectors of σ_A corresponding to eigenvalues $+1$ and -1 respectively. Please note that superscripts $^+$ will appear, but superscript † for adjoint will not appear again. (So you will not need to ask Macbeth's question.) We shall have

$$\mathbf{w}_\mathbf{A}^+ \otimes \mathbf{w}_\mathbf{A}^- - \mathbf{w}_\mathbf{A}^- \otimes \mathbf{w}_\mathbf{A}^+ = \lambda \mathcal{W}$$

for some non-zero λ, λ being non-zero because $\mathbf{w}_\mathbf{A}^+$ and $\mathbf{w}_\mathbf{A}^-$ are linearly independent. In fact, $|\lambda| = \sqrt{2}$. Hence,

$$\lambda(\sigma_\mathbf{A} \otimes \sigma_\mathbf{A})\mathcal{W} = \mathbf{w}_\mathbf{A}^+ \otimes (-\mathbf{w}_\mathbf{A}^-) - (-\mathbf{w}_\mathbf{A}^-) \otimes \mathbf{w}_\mathbf{A}^+ = -\lambda\mathcal{W},$$

and so,

$$(\sigma_\mathbf{A} \otimes \sigma_\mathbf{A})\mathcal{W} = -\mathcal{W} \text{ for every unit vector } \mathbf{A}. \tag{E3}$$

The multiplier -1 of $\mathcal{W}$ is the determinant of $\sigma_\mathbf{A}$.

▶ **F. Weird real-world consequences of entanglement.**   We look at entanglement here as if we had never discussed it in connection with entangling circuits.

The mathematics of the last subsection relates to a situation in which a spin-0 particle splits into two spin-$\frac{1}{2}$ particles of opposite spins, the state $\mathcal{W}$ describing the 'spin-state' of the *pair* of particles after splitting. For unit vectors $\mathbf{B}$ and $\mathbf{D}$, we regard $\sigma_\mathbf{B}^{(1)} = \sigma_\mathbf{B} \otimes I$ as representing the state of the first particle in direction $\mathbf{B}$ and $\sigma_\mathbf{D}^{(2)} = I \otimes \sigma_\mathbf{D}$ as representing the spin in direction $\mathbf{D}$ of the second particle.

Of course, as already mentioned, the full quantum state of the system should include features other than spin. However, it is the case that we can here focus just on spins without committing error.

Note that the state $\mathcal{W}$ cannot be written as a single tensor product $\mathbf{s} \otimes \mathbf{t}$ which would represent the situation 'first particle in state $\mathbf{s}$, second in state $\mathbf{t}$': the state $\mathcal{W}$ is **entangled**. Hence, *when the quantum system is in state $\mathcal{W}$, neither particle has itself a well-defined quantum state. The two particles must be considered as an inseparable pair* – as we have seen before and as we shall now examine again, *a measurement made on one of the particles will affect the other.*

Imagine that the two particles drift apart, and that Observer 1 measures the spin of the first particle in direction A. Now,

$$\mathbf{w}_\mathbf{A}^+ \otimes \mathbf{w}_\mathbf{A}^+, \ \mathbf{w}_\mathbf{A}^+ \otimes \mathbf{w}_\mathbf{A}^-, \ \mathbf{w}_\mathbf{A}^- \otimes \mathbf{w}_\mathbf{A}^+, \ \mathbf{w}_\mathbf{A}^- \otimes \mathbf{w}_\mathbf{A}^-$$

form an orthonormal basis for $\mathbb{C}^2 \otimes \mathbb{C}^2$. We have

$$\mathcal{W} = \lambda^{-1}\left(\mathbf{w}_\mathbf{A}^+ \otimes \mathbf{w}_\mathbf{A}^- - \mathbf{w}_\mathbf{A}^- \otimes \mathbf{w}_\mathbf{A}^+\right),$$

Hence, the vector

$$\mathcal{W} - \lambda^{-1}\mathbf{w}_\mathbf{A}^+ \otimes \mathbf{w}_\mathbf{A}^- = \lambda^{-1}\mathbf{w}_\mathbf{A}^- \otimes \mathbf{w}_\mathbf{A}^+$$

is orthogonal to $\mathbf{w}_\mathbf{A}^+ \otimes \mathbf{w}_\mathbf{A}^+$ and to $\mathbf{w}_\mathbf{A}^+ \otimes \mathbf{w}_\mathbf{A}^-$. Hence $\lambda^{-1}\mathbf{w}_\mathbf{A}^+ \otimes \mathbf{w}_\mathbf{A}^-$ is the orthogonal projection of $\mathcal{W}$ onto the space $[\mathbf{w}_\mathbf{A}^+] \otimes \mathbb{C}^2$ of eigenvectors of $\sigma_\mathbf{A}^{(1)}$ corresponding to eigenvector 1. Especially then, if Observer 1 obtains a result $+1$ from her measurement of

$\sigma_{\mathbf{A}}^{(1)}$, then the state of the system immediately jumps to the unentangled state $\mathbf{w}_{\mathbf{A}}^+ \otimes \mathbf{w}_{\mathbf{A}}^-$, so that Particle 2 now has definite state $\mathbf{w}_{\mathbf{A}}^-$ (modulo phase); and measurement of $\sigma^{(2)}$ gives definite result -1. *Performing the measurement of $\sigma_{\mathbf{A}}$ on Particle 1 fixes the state of Particle 2, irrespective of how far away it is.*

This relates to the EPR–Bohm paradox, Bohm's version of the Einstein–Podolsky–Rosen paradox.

▶ **Fa. Non-locality.** This curious entanglement of particles is not even an effect that travels at the speed of light. Observers 1 and 2 using the same **A** will get opposite results even if their acts of measurement are 'space-like separated' so that no signal travelling at the speed of light can alert either particle, before its own spin is measured, of the result – or the direction – of the measurement on the other particle. This strange non-locality phenomenon does not, however, enable Observer 1 to send a meaningful signal to Observer 2 at a speed greater than that of light. Remember, for instance, that there is no way of detecting that the spin of a particle has become fixed in some direction. [[*Note.* According to Relativity Theory, it is meaningless to say of two space-like separated events that one occurs before the other.]]

Various papers appear on the Internet and in journals claiming faster than light communication (get your search engine to look for 'FTL' or 'faster than light'), for example, the FTL communication claimed by Nimtz of Mozart's G minor 40th Symphony, a great choice of signal whether the Physics is right or not.

G. Reduction to 'two-dimensional' situation.
To proceed further, we shall restrict attention to the case when measurements of spins are always in directions parallel to the (z, x) plane. We shall write (for a single particle)

$$\sigma_\alpha := (\cos\alpha)\sigma_z + (\sin\alpha)\sigma_x = \begin{pmatrix} \cos\alpha & \sin\alpha \\ \sin\alpha & -\cos\alpha \end{pmatrix}. \tag{G1}$$

Ga. Exercise. Define

$$\mathbf{u}_\beta^+ = \begin{pmatrix} \cos\beta \\ \sin\beta \end{pmatrix}, \qquad \mathbf{u}_\beta^- = \begin{pmatrix} -\sin\beta \\ \cos\beta \end{pmatrix}.$$

Show that

$$\sigma_\alpha \mathbf{u}_\beta^+ = +\mathbf{u}_{\alpha-\beta}^+, \qquad \sigma_\alpha \mathbf{u}_\beta^- = -\mathbf{u}_{\alpha-\beta}^-,$$

so that the eigenvectors of σ_α corresponding to $+1$ and -1 are

$$\mathbf{w}_\alpha^+ = \mathbf{u}_{\frac{1}{2}\alpha}^+, \qquad \mathbf{w}_\alpha^- = \mathbf{u}_{\frac{1}{2}\alpha}^-.$$

Recall that the convention regarding such equations as those below is that one first reads the top sign throughout, then, separately, with the bottom sign throughout; one does not switch between top and bottom signs in mid-equation. It is *not* like reading ± 1 as 'either $+1$ or -1'.

Show that we have for same signs:

$$\mathbb{P}_{\mathcal{W}} \left(\sigma_\alpha^{(1)} \text{ is measured as } \pm 1; \ \sigma_\beta^{(2)} \text{ is measured as } \pm 1 \right)$$
$$= \langle \mathcal{W}, \mathbf{w}_\alpha^\pm \otimes \mathbf{w}_\beta^\pm \rangle = \tfrac{1}{2} \sin^2 \tfrac{1}{2}(\alpha - \beta),$$

and for opposite signs:

$$\mathbb{P}_{\mathcal{W}}\left(\sigma_{\alpha}^{(1)} \text{ is measured as } \pm1; \; \sigma_{\beta}^{(2)} \text{ is measured as } \mp1\right)$$
$$= \left\langle \mathcal{W}, \mathbf{w}_{\alpha}^{\pm} \otimes \mathbf{w}_{\beta}^{\mp} \right\rangle = \tfrac{1}{2}\cos^2 \tfrac{1}{2}(\alpha - \beta).$$

Deduce that

$$\mathbb{E}_{\mathcal{W}}\sigma_{\alpha}^{(1)}\sigma_{\beta}^{(2)} = -\cos(\alpha - \beta).$$

Gb. Exercise. Consider the 1-particle situation. Suppose that the true initial state of our particle is $\mathbf{w}_{\gamma}^{+}$. Show that if σ_{α} is measured before σ_{β}, then

$$\mathbb{P}_{\gamma}\left(\sigma_{\alpha} \text{ is measured first as } 1; \; \sigma_{\beta} \text{ is then measured as } 1\right)$$
$$= \cos^2 \tfrac{1}{2}(\gamma - \alpha)\cos^2 \tfrac{1}{2}(\alpha - \beta).$$

▶ **H. Polarization of photons.** Experimental tests associated with entanglement have been done with photons (which are spin-1 particles, though of a peculiar type), not with spin-$\frac{1}{2}$ particles. The necessary modifications are easily made however, and we can utilize our existing notation for the mathematical symbols σ_{α}, $\mathbf{u}_{\alpha}^{+}$, etc.

We first consider a beam of light, moving parallel to the y-axis, which is plane-polarized (or linearly-polarized) in that the electric field (light is an electromagnetic wave) always points in a fixed direction perpendicular to the y-axis. This fixed direction corresponds to an angle α in the (z, x) plane. If we hold a piece of polaroid in the (z, x) plane with polarization direction at angle β, then a fraction $\cos^2(\alpha - \beta)$ of the intensity will pass through the polaroid, and a fraction $\sin^2(\alpha - \beta)$ will be absorbed.

For a single photon of that light, the probability that it will pass through the polaroid is $\cos^2(\alpha - \beta)$.

Now for the Quantum version. The polarization of a single photon in direction β in the (z, x) plane is represented by the observable

$$\rho_{\beta} = \sigma_{2\beta}$$

with eigenvectors $\mathbf{u}_{\beta}^{+}, \mathbf{u}_{\beta}^{-}$ corresponding to eigenvalues $1, -1$ respectively. If the photon is linearly polarized at angle α in the sense that its state is $\mathbf{u}_{\alpha}^{+}$, and if a measurement of ρ_{β} is made on it, the result is $+1$ with probability $\cos^2(\alpha - \beta)$, this because

$$\langle \mathbf{u}_{\alpha}, \mathbf{u}_{\beta}^{+} \rangle = \cos(\alpha - \beta), \qquad \langle \mathbf{u}_{\alpha}, \mathbf{u}_{\beta}^{-} \rangle = \sin(\alpha - \beta).$$

So a measurement $+1$ corresponds to 'going through the polarizer', a measurement -1 to 'being absorbed'.

It is very important that

$$\rho_{\alpha}\rho_{\beta} \neq \rho_{\beta}\rho_{\alpha} \quad \text{unless } \alpha - \beta \text{ is a multiple of } \tfrac{1}{2}\pi. \tag{H1}$$

The entangled state

$$\mathcal{U} := \frac{1}{\sqrt{2}}(\mathbf{u}_\theta^+ \otimes \mathbf{u}_\theta^+ + \mathbf{u}_\theta^- \otimes \mathbf{u}_\theta^-) = \begin{pmatrix} 1 \\ 0 \\ 0 \\ 1 \end{pmatrix}, \quad \text{for every } \theta,$$

represents two photons with the same linear polarization:

$$(\rho_\theta \otimes \rho_\theta)\mathcal{U} = +\mathcal{U}.$$

However, we shall continue to work with the opposite-polarization (that is, now, perpendicular-polarization) state $\mathcal{W}$. From Exercise 481Ga, we have

$$C(\alpha, \beta) := \mathbb{E}_\mathcal{W}\rho_\alpha^{(1)}\rho_\beta^{(2)} = -\cos 2(\alpha - \beta). \tag{H2}$$

Note that this is correctly -1 if $\alpha = \beta$ and 1 if $\alpha - \beta = \frac{1}{2}\pi$.

[[*Remark.* It is true that $\mathcal{W}$ looks a more fermionic state than bosonic in that interchanging the identities of the two particles changes $\mathcal{W}$ to $-\mathcal{W}$. However, $\mathcal{W}$ is not the full quantum state of the system, which would involve positions, momenta, etc.]]

One can create an opposite(perpendicular)-state photon pair by using a certain type of crystal to split an incoming photon into two perpendicularly-polarized photons of lower energy (lower frequency).

Further note on polarization. It is possible to have elliptically-polarized light in which the electric field 'spirals' around the direction of the beam. This idea can be incorporated into our Quantum picture. For example, right-circular polarization is represented by the vector $(1, i)^T$, which, not too surprisingly, is the eigenvector of σ_y corresponding to eigenvalue 1. But we concentrate on the 'plane-polarized' situation.

▶ **I. Hidden Variables: the difficulties.** We work with the opposite-polarization state $\mathcal{W}$. *How do particles* 1 *and* 2 *contrive that if each is tested with a polarizer at the same angle* β, *then precisely one will go through, this independently of the value* β? It might seem at first sight that from the moment of their creation, for every β, Particle 1 had hidden within it the value $r_\beta^{(1)}$ of a Random Variable $R_\beta^{(1)}$ which would be the measured value of ρ_β if the chosen direction of measurement is β; and that Particle 2 has hidden within it the realization $r_\beta^{(2)} = -r_\beta^{(1)}$ for the same purpose. This is a Very Naive Hidden-Variables (VNHV) picture of the quantum world as having a classical one hidden behind it. It is a very strange picture: the function $\beta \mapsto r_\beta^{(1)}$ from S^1 to $\{-1, 1\}$ is either constant or is discontinuous. [[Of course, similar remarks apply to the '3-dimensional' case of spin-$\frac{1}{2}$ particles.]]

J.S. Bell's idea was to derive **inequalities** which would have to be satisfied if the VNHV picture is true, inequalities in conflict with identities which follow from (H2). Such inequalities (and the VNHV theory) could therefore be expected to be 'shot down' by experiment, and brilliant experiments of Aspect and others provide convincing demonstration that it is indeed Quantum Theory which wins. The particular inequality which we consider, the **Bell–CHSH inequality** was produced independently by Bell and by Clauser, Horne, Shimony and Holt. It provides the case tested in actual experiments.

The essential insight is to note that if $r_0^{(1)}, r_{\frac{1}{4}\pi}^{(1)}, r_{\frac{1}{8}\pi}^{(2)}$, and $r_{-\frac{1}{8}\pi}^{(2)}$ are numbers taking values in the two-point set $\{-1, 1\}$, then

$$\left(r_0^{(1)} + r_{\frac{1}{4}\pi}^{(1)}\right) r_{\frac{1}{8}\pi}^{(2)} + \left(r_0^{(1)} - r_{\frac{1}{4}\pi}^{(1)}\right) r_{-\frac{1}{8}\pi}^{(2)}$$

can only take the value $+2$ or -2. Now, the expectation of an RV taking values in $\{-2, 2\}$, must lie within $[-2, 2]$. Hence, if the VNHV theory is true, we must have the **Bell–CHSH inequality**

$$\left| \mathbb{E}_W \left[\left(R_0^{(1)} + R_{\frac{1}{4}\pi}^{(1)}\right) R_{\frac{1}{8}\pi}^{(2)} + \left(R_0^{(1)} - R_{\frac{1}{4}\pi}^{(1)}\right) R_{-\frac{1}{8}\pi}^{(2)} \right] \right| \; \leq \; 2, \tag{I1}$$

Thus, VNHV leads to the result

$$|C(0, \tfrac{1}{8}\pi) + C(\tfrac{1}{4}\pi, \tfrac{1}{8}\pi) + C(0, -\tfrac{1}{8}\pi) - C(\tfrac{1}{4}\pi, -\tfrac{1}{8}\pi)| \; \leq \; 2. \tag{I2}$$

However, it is immediate from 483(H2) that Quantum Theory yields the identity

$$C(0, \tfrac{1}{8}\pi) + C(\tfrac{1}{4}\pi, \tfrac{1}{8}\pi) + C(0, -\tfrac{1}{8}\pi) - C(\tfrac{1}{4}\pi, -\tfrac{1}{8}\pi) \; = \; -2\sqrt{2}.$$

contradicting the Bell-CHSH inequality of VNHV theory.

But, says a less naive Hidden-Variables advocate, "Because of 482(H1), we cannot simultaneously measure all the observables of which expectations feature in (I2). You must tell me what values of α and β are to be used in some experiment. This information should be built into the whole system from the start. Indeed, Exercise 481Ga tells me that, for the given α and β, from the moment of their creation, the two RVs $R_\alpha^{(1)}$ and $R_\beta^{(2)}$ have joint probability mass function satisfying

$$\mathbb{P}\left(R_\alpha^{(1)} = 1; R_\beta^{(2)} = 1\right) \; = \; \mathbb{P}\left(R_\alpha^{(1)} = -1; R_\beta^{(2)} = -1\right) \; = \; \tfrac{1}{2}\sin^2(\alpha - \beta),$$

$$\mathbb{P}\left(R_\alpha^{(1)} = 1; R_\beta^{(2)} = -1\right) \; = \; \mathbb{P}\left(R_\alpha^{(1)} = -1; R_\beta^{(2)} = 1\right) \; = \; \tfrac{1}{2}\cos^2(\alpha - \beta).$$

You have convinced me that I cannot arrange this simultaneously for all α and β. But I need only arrange it for the particular pair (α, β) used in the experiment."

An official spokesperson replies: "Yes, but I haven't told you the main thing about the brilliant Aspect experiment, and the myriad other experiments which confirm his findings. You see, on each experiment, the decision about which directions α and β were to be used was not made until *after* the creation of the entangled pair. Before its own polarization is tested, neither photon could learn from information travelling at the speed of light of the angle of measurement of polarization of the other photon. So, we must have non-locality."

To which the Hidden-Variables advocate can of course reply: "I admit that that seems very disturbing for my attempt to give a realistic picture of what is happening. But you now have to tell me the full details of how the choice of directions is made ... "

And there we leave them.

Ia. A way of thinking about the quantum Bell–CHSH identity. For a single photon,

$$\rho_\alpha = (\cos 2\alpha)\rho_0 + (\sin 2\alpha)\rho_{\frac{1}{4}\pi} \text{ whence } \sqrt{2}\rho_{\frac{1}{8}\pi} = \rho_0 + \rho_{\frac{1}{4}\pi}.$$

For our entangled pair of photons in state $\mathcal{W}$, with probability 1,

$$\left(\rho_0^{(1)} + \rho_{\frac{1}{4}\pi}^{(1)}\right)\rho_{-\frac{1}{8}\pi}^{(2)} = \sqrt{2}\rho_{\frac{1}{8}\pi}^{(1)}\rho_{-\frac{1}{8}\pi}^{(2)} = -\sqrt{2},$$

$$\left(\rho_0^{(1)} - \rho_{\frac{1}{4}\pi}^{(1)}\right)\rho_{\frac{1}{8}\pi}^{(2)} = \sqrt{2}\rho_{-\frac{1}{8}\pi}^{(1)}\rho_{\frac{1}{8}\pi}^{(2)} = -\sqrt{2},$$

whence it is reasonable to claim that in state $\mathcal{W}$, the mathematical observable described by the self-adjoint operator

$$\left(\rho_0^{(1)} + \rho_{\frac{1}{4}\pi}^{(1)}\right)\rho_{\frac{1}{8}\pi}^{(2)} + \left(\rho_0^{(1)} - \rho_{\frac{1}{4}\pi}^{(1)}\right)\rho_{-\frac{1}{8}\pi}^{(2)}$$

is, with probability 1, equal to $-2\sqrt{2}$. (It is hard to imagine that this mathematical observable could be measured in the real world.) Since the expectation of an RV which is equal to $-2\sqrt{2}$ with probability 1, is obviously $-2\sqrt{2}$, we see a motivation for the Bell–CHSH identity.

For further discussion, see Subsection 491A.

10.5 Spin and the Dirac equation

Until now, spin has just been part of a mathematical 'game', so we should take a brief look at some Physics which greatly clarifies its real rôle. Anyone writing on this must feel tempted to include the lovely quote from Dirac:

> "A great deal of my work is just playing with equations and seeing what they give. I don't suppose that applies so much to other physicists; I think it's a peculiarity of myself that I like to play about with equations, just looking for beautiful mathematical relations which maybe don't have any physical meaning at all. Sometimes they do."

In this book, we have already seen several pieces of fine Mathematics: the Central Limit Theorem, the associated asymptotic results for MLEs and for the Likelihood Ratio, the F-test theorem for Linear Models, the Ergodic Theorem for the Gibbs sampler, the Strong Law, the Martingale-Convergence Theorems, the Universality result for quantum gates. Now, the Mathematics surrounding the Dirac equation matches any of these (though not some of the Mathematics mentioned at the end of Chapter 9); and as for the 'Sometimes they do', well, the equation signalled one of the most dramatic developments in the history of Physics.

Inevitably, the true picture is rather more complicated than the one I now present, but as a first approximation, this will do, I think.

Dirac found a remarkable modification of Schrödinger's equation for a free particle (one on which no force acts) which is compatible with Special Relativity Theory (which subject I do *not* assume that you know). Dirac's equation is not however compatible with the law of conservation of angular momentum *unless* one postulates the existence of an **intrinsic angular momentum** or **spin** for the particle which miraculously 'balances things out'. This intrinsic angular momentum has no classical analogue.

Let me mention a little of the rest of the story, even though, sadly, we do not study it in this book. In analogous fashion to that in which we are forced to introduce intrinsic angular momentum, consideration of the movement of an electron in an electro-magnetic field forces us to postulate the existence of an **intrinsic magnetic moment** of the electron, which again corresponds to (a constant multiple of) its spin. This is stunning enough, but Dirac's equation also predicts the existence of **anti-particles**; and the positron, the anti-particle of the electron, was discovered by Anderson just a few years after Dirac produced his equation. And there is much more to the story: it leads naturally on to quantum fields, etc. So the Dirac equation is truly wonderful.

Notation. Since we are now considering Physics,

- the modified Planck constant $\hbar$ will no longer be considered to be 1;

- c will now denote the speed of light (not $2^{-\frac{1}{2}}$);

- a typical wave function will be denoted by ψ, not $\mathbf{v}$, so that leaves $\mathbf{v}$ free to denote a velocity;

- the Hamiltonian will be denoted by H (not $\mathcal{H}$), the Hadamard gate no longer being of interest to us;

- Schrödinger's equation will be written

$$\mathrm{i}\hbar\frac{\partial\psi}{\partial t} = H\psi;$$

- the position of a particle moving on $\mathbb{R}$ will be represented by the observable Q, where $(Q\psi)(x) = x\psi(x)$ and its momentum by the observable $P := -\mathrm{i}\hbar\mathrm{d}/\mathrm{d}x$.

I shall call Q quantum position, P quantum momentum, and H quantum energy.

▶ **A. The Heisenberg picture of evolution of observables.** Up to now, we have used the **Schrödinger picture** in which the *state* evolves according to Schrödinger's equation

$$\mathrm{i}\hbar\frac{\partial\psi}{\partial t} = H\psi, \quad \text{so that } \psi(t) = \mathrm{e}^{-\mathrm{i}tH/\hbar}\psi(0).$$

In the Schrödinger picture, a physical observable such as the momentum of the particle is represented by the same operator A at all times.

In the equivalent **Heisenberg picture**, the state ψ remains fixed, and the *observable* evolves. To have compatibility, the expectation of an observable at time t has to be the same in the two pictures, so we must have

$$\text{(Schrödinger)} \ \langle\psi(t), A(0)\psi(t)\rangle = \langle\psi(0), A(t)\psi(0)\rangle \ \text{(Heisenberg)}.$$

Thus, we need

$$\langle \psi(0),\, A(t)\psi(0) \rangle = \langle e^{-itH/\hbar}\psi(0),\, A(0)e^{-itH/\hbar}\psi(0) \rangle$$
$$= \langle \psi(0),\, e^{itH/\hbar}A(0)e^{-itH/\hbar}\psi(0) \rangle.$$

Heisenberg therefore postulates that

$$A(t) = e^{itH/\hbar}A(0)e^{-itH/\hbar}$$

so that on 'differentiating by parts',

$$\frac{\mathrm{d}}{\mathrm{d}t}A(t) = \frac{iH}{\hbar}e^{itH/\hbar}A(0)e^{-itH/\hbar} + e^{itH/\hbar}A(0)e^{-itH/\hbar}\frac{(-iH)}{\hbar}.$$

Thus, we are led to Heisenberg's equation

$$i\hbar\frac{\mathrm{d}}{\mathrm{d}t}A(t) = [A(t), H] := A(t)H - HA(t). \tag{A1}$$

We note especially that *if $A(0)$ commutes with H, then $A(t) = A(0)$ so that $A(0)$ is an invariant of the motion.* In particular, $H(t) = H$, the law of conservation of quantum energy.

Of course, the Schrödinger and Heisenberg pictures only apply to situations where no measurement is made, so interpretation is again rather bizarre.

B. Free particle: non-relativistic theory. For a particle of mass m_0 moving freely (that is, not subject to any force) at velocity v on $\mathbb{R}$, Newton tells us that its classical momentum is $p = mv$ and that its classical energy is $\frac{1}{2}m_0v^2 = p^2/(2m_0)$. Force is rate of change of momentum, so, since our particle moves freely, $\dot{p} = \mathrm{d}p/\mathrm{d}t = 0$, and p is constant, the law of conservation of classical momentum.

Since the Hamiltonian is quantum energy, we write

$$H = \frac{P^2}{2m_0} = -\frac{\hbar^2}{2m_0}\frac{\mathrm{d}^2}{\mathrm{d}x^2}.$$

We take $P(0) = P, Q(0) = Q$. Since P obviously commutes with H, we have $P(t) = P$, the law of conservation of quantum momentum. In regard to the quantum position $Q(t)$ at time t, it is natural to guess from Newton's theory that

$$Q(t) = Q + m_0^{-1}tP.$$

This checks with

$$[Q, H]\psi = \frac{\hbar^2}{2m_0}\left[-x\frac{\mathrm{d}^2\psi}{\mathrm{d}x^2} + \frac{\mathrm{d}^2}{\mathrm{d}x^2}(x\psi) \right] = \frac{\hbar^2}{m_0}\frac{\mathrm{d}\psi}{\mathrm{d}x} = i\hbar m_0^{-1}P\psi.$$

For a particle of mass m_0 moving freely at velocity $\mathbf{v}$ in $\mathbb{R}^3$, Newton says that $\mathbf{p} = m\mathbf{v}$ and the lack of force implies that $\dot{\mathbf{p}} = \mathbf{0}$. Since the 'orbital' angular momentum

ℓ of the particle about the origin was defined by Newton to be $\mathbf{q} \times \mathbf{p}$, where $\mathbf{q}$ is the particle's position and $\times$ the vector product, we have

$$\frac{\mathrm{d}\ell}{\mathrm{d}t} = \mathbf{q} \times \dot{\mathbf{p}} + \dot{\mathbf{q}} \times \mathbf{p} = \mathbf{0},$$

because $\dot{\mathbf{p}} = \mathbf{0}$ and $\dot{\mathbf{q}}$ is parallel to $\mathbf{p}$.

In the quantum context, the quantum position is the 3-component vector observable $\mathbf{Q} = (Q_x, Q_y, Q_z)^T$ and the quantum momentum is $\mathbf{P} = (P_x, P_y, P_z)^T$, where for $\psi : \mathbb{R}^3 \to \mathbb{C}$, we have

$$(Q_x\psi)(x, y, z) = x\psi(x, y, z), \qquad (P_x\psi)(x, y, z) = -\mathrm{i}\hbar\frac{\partial\psi}{\partial x},$$

etc. The quantum orbital angular momentum is the 3-component vector observable

$$\begin{aligned} \mathbf{L} &= \mathbf{Q} \times \mathbf{P} = (L_x, L_y, L_z)^T \\ &= -\hbar\mathrm{i}(y\partial_z - z\partial_y, \; z\partial_x - x\partial_z, \; x\partial_y - y\partial_x)^T, \end{aligned} \tag{B1}$$

where $\partial_x = \partial/\partial x$ as usual. Of course, H is now

$$\frac{1}{2m_0}\left(P_x^2 + P_y^2 + P_z^2\right) = -\frac{\hbar^2}{2m_0}\Delta,$$

where Δ is the familiar Laplace operator

$$\Delta = \frac{\partial^2}{\partial x^2} + \frac{\partial^2}{\partial y^2} + \frac{\partial^2}{\partial z^2}.$$

You should check that $[L_x, H] = 0$, so that in the Heisenberg picture, $\mathbf{L}(t) = \mathbf{L}(0) = \mathbf{L}$, and $\mathbf{L}$ is an invariant of the motion, yielding the law of conservation of quantum angular momentum.

If $\mathbf{k}$ is the unit vector along the z-axis, then

$$R(t, \mathbf{k})\begin{pmatrix} x \\ y \\ z \end{pmatrix} = \begin{pmatrix} x\cos t - y\sin t \\ x\sin t + y\cos t \\ z \end{pmatrix},$$

and

$$\frac{\partial}{\partial t}\psi(x\cos t - y\sin t, \; x\sin t + y\cos t, \; z)\big|_{t=0} = (x\partial_y - y\partial_x)\psi.$$

This points to the perfect ('Lie-algebra') tie-up between angular momentum and our earlier study of SO(3), but spin as *intrinsic* angular momentum is something different.

▶ **C. Energy in Special Relativity Theory.** Suppose that a free particle has 'rest mass' m_0: this is the mass of the particle as measured by an observer at rest relative to the particle. (If Observer 1 reckons that the particle is moving freely, he will regard the particle as having constant velocity, and an Observer 2, moving at some constant velocity relative to Observer 1, will also believe the particle to be moving freely.)

For a particle of rest mass m_0 moving at velocity $\mathbf{v}$ relative to us, we regard it as having

- mass $m = m_0 \left(1 - \|\mathbf{v}\|^2/c^2\right)^{-\frac{1}{2}}$,

- momentum $\mathbf{p} = m\mathbf{v}$,

- energy (wait for it!) $E = mc^2$.

Juggling with the algebra produces (with the strange $\pm$ discussed later)

$$E^2 = m_0^2 c^4 + c^2 \|\mathbf{p}\|^2, \qquad E = \pm\sqrt{m_0^2 c^4 + c^2 \|\mathbf{p}\|^2}. \tag{C1}$$

▶▶ **D. The Dirac equation.** An essential property of the original Schrödinger equation is that because it is linear and of first order in ∂_t and H is self-adjoint, then $\|\psi(t)\|^2 = \|\psi(0)\|^2$, yielding conservation of probability densities. Now in Special Relativity, space and time are treated on the same footing. Dirac therefore sought an equation which is linear and of first order in all partial derivatives. The obstacle to achieving this is obviously caused by the square root in equation (C1).

However, we already know that the anticommutation relations

$$\{\sigma_r, \sigma_s\} := \sigma_r \sigma_s + \sigma_s \sigma_r = 2\delta_{rs} I_2, \qquad r, s \in \{x, y, z\},$$

(where I_2 is the identity 2×2 matrix) for the Pauli matrices imply that for numbers p_x, p_y, p_z we have

$$(p_x \sigma_x + p_y \sigma_y + p_z \sigma_z)^2 = \|\mathbf{p}\|^2 I_2,$$

so that $p_x \sigma_x + p_y \sigma_y + p_z \sigma_z$ may be regarded as a 'square root' of $\|\mathbf{p}\|^2$.

Dirac therefore had the idea of writing

$$\left(m_0^2 c^4 + c^2 \|\mathbf{p}\|^2\right) I_n = \left(m_0 c^2 \alpha_t + c p_x \alpha_x + c p_y \alpha_y + c p_z \alpha_z\right)^2$$

where, for some n, the α_r are postulated to be self-adjoint $n \times n$ matrices satisfying the anticommutation relations

$$\{\alpha_r, \alpha_s\} := \alpha_r \alpha_s + \alpha_s \alpha_r = 2\delta_{rs} I_n \qquad (r, s \in \{t, x, y, z\}), \tag{D1}$$

and thus producing the Dirac equation

$$i\hbar \frac{\partial \psi}{\partial t} = H_D \psi, \qquad H_D := m_0 c^2 \alpha_t + c P_x \alpha_x + c P_y \alpha_y + c P_z \alpha_z, \tag{D2}$$

where $\psi : \mathbb{R}^3 \to \mathbb{C}^n$. (Note that H_D is (formally) self-adjoint because the α's are, the P's are, and the α's commute with the P's.) We assume for now that such matrices α_r exist.

I skip discussion of the fact that the Dirac equation transforms (from one observer to another) in the appropriate ('Lorentz covariant') manner for Relativity Theory. See references at the end of this section.

▶▶ **E. Electron spin as intrinsic angular momentum.** Once more, let **L** be the orbital angular momentum given by 488(B1). Then

$$[L_x, H_D] = -i\hbar\{(y\partial_z - z\partial_y)c(\alpha_y\partial_y + \alpha_z\partial_z) - c(\alpha_y\partial_y + \alpha_z\partial_z)(y\partial_z - z\partial_y)\}$$
$$= i\hbar c(\alpha_y\partial_z - \alpha_z\partial_y),$$

so the law of conservation of quantum orbital angular momentum about the origin *fails* for our free particle. This is a situation which **must** be remedied.

But now define $\mathbf{S} = (S_x, S_y, S_z)^T$ to be the 3-component observable of self-adjoint matrices

$$\mathbf{S} = -\tfrac{1}{2}i\hbar(\alpha_y\alpha_z, \ \alpha_z\alpha_x, \ \alpha_x\alpha_y).$$

(Note that $\alpha_y\alpha_z = \tfrac{1}{2}[\alpha_y, \alpha_z]$.) Then

$$[S_r, \alpha_t] = [S_r, \alpha_r] = 0 \quad (r \in \{x, y, z\}),$$
$$[S_x, \alpha_y] = i\hbar\alpha_z \quad \text{and likewise for cyclic permutations of } x, y, z,$$

whence $[S_x, H_D] = -i\hbar c(\alpha_y\partial_z - \alpha_z\partial_y)$, and so

$$[T_r, H_D] = 0 \quad (r \in \{x, y, z\}), \quad \text{where } \mathbf{T} := \mathbf{L} + \mathbf{S}.$$

Thus, **T** is an invariant of the motion determined by the Dirac Hamiltonian H_D, and *this* is the true law of conservation of quantum angular momentum. Adding in the intrinsic angular momentum, or spin, **S**, has saved the day, assuming that the α's exist.

So, *how do we find the α's?* This question was answered by Clifford in 1880! We take $n = 4$ and

$$\alpha_t = \sigma_z \otimes I, \quad \alpha_r = \sigma_z \otimes \sigma_r, \quad (r \in \{x, y, z\});$$

indeed, this is essentially the only choice (see below).

You check that these α's have all the required properties. The variable **S** is exactly what we had in our SU(2) picture except for the mysterious (but understood) '$I\otimes$'.

F. A bizarre feature of the Dirac equation. Since $[P_x, H_D] = 0$, the law of conservation of momentum holds for the Dirac equation.

But what does the equation say about the position of the particle? Because the orbital angular momentum is not preserved for the Dirac equation, it *cannot* be the case now that $\dot{\mathbf{Q}}(t)$ is always a scalar multiple of $\mathbf{P}(t)$, even though $\dot{\mathbf{q}}(t)$ *is* a scalar multiple of $\mathbf{p}(t)$ for our free particle in Special Relativity. Check that we have

$$[Q_x, H_D] = i\hbar c\alpha_x,$$

whence,

$$\frac{dQ_x(t)}{dt} = c\alpha_x \quad \text{when } t = 0.$$

Now, $c\alpha_x$ is a *mathematical* observable which can only be 'measured' as $-c$ or c, so things are truly rather bizarre. Note too that $[\alpha_x, H_D] \neq 0$, so that α_x is not invariant under H.

For discussion of this point, and of the associated phenomenon of Zitterbegewung ('jittering'), see Thaller [225] and other books mentioned at the end of the next subsection.

G. What next? There I must end my bit of the 'Dirac' story.

You are probably intrigued that the quantum energy is

$$H_D = m_0 c^2 \alpha_t + \cdots,$$

where α_t can only be measured as $+1$ or -1. Does this point to the existence of negative energy? In classical Special Relativity, one has to take the positive square-root sign in 489(C1). But it seems that with quantum theory, all things are possible.

I cannot do better than hand you over to my Swansea colleague David Olive for a better geometric picture of what we have done, for the remarkable essential uniqueness of the choice of the α's, and for some of the still more fascinating remainder of the story. See Olive [179].

For extensive studies of the Dirac equation (including its Lorentz covariance and its bizarre behaviour in regard to position observables), see Thaller [225], Bjorken and Drell [24], Davydov [56], Messiah [163]. A fine account of some quantum-field theory by Fradkin [82] is available on the Web. See di Francesco, Mathieu and Sénéchal [64] for the definitive version.

10.6 Epilogue

▶▶ **A. Quantum Theory and 'Realism'.**   As the quant-ph website and many other web sites reveal, many people refuse to accept non-locality, and continually put forward 'realistic and local explanations' (some more realistic than others!) of the results obtained by Aspect and others. Some of the objections in regard to 'detection' in such experiments might be considered definitively countered by Rowe *et al.* [203].

However, most physicists believe that we have to accept that Quantum Theory just *is* completely counter-intuitive. I believe that the Mathematics of self-adjoint operators, etc, is all there is to understand. That Mathematics has never been faulted by experiment. Of course, it will be very interesting to see Penrose's development of his gravity theory of R-mode quantum mechanics, and perhaps even more interesting to see what M-theory eventually has to say about it all; but many mysteries will remain if we try to explain things in ordinary language.

A small part of me would like to see a more 'sensible' explanation of the weird phenomena of entanglement, etc.   Some physicists and mathematicians with deep understanding of the orthodox theory have sought such explanations. Amongst the best known attempts are the Bohm and Bohm–Hiley theory ([25, 26]) of *quantum potential*, and Nelson's theory [171] of *stochastic mechanics*. These theories agree with Schrödinger's equation for a particle moving under the influence of some force: indeed the quantum potential and the stochastic mechanics are derived from Schrödinger's equation.

But the Bohm(–Hiley) theory gives the particle a classical position and velocity, and the Nelson theory gives it a classical position. So these are hidden-variable theories, but each postulates some kind of underlying quantum fluctuation as an *extra*, non-classical, force. Influences are not local in these theories.

Feynman's *path-integral formulation* [77] is another way, familiar to probabilists because of the 'Feynman–Kac formula', of thinking about Schrödinger's equation. Everett's (in)famous *many-worlds* interpretation is something in which I do not believe: it is the least parsimonious theory ever! I should not be so flippant: many people who have thought long and hard about Quantum Theory, especially those in Quantum Computing, are tending to embrace the many-worlds theory. Sorry, folks: I cannot accept it.

There are a number of attempts to see Quantum Theory in terms of Quantum Information Theory. See, for example, Zeilinger [239] and Bouwmeester, Ekert and Zeilinger [28], and a forthcoming book by Kelbert and Suhov.

Fundamentally, I prefer to stick to the orthodox theory, regarding Mathematics as the correct language of Nature, and not needing to have it interpreted. The operator formulation applies in exactly the same way to all situations: to the finite-dimensional situation for Quantum Computing exactly as for the position-momentum case; and I consider it very much more elegant than any of its 'realistic' alternatives. Nature rejoices in Pure Mathematics: in group representations, Clifford algebras, etc. She sees SO(3) as an abstract group, utilizing all its representations, not just the SO(3) and SU(2) representations we have studied. I am sure that She loves the Parseval duality between position and momentum which 'realistic' theories lose.

Thought. It always puzzles me that people want things 'explained' in English or Chinese or some other ethnic language, when such languages are extremely crude compared with that of Mathematics in situations where the latter applies. Why try to 'improve' on a perfect mathematical picture such as the 'orthodox' one of the quantum world? People will presumably accept that I cannot describe Beethoven's C sharp minor String Quartet in words. [[Incidentally, I do worry about Lewis Thomas's idea, quoted in Dawkins [57], that we should advertise our achievements on Earth by broadcasting into space Bach and only Bach. What makes Bach's music great is what *cannot* be explained by its 'mathematics', and an alien civilization aware that the music is the creation of intelligent life, is unlikely to be able to appreciate that. I can easily program the computer to write a fugue with 100 parts, with complex stretti and cancrisans statements galore, and all the other tricks of the art of fugue. But it won't be music. Far better to continue broadcasting the fact that we know the Riemann Hypothesis!]]

Decoherence.   It would be quite wrong to leave this topic without mentioning *decoherence*, which recognizes the rôle played by the *environment*, which constantly 'observes' a quantum system. The theory of decoherence answers two key questions. Given the weird phenomena of the quantum world, why does the everyday world generally act in a sane way? (If the environment is continually observing the world at microscopic level, why aren't states jumping to eigenstates of all sorts of different operators all the time?) On the other hand, we need to answer the question: how can it sometimes happen that we *are* able to witness quantum phenomena on a 'macroscopic' scale? Read Zurek [240], and Giulini, Joos, Kiefer, Kupsch, Stamatescu and Zeh [97].

B. And finally ... . I wrote much of my account of entanglement at Bath, finishing that first draft on 3rd July 1997. I went down to the library at Bradford on Avon that morning to find a novel to free my mind from this topic. On the library shelves, however, I happened to spot the novel *The Schrödinger Cat Trilogy* by Robert A Wilson [236]. This extraordinary book was written in 1979, three years before the *famous* Aspect experiment. As mentioned in the quotation, Clauser and Aspect had already done some important experiments, as had Wilson, Lowe and Butt ([237]).)

The following (except for bits within [ ] brackets) are verbatim quotations from the book (the punctuation is verbatim, too!):

> "It wasn't Einstein", Williams was droning along, "and it wasn't Heisenberg or dear old Schrödinger who drove the last nail in the coffin of common sense. It was John S. Bell who published his memorable Theorem in 1964, nearly twenty years ago", and blah, blah, blah. [*Hugo's fantasy about a different kind of entanglement with the girl in front follows here, but is not relevant to our thinking now.*]
>
> Williams continues to transmit to bored faces:
>
> "Bell's Theorem basically deals with non-locality. That is, it shows no local explanation can account for the known facts of quantum mechanics. Um perhaps I should clarify that. A local explanation is one that assumes that things seemingly separate in space and time are really separate. Um? Yes. It assumes, that is to say, that space and time are independent of our primate nervous systems. Do I have your attention, class?
>
> But Bell is even more revolutionary. He offers us two choices if we try to keep locality, [*etc, etc.* ...] Yes, Mr Naranga?"
>
> [*Mr Naranga asks if this topic is going to be on the examination.*]
>
> "No you needn't worry about that Mr Naranga we wouldn't dream of asking anything hard on the examination I believe the last examination with a hard question at this university was in the survey of mathematics course in 1953 yes Mr Lee?"
>
> [*Mr Lee asks if it is possible that the quantum connection is not immediate and unmitigated.*]
>
> "Ah Mr Lee how did you ever land at this university there are times when I suspect you of actually seeking an education but I'm afraid your canny intellect has run aground. Recent experiments by Clauser and Aspect shut that door forever. The quantum connection is immediate, unmitigated, and I might say omnipresent as the Thomist God."
>
> [*Later in the novel*] Bohr also added nearly as much to quantum theory as Planck, Einstein, or Schrödinger, and his model of the atom [...]. Bohr himself, however, had never believed it; nor had he believed any of his other theories. Bohr invented what is called the Copenhagen interpretation, which

holds in effect that a physicist shouldn't believe anything but measurements in his laboratory. Everything else – the whole body of mathematics and the theory relating one measurement to another – Bohr regarded as a model for how the human mind works, not of how the universe works. Williams loved Bohr for the Copenhagen interpretation, which had made it possible for him to study physics seriously, even devoutly, without believing a word of it.

Robert A Wilson

What more can I say?!

APPENDIX A

SOME PREREQUISITES

and ADDENDA

Appendix A1. 'O' notation

For functions f and g on $\mathbb{N}$, we write $f(n) = O(g(n))$ if the ratio $f(n)/g(n)$ remains bounded as $n \to \infty$, that is, if for some constant K and some integer n_0, we have $|f(n)| \leq K|g(n)|$ for $n \geq n_0$. Thus,

$$\ln(n) = O(n), \quad n^4 = O(e^n), \quad \sin(n) = O(1), \qquad (n \to \infty).$$

For functions f and g on $(0, \infty)$ we write $f(x) = O(g(x))$ as $x \downarrow 0$ if for some $b > 0$ we have $|f(x)| \leq K|g(x)|$ for $0 \leq x \leq b$. The modification for the situation where $x \to 0$ rather than $x \downarrow 0$ is now obvious. Examples:

$$x^2 = O(x), \quad \sin(x) = O(x), \qquad (x \to 0).$$

We define $f(x) = O(g(x))$ as $x \to \infty$ in the obviously analogous way.

We can write $O(n) = O(n^2)$ in that if $f(n) = O(n)$, then $f(n) = O(n^2)$. But we cannot write $O(n^2) = O(n)$.

We have (for example) $-3O(n) = O(n)$. We have $O(n) + O(n) = O(n)$ in that if $f(n) = O(n)$ and $h(n) = O(n)$, then $(f + h)(n) = O(n)$; but we would have to be extremely careful about trying to extend this idea to infinite sums.

Appendix A2. Results of 'Taylor' type

I am perfectly happy if you assume throughout the following that f has derivatives of all orders on $[-a, a]$. I play things as I do (still making stronger assumptions than necessary) only because the following is the kind of thing beloved of some analysts and therefore featuring in some courses on Analysis which you may be enjoying.

Suppose that for some $a > 0$, f is a continuous function on $[-a, a]$ for which $f'(0)$ exists. Since $\{f(y) - f(0)\}/y \to f'(0)$ and f is continuous, it is obvious that for some K_0 we have $|f(y) - f(0)| \leq K_0|y|$ for $y \in [-a, a]$. In other words,

$$f(y) = f(0) + O(y) \qquad (y \to 0).$$

Suppose now that $f'(x)$ exists and is continuous on $[-a, a]$ and that $f''(0)$ exists. Then, on applying the result just obtained to f', we have

$$f'(y) \;=\; f'(0) + \mathrm{O}(y) \qquad (y \to 0). \tag{ApA 2.1}$$

But, for $x > 0$, we can integrate this over $[0, x]$ to obtain

$$f(x) \;=\; f(0) + xf'(0) + \mathrm{O}(x^2), \qquad (x \to 0). \tag{ApA 2.2}$$

Remember what's happening here in the integration. Concentrate first on $y > 0$. The meaning of (ApA 2.1) is that for some $b > 0$ and some K_1, the $\mathrm{O}(y)$ term is dominated on the y-interval $[0, b]$ by $K_1 y$, and so, for $0 \le x \le b$ its integral over $[0, x]$ is dominated by $\frac{1}{2}K_1 x^2$. The case when $y < 0$ is, of course, strictly analogous.

If f'' exists and is continuous on $[-a, a]$ and $f'''(0)$ exists, then, by applying result (ApA 2.2) to f', we have

$$f'(y) = f'(0) + yf''(0) + \mathrm{O}(y^2), \qquad (y \to 0),$$

and we can integrate this to get

$$f(x) = f(0) + xf'(0) + \tfrac{1}{2}x^2 f''(0) + \mathrm{O}(x^3) \qquad (x \to 0). \tag{ApA 2.3}$$

In particular,

$$\ln(1 + x) = x - \tfrac{1}{2}x^2 + \mathrm{O}(x^3), \qquad (x \to 0). \tag{ApA 2.4}$$

Appendix A3. 'o' notation

For functions f and g on $\mathbb{N}$, we write $f(n) = \mathrm{o}(g(n))$ if the ratio $f(n)/g(n) \to 0$ as $n \to \infty$. Examples:

$$\ln(n) = \mathrm{o}(n), \quad n^4 = \mathrm{o}(e^n), \quad \sin(n) \ne \mathrm{o}(1), \qquad (n \to \infty).$$

Similarly,

$$x^2 = \mathrm{o}(x), \quad \sin(x) \ne \mathrm{o}(x), \qquad (x \to 0).$$

Appendix A4. Countable and uncountable sets

An *infinite* set S is called countable (or countably infinite) if S may be put in one-one correspondence with the set $\mathbb{N} = \{1, 2, 3, \ldots\}$ of natural numbers; that is, if we can find a way of labelling S as the set of points in a sequence:

$$S \;=\; \{s_1, s_2, s_3, \ldots\};$$

otherwise, S is called uncountable. We shall also call *finite* sets countable, so that every set is either countable or uncountable.

The set $\mathbb{Q}$ of all rational numbers is countable. We see that $\mathbb{Q} \cap (0, 1)$ consists of the points

$$\tfrac{1}{2}, \tfrac{1}{3}, \tfrac{2}{3}, \tfrac{1}{4}, \tfrac{3}{4}, \tfrac{1}{5}, \tfrac{2}{5}, \tfrac{3}{5}, \tfrac{4}{5}, \tfrac{1}{6}, \tfrac{5}{6}, \ldots.$$

Call the above sequence $q_1, q_2, q_3, \ldots$, and we then have

$$\mathbb{Q} \cap (0, \infty) = \{1, q_1, q_1^{-1}, q_2, q_2^{-1}, \ldots\}.$$

It is now obvious that $\mathbb{Q}$ is countable.

The set $\mathbb{R}$ is uncountable. Indeed, its subset H consisting of the irrationals in $(0, 1)$ is already uncountable. Any point of H may be written uniquely as

$$x = \mathrm{CF}(a_1, a_2, a_3, \ldots) := \cfrac{1}{a_1 + \cfrac{1}{a_2 + \cfrac{1}{a_3 + \cfrac{1}{\ddots}}}}, \qquad a_n = a_n(x) \in \mathbb{N},$$

in the sense that the fraction $p_n(x)/q_n(x)$ (in lowest terms) obtained by truncating the fraction at the nth stage converges to x. Thus H is in one-one correspondence with the set of all sequences of elements of $\mathbb{N}$. Suppose that H could be labelled as $H = \{x_1, x_2, x_3, \ldots\}$ and that $x_j = CF(a_{j1}, a_{j2}, a_{j3}, \ldots)$. Define

$$x := CF(a_{11} + 1, a_{22} + 1, a_{33} + 1, \ldots).$$

Then, for every n, x differs from x_n in the nth place, so $x \notin H$, a contradiction. Thus H is uncountable.

In connection with the Strong Law of Large Numbers and the whole philosophy of Probability, it is important that *the set of all strictly increasing sequences of positive integers is uncountable*. This italicized result is true because being given a strictly increasing sequence

$$b_1, b_2, b_3, \ldots$$

is the same as being given the sequence

$$c_1, c_2, c_3, \ldots, \quad \text{where } c_1 = b_1, \ c_2 = c_2 - c_1, \ c_3 = b_3 - b_2, \ldots,$$

so the set of strictly increasing sequences of positive integers is in one-one correspondence with the set of all sequences of positive integers; and we already know that the latter is uncountable from the continued-fraction story.

Note. One of the standard theorems in any elementary course on set theory is that *a countable union $\bigcup C_n$ of countable sets C_n is countable*. But the standard proof (by Cantor) assumes that each C_n is not merely count**able** but count**ed**, that is, already equipped with a labelling as a sequence. In many applications, there is some natural labelling of each C_n, and then Cantor's argument applies. However, in the abstract, where we merely know that there *exists* a labelling of each C_n as a sequence, the standard theorem relies on the Axiom of Choice. See below. Cohen has shown that one can have a system consistent with all the 'other' rules of set theory in which the standard theorem is false, and that indeed one can have such a system in which $\mathbb{R}$ (which is definitely uncountable, as we have seen) is a countable union of countable sets.

Interesting Digression. Continued fractions give the best rational approximations to x. For every x in H, the following is true: for every n, we have

$$\left| \frac{p_n(x)}{q_n(x)} - x \right| < \frac{1}{K q_n(x)^2} \qquad \text{(ApA 4.1)}$$

with $K = 1$; and amongst any two consecutive values of n, at least one satisfies (ApA 4.1) with $K = 2$. 'Conversely', for every x in H, any fraction p/q such that

$$\left| \frac{p}{q} - x \right| < \frac{1}{K q^2} \qquad \text{(ApA 4.2)}$$

with $K = 2$ is one of the continued-fraction approximations. To round off the story: for every x in H, amongst any three consecutive values of n, at least one satisfies (ApA 4.1) with $K = \sqrt{5}$; but for the particular value

$$x = \mathrm{CF}(1, 1, 1, \ldots) = \tfrac{1}{2}(\sqrt{5} - 1),$$

for any $K > \sqrt{5}$, only finitely many fractions p/q satisfy (ApA 4.2). See Hardy and Wright [108].

These results suggest investigating the rate of growth of $q_n(x)$. An amazing result of Lévy and Khinchine states that if we choose a point X in $[0, 1]$ uniformly at random, then ($\mathbb{P}(X \in H) = 1$ and) with probability 1,

$$\sqrt[n]{q_n(X)} \to \exp\left(\frac{\pi^2}{12 \ln 2} \right);$$

but that is another story from Probability. See, for example, Billingsley [21].

Appendix A5. The Axiom of Choice

The discussion here is heuristic, and does not go into axiomatizations of the whole of set theory.

The 'Axiom of Choice' is usually assumed in Mathematics. (Recall that it is needed for the Banach–Tarski Paradox.) It says the following: *the Cartesian product of a collection of non-empty sets is non-empty.*

Let's give an equivalent statement: *Let C be a collection of nonempty sets. Then there exists a function f on C such that, for each set S in the collection, $f(S) \in S$. In other words, we can make a simultaneous choice, choosing one element from each set S in C.* (The described functions f are exactly what make up the Cartesian product.)

Bertrand Russell liked to say: "To choose one sock from each of infinitely many pairs of socks requires the Axiom of Choice, but for shoes the Axiom is not needed." The point here is that there is a well-defined particular element in a pair of shoes, the left one; but the same is not true for socks.

If, for example, one takes a countable union of 'abstract' countably infinite sets and wishes to apply Cantor's argument, then one has to make a simultaneous choice, choosing one labelling for each set from the (uncountably many) different labellings of that set.

For a useful site about the Axiom of Choice, see

http://www.math.vanderbilt.edu/~schectex/ccc/choice.html

prepared by Eric Schecter.

Appendix A6. A non-Borel subset of $[0, 1]$

Return to the continued-fraction story. Let V be the set of irrational x in $[0, 1]$ for which there exists an increasing subsequence $(n_r(x))$ of $\mathbb{N}$ such that $a_{n_r(x)}(x)$ divides $a_{n_{r+1}(x)}(x)$ for every r. Then V is not a Borel subset of $[0, 1]$. Again, the crucial point is that the set of all subsequences of the sequence of positive integers is uncountable; but to *prove* that V is not Borel is *very* difficult. The set V is 'Lebesgue measurable' with measure 1.

Appendix A7. Static variables in 'C'

Static variables cannot be changed except via functions in the file where they are set up. Here I give a header file H.h, a 'module' file M.c which defines the function up(), and a program file P.c which defines a function down() and uses it within main(). Because the variable a in M.c is static, its value is preserved within M.c. The variable a in P.h is different. When up() is called, it uses the values of a and C2 within M.c, keeping them fixed between 'visits to M.c'; whereas down() deals with the values of a and C2 in P.c. This way, you cannot get into trouble by using the same symbol in different files. It *is* a bit like Object Oriented Programming.

```
/*  Header H.h    defines C1=1 */
#if defined H_h    /*    Three of four lines      */
#else              /*  to avoid DOUBLE definition  */
#define H_h        /* of C1, once in M.c, once in P.c */

#define C1 1       /* C1 to be used by both M.c and P.c */
void up();         /* optional reminder of function in module */

#endif             /* The 'fourth line of four'  */
------------------------------------
/* Module M.c     cc -c M.c -o M.o  */
#include "H.h"
#include <stdio.h>
static const C2 = 2;
static int a = 0;
void up(){
  a = a + C2 + C1; printf("%4d", a);
}
------------------------------------
/* Program P.c       cc P.c M.o -o P  */
```

```
#include "H.h"
const C2 = -4;
int a = 0;
void down(){
  a = a + C1 + C2; printf("%4d\n",a);
}
int main(){
  int i;
  for(i=0; i< 3; i++){up(); down();}
  return 0;
}
----------------------------------------
Results:
   3  -3
   6  -6
   9  -9
----------------------------------------
```

Appendix A8. A 'non-uniqueness' example for moments

Let h be an even infinitely differentiable function on $\mathbb{R}$ which (for some δ and a with $0 < \delta < a$) is zero on $(-\infty, -a] \cup [-\delta, \delta] \cup [a, \infty)$ and non-zero elsewhere. Let g be the real function

$$g(\alpha) := \int_{\mathbb{R}} e^{i\alpha x} h(x) \, dx.$$

Then (justification being standard Fourier theory)

$$h(x) = \frac{1}{2\pi} \int_{\mathbb{R}} e^{-ix\alpha} g(\alpha) \, d\alpha,$$

and

$$h^{(n)}(x) = \frac{1}{2\pi} \int_{\mathbb{R}} (-i\alpha)^n e^{-ix\alpha} g(\alpha) \, d\alpha.$$

But $h^{(n)}(0) = 0$ for $(n = 0, 1, 2, \ldots)$, so that

$$\int_{\mathbb{R}} x^n g(x) \, dx = \int_{\mathbb{R}} \alpha^n g(\alpha) \, d\alpha = 0.$$

It is now clear that we can choose a constant K such that Kg^+ and Kg^- are pdfs with the same cumulants but which, since they have different supports, correspond to different distributions.

Simon ([213]) treats the moment problem with characteristic elegance. The above example was suggested by his paper. You will find there many other references for 'moment problems'.

Appendix A9. Proof of a 'Two-Envelopes' result

Result 400(M4) – under the 'bounded-ratio' assumption – is an immediate consequence of the following result.

Suppose that $(X_1, Y_1), (X_2, Y_2), (X_3, Y_3), \ldots$ are IID RVs with values in $(0, \infty)^2$ each with the same distribution as a $(0, \infty)^2$-valued RV (X, Y). Suppose that for some finite constant K and some constant c, we have, with probability 1,

$$Y \leq KX, \quad and \, \mathbb{E}(Y \mid X) \geq cX.$$

Then, for some absolute constant L with $c \leq L \leq K$,

$$\limsup \frac{Y_1 + Y_2 + \cdots + Y_n}{X_1 + X_2 + \cdots + X_n} = L, \quad a.s..$$

Proof. By the 'Independence Golden Rule' 402(Nb),

$$\mathbb{E}(Y_1 \mid X_1, X_2) = \mathbb{E}(Y_1 \mid X_1) \geq cX_1, \quad \text{a.s.,}$$

whence,

$$\mathbb{E}(Y_1 + Y_2 \mid X_1, X_2) \geq c(X_1 + X_2), \quad \text{a.s..}$$

Hence by the 'taking out what is known' Golden Rule 389(E1),

$$\mathbb{E}\left(\frac{Y_1 + Y_2}{X_1 + X_2} \;\middle|\; X_1, X_2\right) = \frac{1}{X_1 + X_2}\mathbb{E}(Y_1 + Y_2 \mid X_1, X_2) \geq c, \text{ a.s.,}$$

so that by obvious extension and the main Golden Rule 390F2,

$$\mathbb{E}\left(\frac{Y_1 + Y_2 + \cdots + Y_n}{X_1 + X_2 + \cdots + X_n}\right) \geq c.$$

The remainder of the proof uses results found (for example) in [W]. *Because $Y \leq KX$,* the ratio in the last equation above is bounded within $[0, K]$. We may therefore use the 'Reverse Fatou Lemma' [W; 5.4(b)] to deduce that

$$\mathbb{E}\limsup\left(\frac{Y_1 + Y_2 + \cdots + Y_n}{X_1 + X_2 + \cdots + X_n}\right) \geq \limsup\mathbb{E}\left(\frac{Y_1 + Y_2 + \cdots + Y_n}{X_1 + X_2 + \cdots + X_n}\right) \geq c.$$

But the lim sup is in the tail σ-algebra [W; 4.10 – 4.12] of the independent sequence $(X_1, Y_1), (X_2, Y_2), (X_3, Y_3), \ldots$, whence, by Kolmogorov's Zero-One Law [W; 4.11, 14.3], the lim sup is a deterministic constant L (possibly infinity). That $c \leq L \leq K$ is now obvious. □

APPENDIX B

DISCUSSION OF SOME SELECTED EXERCISES

'Discussion' signifies that this chapter contains further hints for certain exercises and, in the case of just a few of them, some extra information to 'fill out the picture'.

I regard reading the solution to a problem without having tried really hard to do it as a learning opportunity missed. I therefore hope that you will consult this chapter mainly for confirmation of your own answers and for the occasional extra discussion.

Section 1.1

13Oa. In (b), Prob(ii)$= 3n(n-1)/n^3$. The answer to (c) is

$$\frac{n\binom{r}{2}\left(^{n-1}P_{r-2}\right)}{n^r} = \binom{r}{2}\frac{1}{n}\cdot\frac{n-1}{n}\cdot\frac{n-2}{n}\cdots\frac{n-r+2}{n},$$

the left-hand side by 'counting', the right by 'conditioning'.

It is intuitively obvious that any tendency of birthdays not to be spread uniformly throughout the year will increase the chance that some two out of r people will have the same birthday. Suppose that for the huge underlying population the actual proportion of birthdays falling on the ith day in the year is p_i. Then, the probability that r people chosen randomly from the whole population will have all birthdays different will be

$$\sum^{*} p_{i_1} p_{i_2} \cdots p_{i_r}$$

the $*$ signifying that the sum is over all $^n P_r$ r-tuples $(i_1, i_2, \ldots, i_r)$ of distinct values. This function can be shown to take its maximum value when each p_i is $1/n$. If you know about Lagrange multipliers, you will be able to prove this.

16Ra. Using the binomial theorem and ignoring 0.01^k for $k \geq 2$, we have

$$1 - 0.99^x + \frac{1}{x} = 1 - (1 - 0.01)^x + \frac{1}{x} \approx 1 - [1 - x \times 0.01] + \frac{1}{x} = \frac{x}{100} + \frac{1}{x}.$$

The last expression is minimized when $1/100 - 1/x^2 = 0$, so $x = 10$. A calculator shows that $x = 11$ is very slightly better for the correct expression, giving expected number of tests per person equal to 0.1956, an 80.44% saving.

Section 1.2
22Gb. God tosses a coin N times per minute. A customer arrives when God gets a Tail. On average λ people arrive per minute, so, for the coin,

$$q := \mathbb{P}(\text{Tail}) = \frac{\lambda}{N}, \qquad p := \mathbb{P}(\text{Head}) = 1 - q = 1 - \frac{\lambda}{N}.$$

An interval c without an arrival corresponds to $n = Nc$ Heads in a row.

22Gc. (a) We wait for the first Head. The *next* result (H or T with probability $\frac{1}{2}$ each) decides whether or not HH or HT occurs first. Thus, the answer is $\frac{1}{2}$.
(b) The only way in which HH precedes TH is if the first two tosses produce HH. Think about it! The answer is $\frac{3}{4}$.

22Gd. What is the chance that the first toss after the first Head produces a Head? What does that tell you about a key case?

Section 2.2
41G. $\mathbb{P}(\text{exactly one professor gets the right hat}) = p_{0,n-1}$. The answer to Part (b) is $(1/r!)p_{0,n-r}$.

Section 3.2
54Ea. $f_R(r) = F'_R(r) = 2r$. This makes sense. We should have $f_R(r) \propto r$ because a circle of radius r has perimeter proportional to r; and of course if $f_R(r) \propto r$, then we must have $f_R(r) = 2r$ because f has to integrate to 1 over $[0, 1]$.

56G. The needle crosses the line if and only if $Y < \sin \Theta$. Buffon did the first Monte-Carlo calculation!

Section 3.3
58Dc. The answer is 'obviously' $1\frac{1}{2}$ times the expected number of Heads on the first two tosses, and so is $3/2$. *You* check that this tallies with the formula for p_X at Exercise 52Da.

Section 3.4
64Ja. Let the points be (X, Y) and (U, V). Then $\mathbb{E}(XU) = 0$ by symmetry, and, from Exercise 54Ea, $\mathbb{E}(X^2 + Y^2) = \frac{1}{2}$, giving the answer.

70Jb. I do give the full solution to this important problem. We use throughout the fact that $\mathbb{E}(W^2) = \mu_W^2 + \sigma_W^2$.
Part (a). We have MSE $= (\mu_Y - c)^2 + \sigma_Y^2$, and the desired result is immediate.
Part (b). Now,

$$\begin{aligned}
\text{MSE} &= \{\mu_Y - (\alpha\mu_X + \beta)\}^2 + \text{Var}(Y - \alpha X) \\
&= \{\mu_Y - (\alpha\mu_X + \beta)\}^2 + \sigma_Y^2 - 2\alpha\rho\sigma_X\sigma_Y + \alpha^2\sigma_X^2.
\end{aligned}$$

Clearly, we must take $\beta = \mu_Y - \alpha\mu_X$, and then

$$\text{MSE} = \sigma_Y^2 - 2\alpha\rho\sigma_X\sigma_Y + \alpha^2\sigma_X^2 = (\alpha\sigma_X - \rho\sigma_Y)^2 + \sigma_Y^2\left(1 - \rho^2\right),$$

so that we choose $\alpha = \rho\sigma_Y/\sigma_X$ and then $\text{MSE} = \sigma_Y^2(1-\rho^2)$. If $\rho = \pm 1$, then $\text{MSE} = 0$, so that, with probability 1, $Y = \alpha X + \beta$ for the chosen α, β.

You could, of course, have done

$$\frac{\partial}{\partial\alpha}(\text{MSE}) = 0 = \frac{\partial}{\partial\beta}(\text{MSE}).$$

70Ka. We can imagine that someone just gave out the hats at random. Thus, whenever $i \neq j$, we have $\mathbb{E}(X_i X_j) = \mathbb{E}(X_1 X_2)$. Note that since $X_i = 0$ or 1, we have $X_1^2 = X_1$ and

$$\mathbb{E}(X_1 X_2) = \mathbb{P}(X_1 X_2 = 1) = \mathbb{P}(X_1 = 1, X_2 = 1).$$

If one professor has the right hat, then it is more likely that another has.

71Kd. Show that $\rho(Y, Z) \geq h(\alpha)$ and choose the best α.

Section 4.1

76E. The intuitive idea for the later part is that for final configurations for which i is isolated, 'what happens to the left of i is independent of what happens to the right of i'. Think about this. This leads to

$$p_{i,n} = p_{i,i}p_{n-i+1,n-i+1} = p_i p_{n-i+1},$$

the $p_{n-i+1,n-i+1}$ being the probability that i ends up isolated if we start with the row $i, i+1, \ldots, n$ of sites. Of course, it is false that the probability that $[i-1, i]$ ends up occupied is $(1-p_i)p_{n-i+1}$, for now 'there *is* interference between the two sides of i'.

I have warned you that one of the most common mistakes is to assume independence when it is not present. Because you may have niggling doubts about the 'partial independence' claimed above, I give the following separate argument which will surely clinch the matter for you.

Decomposing the event F that site i ends up isolated according to which of the positions (compatible with F)

$$[1, 2], \quad [2, 3], \quad \ldots, \quad [i-2, i-1],$$
$$[i+1, i+2], \quad [i+2, i+3], \quad \ldots, \quad [n-1, n],$$

the first driver occupies, we find that

$$(n-1)p_{i,n} = (p_{i-2,n-2} + \cdots + p_{1,n-i+1}) + (p_{i,i} + p_{i,i+1} + \cdots + p_{i,n-2}).$$

Let $\mathbf{S}(k)$ be the proposition that for $1 \leq m \leq k$ and $1 \leq i \leq m$, $p_{i,m} = p_i p_{m-i+1}$. Then $\mathbf{S}(1)$ is true. If we assume that $\mathbf{S}(n-1)$ is true then we have, using the earlier recurrence relation for the p_k, for $1 \leq i \leq n$,

$$(n-1)p_{i,n} = (p_{i-2} + \cdots + p_1)p_{n-i+1} + p_i(p_1 + \cdots + p_{n-i+1})$$
$$= (i-1)p_i p_{n-i+1} + p_i(n-i)p_{n-i+1},$$

so that $p_{i,n} = p_i p_{n-i+1}$ and $\mathbf{S}(n)$ is true. Hence $\mathbf{S}(n)$ is true for all n by induction.

90Pa. Here's a full solution. We make the table H(i).

H {John, Mary}	$\mathbb{P}(H)$ =	$\mathbb{P}(H \cap K)$ =	$\mathbb{P}(H \cap K \cap S_1)$ =	$\mathbb{P}(H \cap K \cap S_1 \cap E_2)$ =
{aa,aa}	p^4	0	0	0
{aa,aA}	$4p^3q$	0	0	0
{aa,AA}	$2p^2q^2$	0	0	0
{aA,aA}	$4p^2q^2$	$4p^2q^2$	$3p^2q^2$	$\frac{3}{4}p^2q^2$
{aA,AA}	$4pq^3$	$4pq^3$	$4pq^3$	0
{AA,AA}	q^4	q^4	q^4	0

Table H(i): Re: children of John and Mary

For the $\{aA, aA\}$ row, for instance, we have

$$\mathbb{P}(K \mid H) = 1, \quad \mathbb{P}(S_1 \mid H \cap K) = \tfrac{3}{4}, \quad \mathbb{P}(E_2 \mid H \cap K \cap S_1) = \mathbb{P}(E_2 \mid H) = \tfrac{1}{4},$$

and the General Multiplication Rule allows us to fill in the entries.

Hence,

$$\mathbb{P}(K) = 4p^2q^2 + 4pq^3 + q^4 = q^2(4p^2 + 4pq + q^2)$$
$$= q^2(2p + q)^2 = q^2(1 + p)^2.$$

Similarly,

$$\mathbb{P}(K \cap S_1) = 3p^2q^2 + 4pq^3 + q^4 = q^2(3p^2 + 4pq + q^2)$$
$$= q^2(3p + q)(p + q) = q^2(1 + 2p).$$

Thus, the answer to Part (a) is

$$\mathbb{P}(S_1 \mid K) = \frac{1 + 2p}{(1 + p)^2}.$$

The answer to Part (b) is clearly $\tfrac{3}{4}$ because, given that their first child is exceptional and that John and Mary themselves are standard, they must both be aA.

We have $\mathbb{P}(K \cap S_1 \cap E_2) = \tfrac{3}{4}p^2q^2$, whence the answer to Part (c) is

$$\mathbb{P}(E_2 \mid K \cap S_1) = \frac{\tfrac{3}{4}p^2q^2}{q^2(1 + 2p)} = \frac{3p^2}{4(1 + 2p)}.$$

Section 4.2
97D. Convince yourself that we can assume that A beats B, C beats D, and A beats C

without affecting the answer. [[First label the people as $-2, -1, 0, 1, 2$, and let $W(i, j)$ be the winner of the 'i versus j' game. Let

$$A = W(W(-2, -1), W(1, 2)), \quad B = 2/A, \quad C = W(-A, -B), \quad D = 2/C,$$

to label the people as A,B,C,D,E. This nonsense cannot affect the probabilities for the remaining games.]]

	A	B	C	D	E
A	–	W	W		
B	L	–			
C	L		–	W	
D			L	–	
E					–

Table H(ii): 'Five-Nations Table'

The W in the '(A,B)' position in Table H(ii) signifies that A Wins against B, so B Loses against A: the table is antisymmetric. There are $2^7 = 128$ ways of completing the table, all equiprobable by independence. But, as you can check, there are only 3 ways of completing the table such that each team wins two games.

100J. In the last part, $\mathbb{P}(B_p) = p^{-2s}$. The event $\bigcap(B_p^c)$ occurs if and only if for no prime p is it true that p divides both X and Y, that is, if and only if $H = 1$. Thus,

$$\mathbb{P}(H = 1) = \prod(1 - p^{-2s}) = \zeta(2s)^{-1}.$$

You argue that $\mathbb{P}(H = m) = m^{-2s}\mathbb{P}(H = 1)$.

Section 4.3

103C. With probability 1, precisely one of $X_1, X_2, \ldots, X_n$ is the largest, and, by symmetry, each has chance $1/n$ of being the largest. Hence $\mathbb{P}(E_n) = 1/n$.

Consider $\mathbb{P}(E_1 \cap E_3 \cap E_6)$, remembering that if $E_1 \cap E_3 \cap E_6$ occurs, then E_4 (for example) may also occur. Think of how many ways the X_k's may be arranged in increasing order consistent with $(E_1 \cap E_3 \cap E_6)$. We must have

$$X_1 < X_3 < X_6.$$

There are 2 places where we can 'insert X_2': we have

$$\text{either } X_1 < X_2 < X_3 < X_6 \quad \text{or} \quad X_2 < X_1 < X_3 < X_6.$$

(We cannot have $X_2 > X_3$ or else E_3 would not occur.) For each of the two arrangements above, there are 4 places where we can insert X_4. Then there will be 5 places where we can insert X_5, so that

$$\mathbb{P}(E_1 \cap E_3 \cap E_6) = \frac{2 \times 4 \times 5}{6!} = 1 \times \frac{1}{3} \times \frac{1}{6} = \mathbb{P}(E_1)\mathbb{P}(E_3)\mathbb{P}(E_6).$$

You can now see why the desired result is true.

103D. I leave you to relate this to the previous exercise!

Discussion. (This discussion, which assumes a number of results, illustrates that even a simple probability model can be very rich in terms of the mathematics it supports.)

The situation where, as here, $\mathbb{P}(L_1 = \infty) = 0$ but in which the tail of the distribution of L_1 is so large that $\mathbb{E}(L_1) = \infty$, is surprisingly common: we shall see several more examples. Suppose that one repeated the *whole* experiment, with strings of convoys for each experiment. Let $L_1(k)$ be the length of the *first* convoy in the kth *experiment*. The fact that $\mathbb{E}(L_1) = \infty$ means that instead of 'stabilizing to a finite value', the 'average length of the first convoy over the first K experiments',

$$A_1(K) := \frac{L_1(1) + L_1(2) + \cdots + L_1(K)}{K}$$

converges to infinity as $K \to \infty$. It follows from the Second Borel–Cantelli Lemma 98E that, with probability 1 there will be infinitely many values of K for which $L_1(K) > K \ln(K)$, and $A_1(K)$ is clearly greater than $\ln(K)$ for those values.

Of much greater interest is what happens to the sequence L_1, L_2, ... of lengths of *successive* convoys in a single experiment. With probability 1,

$$\frac{C(n)}{\ln(n)} \to 1,$$

where $C(n)$ is the number of convoys formed by the first n cars. *Roughly speaking*, therefore, $L_1 + L_2 + \cdots + L_n \approx e^n$. Convoys are tending (rapidly) to get longer and longer because they are headed by slower and slower drivers.

116Ya. As I said, this is very instructive from the point of view of understanding 'about ω', but it *is* quite tricky. Recall that we assume that F is continuous and strictly increasing.

Let $x_{i,r}$, where $r \in \mathbb{N}$ and $0 < i < r$, be the unique solution of the equation $F(x_{i,r}) = i/r$. Then, because there are only countably many $x_{i,r}$ (where $r \in \mathbb{N}$ and $0 < i < r$), the set G of those ω for which

$$F_n(x_{i,r}, \mathbf{Y}(\omega)) \to F(x_{i,r}) \quad \text{whenever } r \in \mathbb{N} \text{ and } 0 < i < r,$$

is an event of probability 1. We need to prove that for $\omega \in G$,

$$F_n(x; \mathbf{Y}(\omega)) \to F(x) \quad \text{uniformly over } x \in \mathbb{R}.$$

So, let $\omega \in G$, and let $\varepsilon > 0$ be given. Choose an integer r_0 such that $r_0 > 2/\varepsilon$. The set of points $\{x_{1,r_0}, x_{2,r_0}, \ldots, x_{r_0-1,r_0}\}$ is finite, so, by definition of G, there will exist $n_0(\omega)$ such that for $n \geq n_0(\omega)$, we have

$$\left| F_n(x_{i,r_0}; \mathbf{Y}(\omega)) - F(x_{i,r_0}) \right| < \tfrac{1}{2}\varepsilon \quad \text{whenever } 0 < i < r_0 \text{ and } n \geq n_0(\omega).$$

We now make the obvious definitions $x_{0,r_0} = -\infty$, $x_{r_0,r_0} = \infty$. 'Every kind of F' is 0 at $-\infty$ and 1 at ∞. For any $x \in \mathbb{R}$, we have for some i with $1 \le i \le r_0$,

$$
\begin{array}{ccccc}
x_{i-1,r_0} & \le & x & \le & x_{i,r_0} \\
F_n(x_{i-1,r_0}; \mathbf{Y}(\omega)) & \le & F_n(x; \mathbf{Y}(\omega)) & \le & F_n(x_{i,r_0}; \mathbf{Y}(\omega)) \\
F(x_{i-1,r_0}) & \le & F(x) & \le & F(x_{i,r_0}).
\end{array}
$$

So, for $n \ge n_0(\omega)$, we have (with the lack of moduli being deliberate)

$$
\begin{aligned}
F_n(x; \mathbf{Y}(\omega)) - F(x) &\le F_n(x_{i,r_0}; \mathbf{Y}(\omega)) - F(x_{i-1,r_0}) \\
&= F_n(x_{i,r_0}; \mathbf{Y}(\omega)) - F(x_{i,r_0}) + F(x_{i,r_0}) - F(x_{i-1,r_0}) \\
&< \tfrac{1}{2}\varepsilon + r_0^{-1} = \varepsilon,
\end{aligned}
$$

and, by an analogous argument,

$$
F_n(x; \mathbf{Y}(\omega)) - F(x) > -\varepsilon.
$$

The result is proved.

To do the general case is a bit messy. There *is* a neat trick, but it would not be helpful to present it here.

Section 4.4

123Ga. Let $x = \mathbb{P}_O(\text{hit A})$. Then $x = \tfrac{1}{4} + \tfrac{3}{4}x^2$, so that $x = 1$ or $x = \tfrac{1}{3}$. *You* use the argument at 119Da to prove that $x = \tfrac{1}{3}$. (That x cannot be 1 can also be seen from the argument at the end of this solution.

Next,

$$
y := \mathbb{P}_O(\text{hit A before B}) = \tfrac{1}{4}.1 + \tfrac{1}{4}.0 + \tfrac{1}{2}\mathbb{P}_C(\text{hit A before B}) = \tfrac{1}{4} + \tfrac{1}{2}xy,
$$

so that $y = 3/10$. We have

$$
\mathbb{P}_O(\text{hit both A and B}) = 2y \times x^2 = \tfrac{1}{15}.
$$

Now let $f_i := \mathbb{P}_i(\text{hit H before either A or B})$. Then,

$$
f_O = \mathbb{P}_O(\text{hit D before either A or B})f_D.
$$

But

$$
r := \mathbb{P}_O(\text{hit D before either A or B}) = \tfrac{1}{4} + \tfrac{1}{4}xr,
$$

and $r = 3/11$. The remaining equation needed to find f_O is that

$$
f_D = \tfrac{1}{4} + \tfrac{1}{4}f_O + \tfrac{1}{2}xf_D.
$$

If the current reduced word has length at least 1, then its length will increase by W after the next step, where $\mathbb{P}(W = 1) = \tfrac{3}{4}$ and $\mathbb{P}(W = -1) = \tfrac{1}{4}$. So, the answer regarding c is that $c = \mathbb{E}(W) = \frac{1}{2}$. But do give some thought to the difficulty that the Random Walk may visit O (the word of length 0) several times. The length forms a reflecting Random Walk with $p = \frac{3}{4}$ on the set $\mathbb{Z}^+$.

126La. To prove that

$$\mathbb{P}(\text{B is never in the lead}) = 1 - \frac{b}{a+1},$$

is actually easier than the first part.

Section 4.6

139Ic. Here's my program. This one *does* use pointers as a neat way of choosing (X, Y) in the disc via `InDisc(&X,&Y)`, `&X` being the address at which X is stored. Sorry!

```
/* SimDisc.c
 cc SimDisc.c RNG.o -o SimDisc -lm
*/
#include <stdio.h>
#include "RNG.h"
#define N 100000
double ssq(double a, double b){return a*a + b*b;}
void InDisc(double *x, double *y){
  do{*x = 2.0*Unif()-1; *y = 2.0*Unif()-1;} while (ssq(*x, *y) >1);
}
int main(){ long i;
  double sumDsq = 0.0, X,Y,U,V;
  setseeds();
  for(i=0; i<N; i++){
     InDisc(&X,&Y); InDisc(&U,&V); sumDsq += ssq(X-U, Y-V);
  }
  ShowOldSeeds();
  printf("\nEstimated mean of D squared = %6.4f\n", sumDsq/N);
  return 0;
}
/*RESULTS:
  Generator used was Wichmann-Hill(3278, 29675,9671)
  Estimated mean of D squared = 1.0012
*/
```

Section 5.3
149Da. We have

$$\Gamma(\tfrac{1}{2}) = \int_0^\infty x^{-\frac{1}{2}} e^{-x}\, \mathrm{d}x = \cdots = \sqrt{\pi} \int_{-\infty}^\infty \varphi(y)\mathrm{d}y,$$

on putting $x = \frac{1}{2}y^2$ and using the symmetry of φ.

150Eb. We have, for $\alpha < \lambda$,

$$\int_0^\infty e^{\alpha x} \frac{\lambda^K x^{K-1} e^{-\lambda x}}{\Gamma(K)}\, dx = \text{what?} \times \int_0^\infty \frac{(\lambda - \alpha)^K x^{K-1} e^{-(\lambda-\alpha)x}}{\Gamma(K)}\, dx.$$

Section 6.2
177Da. We have, with roughly 95% probability,

$$|\overline{Y} - \lambda| \le 2\sqrt{\frac{\lambda}{n}}.$$

You can say that you are 95% confident that

$$(\lambda - \overline{y}_{\text{obs}})^2 \le 4\lambda/n$$

and that λ therefore lies between the roots of the obvious quadratic, or, in more rough-and-ready fashion, use

$$\left[\overline{y}_{\text{obs}} - 2\sqrt{\frac{\overline{y}_{\text{obs}}}{n}},\ \overline{y}_{\text{obs}} + 2\sqrt{\frac{\overline{y}_{\text{obs}}}{n}}\right].$$

178Eb. For the ith individual chosen, let

$$X_i := \begin{cases} 1 & \text{if the person says he/she will vote for Party 1,} \\ 0 & \text{otherwise,} \end{cases}$$

with Y_i the analogous Variable for Party 2. Then

$$\mathbb{E}(Y_i - X_i) = p_2 - p_1,$$
$$\mathbb{E}\left[(Y_i - X_i)^2\right] = \mathbb{E}(Y_i^2) + \mathbb{E}(X_i^2) - 2\mathbb{E}(X_i Y_i)$$
$$= p_2 + p_1 - 0,$$

and

$$\text{Var}(Y_i - X_i) = p_2 + p_1 - (p_1 - p_2)^2.$$

Note the sense of this when $p_1 = 0$. The variables $Y_1 - X_1, Y_2 - X_2, \ldots$ are IID and so

$$\frac{S_2 - S_1}{n} \approx \text{N}\left(p_2 - p_1,\ \frac{p_2 + p_1 - (p_1 - p_2)^2}{n}\right).$$

As a rough and ready guide, you could pretend in estimating the variance that p_i is $S_i(\omega^{\text{act}})/n$.

Section 6.3
186Ga. We have

$$\mathbb{P}(M \leq m) = \mathbb{P}(Y_1 \leq m; \ldots; Y_n \leq m) = \left(\frac{m}{\theta}\right)^n,$$

which tells Zeus how to choose M. Tyche lets $Z_1 = M$ and then chooses $n - 1$ Variables $Z_2, Z_3, \ldots, Z_n$ independently, each with the U$[0, M]$ distribution. She then lets $Y_1, Y_2, \ldots, Y_n$ be a random permutation of $Z_1, Z_2, \ldots, Z_n$.

Section 6.4
188Bc. Let $T = \tau(Y)$. Then the desired unbiasedness property says that

$$\frac{1}{e^\theta - 1} \sum_{y=1}^\infty \tau(y) \frac{\theta^y}{y!} = (1 - e^{-\theta}),$$

so that

$$\sum_{y=1}^\infty \tau(y) \frac{\theta^y}{y!} = 2\cosh\theta - 2 = 2\sum_{n=1}^\infty \frac{\theta^{2n}}{(2n)!}.$$

Section 6.5
198Fa. We have

$$\int \ln\left(\frac{f(h(z))h'(z)}{g(h(z))h'(z)}\right) f(h(z))h'(z)\, dz$$

$$= \int \ln\left(\frac{f(h(z))}{g(h(z))}\right) f(h(z))h'(z)\, dz = \int \ln\left(\frac{f(y)}{g(y)}\right) f(y)\, dy,$$

the last by change of variables.

199Ha. We have

$$-\mathbb{E}_0\left(\frac{h'(Y)}{h(Y)} \frac{\varphi'(Y)}{\varphi(Y)}\right) = \mathbb{E}_0\left(\frac{h'(Y)}{h(Y)}Y\right) = \int \frac{h'(y)}{h(y)} y h(y)\, dy = \cdots.$$

Section 6.6
206Gd. By differentiation,

$$\partial_\gamma f^*(Y \mid \gamma) = -\frac{1}{\gamma}\left[1 + \frac{Y - \theta_0}{\gamma}(g^*)'\left(\frac{Y - \theta_0}{\gamma}\right)\right].$$

But if Y has pdf $f(y \mid \gamma)$ and $Z := (Y - \theta_0)/\gamma$, so that $Y = h(Z) := \theta_0 + \gamma Z$, then

$$f_Z(z) = f_Y(h(z))h'(z) = \gamma^{-1}g(z)\gamma = g(z).$$

213La. We have

$$\mathbb{P}(N = k; \Theta \in d\theta; \mathbf{Y} \in d\mathbf{y}) = p(k)\pi_k(\theta)\mathrm{lhd}(\theta; \mathbf{y})d\theta d\mathbf{y}.$$

It is all easy if you keep a cool head. **212Kc.** For a Frequentist, $Y_{n+1} - \overline{Y}$ (where $\overline{Y}$ is the average of $Y_1, Y_2, \ldots, Y_n$) has the normal distribution of mean 0 and known variance $\sigma_0^2(n^{-1} + 1)$. **221Pb.** Find the appropriate inequality which must be satisfied by a mixture.

Section 7.4
263Ea. You check that (with $t := \min y_i$ and $t + a = \max y_i$)

$$\mathrm{lhd}(\theta; \mathbf{y}) = I_{[\theta - \frac{1}{2}, \theta + \frac{1}{2}]}(t) I_{[\theta - \frac{1}{2}, \theta + \frac{1}{2}]}(t + a),$$

so that (T, A) is Sufficient for θ. You may assume that this pair is Minimal Sufficient (how could it not be?). Clearly A has the distribution of the range of an IID sample, each $U[0, 1]$.

So, let $U_1, U_2, \ldots, U_n$ be IID each $U[0, 1]$. Let

$$V := \min U_i, \quad W = \max U_i.$$

(*Note.* Whether we deal with open or closed intervals is irrelevant in the following discussion.) Then, for $0 < v < w < 1$,

$$\mathbb{P}(V > v, \ W < w) = \mathbb{P}(U_i \in (v, w) \text{ for every } i) = (w - v)^n.$$

Hence,

$$\mathbb{P}(V > v; \ W \in dw) = \left\{ \frac{\partial}{\partial w}(w - v)^n \right\} dw = n(w - v)^{n-1} dw,$$

$$\mathbb{P}(V \in dv; \ W \in dw) = -\left\{ \frac{\partial}{\partial v} n(w - v)^{n-1} \right\} dv dw = n(n - 1)(w - v)^{n-2} dv dw,$$

$$f_{V,W}(v, w) = n(n - 1)(w - v)^{n-2} I_{\{0 < v < w < 1\}}.$$

Let $B = W - V$, $C = V$. Then $v = c$, $w = b + v$, the Jacobian is 1 and so

$$f_{B,C}(b, c) = n(n - 1)b^{n-2} I_{\{0 < b < 1, \ 0 < c < 1 - b\}},$$

whence

$$f_B(b) = \left(\int_0^{1-b} f_{B,C}(b, c) dc \right) I_{(0,1)}(b) = n(n - 1)b^{n-2}(1 - b) I_{(0,1)}(b),$$

giving the required pdf of A.

Back to $T, A, \mathbf{Y}$. We now know about Stage 1 where Tyche chooses A. Since the conditional probability of C given B is $U[0, 1 - B]$, the conditional distribution of $T - (\theta - \frac{1}{2})$ given A is $U[0, 1 - A]$. Thus, conditionally on $A = a^{\mathrm{obs}}$, T is uniform on $[\theta - \frac{1}{2}, \theta + \frac{1}{2} - a^{\mathrm{obs}}]$; and we now understand Stage 2. For Stage 3, Tyche chooses numbers $z_1^{\mathrm{act}}, z_2^{\mathrm{act}}, \ldots, z_{n-2}^{\mathrm{act}}$ independently and uniformly on $(t^{\mathrm{obs}}, t^{\mathrm{obs}} + a^{\mathrm{obs}})$ and then lets $y_1^{\mathrm{obs}}, y_2^{\mathrm{obs}}, \ldots, y_n^{\mathrm{obs}}$ be a random permutation of $t^{\mathrm{obs}}, t^{\mathrm{obs}} + a^{\mathrm{obs}}, z_1^{\mathrm{act}}, z_2^{\mathrm{act}}, \ldots, z_{n-2}^{\mathrm{act}}$.

Section 8.3

327La. We have

$$H_0 : \boldsymbol{\mu} = \mu\mathbf{1} \otimes \mathbf{1} + \mathbf{1} \otimes \boldsymbol{\beta} + \sigma\mathbf{G},$$
$$H_A : \boldsymbol{\mu} = \mu\mathbf{1} \otimes \mathbf{1} + \boldsymbol{\alpha} \otimes \mathbf{1} + \mathbf{1} \otimes \boldsymbol{\beta} + \sigma\mathbf{G},$$
$$U = [\mathbf{1}] \otimes [\mathbf{1}] + [\mathbf{1}] \otimes [\mathbf{1}]^{\perp} = [\mathbf{1}] \otimes \mathbb{R}^K, \quad P_U = P_{\text{mean}} + P_{\boldsymbol{\mathcal{B}} \text{ eff}},$$
$$W = U + [\mathbf{1}]^{\perp} \otimes [\mathbf{1}], \quad Z = [\mathbf{1}]^{\perp} \otimes [\mathbf{1}], \quad Z \perp U, \quad P_Z = P_{\boldsymbol{\mathcal{A}} \text{ eff}}.$$

336Pa. If we change notation and think of the errors for successive months as $\varepsilon_1, \varepsilon_2, \varepsilon_3, \ldots$, then it might be better to use an IID sequence $\eta_1, \eta_2, \eta_3, \ldots$ each $N(0, \sigma^2)$, let $c \in (0, 1)$ be an unknown parameter, and then let

$$\varepsilon_1 = \eta_1, \quad \varepsilon_n = c\varepsilon_{n-1} + \sqrt{1 - c^2}\, \eta_n,$$

the sort of **autoregressive model** used in books on the theory and practice of **Time Series**. Chatfield [38] and Brockwell and Davis [33] are fine introductions. Box and Jenkins [29], Brillinger [32], Brockwell and Davis [34], and Priestley [188] are classics. For important new ideas, see Tong [229].

Our geometry can cope with the time-series model, but at the cost of considerable complexity. Of course, WinBUGS can deal with it.

APPENDIX C

TABLES

The following tables are included.

- Table of Distribution Function Φ of the standard normal $N(0,1)$ distribution, calculated as explained in Subsection 158C.

- Table of upper percentage points for t_ν, Student's t distribution with ν degrees of freedom, listing t such that $\mathbb{P}(T > t) = \eta$ or $\mathbb{P}(|T| > t) > \eta$ for various values of η and ν, where $T \sim t_\nu$. This table was calculated as explained in Subsection 253G.

- Table of upper percentage points for the χ^2 distribution with ν degrees of freedom, calculated as explained at 149Ea, using the fact that $\chi^2_\nu = \text{Gamma}(\frac{1}{2}\nu, \text{rate } \frac{1}{2})$.

- Table of upper 5% points for Fisher's $F_{r,s}$ distribution, calculated as explained at 252F and 210Id.

I am reasonably sure that the tables are correct!

Table of Distribution Function Φ
of the standard normal $N(0, 1)$ distribution

	0.00	0.01	0.02	0.03	0.04	0.05	0.06	0.07	0.08	0.09
0.0	0.5000	0.5040	0.5080	0.5120	0.5160	0.5199	0.5239	0.5279	0.5319	0.5359
0.1	0.5398	0.5438	0.5478	0.5517	0.5557	0.5596	0.5636	0.5675	0.5714	0.5753
0.2	0.5793	0.5832	0.5871	0.5910	0.5948	0.5987	0.6026	0.6064	0.6103	0.6141
0.3	0.6179	0.6217	0.6255	0.6293	0.6331	0.6368	0.6406	0.6443	0.6480	0.6517
0.4	0.6554	0.6591	0.6628	0.6664	0.6700	0.6736	0.6772	0.6808	0.6844	0.6879
0.5	0.6915	0.6950	0.6985	0.7019	0.7054	0.7088	0.7123	0.7157	0.7190	0.7224
0.6	0.7257	0.7291	0.7324	0.7357	0.7389	0.7422	0.7454	0.7486	0.7517	0.7549
0.7	0.7580	0.7611	0.7642	0.7673	0.7704	0.7734	0.7764	0.7794	0.7823	0.7852
0.8	0.7881	0.7910	0.7939	0.7967	0.7995	0.8023	0.8051	0.8078	0.8106	0.8133
0.9	0.8159	0.8186	0.8212	0.8238	0.8264	0.8289	0.8315	0.8340	0.8365	0.8389
1.0	0.8413	0.8438	0.8461	0.8485	0.8508	0.8531	0.8554	0.8577	0.8599	0.8621
1.1	0.8643	0.8665	0.8686	0.8708	0.8729	0.8749	0.8770	0.8790	0.8810	0.8830
1.2	0.8849	0.8869	0.8888	0.8907	0.8925	0.8944	0.8962	0.8980	0.8997	0.9015
1.3	0.9032	0.9049	0.9066	0.9082	0.9099	0.9115	0.9131	0.9147	0.9162	0.9177
1.4	0.9192	0.9207	0.9222	0.9236	0.9251	0.9265	0.9279	0.9292	0.9306	0.9319
1.5	0.9332	0.9345	0.9357	0.9370	0.9382	0.9394	0.9406	0.9418	0.9429	0.9441
1.6	0.9452	0.9463	0.9474	0.9484	0.9495	0.9505	0.9515	0.9525	0.9535	0.9545
1.7	0.9554	0.9564	0.9573	0.9582	0.9591	0.9599	0.9608	0.9616	0.9625	0.9633
1.8	0.9641	0.9649	0.9656	0.9664	0.9671	0.9678	0.9686	0.9693	0.9699	0.9706
1.9	0.9713	0.9719	0.9726	0.9732	0.9738	0.9744	0.9750	0.9756	0.9761	0.9767
2.0	0.9772	0.9778	0.9783	0.9788	0.9793	0.9798	0.9803	0.9808	0.9812	0.9817
2.1	0.9821	0.9826	0.9830	0.9834	0.9838	0.9842	0.9846	0.9850	0.9854	0.9857
2.2	0.9861	0.9864	0.9868	0.9871	0.9875	0.9878	0.9881	0.9884	0.9887	0.9890
2.3	0.9893	0.9896	0.9898	0.9901	0.9904	0.9906	0.9909	0.9911	0.9913	0.9916
2.4	0.9918	0.9920	0.9922	0.9925	0.9927	0.9929	0.9931	0.9932	0.9934	0.9936
2.5	0.9938	0.9940	0.9941	0.9943	0.9945	0.9946	0.9948	0.9949	0.9951	0.9952
2.6	0.9953	0.9955	0.9956	0.9957	0.9959	0.9960	0.9961	0.9962	0.9963	0.9964
2.7	0.9965	0.9966	0.9967	0.9968	0.9969	0.9970	0.9971	0.9972	0.9973	0.9974
2.8	0.9974	0.9975	0.9976	0.9977	0.9977	0.9978	0.9979	0.9979	0.9980	0.9981
2.9	0.9981	0.9982	0.9982	0.9983	0.9984	0.9984	0.9985	0.9985	0.9986	0.9986
3.0	0.9987	0.9987	0.9987	0.9988	0.9988	0.9989	0.9989	0.9989	0.9990	0.9990
3.1	0.9990	0.9991	0.9991	0.9991	0.9992	0.9992	0.9992	0.9992	0.9993	0.9993

Example. We have $\Phi(1.63) = \Phi(1.6 + 0.03) = 0.9484$.
One can use linear interpolation:

$$\Phi(1.634) \approx \Phi(1.63) + 0.4\left\{\Phi(1.64) - \Phi(1.63)\right\},$$

not that such accuracy is ever relevant in practice! 'Tail percentage points' may be read off from the last row of the 't' table, since the t_∞ and $N(0, 1)$ distributions are the same.

Upper percentage points for t_ν, Student's t-distribution with ν degrees of freedom

$\mathbb{P}(|T| > t)$	50.0%	20.0%	10.0%	5.0%	2.0%	1.0%	0.2%
$\mathbb{P}(T > t)$	25.0%	10.0%	5.0%	2.5%	1.0%	0.5%	0.1%
ν	t	t	t	t	t	t	t
1	1.000	3.08	6.31	12.71	31.82	63.66	318.31
2	0.816	1.89	2.92	4.30	6.96	9.92	22.33
3	0.765	1.64	2.35	3.18	4.54	5.84	10.21
4	0.741	1.53	2.13	2.78	3.75	4.60	7.17
5	0.727	1.48	2.02	2.57	3.36	4.03	5.89
6	0.718	1.44	1.94	2.45	3.14	3.71	5.21
7	0.711	1.41	1.89	2.36	3.00	3.50	4.79
8	0.706	1.40	1.86	2.31	2.90	3.36	4.50
9	0.703	1.38	1.83	2.26	2.82	3.25	4.30
10	0.700	1.37	1.81	2.23	2.76	3.17	4.14
11	0.697	1.36	1.80	2.20	2.72	3.11	4.02
12	0.695	1.36	1.78	2.18	2.68	3.05	3.93
13	0.694	1.35	1.77	2.16	2.65	3.01	3.85
14	0.692	1.35	1.76	2.14	2.62	2.98	3.79
15	0.691	1.34	1.75	2.13	2.60	2.95	3.73
16	0.690	1.34	1.75	2.12	2.58	2.92	3.69
17	0.689	1.33	1.74	2.11	2.57	2.90	3.65
18	0.688	1.33	1.73	2.10	2.55	2.88	3.61
19	0.688	1.33	1.73	2.09	2.54	2.86	3.58
20	0.687	1.33	1.72	2.09	2.53	2.85	3.55
21	0.686	1.32	1.72	2.08	2.52	2.83	3.53
22	0.686	1.32	1.72	2.07	2.51	2.82	3.51
23	0.685	1.32	1.71	2.07	2.50	2.81	3.49
24	0.685	1.32	1.71	2.06	2.49	2.80	3.47
25	0.684	1.32	1.71	2.06	2.49	2.79	3.45
30	0.683	1.31	1.70	2.04	2.46	2.75	3.39
35	0.682	1.31	1.69	2.03	2.44	2.72	3.34
40	0.681	1.30	1.68	2.02	2.42	2.70	3.31
45	0.680	1.30	1.68	2.01	2.41	2.69	3.28
50	0.679	1.30	1.68	2.01	2.40	2.68	3.26
55	0.679	1.30	1.67	2.00	2.40	2.67	3.25
60	0.679	1.30	1.67	2.00	2.39	2.66	3.23
65	0.678	1.29	1.67	2.00	2.39	2.65	3.22
70	0.678	1.29	1.67	1.99	2.38	2.65	3.21
75	0.678	1.29	1.67	1.99	2.38	2.64	3.20
∞	0.678	1.28	1.64	1.96	2.33	2.58	3.09

Example. If T has the t_{16} distribution, then the value t such that $\mathbb{P}(T > t) = 2.5\%$ (equivalently such that $\mathbb{P}(|T| > t) = 5\%$) is $t = 2.12$.

Upper percentage points for $\chi_\nu^2 = \mathrm{Gamma}(\frac{1}{2}\nu, \text{mean } 2)$,
the χ^2 distribution with ν degrees of freedom

ν	99.0%	97.5%	95.0%	50.0%	10.0%	5.0%	2.5%	1.0%
1	(0.0002)	(0.001)	(0.004)	0.45	2.71	3.84	5.02	6.63
2	(0.020)	(0.051)	(0.103)	1.39	4.61	5.99	7.38	9.21
3	(0.115)	(0.216)	(0.352)	2.37	6.25	7.81	9.35	11.34
4	(0.297)	(0.484)	(0.711)	3.36	7.78	9.49	11.14	13.28
5	(0.554)	(0.831)	(1.145)	4.35	9.24	11.07	12.83	15.09
6	(0.872)	1.24	1.64	5.35	10.64	12.59	14.45	16.81
7	1.24	1.69	2.17	6.35	12.02	14.07	16.01	18.48
8	1.65	2.18	2.73	7.34	13.36	15.51	17.53	20.09
9	2.09	2.70	3.33	8.34	14.68	16.92	19.02	21.67
10	2.56	3.25	3.94	9.34	15.99	18.31	20.48	23.21
11	3.05	3.82	4.57	10.34	17.28	19.68	21.92	24.73
12	3.57	4.40	5.23	11.34	18.55	21.03	23.34	26.22
13	4.11	5.01	5.89	12.34	19.81	22.36	24.74	27.69
14	4.66	5.63	6.57	13.34	21.06	23.68	26.12	29.14
15	5.23	6.26	7.26	14.34	22.31	25.00	27.49	30.58
16	5.81	6.91	7.96	15.34	23.54	26.30	28.85	32.00
17	6.41	7.56	8.67	16.34	24.77	27.59	30.19	33.41
18	7.01	8.23	9.39	17.34	25.99	28.87	31.53	34.81
19	7.63	8.91	10.12	18.34	27.20	30.14	32.85	36.19
20	8.26	9.59	10.85	19.34	28.41	31.41	34.17	37.57
21	8.90	10.28	11.59	20.34	29.62	32.67	35.48	38.93
22	9.54	10.98	12.34	21.34	30.81	33.92	36.78	40.29
23	10.20	11.69	13.09	22.34	32.01	35.17	38.08	41.64
24	10.86	12.40	13.85	23.34	33.20	36.42	39.36	42.98
25	11.52	13.12	14.61	24.34	34.38	37.65	40.65	44.31
30	14.95	16.79	18.49	29.34	40.26	43.77	46.98	50.89
35	18.51	20.57	22.47	34.34	46.06	49.80	53.20	57.34
40	22.16	24.43	26.51	39.34	51.81	55.76	59.34	63.69
45	25.90	28.37	30.61	44.34	57.51	61.66	65.41	69.96
50	29.71	32.36	34.76	49.33	63.17	67.50	71.42	76.15
55	33.57	36.40	38.96	54.33	68.80	73.31	77.38	82.29
60	37.49	40.48	43.19	59.33	74.40	79.08	83.30	88.38
65	41.44	44.60	47.45	64.33	79.97	84.82	89.18	94.42
70	45.44	48.76	51.74	69.33	85.53	90.53	95.02	100.43
75	49.48	52.94	56.05	74.33	91.06	96.22	100.84	106.39
80	53.54	57.15	60.39	79.33	96.58	101.88	106.63	112.33
90	61.75	65.65	69.13	89.33	107.57	113.15	118.14	124.12
100	70.06	74.22	77.93	99.33	118.50	124.34	129.56	135.81

(Brackets surround entries which are 'misaligned' through being given to more than 2 decimal places.) **Example.** If H has the χ_{13}^2 distribution with $\nu = 13$ degrees of freedom, then the value c such that $\mathbb{P}(H > c) = 5\%$ is 22.36. **Important Note.** See 164Ka.

Upper 5% points for the $F((num) \nu_1, (den) \nu_2)$ distribution

	ν_1									
	1	2	3	4	5	6	7	8	9	10
1	161.45	199.50	215.71	224.58	230.16	233.99	236.77	238.88	240.54	241.88
2	18.51	19.00	19.16	19.25	19.30	19.33	19.35	19.37	19.38	19.40
3	10.13	9.55	9.28	9.12	9.01	8.94	8.89	8.85	8.81	8.79
4	7.71	6.94	6.59	6.39	6.26	6.16	6.09	6.04	6.00	5.96
5	6.61	5.79	5.41	5.19	5.05	4.95	4.88	4.82	4.77	4.74
6	5.99	5.14	4.76	4.53	4.39	4.28	4.21	4.15	4.10	4.06
7	5.59	4.74	4.35	4.12	3.97	3.87	3.79	3.73	3.68	3.64
8	5.32	4.46	4.07	3.84	3.69	3.58	3.50	3.44	3.39	3.35
9	5.12	4.26	3.86	3.63	3.48	3.37	3.29	3.23	3.18	3.14
ν_2 10	4.96	4.10	3.71	3.48	3.33	3.22	3.14	3.07	3.02	2.98
11	4.84	3.98	3.59	3.36	3.20	3.09	3.01	2.95	2.90	2.85
12	4.75	3.89	3.49	3.26	3.11	3.00	2.91	2.85	2.80	2.75
13	4.67	3.81	3.41	3.18	3.03	2.92	2.83	2.77	2.71	2.67
14	4.60	3.74	3.34	3.11	2.96	2.85	2.76	2.70	2.65	2.60
15	4.54	3.68	3.29	3.06	2.90	2.79	2.71	2.64	2.59	2.54
20	4.35	3.49	3.10	2.87	2.71	2.60	2.51	2.45	2.39	2.35
30	4.17	3.32	2.92	2.69	2.53	2.42	2.33	2.27	2.21	2.16
60	4.00	3.15	2.76	2.53	2.37	2.25	2.17	2.10	2.04	1.99
120	3.92	3.07	2.68	2.45	2.29	2.18	2.09	2.02	1.96	1.91
∞	3.84	3.00	2.60	2.37	2.21	2.10	2.01	1.94	1.88	1.83

	ν_1									
	11	12	13	14	15	20	30	60	120	∞
1	242.98	243.91	244.69	245.36	245.95	248.01	250.10	252.20	253.25	254.31
2	19.40	19.41	19.42	19.42	19.43	19.45	19.46	19.48	19.49	19.50
3	8.76	8.74	8.73	8.71	8.70	8.66	8.62	8.57	8.55	8.53
4	5.94	5.91	5.89	5.87	5.86	5.80	5.75	5.69	5.66	5.63
5	4.70	4.68	4.66	4.64	4.62	4.56	4.50	4.43	4.40	4.36
6	4.03	4.00	3.98	3.96	3.94	3.87	3.81	3.74	3.70	3.67
7	3.60	3.57	3.55	3.53	3.51	3.44	3.38	3.30	3.27	3.23
8	3.31	3.28	3.26	3.24	3.22	3.15	3.08	3.01	2.97	2.93
9	3.10	3.07	3.05	3.03	3.01	2.94	2.86	2.79	2.75	2.71
ν_2 10	2.94	2.91	2.89	2.86	2.85	2.77	2.70	2.62	2.58	2.54
11	2.82	2.79	2.76	2.74	2.72	2.65	2.57	2.49	2.45	2.40
12	2.72	2.69	2.66	2.64	2.62	2.54	2.47	2.38	2.34	2.30
13	2.63	2.60	2.58	2.55	2.53	2.46	2.38	2.30	2.25	2.21
14	2.57	2.53	2.51	2.48	2.46	2.39	2.31	2.22	2.18	2.13
15	2.51	2.48	2.45	2.42	2.40	2.33	2.25	2.16	2.11	2.07
20	2.31	2.28	2.25	2.22	2.20	2.12	2.04	1.95	1.90	1.84
30	2.13	2.09	2.06	2.04	2.01	1.93	1.84	1.74	1.68	1.62
60	1.95	1.92	1.89	1.86	1.84	1.75	1.65	1.53	1.47	1.39
120	1.87	1.83	1.80	1.78	1.75	1.66	1.55	1.43	1.35	1.25
∞	1.79	1.75	1.72	1.69	1.67	1.57	1.46	1.32	1.22	1.00

APPENDIX D

A SMALL SAMPLE OF THE LITERATURE

These days, there are so many brilliant young (and youngish) probabilists and statisticians around that I am not going to single any out. Tributes to people who helped develop the fields are more 'classical'. But, of course, many of the books mentioned are modern.

The Bibliography is fairly extensive, though still a small fraction of what it could/should have been. For everyone, there are many books in the Bibliography of great interest. And for those determined to strive, to seek, to find, but not to yield, the Bibliography contains – for inspiration – some books and papers (several of which were mentioned in the main text) which are at a significantly more advanced level than this book.

Probability

In the early days, Probability Theory was developed by Cardano, Galileo, Pascal, de Moivre, two Bernoullis, Bayes, Laplace, Gauss, Markov, Tchebychev. In the 20th century, the Measure Theory of Borel and Lebesgue allowed Borel, Wiener, Kolmogorov [140], Doob [66] and others to put Probability on a rigorous basis and to extend its scope dramatically. Lévy [151, 152] provided much of the intuition and inspiration.

Volume 1 of Feller's book [75], a little more challenging than this present one, will always hold a special place in the hearts of probabilists. It *is* quite challenging, and should perhaps be read after, or in conjunction with, this one. Pitman [184] is a good route-in as prerequisite for the Probability in this book, should you need one.

At about the same level as this book are Grimmett and Stirzaker [103], Volume 1 of Karlin and Taylor [125], Ross [201, 202].

There are a lot of books on the measure-theoretic foundations of Probability. Neveu [172] has real class and French elegance, Breiman [30] is one of the all-time greats, and Williams [235] is mischievous.

Other favourites on Probability: Chow and Teicher [41], Chung [43], Durrett [69], Fristedt and Gray [85], Laha and Rohatgi [146], Rényi [193]. The sad death of E T Jaynes means that only some of a fascinating book is complete. See [117]. (Skip his remarks about Feller, though.)

Genetics.   Elizabeth Thompson, daughter of Edward who so inspired many of us 'tutees' at Oxford, is a very considerable authority on Pedigree Analysis in Genetics. See [227, 228]. I have already enthused about [101] which shows in particular that there is much more to Genetics than Maths. Richard Dawkins' books have also had my very strong recommendation.

Fisher wrote one of the most important books on Genetics since Darwin, now updated as [80], work further developed by W D Hamilton and others. Genetics has provided motivation for much Probability of a type too advanced for this book: measure-valued diffusions, etc. Find out on the Web what people such as Donald Dawson, Peter Donnelly, Stewart Ethier, Steven Evans, Thomas Kurtz, Edwin Perkins, ..., are doing. The FKPP equation of Fisher and of Kolmogorov, Petrovskii and Piscunov was a big motivating factor for branching processes in Probability and for much work in Analysis. Interestingly, the probabilistic methods of Bramson and others have here been able to beat the analytical at their own game.

Of course, this is the era of the Human Genome Project about which you can find much information via the website: http://www.ornl.gov/hgmis/ .

Random walks, Markov chains, stochastic processes. For these important topics, see Feller [75], Kemeny and Snell [131], Norris [176], Spitzer [216]. Markov chains play an important part in queueing theory; see Cox and Smith [52], and Kelly's fascinating book [130]. For stochastic processes generally and applications, see Brémaud [31], Çinlar [45], Cox and Miller [50], Volume 2 of Karlin and Taylor [125].

The theory of **martingales** is due principally to Doob [66], with some key contributions from Lévy and Meyer. The books, Neveu [173], Hall and Heyde [107], Williams [235] and Rogers and Williams [199] give an idea of the scope, Dellacherie and Meyer [60] the definitive account. Martingales form the foundation for the very important technique of **stochastic calculus** created by the great Japanese mathematician Itô. Øksendal [178] is a nice introduction, as are Chung and (Ruth) Williams [44] and Durrett [70]. Revuz and Yor [194], Karatzas and Shreve [123], and Rogers and Williams [199], which also begin at the beginning, take the theory further. For references for martingales in finance, see Subsection 421K.

I became *really* enthusiastic about some topics in Probability, especially Statistical Mechanics, at the end of Chapter 9. It will surprise some people to realize that the topics I find most interesting (and have for a long time found most interesting) in Probability are ones which do *not* have to do with martingales! But of course, such topics are too hard for me to have been able to do anything with them, in spite of determined efforts.

Frequentist Statistics

Early Frequentist Statistics owed a lot to Galton, two Pearsons, 'Student' (W S Gosset), Neyman. The greatest of statisticians was Sir Ronald Fisher, and key work originated by him on ANOVA was developed by Yates and coworkers at the Rothamsted Experimental Station. As stated earlier, we owe several key concepts – sufficiency, ancillarity, likelihood, F tests, ... – to Fisher. See [79] for some of his ideas. Huge contributions to Statistics were later made by Bartlett, Cox, Tukey, Wald, Wilks and others.

A great introductory book on Frequentist Theory is Freedman, Pisani and Purves [83]: how that book makes you think! Amongst classics are Kendall and Stuart (now modernized with the help of Ord and Arnold as [132, 133]), Cox and Hinkley [49] (which also has a substantial Bayesian section), Lehmann [149, 150]. I always liked Silvey [212] as a succinct account. See also Casella and Berger [37], Garthwaite and Jolliffe [88], Kalbfleisch [122].

On **regression**, see Draper and Smith [68], Montgomery and Peck [167], Myers [170], Rousseeuw and Leroy [200], Ryan [205] and Weisberg [234]. See further references under 320Fc. On **ANOVA**, which cannot really be separated from **Experimental Design**, see Cochran and Cox [46], Cox [48], Fisher and Bennett [79], Huitema [113], Mead [161], Snedecor and Cochran [214], Scheffé [207]. Cox and Reid [51] is an important recent book which, amongst many other things, explains how Galois fields and finite geometries feature in Experimental Design. Rosemary Bailey has in preparation a book showing how several other pieces of interesting mathematics also feature there.

The theory of Linear Models expands to Multivariate Analysis (see Anderson [5], Chatfield and Collins [39], Everitt and Dunn [73], Fahrmeir and Tutz [74], Johnson and Wichern [121], Krzanowski and Marriott [143], Manly [156], Mardia, Kent and Bibby [157], Murtagh and Heck [169], Scott [210], Seber [211], for example). It is interesting to look at a book such as Tabachnick and Fidell [223], written for social scientists, and therefore with a somewhat different emphasis. The fact that its Contents section runs to 21 pages listing chapters on Multiple Regression, Canonical Correlation, Multiway Frequency Analysis, Analysis of Covariance (ANCOVA), Multivariate Analysis of Covariance (MANCOVA), Discriminant Factor Analysis, Principal Components and Factor Analysis, Structural Equation Modelling, etc, gives a clue to the degree of expansion the subject has witnessed. That book surveys some of the Frequentist computing packages available in 1996.

The theory of Generalized Linear Models (see Dobson [65], McCullagh and Nelder [160]) keeps some features of Linear-Model theory, but allows a certain degree of non-linearity and the use of non-normal distributions. GENSTAT, GLIM, S-PLUS are amongst packages which can implement the theory.

I gave references for **time series** at the end of Appendix B.

Bayesian Statistics

That the world has now to a large extent embraced the Bayesian view of Statistics is due to de Finetti [58], Good [99], Jeffreys [118], Savage [206], and, perhaps above all, to the missionary zeal (and good sense) of Lindley. Bayesian Statistics is well served in the literature: see Bernardo and Smith [17], Carlin and Louis [36], Lindley [153], Gelman, Carlin, Stern and Rubin [90], O'Hagan [177]. A nice introduction is provided by Lee [148]. I am very much looking forward to Draper [67]. The series of volumes [16] edited by Bernardo, Berger, Dawid and Smith provide the authoritative way of keeping up to date.

The Bayesian approach was always elegant (if one accepts its philosophy). However, it only became of real practical use for complex models with the advent of MCMC computing. I have illustrated this both with bare-hands programs and via WinBUGS . Some key references were given at the end of Subsection 268C. See also Gamerman [87], Robert [197], Robert and Casella [198]. As is so often the case, we owe the original ideas to physicists. The structure of WinBUGS relies heavily on the theory of graphical models as in Lauritzen [147] (building inevitably on the work of physicists on Markov random fields). I have explained that MLwiN is a user-friendly package capable of doing some 'classical' and some MCMC work.

I have probably been unfair to **Decision Theory** (Frequentist and Bayesian), but I cannot get to like it. See Berger [14], Chernoff and Moses [40], the relevant chapter of Cox and Hinkley [49], de Groot [59], the very influential Ferguson [76], French and Insua [84].

Some references for Differential Geometry in Statistics were given at 380Na.

On the question of **model choice**, see Akaike [2], Burnham and Anderson [35], Christensen [42] Gelfand and Ghosh [89], Mallows [155], Schwarz [209], Spiegelhalter, Best and Carlin [215], and references contained in these publications. See also the discussion at 320Fc for the case of Regression.

Quantum Theory

Amongst the great names of early Quantum Theory are Planck, Einstein, Schrödinger, Heisenberg, Bohr, Dirac, and, of course, though less early, Feynman. The rigorous probabilistic theory is due principally to von Neumann.

Feynman's place as the greatest expositor of Science is likely to remain: see Volume III of [78] and the magical [77] which presents the alternative 'path-integral' formulation of standard Quantum Theory. I did greatly enjoy the popular books, Greene [100] and Penrose [183]. Just look at the magic-dodecahedron version of entanglement in the latter.

As a guide to the mathematical theory, one of my favourites is Isham [115], and I am fond of the classic accounts, Davydov [56], Messiah [163], Merzbacher [162], Schiff [208] . For the full mathematical theory, see the great Reed and Simon [191], and Mackey [154]. For references for the Dirac equation, see the end of Section 10.5.

For some connections between Quantum Theory and Probability and Statistics, see Helstrom [111], Holevo [112], Streater [219], and current work of Barndorff-Nielsen and Gill which you may track via the Web. The books by Meyer [165] and Parthasarathy [181] are of much interest to probabilists as they present the Hudson–Parthasarathy stochastic calculus, a non-commutative version of Itô's.

Quantum computing

The idea of quantum computing goes back to Manin and, later and independently, Feynman. Deutsch made a striking contribution, but what made the subject really take off was Schor's algorithm (which is presented in the Preskill, Gruska, and Nielsen and Chuang references now to be mentioned.)

Most of what I have learnt about quantum computing has been from Preskill's marvellous notes [187], from Gruska's fine book [105], and from very many papers under Quantum Physics at the Los Alamos National Laboratory archive site

`http://www.arXiv.org` mirrored as `http://uk.arXiv.org`

and mirrored at other sites too. I refer to these as quant-ph. I don't myself add quant-ph to the site address because there are lots of other interesting things on the site.

After I had essentially completed my account, a superb book, Nielsen and Chuang [174], appeared. Read it. Also read Bouwmeester, Ekert and Zeilinger [28].

S-PLUS, R, Minitab, LaTeX, Postscript, C, Emacs, WinEdt

As already remarked, S-PLUS (see Krause and Olson [142], Venables and Ripley [232]) is the most popular package amongst academic statisticians, and R, of which a free version is available ([190]), contains many features of S-PLUS. For Minitab, see [204].

For LaTeX (Lamport [144]), I found Kopka and Daly [141] very useful. I had earlier learnt TeX from the master in Knuth [139]. I found Postscript ([1]) fun and, after a time, surprisingly easy to use, sometimes via a C-to-Postscript converter I wrote.

I learnt 'C' from Kernighan and Richie [136] and Kelley and Pohl [129]. I did once learn something of Object Oriented Programming but haven't found it of any real use for my work. (OOPs, I'll get some flak for that! But static-variable modules seem to do things much more neatly.)

I used the (free) Emacs and (very-low-cost) WinEdt (correct spelling!) to prepare the LaTeX file. Both are excellent, and both downloadable from the Web. I particularly like WinEdt's 'live' correction of spelling.

Bibliography

[1] Adobe Systems, Inc, *Postscript Language Reference Manual* (second edition), Addison Wesley, Reading, Mass., 1993.

[2] Akaike, H., Prediction and Entropy, Chapter 1 of [7].

[3] Aldous, D., *Probability Approximations via the Poisson Clumping Heuristic*, Springer-Verlag, Berlin, Heidelberg, New York, 1989.

[4] Amari, S., *Differential-Geometrical Methods in Statistics*, Springer-Verlag, Berlin, Heidelberg, New York, 1985.

[5] Anderson, T.W., *An Introduction to Multivariate Statistical Analysis*, Wiley, New York, 1958.

[6] Athreya, K.B. and Ney, P., *Branching Processes*, Springer-Verlag, Berlin, Heidelberg, New York, 1989.

[7] Atkinson, A.C. and Fienberg, S.E. (editors), *A Celebration of Statistics : the ISI Centenary Volume,* Springer-Verlag, Berlin, Heidelberg, New York, 1985.

[8] Barndorff-Nielsen, O.E. and Cox, D.R., *Asymptotic Techniques for Use in Statistics*, Chapman and Hall, London, 1989.

[9] Barndorff-Nielsen, O.E. and Cox, D.R., *Inference and Asymptotics*, Chapman and Hall, London, 1994.

[10] Barnett, V. and Lewis, T.: *Outliers in Statistical Data*, Wiley, Chichester, New York, 1984.

[11] Barron, A.R., Entropy and the Central Limit Theorem, *Ann. Prob.* **14**, 336–343, 1986.

[12] Baxter, R.S., *Exactly Solvable Models in Statistical Mechanics*, Academic Press, New York, 1982.

[13] Baxter, R.J. and Enting, I.G., 399th solution of the Ising model, *J. Phys. A.*, **11**, 2463–2473, 1978.

[14] Berger, J.O., *Statistical Decision Theory and Bayesian Analysis*, second edition, Springer-Verlag, Berlin, Heidelberg, New York, 1985.

[15] Bernardo, J.M. and Berger, J.O., On the development of reference priors, in *Volume 4* of [16], 1992.

[16] Bernardo, J.M., Berger, J.O., Dawid, A.P., and Smith, A.F.M. (editors), *Bayesian Statistics*, several volumes, Oxford University Press.

[17] Bernardo, J.M. and Smith, A.F.M., *Bayesian Theory*, Wiley, Chichester, New York, 1994. Further volumes are due.

[18] Besag, J., The statistical analysis of dirty pictures (with discussion), *J. Royal Stat. Soc., B*, **48**, 259–302, 1986.

[19] Besag, J., Green, P., Higdon, D., and Mengerson, K., Bayesian computation and stochastic systems (with discussion), *Statistical Science*, **10**, 3–66, 1995.

[20] Besag, J. and Higdon, D., Bayesian analysis of agricultural field experiments, *J. Royal Stat. Soc.*, **61**, 691–746, 1999.

[21] Billingsley, P., *Ergodic Theory and Information*, Wiley, Chichester, New York, 1965.

[22] Billingsley, P., *Convergence of Probability Measures*, Wiley, Chichester, New York, 1968.

[23] Bingham, N.H. and Kiesel, R., *Risk-neutral Valuation: Pricing and Hedging of Financial Derivatives*, Springer-Verlag, Berlin, Heidelberg, New York, 1998.

[24] Bjorken, J.D. and Drell, S.D., *Relativistic Quantum Mechanics*, McGraw–Hill, New York, Toronto, 1964.

[25] Bohm, D., *Wholeness and the Implicate Order*, Routledge (Taylor and Francis), London, New York, 1980.

[26] Bohm, D. and Hiley, B.J., *The Undivided Universe*, Routledge (Taylor and Francis), London, New York, 1993, 1995.

[27] Borcherds, R.E., What is moonshine?, *Documenta Mathematica (extra volume of) Proc. Intern. Cong. Math. , 1998*, available from `http://xxx.lanl.gov/abs/math.QA/9809110`.

[28] Bouwmeester, D., Ekert, A.K. and Zeilinger, A., *The Physics of Quantum Information. Quantum Cryptography, Quantum Teleportation, Quantum Computation*, Springer-Verlag, Berlin, Heidelberg, New York, 2000.

[29] Box, G.E.P. and Jenkins, G. M., *Time Series Analysis, Forecasting and Control*, Holden–Day, San Francisco, 1970.

[30] Breiman, L., *Probability*, SIAM, Philadelphia, 1992 (originally, Addison–Wesley, Reading, Mass., 1968).

[31] Brémaud, P., *An Introduction to Probabilistic Modelling*, Springer-Verlag, Berlin, Heidelberg, New York, 1988.

[32] Brillinger, D.R., *Time Series: Data Analysis and Theory*, Holt, New York, 1975.

[33] Brockwell, P.J. and Davis, R.A., *An Introduction to Time Series and Forecasting*, Springer-Verlag, Berlin, Heidelberg, New York, 1997.

[34] Brockwell, P.J. and Davis, R.A., *Time Series: Theory and Methods*, Springer-Verlag, Berlin, Heidelberg, New York, 1987.

[35] Burnham, K.P. and Anderson, D.R., *Model Selection and Inference*, Springer, Berlin, Heidelberg, New York, 1998.

[36] Carlin, B.P. and Louis, T.A., *Bayes and Empirical Bayes Methods for Data Analysis*, Chapman and Hall, London, 2000.

[37] Casella, G. and Berger, R.L., *Statistical Inference*, Wadsworth, Belmont, CA, 1990.

[38] Chatfield, C., *The Analysis of Time Series: Theory and Practice*, Chapman and Hall, London, 1975.

[39] Chatfield, C. and Collins, A.J., *Introduction to Multivariate Analysis*, Chapman and Hall, London, 1980.

[40] Chernoff, H. and Moses, L., *Elementary Decision Theory*, Dover, Mineola, NY, 1987.

[41] Chow, Y.S. and Teicher, H., *Probability Theory: Independence, Interchangeability, Martingales* (second edition), Springer-Verlag, Berlin, Heidelberg, New York, 1988.

[42] Christensen, R., *Analysis of Variance, Design, and Regression: Applied Statistical Methods*, Chapman and Hall, London, 1996.

[43] Chung, K.L., *A Course in Probability Theory*, Academic Press, New York, 1974.

[44] Chung, K.L. and Williams, R.J., *Introduction to Stochastic Integration*, Birkhäuser, Boston, 1983.

[45] Çinlar, E., *Introduction to Stochastic Processes*, Prentice–Hall, Englewood Cliffs, NJ, 1975.

[46] Cochran, W.G. and Cox, G.M., *Experimental Designs* (second edition), Wiley, New York, 1957.

[47] Cox, D.A., *Primes of the form $x^2 + ny^2$*, Wiley, New York, 1989.

[48] Cox, D.R., *Planning of Experiments*, Wiley, New York, 1992.

[49] Cox, D.R. and Hinkley, D.V., *Theoretical Statistics*, Chapman and Hall, London, 1974. *Problems and Solutions in Theoretical Statistics*, Chapman and Hall, London, 1978.

[50] Cox, D.R. and Miller, H.D., *The Theory of Stochastic Processes*, Chapman and Hall, London, 1965.

[51] Cox, D.R. and Reid, N., *The Theory of the Design of Experiments*, Chapman and Hall, London, 2000.

[52] Cox D.R. and Smith, W.L., *Queues*, Methuen, London, 1961.

[53] Daniels, H.E., Saddlepoint approximations in statistics, *Ann. Math. Statist.*, **25**, 631–650, 1954.

[54] Davis, B., Reinforced random walk, *Probab. Th. Rel. Fields* **84**, 203–229, 1990.

[55] Davison, A.C. and Hinkley, D.V., *Bootstrap Methods and Their Application*, Cambridge University Press, 1997.

[56] Davydov, A.S., *Quantum Mechanics*, Pergamon, Oxford, 1965.

[57] Dawkins, R., *River out of Eden*, Weidenfeld and Nicholson, London, 1995.

[58] de Finetti, B., *Theory of Probability: a Critical Introductory Treatment* (two volumes), Wiley, Chichester, New York, 1974.

[59] de Groot, M., *Optimal Statistical Decisions*, McGraw–Hill, New York, 1970.

[60] Dellacherie, C. and Meyer, P.-A., *Probabilités et Potentiel* (several volumes), Hermann, Paris. English translation, *Probability and Potential*, published by North–Holland, Amsterdam.

[61] Deuschel, J.-D. and Stroock, D.W., *Large Deviations*, Academic Press, Boston, 1989.

[62] Diaconis, P., Recent progress on de Finetti's notion of exchangeability, in *Volume 3* of [16].

[63] Diaconis, P., *Group Representations in Probability and Statistics*, Institute of Math. Stat., Hayward, California, 1988.

[64] di Francesco, P., Mathieu, P. and Sénéchal, D., *Conformal Field Theory* (two volumes), Springer, Berlin, Heidelberg, New York, 1997.

[65] Dobson, A.J., *An Introduction to Statistical Modelling*, Chapman and Hall, London, 1983.

[66] Doob, J.L., *Stochastic Processes*, Wiley, New York, 1953.

[67] Draper, D., *Bayesian Hierarchical Modeling* (to appear).

[68] Draper, N.R. and Smith, H., *Applied Regression Analysis*, Wiley, New York, 1981.

[69] Durrett, R., *Probability: Theory and Examples*, second edition, Duxbury Press, Pacific Grove CA, 1996.

[70] Durrett, R., *Stochastic Calculus: A Practical Introduction*, CRC Press, London, 1996.

[71] Efron, B. and Tibshirani, R.J., *An Introduction to the Bootstrap*, Chapman and Hall, New York, 1993.

[72] Einstein, A., Podolsky, B. and Rosen, N, Can quantum-mechanical description of physical reality be considered complete?, *Phys. Rev.* **41**, 777–780, 1935.

[73] Everitt, B.S. and Dunn, G., *Applied Multivariate Data Analysis*, Edward Arnold, London, 1992.

[74] Fahrmeir, L. and Tutz, G., *Multivariate Statistical Modelling based on Generalized Linear Models*, Springer-Verlag, Berlin, Heidelberg, New York, 1994.

[75] Feller, W., *Introduction to Probability Theory and its Applications, Vol.1*, Wiley, New York, 1957.

[76] Ferguson, T.S., *Mathematical Statistics: A Decision Theoretic Approach*, Academic Press, New York, 1967.

[77] Feynman, R.P., *QED: The Strange Story of Light and Matter*, Princeton University Press, 1985; Penguin, London, 1990.

[78] Feynman, R.P., Leighton, R.B., Sands, M., *The Feynman Lectures on Physics*, 2 volumes, Addison–Wesley, Reading, Mass., 1979.

[79] Fisher, R.A. and Bennett, J.H. (editor), *Statistical Methods, Experimental Design, and Scientific Inference*, Oxford University Press, 1990.

[80] Fisher, R.A. and Bennett, J.H. (editor), *The Genetical Theory of Natural Selection: A Complete Variorum Edition*, Oxford University Press, 1990.

[81] Flury, B., *A First Course in Multivariate Statistics*, Springer-Verlag, Berlin, Heidelberg, New York, 1997.

[82] Fradkin, E., *Physics 483 Course on Quantum Field Theory* at University of Illinois,
`http://w3.physics.uiuc.edu/ efradkin/phys483/physics483.html`

[83] Freedman, D., Pisani, R. and Purves, R., *Statistics*, third edition, Norton, New York, London, 1998.

[84] French, S. and Insua, D.R., *Statistical Decision Theory*, Arnold, London, 2000.

[85] Fristedt, B., and Gray, L., *A Modern Approach to Probability Theory and its Applications*, Springer-Verlag, Berlin, Heidelberg, New York, 1996.

[86] Frieden, B.R., *Physics from Fisher Information*, Cambridge University Press, 1998.

[87] Gamerman, D., *Markov Chain Monte Carlo: Stochastic Simulation for Bayesian Inference*, Chapman and Hall, London, 1997.

[88] Garthwaite, P.H. and Jolliffe, I.T., *Statistical Inference*, Prentice Hall, Englewood Cliffs, NJ, 1995.

[89] Gelfand, A.E. and Ghosh, S.K., Model choice: a minimum posterior predictive loss approach, *Biometrika*, **85**, 1–1, 1998.

[90] Gelman, A., Carlin, J.B., Stern, H.S. and Rubin, D.B., *Bayesian Data Analysis*, Chapman and Hall, London, 1995.

[91] Geman, S. and Geman, D., Stochastic relaxation, Gibbs distributions and the Bayesian restoration of images, *IEEE Trans. Pattn. Anal. Mach. Intel.*, **6**, 721–741, 1984.

[92] Gershenfeld, N. and Chuang, I.L., Quantum computing with molecules, *Sci. Am.*, June, 1998.

[93] Gilks, W.R., Full conditional distributions, in [94].

[94] Gilks, W.R., Richardson, S. and Spiegelhalter, D., *Markov Chain Monte Carlo in Practice*, Chapman and Hall, London, 1996.

[95] Gilks, W.R. and Roberts, G.O., Strategies for improving MCMC, in [94].

[96] Gilmour, S.G., The interpretation of Mallows's C_p statistic, *J. Royal Stat. Soc., Series D (The Statistician)*, **45**, 49–56, 1996.

[97] Giulini, D., Joos, E., Kiefer, C., Kupsch, J., Stamatescu, I.-O. and Zeh, H.D., *Decoherence and the Appearance of a Classical World in Quantum Theory*, Springer-Verlag, Berlin, Heidelberg, New York, 1996.

[98] Goddard, P., The work of Richard Ewen Borcherds, *Documenta Mathematica (extra volume of) Proc. Intern. Cong. Math. 1998*, 99–108, available from http://xxx.lanl.gov/abs/math.QA/9808136

[99] Good, I.J., *Probability and the Weighing of Evidence*, Griffin, London, 1950.

[100] Greene, B., *The Elegant Universe*, Vintage, Random House, London, 2000. (originally published by Jonathan Cape, 1999).

[101] Griffiths, A.J.F., Lewontin, R.C., Gelbart, W.M.G., and Miller, J., *Modern Genetic Analysis*, Freeman, New York, 1999.

[102] Grimmett, G., *Percolation*, second edition, Springer-Verlag, Berlin, Heidelberg, New York, 1999.

[103] Grimmett, G. and Stirzaker, D.R., *Probability and Random Processes*, Oxford University Press, 1992.

[104] Grover, L., A fast quantum mechanical algorithm for database search, *Proc. 28th ACM Symp. Theory Comp.*, 212–219, ACM Press, New York, 1996.

[105] Gruska, J., *Quantum Computing*, McGraw–Hill, London, 1999.

[106] Hall, P., *Introduction to the Theory of Coverage Processes*, John Wiley, New York, 1988.

[107] Hall, P. and Heyde, C.C., *Martingale Limit Theory and its Applications*, Academic Press, New York, 1980.

[108] Hardy, G.H. and Wright, E.M., *An Introduction to the Theory of Numbers*, fourth edition, Oxford University Press, 1960.

[109] Harris, T.E., *The Theory of Branching Processes*, Springer-Verlag, Berlin, Heidelberg, New York, 1963.

[110] Hastings, W.K., Monte Carlo sampling methods using Markov chains and their applications, *Biometrika*, **57**, 97–109, 1970.

[111] Helstrom, C.W., *Quantum Detection and Estimation Theory*, Academic Press, New York, 1976.

[112] Holevo, A.S., *Probabilistic and Statistical Aspects of Quantum Theory*, North–Holland, Amsterdam, 1982.

[113] Huitema, B.E., *The Analysis of Variance and Alternatives*, Wiley, New York, 1980.

[114] Hunt, P.J. and Kennedy, J.E., *Financial Derivatives in Theory and Practice*, Wiley, Chichester, New York, 2000.

[115] Isham, C.J., *Lectures on Quantum Theory*, Imperial College Press, London, and World Scientific, 1995.

[116] Itzykson, C. and Drouffe, J.M., *Statistical Field Theory* (two volumes), Cambridge University Press, 1991.

[117] Jaynes, E.T., *Probability Theory: The Logic of Science*, http://omega.albany.edu:8008/JaynesBook

[118] Jeffreys, H., *Theory of Probability*, Oxford University Press, 1961.

[119] Jennison, C. and Turnbull, B.W., *Group Sequential Methods with Applications to Clinical Trials*, Chapman and Hall, London, 2000.

[120] Johnson, N.L. and Kotz, S., *Distributions in Statistics* (several volumes), Wiley, New York, 1969–.

[121] Johnson, R.A. and Wichern, D.W., *Applied Multivariate Statistical Analysis* (fourth edition), Prentice–Hall, Englewood Cliffs, NJ, 1998.

[122] Kalbfleisch, J.G., *Probability and Statistical Inference: Volume 1: Statistical Inference*, Springer-Verlag, Berlin, Heidelberg, New York, 1985.

[123] Karatzas, I. and Shreve, S.E., *Brownian Motion and Stochastic Calculus*, Springer-Verlag, Berlin, Heidelberg, New York, 1987.

[124] Karatzas, I. and Shreve, S.E., *Methods of Mathematical Finance*, Springer-Verlag, Berlin, Heidelberg, New York, 1998.

[125] Karlin, S. and Taylor, H.M., *A First Course in Stochastic Processes*, Academic Press, New York, 1975. *A Second Course in Stochastic Processes*, Academic Press, New York, 1982.

[126] Kass, R.E. and Vos, P.W., *Geometrical Foundations of Asymptotic Inference*, Wiley, New York, 1997.

[127] Kasteleyn, P.W., The statistics of dimers on a lattice, I, The number of dimer arrangements on a quadratic lattice, *Physica*, **27**, 1209–1225, 1961.

[128] Kaufman, B., Crystal statistics II: partition function evaluated by spinor analysis, *Phys. Rev.*, **76**, 1232–1243, 1949.

[129] Kelley, A. and Pohl, I., *A Book on C*, second edition, Benjamin/Cummings, Redwood City, CA, 1990.

[130] Kelly, F.P., *Reversibility and Stochastic Networks*, Wiley, Chichester, New York, 1979.

[131] Kemeny, J.G. and Snell, J.L., *Finite Markov Chains*, Van Nostrand, Princeton, NJ, 1959.

[132] Kendall, M.G., Stuart, A. and Ord, J.K., ('Kendall's Advanced Theory of Statistics, Volume 1'), *Distribution Theory*, Arnold, London, 1994.

[133] Kendall, M.G., Stuart, A., Ord, J.K. and Arnold, S., ('Kendall's Advanced Theory of Statistics, Volume 2A'), *Classical Inference and the Linear Model*, Arnold, London, 1999.

[134] Kendall, D.G., Branching processes since 1873, *J. London Math. Soc.*, **41**, 385–406, 1966.

[135] Kendall, D.G., The genealogy of genealogy: branching processes before (and after) 1873, *Bull. London Math. Soc.* **7**, 225–253, 1975.

[136] Kernigan, B.W. and Richie, D.M., *The C Programming Language* (second edition), Prentice Hall, Englewood Cliffs, NJ, 1988.

[137] Kingman, J.F.C., Martingales in the OK Corral, *Bull. London Math. Soc.* **31**, 601–606, 1999.

[138] Knill, E., Laflamme, R. and Milburn, G.J., A scheme for efficient quantum computation with linear optics, *Nature* **409**, 46–52, 2001.

[139] Knuth, D.E., *The TEXbook*, Addison Wesley, Reading, Mass., 1987.

[140] Kolmogorov, A. N. *Foundations of the Theory of Probability* (translated from the German), second edition, Chelsea, New York, 1956.

[141] Kopka, H. and Daly, P.W., *A Guide to LATEX2e: Document Preparation for Beginners and Advanced Users*, second edition, Addison–Wesley, Harlow and Reading, Mass., 1995.

[142] Krause, A. and Olson, M., *The Basics of S and S-PLUS*, Springer-Verlag, Berlin, Heidelberg, New York, 1997.

[143] Krzanowski, W.J. and Marriott, F.H.C., *Multivariate Analysis* (two volumes), Edward Arnold, London, Wiley, New York, 1994.

[144] Lamport, L., *LATEX – A Document Preparation System*, second edition, Addison Wesley, Reading, Mass., 1994.

[145] Langlands, R., Pouliot, Ph. and Saint–Aubin, Y., Conformal invariance in two-dimensional percolation theory, *Bull. Amer. Math. Soc.*, **30**, 1–61, 1994.

[146] Laha, R. and Rohatgi, V., *Probability Theory*, Wiley, New York, 1979.

[147] Lauritzen, S., *Graphical Models*, Oxford University Press, 1996.

[148] Lee, P.M., *Bayesian Statistics*, Arnold, London, 1997.

[149] Lehmann, E.L., *Testing Statistical Hypotheses*, second edition, Springer-Verlag, Berlin, Heidelberg, New York, 1997.

[150] Lehmann, E.L., *Theory of Point Estimation*, second edition, Springer-Verlag, Berlin, Heidelberg, New York, 1998.

[151] Lévy, P., *Théorie de l'Addition des Variables Aléatoires*, Gauthier–Villars, Paris, 1954.

[152] Lévy, P., *Processus Stochastiques et Mouvement Brownien*, Gauthier–Villars, Paris, 1965.

[153] Lindley, D.V., *Introduction to Probability and Statistics from a Bayesian Viewpoint* (two volumes), Cambridge University Press, 1965.

[154] Mackey, G.W., *The Mathematical Foundations of Quantum Mechanics*, Benjamin, New York, 1963.

[155] Mallows, C.L., Some comments on C_p, *Technometrics*, **15**, 661–675, 1973.

[156] Manly, B.F.J., *Multivariate Statistical Methods: A Primer*, Chapman and Hall, London, 1994.

[157] Mardia, K.V., Kent, J.T. and Bibby, J.M., *Multivariate Analysis*, Academic Press, New York, 1980.

[158] Marriott, P. and Salmon, M (editors), *Applications of Differential Geometry to Econometrics*, Cambridge University Press, 2000.

[159] McCullagh, P., *Tensor Methods in Statistics*, Chapman and Hall, London, 1987.

[160] McCullagh, P. and Nelder, J.A., *Generalized Linear Models*, Chapman and Hall, London, 1983.

[161] Mead, R., *The Design of Experiments: Statistical Principles for Practical Application*, Cambridge University Press, 1988.

[162] Merzbacher, E., *Quantum Mechanics*, Wiley, New York, 1970.

[163] Messiah, A., *Quantum Mechanics* (two volumes), Dunod, Paris; North–Holland, Amsterdam, Wiley, New York.

[164] Metropolis, N., Rosenbluth, A.W., Rosenbluth, M.N., Teller, A.H. and Teller, E., Equations of state calculations by fast computing machine, *J. Chem. Phys.*, **21**, 1087–1091, 1953.

[165] Meyer, P.-A., *Quantum Theory for Probabilists*, Springer Lecture Notes, Berlin, Heidelberg, New York, 1993.

[166] MLwiN is available at a cost from the site http://www.ioe.ac.uk/mlwin/.

[167] Montgomery, D.C. and Peck, E.A., *Introduction to Linear Regression Analysis*, Wiley, New York, 1992.

[168] Murray, M. K. and J. W. Rice, *Differential Geometry and Statistics*, Chapman and Hall, London, 1993.

[169] Murtagh, F. and Heck, A., *Multivariate Data Analysis*, Kluwer, Dordrecht, 1987.

[170] Myers, R.H., *Classical and Modern Regression with Applications* (second edition), PWS-Kent, Boston, 1990.

[171] Nelson, E. *Quantum Fluctuations*, Princeton University Press, 1985.

[172] Neveu, J., *Bases Mathématiques Du Calcul des Probabilités*, Masson, Paris 1967. English translation: *Mathematical Foundations of the Theory of Probability*, Holden–Day, San Francisco.

[173] Neveu, J., *Discrete-parameter Martingales*, North–Holland, Amsterdam, 1975.

[174] Nielsen, M.A. and Chuang, I.L., *Quantum Theory and Quantum Information*, Cambridge University Press, 2000.

[175] Nobile, A. and Green, P.J., Bayesian analysis of factorial experiments by mixture modelling, *Biometrika*, **87**, 15–35, 2000.

[176] Norris, J.R., *Markov Chains*, Cambridge University Press, 1997.

[177] O'Hagan, A., *Bayesian Inference* ('Kendall's Advanced Theory of Statistics, Volume 2B'), Arnold, London, 1994.

[178] Øksendal, B., *Stochastic Differential Equations*, Springer-Verlag, Berlin, Heidelberg, New York, 1985.

[179] Olive, D.I., The relativistic electron, in *Electron, a Centenary Volume* (edited by M. Springford), Cambridge University Press, 1997.

[180] Onsager, L., Crystal statistics, I: a two-dimensional model with an order-disorder transition, *Phys. Rev.*, **65**, 117–149, 1944.

[181] Parthasarathy, K.R., *An Introduction to Quantum Stochastic Calculus*, Birkhäuser, Basel, Boston, 1992.

[182] Pemantle, R., Vertex-reinforced random walk. *Prob. Theor. and Rel. Fields* **92**, 117–136, 1990.

[183] Penrose, R., *Shadows of the Mind: a Search for the Missing Science of Consciousness*, Oxford University Press, 1994.

[184] Pitman, J.W., *Probability*, Springer-Verlag, Berlin, Heidelberg, New York, 1993.

[185] Pólya, G, *How to Solve it*, Penguin, London, 1990.

[186] Pólya, G., *Mathematics and Plausible Reasoning: 1. Induction and Analogy in Mathematics*, Princeton University Press, 1954.

[187] Preskill, J., *Physics 229: Advanced Mathematical Methods of Physics – Quantum Computing and Quantum Information*, California Institute of Technology, 1998. Available at http://www.theory.caltech.edu/people/preskill/ph229.

[188] Priestley, M., *Spectral Analysis and Time Series* (two volumes), Academic Press, London, New York, 1994.

[189] Propp, J.G. and Wilson, D.B., Exact sampling with coupled Markov chains and applications to statistical mechanics, *Random Structures and Algorithms*, **9**, 223–252, 1996.

[190] R is available free from `http://cran.r-project.org`.

[191] Reed, M. and Simon, B., *Methods of Modern Mathematical Physics* (four volumes), Academic Press, New York.

[192] Reid, N., Saddlepoint methods and statistical inference (with discussion), *Stat. Sci.*, **3**, 213–238, 1988.

[193] Rényi, A., *Probability Theory*, North–Holland, Amsterdam, 1970.

[194] Revuz, D. and Yor, M., *Continuous Martingales and Brownian Motion*, Springer-Verlag, Berlin, Heidelberg, New York, 1991.

[195] Ripley, B.D., Modelling spatial patterns, *J. Royal Statist. Soc., Ser. B*, **39** 172–212, 1977.

[196] Ripley, B.D., *Stochastic Simulation*, Wiley, Chichester, New York, 1987.

[197] Robert, C.P., *Discretization and MCMC Convergence Assessment*, Springer-Verlag, Berlin, Heidelberg, New York, 1998.

[198] Robert, C.P. and Casella, G., *Monte Carlo Statistical Methods*, Springer-Verlag, Berlin, Heidelberg, New York, 1999.

[199] Rogers, L.C.G. and Williams, D., *Diffusions, Markov Processes, and Martingales* (two volumes), Cambridge University Press, 2000 (originally published by Wiley).

[200] Rousseeuw, P.J. and Leroy, A.M.: *Robust Regression and Outlier Detection*, Wiley, Chichester, New York, 1987.

[201] Ross, S.M., *Stochastic Processes*, Wiley, New York, 1983.

[202] Ross, S.M., *Introduction to Probability Models*, Academic Press, 1981.

[203] Rowe, M.A., Kielpinski, D., Meyer, V., Sackett, C.A., Itano, W.M., Monroe, C. and Wineland, D.J., Experimental violation of a Bell's inequality with efficient detection, *Nature*, **409**, 791–793, February, 2001.

[204] Ryan, B.F. and Joiner, B.L., *MINITAB Handbook*, Duxbury Press, Pacific Grove, Ca, 2000.

[205] Ryan, T.P., *Modern Regression Methods*, Wiley, New York, 1996.

[206] Savage, L.J., *The Foundations of Statistics* (second edition), Dover, New York, 1972. (First edition, Wiley, 1954.)

[207] Scheffé, H., *The Analysis of Variance*, Wiley, New York, 1958.

[208] Schiff, L.I., *Quantum Mechanics*, second edition, McGraw–Hill, New York, 1955.

[209] Schwarz, G., Estimating the dimension of a model, *Ann. Stat.*, **6**, 461–464, 1978.

[210] Scott, D.W., *Multivariate Density Estimation: Theory, Practice, and Visualization*, Wiley, New York, 1992.

[211] Seber, G.A.F., *Multivariate Observations*, Wiley, New York, 1984.

[212] Silvey, S.D., *Statistical Inference*, Chapman and Hall, London, 1970.

[213] Simon, B., The classical moment problem as a self-adjoint finite difference operator, *Advances in Math.* **137**, 82–203, 1998.

[214] Snedecor, G.W. and Cochran, W.G., *Statistical Methods*, Iowa State University Press, Ames, 1980.

[215] Spiegelhalter, D., Best, N. and Carlin, B., Bayesian deviance, the effective number of parameters, and the comparison of arbitrarily complex models (to appear), 2001.

[216] Spitzer, F., *Principles of Random Walk*, Van Nostrand, Princeton, NJ, 1964.

[217] Stern, C. and Sherwood, E.R., *The Origin of Genetics: A Mendel Source Book*, Freeman, San Francisco, 1966.

[218] Stigler, S.M., The 1988 Neyman Memorial Lecture: A Galtonian perspective on shrinkage estimators, *Statistical Science*, **5**, 147–155, 1990.

[219] Streater, R.F., Classical and quantum probability, *J. Math. Phys.*, **41**, 3556–3603, 2000.

[220] Streater, R.F. and Wightman, A.S., *PCT, Spin and Statistics, and all that*, Benjamin/Cummings, Reading, Mass., 1978.

[221] Stroock, D.W., *Probability Theory, an Analytic View*, Cambridge University Press, 1993.

[222] Stoyan, D., Kendall, W.S. and Mecke, J., *Stochastic Geometry and its Applications*, Wiley, Chichester, New York, 1987.

[223] Tabachnick, B.G. and Fidell, L.S., *Using Multivariate Statistics*, third edition, HarperCollins, New York, 1996.

[224] Temperley, H.N.V. and Fisher, M.E., Dimer problem in statistical mechanics – an exact result, *Phil. Mag.*, **6**, 1061–1063, 1961.

[225] Thaller, B., *The Dirac Equation*, Springer-Verlag, Berlin, Heidelberg, New York, 1992.

[226] Thompson, C.J., *Mathematical Statistical Mechanics*, Princeton University Press, 1979.

[227] Thompson, E.A., *Pedigree Analysis in Human Genetics*, Johns Hopkins University Press, Baltimore, 1986.

[228] Thompson, E.A. and Cannings, C., *Genealogical and Genetic Structures*, Cambridge University Press, 1981.

[229] Tong, H., *Non–Linear Time Series, A Dynamical System Approach*, Oxford University Press, 1990 .

[230] Tukey, J., The philosophy of multiple comparisons, *Statistical Science*, **6**, 100–116, 1991.

[231] Varadhan, S.R.S., *Large Deviations and Applications*, SIAM, Philadelphia, 1084.

[232] Venables, W.N. and Ripley, B., *Modern Applied Statistics with S–PLUS; Volume 1: Data Analysis*, third edition, Springer-Verlag, Berlin, Heidelberg, New York 1999.

[233] Wagon, S., *The Banach–Tarski Paradox*, Cambridge University Press, 1985.

[234] Weisberg, S., *Applied Linear Regression*, second edition, Wiley, New York, 1985.

[235] Williams, D., *Probability with Martingales*, Cambridge University Press, 1991.

[236] Wilson, R.A., *Schrödinger's Cat Trilogy*, Dell, New York, 1979, Orbit, 1990.

[237] Wilson, A.R., Lowe, J. and Butt, D.K., Measurement of the relative planes of polarisation of annihilation quanta as a function of separation distance, *J. Phys. G Nuclear Physics*, **2**, 613–623, 1976.

[238] `WinBUGS` is available from `http://www.mrc-bsu.cam.ac.uk/bugs/welcome.shtml`.

[239] Zeilinger, A., A foundational principle for quantum mechanics, *Foundations of Physics*, **29**, 631643, 1999.

[240] Zurek, W.H., Decoherence and the transition from quantum to classical, *Physics Today*, **44**, 36–44, 1991.

Index